T0179460

Fundamentals of
Abstract Analysis

Fundamentals of Abstract Analysis

ANDREW M. GLEASON
Harvard University

CRC Press
Taylor & Francis Group
Boca Raton London New York

CRC Press is an imprint of the
Taylor & Francis Group, an **informa** business

Visit the Taylor & Francis Web site at
http://www.taylorandfrancis.com

and the CRC Press Web site at
http://www.crcpress.com

Library of Congress Cataloging-in-Publication Data
Not available at press time.

Cover Design: Hannus Design Associates

PREFACE TO THIS EDITION

This edition of *Fundamentals of Abstract Analysis* differs from the first only by the correction of a few typos and a couple of out-and-out mistakes.

Were I starting from scratch, I would probably write a different book, but not a very different book. During the past 25 years emphases have changed within my own mind as indeed they have in mathematics as a whole. But the set-theoretic organization of mathematics is still standard and its language is the lingua franca of the profession; so I am pleased to be able to offer once again my attempt to explain it to students who are seriously interested in pursuing advanced mathematics.

I want to thank Klaus Peters and Jones and Bartlett Publishers for making this new edition possible.

PREFACE

For many years I have felt the lack of any book which seriously attempts to explain to the beginning student the relation of set-theoretic mathematics to mathematics itself. Considering the large number of books on "modern" mathematics which have appeared in the past few years, this may seem to be an odd statement, but the books that I have seen all explain the axiomatic method, not the set-theoretic point of view.

There are mathematicians who claim that there is no difference between mathematics and set theory, but I believe this claim can be dismissed. No mathematician of my acquaintance would abandon his field if an apparently insurmountable contradiction were discovered in the general concept of subset. Obviously, mathematics has a real content which transcends the inadequacies of our efforts to formalize it. On the other hand, there can be no doubt that the set-theoretic point of view has contributed mightily to our understanding of even the most primitive concepts.

It is unfortunate that the technical devices necessary to maintain an abstract approach often obscure the origins of the problems they are designed to handle. The result has been a widening of the intellectual gap between pure and applied mathematics and, regrettably, a virtual estrangement of pure and applied mathematicians. Those who find the precision of set-theoretic formulations fascinating often lose sight of mathematics itself, while those who are repelled by formalisms often dismiss all abstractions as mere axiom-pushing and turn a blind eye to the insights that abstraction may provide.

The separation begins with the first course adopting the abstract point of view, usually either "modern algebra" or "real variable theory." After a few courses which at least claim to be devoted to problems of direct applicability in the real world our students are often suddenly and unceremoniously transplanted into a course with an abstract point of view with no explanation or apology for a fundamental change in philosophy. They are given the impression that set theory is the ultimate approach to mathematics and that no true mathematician will even consider any question until it has been completely formalized in set-theoretic terms. It is only much later that the student discovers that mathematical research is largely a process of winnowing theorems from a melange of hunches, vague analogies, and geometrical images.

I do not suggest any retreat from abstraction, far from it, but I do believe that our students will find set-theoretic mathematics easier to understand and at the same time more valuable if it is presented with a frank acknowledgement that it is only one of the possible ways to record mathematical ideas. And so I have written this book.

It is a very abstract and highly formalistic book, but at several strategic places I have tried to point out how formalism is related to the elusive "real mathematics" which exists only in our intuition. The value of this book will depend to a considerable extent on the success of these passages.

After the preliminary chapters on logic and sets I have tried in Chapter 4 to sketch with broad strokes the set-theoretic point of view to show what it has to offer to mathematics. In the subsequent chapters I have illustrated these ideas. Starting off with order and equivalence relations, I have tried to increase the level of sophistication gradually, going next to the construction of the real numbers, following a path close to the original route of Dedekind. The treatment of cardinal numbers in Chapter 11 is minimal, covering primarily the few theorems that often come up in analysis. However, this chapter serves as a vehicle for a discussion of the nature of axiomatic set theory and the significance of the axiom of choice, which I hope will clarify the relation of these subjects to mathematics itself. Chapters 12 and 13 develop the theory of limits, infinite sums, and infinite products. Chapter 14 presents a rapid survey of metric space topology. Again I have tried to limit myself to those definitions and theorems which actually come up in analysis.

After fourteen chapters devoted to developing the tools of analysis it seemed important to apply these tools in some significant way. I chose part of the elementary theory of analytic functions of a complex variable. Since the book contains no integration theory at all, the work is based entirely on power series. In this respect it differs sharply from most elementary treatments. Doing analytic function theory without line integrals is rather like fighting with one arm tied, but one can give a significant preview of the general theory nevertheless. The main objectives of Chapter 15 are the modulus principle and the exponential, logarithmic, and circular functions. The work on explicit functions serves as a brief introduction to the idea of analytic continuation and provides an opportunity to discuss the problem of angles. Two proofs of the fundamental theorem of algebra are included, one using the modulus principle, the other using winding numbers (the latter is in the exercises).

The whole book is based on naive set theory, because I feel that an axiomatic approach to set theory cannot be understood by a student until after he has thoroughly digested much of the material in this book from a naive point of view. On the other hand, I believe that every theorem and proof will withstand a translation into the Hilbert-Bernays-Gödel axiomatic system.

As the preceding remarks and the title suggest, the book is intended for the first course in completely rigorous analysis commonly designed for fourth ($\pm$ 1)-year students. As usual in such a course, no technical knowledge is required to study the book, but only an exceptional student will profit from the experience if he has not already acquired a considerable sophistication in mathematics. With some omissions the material can be covered in a semester or even less, but it will take two if an effort is made to discuss the subtle points thoroughly and to study the more advanced exercises.

The exercises at the end of each section have been arranged roughly in increasing order of difficulty. Many of them depend on the results of prior exercises. This makes it desirable for the student to do them all. Some, especially in the later chapters, present substantial extensions of the theory. Particularly at the beginning of a book which attempts to present mathematics *ab initio*, it is hard to find exercises of any apparent significance which rely solely on the theorems and concepts already developed. Consequently, I have included some problems which depend on knowledge and techniques not discussed in the preceding text. Such exercises are marked with an asterisk. Hints and solutions to most of the exercises will be found at the end of the book.

It is a pleasure to record my indebtedness to numerous friends, colleagues, and students. Among them are several whose contributions deserve particular mention. Burton Dreben, Paul Halmos, and Lynn Loomis all made valuable comments on parts of the manuscript. Ethan Bolker showed me a number of places where the text could be made more meaningful by being more concrete and thus counteracted in part my tendency towards over-abstraction. Many thanks are due to my wife, who, although she still insists she didn't understand a single word of it, read the galley proofs aloud and spotted numerous ambiguities and grammatical solecisms, not to mention typographical errors. Walter Stewart was extremely helpful in reading the page proofs and preparing the index and answers.

A.M.G.

Cambridge, Massachusetts
March 1966

CONTENTS

Chapter 1. Sets

1-1. The notion of set 1
1-2. Equality 2
1-3. Parentheses 3
1-4. Membership 3
1-5. The empty set 3
1-6. The list notation 4
1-7. Set inclusion 5

Chapter 2. Logic

2-1. Propositions and logical connectives 8
2-2. Tautologies 10
2-3. The conditional 13
2-4. Propositional schemes and quantifiers 15
2-5. Proof and inference 19
2-6. Set formation 24
2-7. The set-theoretic paradoxes 25
2-8. Dummy variables 26

Chapter 3. The Set-Theoretic Machinery

3-1. Binary set combinations 29
3-2. The power set 37
3-3. Ordered pairs and direct products 37
3-4. Functions 40
3-5. Relations 49
3-6. Indexed unions and intersections 50
3-7. Indexed direct products 53

Chapter 4. Mathematical Configurations

4-1. Structures and configurations 55
4-2. Definitions, postulates, and theorems 59
4-3. Consistency 62
4-4. The classification problem 63

Chapter 5. Equivalence

5-1. Equivalence relations and partitions 65
5-2. Factoring functions 67

Chapter 6. Order

6-1. Order relations 70
6-2. Maps of ordered sets 73

6–3.   Linear order   75
6–4.   Bounds   77
6–5.   Complete ordered sets   79
6–6.   Well-ordering   82

Chapter 7. Mathematical Induction
7–1.   Chains   84
7–2.   Inductive proof   85
7–3.   The natural numbers and inductive definitions   88

Chapter 8. Fields
8–1.   Binary operations   94
8–2.   Fields   96
8–3.   The elementary arithmetic of fields   99
8–4.   Whole numbers and rational numbers   100
8–5.   Ordered fields   104
8–6.   Archimedean ordered fields   107
8–7.   Complete ordered fields   108

Chapter 9. The Construction of the Real Numbers
9–1.   The arithmetic of the natural numbers   112
9–2.   Fractions and rational numbers   116
9–3.   The positive real numbers   121
9–4.   Real numbers   126

Chapter 10. Complex Numbers
10–1.   Complex number systems   130
10–2.   Permanent notation   132
10–3.   Conjugates and absolute values   132
10–4.   Exponents   134

Chapter 11. Counting and the Size of Sets
11–1.   Similarity and dominance   136
11–2.   Finite sets   141
11–3.   Countable sets   143
11–4.   Another form of inductive definition   144
11–5.   The axiom of choice   148
11–6.   Cardinal numbers   155

Chapter 12. Limits
12–1.   Convergent sequences   160
12–2.   Limits and arithmetic   168

12–3.   Infinity and the extended real number system 173
12–4.   Superior and inferior limits 175
12–5.   Criteria for the existence of limits 180
12–6.   Subsequences 182

Chapter 13.   **Sums and Products**

13–1.   Finite sums and products 186
13–2.   Infinite series 191
13–3.   Infinite products 212
13–4.   Numeration and calculation 217

Chapter 14.   **The Topology of Metric Spaces**

14–1.   Metric spaces 223
14–2.   Convergence. 229
14–3.   Closure, closed sets, and open sets 232
14–4.   Continuous functions 239
14–5.   Uniform continuity and uniform convergence 245
14–6.   Homeomorphism 249
14–7.   Complete spaces 253
14–8.   Compact spaces 266
14–9.   Separable spaces 273
14–10.   Connectedness 278

Chapter 15.   **Introduction to Analytic Functions**

15–1.   Differentiation 287
15–2.   Power series 293
15–3.   Analytic functions 298
15–4.   The exponential and circular functions 308
15–5.   The modulus principle 315
15–6.   The logarithm 319
15–7.   Exponents 324
15–8.   Geometric considerations 326

Answers and Solutions. 340

Index of Symbols and Special Notations 395

Index 397

CHAPTER 1

SETS

In this book we propose to develop a portion of mathematics from the beginning. Since the most widely accepted basis for mathematics today is the theory of sets, we shall begin our study with the notion of set. It is possible to treat set theory itself as an abstract mathematical discipline, taking set and membership as undefined notions and laying down axioms to describe the interrelationships of these concepts; however, we shall not do this. We shall rely on intuition for the meaning of the ideas of set and membership and for the validity of the various set-theoretic techniques.

1-1. THE NOTION OF SET

The concept of a set of objects is inherent in our everyday speech whenever we use a collective noun such as *jury* or *team*; hence it can safely be assumed that the reader is familiar with the idea. To clarify somewhat the mathematical idealization of this concept we offer the following definition.

1-1.1. Definition. A *set* is any collection of mathematical objects which is sufficiently well defined to be the subject of logical analysis.

We mean by this that it must be a clear question of fact whether a given object does or does not belong to a set, although we do not demand that the answer to this question be available to us. We may talk of the set of rational numbers, i.e., those numbers which can be expressed as the quotient of two integers. Given a number, it is clearly a question of fact whether or not it can be so expressed, but it may be very difficult to decide in practice. The proofs that $\sqrt{2}$ and π are irrational are both great milestones in the history of mathematics; and there are numbers, such as 2^e, for which the question has not yet been answered. Such difficulties are not, however, regarded as barriers to the consideration of the set of rational numbers.

As one might expect, the sets with which we shall deal will contain only mathematical objects. While there is no *a priori* objection to sets of physical objects, they are of no importance in pure mathematics. Moreover, once physical sets are admitted, there is a temptation to overlook the distinction between the real world and a mathematical model of it, and the results so obtained are often absurd.

Although we cannot now give a precise definition of them, it is convenient to introduce the following sets, from which we can construct many examples which will be familiar to the reader.

N will denote the set of positive integers.
I will denote the set of all integers.
Q will denote the set of rational numbers.
R will denote the set of real numbers.

Formal definitions of these sets will appear much later in Chapter 10.

1-1.2. A set is identified by the objects it contains, regardless of the way it is defined. Thus the set of real numbers between 0 and 1 and the set of real numbers which exceed their squares are the same set in spite of the different criteria for membership.

1-1.3. Because we shall often consider sets whose members are themselves sets and then sets of these, etc., it is convenient to have synonyms for the word *set* to help keep the various levels straight. We shall use the words *class* and *collection* for this purpose. They are entirely synonymous with *set*.

EXERCISES

1. Which of the following are bona fide sets?

The set of all even integers.
The set of all large integers.
The set of irrational numbers which are square roots of integers.
The set of interesting numbers.
The set of points in a given euclidean plane.
The set of circles in a given noneuclidean plane.

2. Was the population of the United States at noon (EST) on January 1, 1960, an odd or an even number? Is this question meaningful? Is the set of inhabitants of the United States at that time a well-defined set?

1-2. EQUALITY

Mathematical objects are sometimes denoted by alphabetical or numerical characters and sometimes by compounds of these, with possibly some not so familiar symbols. We shall always regard these notations as names that we have chosen for things which have an existence (in some abstract world) which is quite independent of nomenclature. It will frequently happen that two different notations refer to the same object. To express this state of affairs we use the word *equals* or the symbol $=$.

1-2.1. When $=$ appears between two symbols (or compound symbols), each of which denotes an object, it means that the symbol (or compound symbol) to the left denotes the same object as the symbol (or compound symbol) to the right. The word *equals* is used in the same way.

The symbol $=$ is rather frequently used in expressions like $1 + 2^2 = 1 + 4 = 5$. These are abbreviations. The example is an abbreviation for the two expressions $1 + 2^2 = 1 + 4$ and $1 + 4 = 5$. Note that "$1 + 4$" is not a command to execute a certain addition, it is a notation for the result.

Once this interpretation of equality is agreed upon the familiar rules of equality are obviously valid. However, we have then renounced the use of the word *equals* in any other context than identity. In classical geometry the word is used to mean variously *has the same length as, has the same area as,* or *has the same volume as;* all such uses we forego.

1-3. PARENTHESES

As we have remarked above, objects may have names which are compound symbols. In order to delineate the boundaries of these compound symbols, parentheses, ( ), and brackets, [ ], are used. Formally this means that whenever a compound symbol is a name of an object, then the same symbol surrounded by parentheses or brackets is also a name of the same object; e.g., in $4 - (5 - 3)$, the compound symbol $(5 - 3)$ is another name for $5 - 3$, or 2. Braces, { }, are often used under this convention, but in this book they are reserved for use in defining sets as explained in Sections 1-6 and 2-5. Parentheses will also be used in the notation for functional values (see Section 3-4.3).

1-4. MEMBERSHIP

The objects from which a set is formed are called its members or elements; in special contexts words like *point* or *number* might be used. The fundamental relationship of membership is denoted by the symbol $\in$.

1-4.1. "$x \in y$" means "x is a member of the set y."

This is frequently read "x is an element of y," "x belongs to y," or "x is in y." When dictated by the grammatical context, inflected readings are used; for example, "Let $x \in y$" might be read "Let x be a member of y."

The contrary relation, nonmembership, is denoted by $\notin$; thus "$x \notin y$" means "x is not a member of y."

1-5. THE EMPTY SET

One set deserves special mention, the *empty* set or *null* set, which is characterized by having no elements at all. There is only one such set because a set is known by its members, and hence it is appropriate that we assign it a permanent name, $\emptyset$, which we will use throughout the book.

1-5.1. $\emptyset$ denotes the empty set.

Perhaps it seems odd to devote special attention to the null set, which is evidently not worthy of much serious thought. There are numerous situations in mathematics where trivial cases arise, and it is a matter of experience that they cause the least difficulty when definitions and notations are devised explicitly to allow for them. We might have defined a set to exclude the possibility of its being empty, but this would prove awkward when we want to form the set of objects common to two different sets: two cases would have to be considered, one when there are common elements and another when there are none. By admitting the empty set this complication is avoided. There are, of course, some disadvantages to our convention, and in certain contexts the empty set is exceptional. Even experienced mathematicians occasionally make the mistake of assuming without proof that a set is not empty. There is no doubt, however, that mathematical exposition is simplified by accepting the null set as a bona fide set.

1-6. THE LIST NOTATION

When we wish to consider a set whose elements a, b, c, etc., can be explicitly listed, we may denote it by $\{a, b, c, \ldots, k\}$. Formally this notation can be used only when there are finitely many elements and we are prepared to write them out in full. However, it is often used informally when it contributes to clarity; thus one might write $\{7, 8, 9, \ldots\}$ for the set of all integers greater than 6. This is permissible only when the writer is prepared to substitute a formally correct notation if challenged.

In connection with this notation, it must be emphasized that there is a distinction between an object x and the set $\{x\}$ which has just that one object as member. Similarly, when a set A appears as an element of another set B, the elements of A are not counted among the elements of B, at least not by virtue of their membership in A. Suppose that $A = \{1, 2\}$ and $B = \{A, 1\}$. Here $2 \in A$, but $2 \notin B$; the fact that $1 \in A$ and $1 \in B$ is a coincidence.

This list notation for a set perforce assigns an order to the members of the set, but it must be understood that this has no significance; the members of a set all stand on the same footing insofar as membership in the set is concerned. In the example above, A and 1 enjoy membership in B without regard to the incidental relation $1 \in A$ or the fact that A precedes 1 in the list for B as we have written it.

An element is not to be counted more than once as a member of a set. Suppose that we define $A = \{a, b, c\}$ and it develops subsequently that $a = b$. We do not regard a as a twofold member of A on this account; a's membership in A is no stronger than c's. Thus $\{(2 + 3), 4, 5\} = \{4, 5\}$.

As tacitly assumed in the example above, there is no objection to forming a set from a mixture of sets and members of those sets. To emphasize this point, let us exhibit some of the sets which can be built up from the empty set. Starting with $\emptyset$, we can form $\{\emptyset\}$, which is certainly not $\emptyset$ because it has a member. Having now two sets at our disposal we can form $\{\{\emptyset\}\}$ and $\{\emptyset, \{\emptyset\}\}$, and these are both new sets. If we repeat this process we will obtain twelve new sets, making a total of sixteen which can be formed from $\emptyset$ and braces, allowing the braces to be not more than three deep.

EXERCISE

List the sixteen sets that can be formed from $\emptyset$, using braces not more than three deep.

1-7. SET INCLUSION

It frequently happens that one set A is part of another, B. In this case, A is said to be a subset of B, or to be included or contained in B, and B is said to contain or include A; rarely, B is called a superset of A. The notation for this relation is $A \subseteq B$ or $B \supseteq A$.

1-7.1. "$A \subseteq B$" or "$B \supseteq A$" means "Every element of A is also an element of B." (Examples: $N \subseteq I, I \subseteq R, R \supseteq N$.)

The following two theorems are immediate consequences of this formal definition.

1-7.2. Theorem. If $A \subseteq B$ and $B \subseteq C$, then $A \subseteq C$.

1-7.3. Theorem. If $A \subseteq B$ and $B \subseteq A$, then $A = B$.

Theorem 1-7.2 is a simple restatement of the obvious fact that if every member of A is a member of B and every member of B is a member of C, then every member of A is a member of C. The hypothesis of 1-7.3 shows that A and B have the same elements; hence the theorem is a restatement of the principle that a set is known by its elements.

Since one set may be a member of another set, we must distinguish carefully between $X \in A$ and $X \subseteq A$. It is possible, although unusual in most areas of mathematics, that both statements are true; for example, given that $A = \{1, \{1\}\}$, then $\{1\} \in A$ and $\{1\} \subseteq A$.

We should note that for any set A, we have $\emptyset \subseteq A$ and $A \subseteq A$. The first of these formulas says, "Every member of the null set is a member of A," which is evidently true, if only in a vacuous way. The second is even more obvious. Both serve to underscore the idea that mathematical definitions are usually worded in the most comprehensive manner and are intended to include trivial cases. Thus the word *part* in ordinary parlance excludes the extreme cases of none and the whole, but the inclusive meaning is more useful mathematically.

There is occasional use for inclusion which excludes equality and for this we introduce the following notation.

1-7.4. "$A \subset B$" or "$B \supset A$" means "$A \subseteq B$ and $A \neq B$."

In other words, "$A \subset B$" means that every element of A is also in B and that B contains, in addition, at least one more element. This is usually expressed verbally by saying that A is a *proper* subset of B. (*Note:* For any set A other than the null set, $\emptyset \subset A$.)

Mathematical writers are divided on the use of the symbol $\subset$. Some, perhaps the majority, use $\subset$ without excluding the possibility of equality. Use of the two symbols $\subset$ and $\subseteq$, which are patterned after the classical symbols $<$ and $\leq$, seems preferable. The negated symbols $\nsubseteq$ and $\not\subset$ are rarely used.

Inclusions are often run together in the same way as equalities. Thus, "$A \subseteq B \subseteq C$" is an abbreviation for "$A \subseteq B$ and $B \subseteq C$." The abbreviation is convenient because it is easy to read off the implied inclusion $A \subseteq C$. The same is true of inclusions directed the other way (e.g., $D \supseteq E \supseteq F$), but one does not ordinarily mix inclusions in opposite directions; "$X \subseteq Y \supseteq Z$" is certainly unusual.

CHAPTER 2

LOGIC

Sets are the objects of modern mathematical study, and logic is the method by which they are studied. We must therefore have some understanding of logic in order to study mathematics.

Before starting our study of logic, it is appropriate to consider more carefully the relation between logic and mathematics. First, we must distinguish between *logic* and the more specialized topic *mathematical logic*. Logic is the study of the processes of reasoning. Mathematical logic is not simply the study of the reasoning processes of mathematics, but the use of mathematical methods to study the reasoning processes of mathematics.

Mathematicians usually interpret their formulas as being statements of fact about objects in some abstract world. Accordingly, their reasoning process is based on the meanings of these formulas. Adopting this point of view, we shall see that the theorems of abstract mathematics are true entirely by virtue of the meanings of the words and symbols involved.

On the other hand, no one has described the subtleties of meaning in terms sufficiently precise to be analyzed mathematically. Hence in mathematical logic one replaces the reasoning process by a set of rules of procedure. With the aid of these rules it is possible to decide objectively whether a mathematical proof is correct (i.e., conforms to the rules) or not. The objective character of such a system is clearly a great advantage, but there are also serious disadvantages. To make such a system workable, a highly formalized language is necessary and proofs are likely to be extremely long. Consequently, only a small part of mathematics has actually been formalized, and mathematicians continue to rely on what is ultimately a subjective process for evaluating proofs. Moreover, the significance of mathematics—both as an intellectual activity and as an applied science—lies in its meaning; hence no system which abandons the domain of meaning is likely to be completely satisfactory.

It is fortunate that partial formalizations of mathematics have been found which in a sense bridge the gap between intuitive reasoning and strict formalization. The idea of set has a certain tangible quality about it which makes it possible for mathematicians to ascribe reality to the objects of their study, and it is possible to interpret other essential ideas of mathematics within set theory without doing too much violence to the intuitive picture. The first-order predicate calculus, which we shall briefly study in this chapter, provides a reasoning system that is quite simple, accords with our intuition, and is easily formalized. Set theory and the predicate calculus together make up a system of mathematical communication that can easily be interpreted in either the formal or the intuitive domain. Almost all of modern mathematics is written in a partially formalized version of this system. The degree of formalization is sufficient to make the standards for proofs effectively objective. At the same time the underlying meanings of mathematical statements are kept clearly in view.

Throughout this book we shall see this semiformal mathematics in action, but we shall consider it only in terms of its meaning. In the present chapter, we shall discuss the basic ideas of the predicate calculus, not from the formal point of view appropriate to mathematical logic, but as a system for the precise recording of elementary logical ideas. Our objective is only to clarify the meanings of the logical words and symbols we shall use and to sketch how we shall use them.

Our first logical convention is that we accept the English language with its ordinary meanings except for those words to which we specifically assign a technical meaning. We shall define only the basic forms of the technical words and assume that the reader will understand the inflected forms when they appear.

2–1. PROPOSITIONS AND LOGICAL CONNECTIVES

A proposition is a statement which is either true or false as a matter of fact, not opinion. Whereas statements of ordinary language frequently fall somewhere between absolute truth and absolute falsehood, we shall be concerned with mathematical statements only, and it is often easy to see that a statement written in mathematical language must be either true or false, even though it may be difficult to tell which. To return to an example given in Chapter 1, the statement, 2^e *is a rational number*, is a proposition, since it is surely either true or false. It is immaterial that no one knows which. The intention of the discussion concerning the proper definition of a set A is that for any given object b, the statement $b \in A$ should be a proposition. Indeed, if we were to carry out a more thorough analysis of the logical structure of mathematics, we should find that propositions of the form $b \in A$ are the only propositions which cannot be dissected into component propositions.

Our definition says that every proposition has one of two truth values, namely, true and false, and this commits us to the so-called two-valued system of logic. Logical systems have been proposed which allow three or more truth values for a proposition. Such systems may prove to be valuable in fields where sharp dichotomies are rare, but it seems difficult to escape from the two-valued system in mathematics. Certain mathematicians, called intuitionists, have seriously proposed that the two-valued system be abandoned, but this view is not wide spread and need not concern us.

One might wonder why mathematicians consider false statements at all; surely our primary concern is with true ones. There are several reasons for considering propositions which are false. Perhaps the strongest reason is the simple fact that there are many statements, such as the one above, which, although identifiable as propositions, have not yet been settled one way or the other. Moreover, false propositions serve an auxiliary function in finding true ones. The technique of argument known as indirect proof consists in introducing a false proposition into the premises and proceeding to a recognizable contradiction. It is possible to build true propositions from false ones: from the false proposition $\frac{1}{3} \in I$ we can make the true one $\frac{1}{3} \notin I$. Although this example is trivial, it points out an important fact. Propositions naturally go in pairs, and within each pair one is true and the other false. Because of the absolute character of the true-false dichotomy, it makes little difference whether we focus on the true or the false member of each pair.

From the beginning, we must understand that logic is not concerned with deciding the truth or falsity of individual propositions unless the propositions are themselves compounded from simpler ones. Thus logic offers us no assistance in deciding the truth value of a proposition such as *it is raining*; this depends on the state of the atmosphere and the subtle question of what we mean by the word *raining*. On the other hand, logic does assert that the compound proposition *either it is raining or it is not raining* is true. Here the proposition is true quite independently of the meaning of the word *raining*; it is true by virtue of the meaning of the words *or* and *not*. The first step in logic is the study of how a few special words, like *and*, *or*, and *not*, affect the meaning of propositions.

When studying logic it is customary to designate by a single letter, for instance p, a proposition whose internal meaning we shall not investigate. In applications, of course, p will be replaced by an explicit proposition. There is no intention that p should be replaced by a true proposition; but we do insist that p be replaced by an actual proposition; that is, a sentence which is either true or false.

If p is a proposition, then *not-p* is the logical opposite of p. When p is explicitly given in words, we usually form *not-p* by inserting (or removing) the word *not* at a suitable point in the text of p and possibly rewording the text slightly to conform to the rules of grammar. When p is given in symbols, we often merely add a / at an appropriate place. Thus, if p stands for $2 \in A$, then *not-p* stands for $2 \notin A$. Precise general rules for how to form *not-p* from p are extremely hard to give. Fortunately, such rules are unnecessary, because we do not propose to examine the internal structure of p. For our purpose, *not-p* is an entirely adequate symbol for the opposite of p. What is essential is that *not-p* stands for a proposition which is false if p is true, and true if p is false.

There are many ways to combine two or more propositions to make a single proposition. The simplest of these are expressed by the words *and* and *or*. The logical meanings of these words are extremely close to the ordinary English meanings.

The word *and* is used in logic in exactly the same way as it is in language. The proposition *p and q* is deemed to be true if both *p* and *q* are true, and false if either or both of the component propositions are false.

The word *or* has two distinct meanings in English. In the sentence, *You may have cake or pie for dessert*, the intended meaning is *cake or pie but not both*. This is called the exclusive meaning. On the other hand, the sentence, *The man we are looking for may be short or fat*, probably is not intended to exclude the possibility that he is both short and fat. This is called the inclusive meaning. The meaning is so often ambiguous that in the legal profession an exclusive *or* is regularly written ... *or* ... *but not both* and the inclusive *or* is often denoted by the barbarism *and/or*. In mathematics and logic the word *or* always has its inclusive meaning; thus *p or q* is deemed to be true if either *p* is true, or *q* is true, or both are true. It frequently happens that the two component propositions are incompatible, so that *p or q* appears to involve the exclusive meaning of *or*, but in reality the exclusion is entirely due to the meanings of *p* and *q* and should not be ascribed to the word *or*.

In the preceding description of the logical use of *and* and *or* there is no requirement that the component propositions be in any way related in meaning. We can write, for example,

> *Washington was the first President of the United States*
> *or the moon is made of green cheese.*

Applying the definitions, we see that this ridiculous sentence is not only a proposition; it is true. If we examine the sentence grammatically, we can find no fault with it, but stylistically it is unacceptable. Similarly, the rules of logic do not forbid the linking of totally unrelated mathematical propositions, but as a matter of style we do not compose a sentence from several mathematical propositions unless they are intrinsically related.

2-2. TAUTOLOGIES

Having defined the meaning of the words, *and*, *or*, and *not* in connection with simple propositions, we can apply them to compound propositions. Consider

$$((p \text{ or } q) \text{ and } r) \text{ or } (p \text{ and } q). \tag{1}$$

Our present interpretation of (1) is that *p*, *q*, and *r* represent definite but, for convenience, unspecified propositions; (1) is then itself a proposition. There is another important interpretation, however. We may regard (1) as a formula which produces a new proposition from any three propositions *p*, *q*, and *r*. These two interpretations are analogous to the two meanings assignable to the expression

$$x^2 + x + 1. \tag{2}$$

First, we may think of x as representing a definite but probably unknown number, in which case (2) represents a number. Secondly, we may think of (2) as a formula which converts a real number x into another real number; that is, as the formula for a function.

Among the formulas similar to (1) there are some which produce a true proposition no matter what propositions are substituted for the letters p, q, and r. Such formulas are called *tautologies*. A simple example is

$$p \quad or \quad (not\text{-}p).$$

We have already discussed the result of replacing p by *it is raining* in this formula. More elaborate examples are

$$((not\text{-}p \; and \; not\text{-}q) \; or \; p) \; or \; not\text{-}(p \; and \; q), \tag{3}$$

$$(r \; or \; not\text{-}p) \; or \; not\text{-}((q \; or \; not\text{-}p) \; and \; (r \; or \; not\text{-}q)). \tag{4}$$

(It is customary to assume that *not-* always applies to the shortest possible block after it. This reduces the number of parentheses required.) The reader should verify that (3) is indeed a tautology by working out its truth value in each of the four possible cases, viz., p and q both true, p true and q false, p false and q true, p and q both false. Similarly, (4) can be verified by working out eight cases. A convenient way to organize this work is shown in Fig. 2-1.

p	q	r	A	B	C	D	E
T	T	T	T	T	T	T	T
T	T	F	F	T	F	F	T
T	F	T	T	F	T	F	T
T	F	F	F	F	T	F	T
F	T	T	T	T	T	T	T
F	T	F	T	T	F	F	T
F	F	T	T	T	T	T	T
F	F	F	T	T	T	T	T

FIG. 2-1. The columns under p, q, and r have been filled out with T's (for true) and F's (for false), so that each of the eight possible combinations occur in the rows. Column A shows in the various cases the truth value of *r or not-p*. Column B shows the truth value of *q or not-p*. Column C shows the truth value of *r or not-q*. Column D is computed directly from B and C and shows the truth value of *B and C*; i.e., of *(q or not-p) and (r or not-q)*. Finally column E shows *A or not-D*; i.e., the compound formula (4) of the text. Since this is true in all cases, (4) is a tautology.

Continuing the previous analogy, a tautology is like the algebraic formula $(x + 1)^2 - (x - 1)(x - 3) - 6x + 3$ which has the value 1 for all values of x.

The importance of tautologies for mathematics lies in the fact that we obtain a true mathematical proposition (i.e., a theorem) by replacing each letter by a mathematical proposition, even if we are in doubt as to the truth value of these replacements. Thus we can replace p in (3) by $2^e \in \mathbf{Q}$ and q by $3^\pi \notin \mathbf{Q}$ and obtain the theorem

$$((2^e \notin \mathbf{Q} \text{ and } 3^\pi \in \mathbf{Q}) \text{ or } 2^e \in \mathbf{Q}) \text{ or } not\text{-}(2^e \in \mathbf{Q} \text{ and } 3^\pi \notin \mathbf{Q}).$$

Propositions derived in this way are said to be tautologically true.

It must be admitted that tautologically true theorems are not very exciting. But we note that such theorems do not owe their truth to any axioms or postulates, the usual remarks about how each statement must be deduced from a prior statement, etc., notwithstanding. We shall point out in Chapter 4 that in abstract mathematics, what are usually called postulates are actually parts of definitions. It will then appear that all the theorems of abstract mathematics are consequences of the meanings of the words and symbols involved in their statements in much the same way as tautologically true propositions are consequences of the meanings of the words *and*, *or*, and *not*.

Closely related to the idea of tautologies is the concept of tautologically equivalent formulas. Suppose that S and T are formulas both involving the letters $p, q, r, \dots$ It may happen that each way of replacing the letters by actual propositions makes the formulas S and T into propositions having the same truth value. (It is understood, of course, that all p's appearing are to be replaced by the same proposition, and similarly for all the q's, all the r's, etc.) In this case, we say that S and T are tautologically equivalent formulas. We also say that the propositions which arise in this manner are tautologically equivalent propositions. For example, *p or q* and *q or p* are tautologically equivalent formulas if p and q are indefinite, and tautologically equivalent propositions if p and q are explicit propositions.

Somewhat less trivial are the pairs

$$not\text{-}(p \text{ and } q) \qquad \text{and} \qquad not\text{-}p \text{ or } not\text{-}q \tag{5}$$

and

$$not\text{-}(p \text{ or } q) \qquad \text{and} \qquad not\text{-}p \text{ and } not\text{-}q. \tag{6}$$

By repeated application of these equivalences, we can replace any proposition compounded with *and*, *or*, and *not* by one in which all the *not*'s apply directly to the basic component propositions. Thus the formal negative of

$$p \text{ or } (q \text{ and } r)$$

can be replaced successively by

$$not\text{-}p \text{ and } not\text{-}(q \text{ and } r)$$

and

$$not\text{-}p \text{ and } (not\text{-}q \text{ or } not\text{-}r).$$

This is important in mathematics, because the *not*'s can then be absorbed into the internal structure of the propositions p, q, r, etc.

2-3. THE CONDITIONAL

A sentence having the form

if p, then q

is called a conditional sentence; the connective *if . . . , then . . .* is called the *conditional* connective. The constituents *p* and *q* are called respectively the antecedent and the consequent.

The conditional is one of the most commonly used connectives in mathematics. At the same time it is one of the biggest sources of confusion. The conditional in English is a more subtle construction than it is in mathematics. It often involves idiomatic meanings or figures of speech. Furthermore, the stylistic requirement that different clauses in the same sentence be connected in meaning is stronger for the conditional connective than it is for simple conjunctions. However, the principal source of confusion is probably a reluctance to interpret conditional sentences as propositions in all situations. For example, consider the sentence

If it is raining, then the streets are wet.

This is apparently intended as a proposition. But it is not always so regarded, for if this statement were made at a time when it was clear to both speaker and listeners that it was not raining, it would not be considered to be either true or false, but only irrelevant. In mathematics the result of connecting two propositions with a conditional is invariably interpreted as a proposition. This means that we must assign it a truth value even when it does not appear to assert anything. Moreover, we shall make the truth value depend only on the truth values of the antecedent and consequent, not on any "relation" between their internal meanings.

2-3.1. In mathematics and logic the conditional proposition

if p, then q

is defined to be true when *p* is false or *q* is true; it is false only when *p* is true and *q* is false. Thus it is by definition tautologically equivalent to *q or not-p*.

This definition has the properties we want. Suppose it is known that *if p, then q* is true. There are then three possible combinations of truth values for the component propositions, namely, *p* and *q* both true, *p* false and *q* true, *p* and *q* both false. (a) These combinations do not permit any immediate inference concerning the truth value of *q*. (b) If *p* is also known to be true, we can infer that *q* is true. (c) But, if *p* is false, no inference can be made concerning *q*. It is easy to see that there is no other way to define the conditional as a propositional connective which has these properties.

Since properties (a), (b), and (c) are shared by most of the nonidiomatic uses of the conditional in English, it appears that the mathematical meaning of the conditional is as close as we can get to the ordinary meaning and still meet the requirement that the truth value of a conditional should depend only on the truth values of its component propositions.

Conditionals occur so frequently in mathematics that it is desirable as a matter of style to have alternative ways of expressing them. Hence we find the compound proposition written *p implies q* or *p only if q*. Often it is expressed *p is a sufficient condition for q* or *q is a necessary condition for p*. Finally the symbol ⇒ is used, as in $p \Rightarrow q$; less commonly $q \Leftarrow p$. All of the last six compounds mean exactly the same thing as *if p, then q*; the choice of which one to use depends only on linguistic convenience. In propositions which are compounded with several conditionals one customarily uses different expressions to keep the levels straight; for example, *if p implies q and q implies r, then p implies r*.

If we express the conditionals in this last formula by their meanings in terms of *or* and *not*, we obtain the tautology 2-2(4). Of course we believe this sentence on the basis of our ordinary language. Hence the fact that it is a tautology may strengthen our belief that we have given an appropriate technical definition of the conditional.

Propositions of the form *p implies q and q implies p* are very common in mathematics. Such a proposition is called a *biconditional*. It is most commonly expressed, *p is a necessary and sufficient condition for q*, or *p if and only if q*. Recently the abbreviation *p iff q* has come into vogue. The symbolic form is $p \Leftrightarrow q$. From the first form given we see that a biconditional proposition is true when the components are both true or both false, and that it is false when the components have different truth values.

It is not difficult to find English sentences in which the word *if* is intended to represent the biconditional; e.g., "A person is called a minor if he is under twenty-one." Until recently this usage has also been common in mathematical definitions, but now most authors use the more precise *if and only if* (or *iff*).

EXERCISES

(*Note.* In mathematics, as in ordinary thought, the process of reasoning is generally directed toward the simplification of complicated situations rather than the other way around. Hence it is comparatively rare to be concerned with a compound proposition as an end in itself when the truth values of its components are known. Thus, although Exercise 1 is an appropriate prelude to Exercises 2 and 3, it represents atypical reasoning.)

1. Assume that the following propositions are true: $x \in A$, $x \notin B$, $x \in C$, $y \in A$, $y \in B$, $y \notin C$, $z \notin A$, $z \notin B$, $z \in C$. Which of the following are true?

(a) $y \notin B$ or $z \in A$.
(b) $x \notin C$ and $y \in B$.
(c) $x \in C$ implies $z \in A$.
(d) ($x \in A$ or $y \in B$) implies $z \notin C$.
(e) $x \notin A$ if and only if $y \in C$.
(f) If $x \in A$ implies $z \in C$, then $y \notin A$ implies $z \in B$.
(g) If $z \in A$ implies $x \in B$, then $y \in B$.
(h) If $x \notin B$, then $y \in C$, $z \in B$, and $x \notin A$.

(i) $x \notin A$ only if $y \notin B$.

(j) $z \in C$ is a necessary condition for $y \in C$.

(k) $z \in C$ is a sufficient condition for $x \in A$.

(l) In order that $z \notin C$, it is necessary and sufficient that $x \in A$ imply $y \notin B$.

2. Suppose it is known that one of the following cases is true:

CASE I. $x \in A$, $x \in B$, $x \notin C$, $y \in A$, $y \notin B$, and $y \in C$;

CASE II. $x \notin A$, $x \in B$, $x \in C$, $y \notin A$, $y \notin B$, and $y \in C$.

Which of the following propositions are certainly true? Which are certainly false?

(a) $x \in C$ if and only if $y \in C$.

(b) $(x \in B$ and $y \in A)$ if and only if $(x \in A$ or $y \in B)$.

(c) If $x \in A$ and $y \in C$, then $x \in C$ or $y \in A$.

(d) $(x \in B$ and $y \in B)$ or $(x \in C$ and $y \in C)$.

(e) not-(if $x \in A$, then $y \in B$) implies $(x \in C$ and $y \in B)$.

3. Suppose that the following propositions are true:

$$\text{if } x \in A \text{ and } y \in B, \text{ then } y \in A;$$
$$\text{if } x \notin A \text{ or } y \in A, \text{ then } y \notin B.$$

What simple conclusion can we draw?

2–4. PROPOSITIONAL SCHEMES AND QUANTIFIERS

The sentence $x^2 = x$ is a proposition if x denotes a particular number, but it is not a proposition, being neither true nor false, if x remains unspecified. Such a sentence, which has the form of a proposition but which contains a symbol of unspecified meaning, is called a *propositional scheme* (sometimes a *propositional function*). We shall confine our attention to the case in which the symbol of unspecified meaning appears only in places where we expect the name of an object. We imagine that we have in mind a set D of objects with the following property: If $b \in D$, and we replace x by the symbol b in each of its occurrences in our scheme, then the resulting sentence is a proposition. The symbol x is called the *variable* of the scheme, and the set D is called the *domain* of the variable. Frequently, the domain will be obvious from context and left unexpressed.

A propositional scheme involving the variable x may be denoted $P(x)$. The proposition resulting from replacing x by the specific name b is then written $P(b)$.

Concerning the symbol x, it is to be understood that it does not refer to anything; its only function is to indicate where the name of an object is to be inserted. It has been very aptly called a "placeholder."

In the remainder of this chapter, we will generally denote variables by the letters x, y, and z. Such a convention could be adopted as a rule. To do so would have certain advantages, but this convention is not often followed by mathematicians; in fact, it is common to find the same symbol appearing in consecutive sentences once as a variable and once as the name of an object. This is often confusing to inexperienced readers, but after one learns to distinguish the two uses, the double significance turns out to be more of a help than a hindrance.

Besides substituting the name of an object for the variable, there are other ways of converting a propositional scheme into a proposition. Two of these are of particular importance.

We may assert that the proposition is true for all permissible substitutions. This proposition is written, in symbols,

$$(\forall x) \quad P(x).$$

The inverted A is usually read, "For all . . ." Other readings are, "For each . . . ," "For every . . . ," or, "For any . . ."

A second method of converting a scheme to a proposition is to assert that it is true for at least one object in the domain of the variable. This is written symbolically

$$(\exists x) \quad P(x).$$

The inverted E is read, "There exists . . . ," "For some . . . ," or, "There is at least one . . ."

These two ways of changing a scheme to a proposition are called *quantification*, and the symbols $\forall$ and $\exists$ are called the *universal* and *existential quantifiers*, respectively.

We can regard the universal quantifier as being an extension of the simple connective *and*. If the domain of the variable x happens to be the two-element set $\{a, b\}$, then $(\forall x) P(x)$ means the same thing as $P(a)$ *and* $P(b)$. More generally, whenever the domain of x is finite, $(\forall x) P(x)$ can be replaced by an expression involving a string of *and*'s instead of the universal quantifier. When the domain of x is infinite, $(\forall x) P(x)$ represents infinitely many statements joined together by *and*'s. Since this cannot be written out, the quantifier is necessary. Similarly, the existential quantifier is an extension of the connective *or*.

In Section 2–2, we found that there may be several different compounds of the same propositions which always have the same truth value. These were the tautologically equivalent compounds. Now that we have quantifiers, which are in effect abbreviations for long strings of *and*'s or *or*'s, we expect analogous equivalences. Corresponding to 2–2(5), we find that

$$not\text{-}\big((\forall x) \ P(x)\big) \quad \text{and} \quad (\exists x) \quad not\text{-}P(x) \qquad (1)$$

have the same truth values no matter what the scheme $P(x)$ is. Similarly,

$$not\text{-}\big((\exists x) \ P(x)\big) \quad \text{and} \quad (\forall x) \quad not\text{-}P(x), \qquad (2)$$

in analogy with 2–2(6). Such equivalent propositions are not, however, referred to as tautologically equivalent.

The equivalence of the propositions of (1) has the following interpretation: To disprove an assertion of a universal truth, i.e., to prove $not\text{-}\big((\forall x) P(x)\big)$, is equivalent to producing a single counter-example, i.e., proving $(\exists x) \ not\text{-}P(x)$.

Similarly, to disprove the existence of any object with a certain property, i.e., to prove $not\text{-}((\exists x)\, P(x))$, is equivalent to proving that every object lacks the property, that is, $(\forall x)\, not\text{-}P(x)$.

When the domain of the variable in a propositional scheme is not clear from the context, it may conveniently be expressed along with the quantifier. Thus we write $(\forall x \in D)$ or $(\exists x \in D)$. For example, the definition of inclusion (1–7.1) expressed with a quantifier is

$$\text{``}A \subseteq B\text{''} \text{ means ``}(\forall x \in A)\ x \in B.\text{''} \tag{3}$$

The latter is equivalent to

$$(\forall x)\quad x \in A \Rightarrow x \in B,$$

where now the variable x has all mathematical objects in its domain.

One particular case of quantification deserves special mention. If the set D should happen to be the empty set, then $(\forall x \in D)\, P(x)$ is true, regardless of the meaning of $P(x)$. We can see that this is the appropriate interpretation by considering the equivalent form $(\forall x)\, x \in \emptyset \Rightarrow P(x)$, which is obviously true. Of course, $(\exists x \in \emptyset)\, P(x)$ is always false. Note that these interpretations are consistent with the equivalences (1) and (2).

Mathematicians do not deliberately quantify schemes over the null set, but they often quantify over a domain whose extent is uncertain. Occasionally such a domain turns out to be empty. By accepting a void domain, we can eliminate many special cases, but there is also a possible pitfall. At first glance, $(\forall x)\, P(x)$ asserts more than $(\exists x)\, P(x)$; hence, we ought to be able to infer $(\exists x)\, P(x)$ whenever we know $(\forall x)\, P(x)$. However, this inference is valid only when we are sure that the domain of the variable is not void.

Propositional schemes involving two or more variables are a natural generalization of those involving only one. A typical mathematical example is $x \in y$. In our discussion we shall use notations such as $P(x, y)$ or $P(x, y, z)$.

The expression $(\forall y)\, P(x, y)$ becomes an ordinary proposition as soon as the variable x is replaced by the name of an object; hence it is a propositional scheme in one variable. We can therefore apply a second quantifier and obtain the propositions $(\exists x)(\forall y)\, P(x, y)$ and $(\forall x)(\forall y)\, P(x, y)$. Similar considerations apply to schemes in more variables.

Altogether there are eight ways to convert $P(x, y)$ into a proposition with the aid of universal and existential quantifiers. The propositions $(\forall x)(\forall y)\, P(x, y)$ and $(\forall y)(\forall x)\, P(x, y)$ always have the same truth value. Similarly for $(\exists x)(\exists y)\, P(x, y)$ and $(\exists y)(\exists x)\, P(x, y)$. We can summarize the situation by saying that the order of successive universal quantifiers or of successive existential quantifiers does not matter. Consequently, when the same quantifier is applied to several variables in a row, particularly when they have the same domain, it is common to abbreviate by writing, for example, $(\forall x, y)$ instead of $(\forall x)(\forall y)$ and $(\exists x, y, z)$ in place of $(\exists x)(\exists y)(\exists z)$.

The order in which quantifiers are applied is important when one is universal and the other is existential. Consider, for example, the propositions

$$(\forall x)(\exists y) \quad x < y$$

and

$$(\exists y)(\forall x) \quad x < y,$$

where the domain for both variables is the set **I.** The first asserts, "Given any integer there is a larger integer," and this is true. The second asserts, "There exists an integer which is larger than every integer," and this is false, because this integer would be larger than itself.

There are other quantifiers which will arise in our work and for which we introduce no symbolism. Thus we may write a statement like, "There is at most one x such that $P(x)$." This can be formalized with the universal quantifier as follows:

$$(\forall x, y) \quad \big(P(x) \text{ and } P(y)\big) \Rightarrow x = y.$$

Most mathematical theorems and definitions are intended to apply in a variety of situations, which means that universal quantifiers are involved. Universal quantifiers which appear at the beginning of a statement, however, are often left unexpressed. For example, (3) above is intended to apply to any sets A and B; hence it might have been written

$$(\forall Y, Z) \quad Y \subseteq Z \Leftrightarrow \big((\forall x \in Y) x \in Z\big).$$

As a general rule, any unquantified symbol appearing in a theorem or definition, except those such as $1, 2, \ldots, \emptyset, \ldots$ which have a fixed significance, should be understood as a variable and universally quantified.

EXERCISES

1. Express in terms of the ordinary quantifiers:

(a) There are at most two x's such that $P(x)$.

(b) There are at least two x's such that $P(x)$.

2. Let $\phi(x, y)$ mean x *is a parent of* y. Let M be the set of all men (living or dead) and W be the set of all women.

The proposition, *a is the mother of b*, can be written $a \in W$ and $\phi(a, b)$, while *a is the brother (possibly half-brother) of b* can be written $a \in M$ and $a \neq b$ and $(\exists x) (\phi(x, a)$ and $\phi(x, b))$.

Because ordinary language is less precise than the language of quantified statements and because there are different ways the biological facts can enter the interpretations of the problems, the answers to the following problems are not all unique.

Express with quantifiers

(a) *a* is the grandfather of *b*.

(c) *a* is the aunt of *b*.

(b) *a* is the grandson of *b*.

(d) *a* and *b* are sisters.

(e) a and b are first cousins.
(g) a is the full brother of b.
(i) Every person has at most one father.
(j) Every person has exactly two parents.
(k) Some person has at least two children.
(l) Every person has a grandfather.
(m) No one is his own grandfather.

(f) a has neither brothers nor sisters.
(h) Every person has a father.

Let $\psi(x, y)$ mean *x is the ancestor of y*. Express $\phi(a, b)$ in terms of $\psi(x, y)$ and quantifiers. Can you express $\psi(a, b)$ in terms of $\phi(x, y)$ and quantifiers?

2-5. PROOF AND INFERENCE

A mathematical proof consists of a sequence of mathematical propositions. We can classify these propositions into three categories: definitions, assumptions, and propositions which are inferred from previous propositions. Propositions of the first two kinds are always clearly identified as such, while those of the third kind are often, but not always, introduced with a word like *hence* or *therefore*.

In practice the logical inferences in a proof are rather simple (that is, the nature of the logical dependence on previous propositions is simple) except when a conscious ellipsis is made to avoid a routine verification. Consequently a detailed examination of all the different kinds of inferences is unnecessary.

We shall consider a few of the inferences that depend on the meanings of the logical connectives. If we have previously established the proposition *p implies q* and the proposition *p*, we may infer *q*. This inference is known as *modus ponens*. We should realize that the inference is not to be justified in terms of the ordinary meaning of *implies*, but rather in terms of the technical meaning given in 2–3.1, since we have agreed to use the word *implies* only in this sense. In the discussion following 2–3.1 we showed that the inference followed from the analysis of cases. By a similar analysis we find that given *p or q* and *not-p*, we may infer *q*.

Another common inference depends on the fact that *p implies q* and *not-q implies not-p* are tautologically equivalent propositions. When we have established either of these propositions, the other follows at once. (These two compounds, incidentally, are known as *contrapositive* forms of one another.) This leads to the following inference: given *p implies q* and *not-q*, we infer *not-p*. This is just *modus ponens* after we replace *p implies q* by *not-q implies not-p*. The equivalence of a conditional with its contrapositive form often figures tacitly in mathematical proofs. Yet the relation between them is not always completely trivial, for it may happen that a conditional and its contrapositive have a distinctly different intuitive flavor when the *not*'s appearing are absorbed into the internal structure of the component propositions.

It is well known that *p implies q* is not tautologically equivalent to *q implies p*. Going from one to the other is not actually a common error except in the context of a biconditional. The proposition *p if and only if q*, as we know, is tautologically equivalent to (*p implies q*) *and* (*q implies p*). To establish a biconditional one often proves the two propositions *p implies q* and *q implies p* separately. One or both of these propositions may be proved in the contrapositive form. Especially when the contrapositive form seems remote from the original implication, it is a natural mistake to prove one of these implications twice in the forms *p implies q* and *not-q implies not-p* and omit the proof of the other altogether. This mistake can be avoided by stating clearly which half of the double implication is established by each part of the proof.

Often a theorem asserts that several propositions have the same truth value. Such theorems are commonly proved by a circular chain of implications. For example, the propositions $p \Rightarrow q$, $q \Rightarrow r$, $r \Rightarrow s$, and $s \Rightarrow p$ are proved separately; then it follows that p, q, r, and s are either all true or all false.

The role of temporary assumptions deserves some discussion. They are commonly used to establish conditional propositions. We introduce a proposition p as an assumption and, after a sequence of steps, deduce q. Finally, we assert *p implies q*. While this procedure agrees with the original meaning of *p implies q*, namely, that the truth of p influences the truth of q, we must justify it for the technical meaning of *implies*. If p is false, then *p implies q* is true by definition; whatever propositions we may have deduced from the assumption p are irrelevant to the desired conclusion. If p is true, then all of the derived propositions, including q, are true; hence, *p implies q* is true. When we use this form of reasoning we must remember that the propositions from the assumption p to the conclusion q are not asserted and must not be referred to in subsequent arguments. Only the final conclusion *p implies q* is asserted.

In the course of an argument several temporary assumptions may appear. It is important to keep track of how the propositions are related to the assumptions. One easy way to do this in writing is to indent each step that is based on a new assumption. We shall illustrate this by giving a proof of the proposition *if p implies q and q implies r, then p implies r*. This is tautologically true; that is, true regardless of the internal meanings of the propositions involved; hence it is unnecessary to specify the particular propositions. We give the reasons in the right-hand column.

(a) Assume *p* implies *q* and *q* implies *r*
(b)　　*p* implies *q*　　　　　　　　　　　　　　　　by (a),
(c)　　*q* implies *r*　　　　　　　　　　　　　　　　by (a),
(d)　　assume *p*
(e)　　　　*q*　　　　　　　　　　　　　　*modus ponens* (b) and (d),
(f)　　　　*r*　　　　　　　　　　　　　　*modus ponens* (c) and (e),
(g)　　*p* implies *r*　　　　　　　　　　　　　　　　(d) through (f),
(h) If *p* implies *q* and *q* implies *r*, then *p* implies *r*　　　(a) through (g).

As we remarked earlier, tautological theorems are not very interesting. Usually the truth of a theorem depends in a significant way on the internal meanings of the propositions involved. A very simple example is provided by the proof of the theorem, *If 2^e is rational, then 4^e is rational.* This is a compound proposition whose truth we can guarantee even though we are in doubt about the truth values of the components.

(i) Assume $2^e \in \mathbf{Q}$

(j) $(2^e)^2 \in \mathbf{Q}$ by a prior theorem, presumably known to the reader, that the product of two rationals is rational;

(k) $(2^e)^2 = 2^{2e}$ by a prior theorem, usually called a law of exponents;

(l) $2^{2e} = (2^2)^e$ law of exponents;

(m) $2^2 = 4$ a theorem of arithmetic;

(n) $(2^e)^2 = 4^e$ by (k) through (m);

(o) $4^e \in \mathbf{Q}$ by (j) and (n);

(p) If $2^e \in \mathbf{Q}$, then $4^e \in \mathbf{Q}$ by (i) through (o).

The theorem which enters the proof in (j) is

$$(\forall x, y)\quad (x \in \mathbf{Q} \ and \ y \in \mathbf{Q}) \Rightarrow xy \in \mathbf{Q}. \tag{1}$$

Presumably this has been proved before we attempt the proof of (p). The domain of the variables in (1) is $\mathbf{R}$ from context. (If we replaced x by $\emptyset$ and y by $\mathbf{R}$, then $\emptyset\mathbf{R} \in \mathbf{Q}$ would be meaningless. Thus (1) is not meaningful unless the domain is such that the indicated product is defined.) Since 2^e is known to be a real number (obviously the theorem presupposes a knowledge of the theory of exponents), we can replace both x and y by 2^e in (1) and get

$$(2^e \in \mathbf{Q} \ and \ 2^e \in \mathbf{Q}) \Rightarrow (2^e)^2 \in \mathbf{Q}. \tag{2}$$

Then (j) follows from (2) and (i) by *modus ponens*. Something rather similar to the foregoing would be required in a strictly formal proof whenever we applied a general theorem to a particular case.

The derivation of (o) from (j) and (n) is immediate when we recall the meaning of the symbol "$=$" (1–2.1). Various rules are often given for the manipulation of equalities. These are indispensable in strictly formal proofs, but quite unnecessary when we base our reasoning on the meaning.

The proofs we have just given, like the other proofs in this section, are far more formally organized than anything we are likely to encounter in practice. We might find (j) through (o) condensed as follows:

$$4^e = (2^2)^e = (2^e)^2$$

and the last, being the square of a rational, is rational.

Most mathematical theorems would be very complicated if written without any abbreviations as combinations of propositional schemes with quantifiers and connectives. Usually the complexity is hidden in a few technical words. Consider, for example, the word *differentiable*. In the context of real-valued functions defined on **R**, "*f* is differentiable" means

$$(\forall x)(\exists y)(\forall \epsilon > 0)(\exists \delta > 0)(\forall h) \quad 0 < |h| < \delta \Rightarrow \left| \frac{f(x + h) - f(x)}{h} - y \right| < \epsilon.$$

Moreover, the five quantifiers do not tell the whole story, because the propositional scheme involves the technical symbols $<$, $-$, $|\ |$, and the standard notation for quotients, all of which are defined in terms of the more fundamental notions of positivity, addition, and multiplication.

Technical words and notations are necessary to make mathematics comprehensible to the human mind. They help us in our thought not merely because they are abbreviations, but also because technical concepts are often defined to mirror concepts, such as continuity, which have direct intuitive meaning. In rigorous mathematics each technical word or notation must be defined in terms of other technical words and notations. In the system we are following, we can trace each definition back to the propositional scheme $x \in y$. Here it stops; the logician says $x \in y$ is undefined, the mathematician "knows" what it means.

Among defined terms it is worth distinguishing the permanent from the temporary, although the distinction is not sharp. Words like *continuous* and symbols like "2" have a permanent significance, even though they may have more than one meaning as contexts vary. Such words and symbols are usually defined formally. On the other hand, it is quite common to introduce temporary notations and sometimes temporary technical words. These appear quite casually in proofs; for example, "Let A be the set of all real numbers that exceed their squares."

In the course of a mathematical proof there is usually much switching back and forth between technical terms and notations and the more primitive terms and notations with which the technical ones are defined. This is one of the most important ways to bring the internal meaning of a proposition into a proof. We shall illustrate this process by proving (1). At the same time we shall find a new use of temporary assumptions.

Recall first the definition of *rational*. To say that a real number a is rational means that it can be expressed as a quotient of two integers. In symbols,

$$a \in \mathbf{Q} \Leftrightarrow (\exists x, y \in \mathbf{I})\, a = x/y. \tag{3}$$

Actually this should be written

$$(\forall z) \quad (z \in \mathbf{Q} \Leftrightarrow (\exists x, y \in \mathbf{I})\, z = x/y), \tag{4}$$

because the definition applies to any real number.

We now prove (1):

(i)   Let a and b be any real numbers.
(ii)   Assume $a \in \mathbf{Q}$ and $b \in \mathbf{Q}$
(iii) $(\exists x, y \in \mathbf{I})\, a = x/y$ (ii) and (4),
(iv)   pick $p, q \in \mathbf{I}$ so that $a = p/q$ by (iii); see below,

(v)	pick $r, s \in \mathbf{I}$ so that $b = r/s$	by argument similar to (iii) and (iv),
(vi)	$ab = (p/q)(r/s) = pr/qs$	from arithmetic,
(vii)	$pr \in \mathbf{I}, qs \in \mathbf{I}$	from arithmetic,
(viii)	$(\exists x, y \in \mathbf{I})\, ab = x/y$	(vi) and (vii),
(ix)	$ab \in \mathbf{Q}$	definition of rational,
(x)	$(a \in \mathbf{Q}$ and $b \in \mathbf{Q}) \Rightarrow ab \in \mathbf{Q}$	(ii) through (ix),
(xi)	$(\forall x, y \in \mathbf{R})(x \in \mathbf{Q}$ and $y \in \mathbf{Q}) \Rightarrow xy \in \mathbf{Q}.$	

The justification of this last step is important. To prove a universally quantified proposition it is sufficient to prove just one case, provided it is a "general" case. In (i) we started with *any* particular real numbers a and b and arrived at (x). Since (x) is free of any assumptions beyond the fact that a and b are real numbers, we can go to the universally quantified proposition (xi). One way to see the validity of this inference is to think of the same argument being repeated for each possible pair of real numbers. Another way is to imagine a malevolent genie who gives us the real numbers a and b. If (xi) were not true, the genie would presumably give us a particular pair of real numbers for which (x) is false. Since we proved (x) in spite of the genie, (xi) must be true.

The step from (iii) to (iv) demands further investigation. Evidently, the knowledge that two integers exist with certain properties is not quite the same thing as actually knowing two such integers. Yet this is essentially what we have claimed. In abstract mathematics we pretend to omniscience and ignore this distinction.

The switch of the letters from a and b to x and y in passing from (x) to (xi) is to emphasize that the letters in (xi) are variables, whereas a and b were supposed to be individual real numbers. Most mathematicians would write (xi) with a and b as the variables to indicate how they are related to (x). Some would say that a and b are variables all along. Then the intermediate steps in the proof are not propositions but propositional schemes. Since a propositional scheme is not a meaningful statement, this interpretation runs counter to the philosophy of this book by moving mathematics away from meaning and towards formal manipulation.

The preceding proofs illustrate in rather simple contexts many of the essential logical steps encountered in mathematics. In the remainder of the book, proofs will be given less formally. The level of formality will be about what one finds in most current research papers. Particularly at the beginning, however, proofs will be very detailed. The amount of detail and formality which will best clarify a given theorem varies from reader to reader. A student should spend some time in recasting proofs at different levels of formality. This may be helpful in understanding individual theorems, and it is certain to improve one's ability to express intuitive ideas in rigorous language.

EXERCISE

Refer to Exercise 2, Section 2–4. Prove formally that (h) $\Rightarrow$ (l).

2–6. SET FORMATION

Given a propositional scheme $P(x)$, we can imagine x being replaced successively by the names of all the members of its domain. In general, we must expect that some of the resulting propositions will be true and some false. Those for which it is true form a certain subset of the domain of x and this subset is denoted by

$$\{x \mid P(x)\}. \tag{1}$$

Thus, $\{x \mid x > 0\}$ is the set of all positive numbers and $\{x \mid (\exists y \in \mathbf{I})\, x = 2y + 1\}$ is the set of all odd integers.

The expression (1) is read, "the set of all x such that $P(x)$." A common variant of this notation is $\{x : P(x)\}$. If we also designate the set so defined by A, we can say that the scheme $P(x)$ serves as a defining criterion for membership in A. In fact,

$$(\forall x)\quad x \in A \Leftrightarrow P(x). \tag{2}$$

There are a number of variations on this set-forming notation which are almost self-explanatory. The following are typical. We may write

$$\{x \in D \mid P(x)\}$$

if we wish to specify or restrict the domain of the variable x. Again

$$\{3x \mid x \in \mathbf{I}\}$$

is a lucid abbreviation for $\{y \mid (\exists x \in \mathbf{I})\, y = 3x\}$.

Our notation makes a set correspond to each criterion of membership. Conversely, every set arises in this way, for if B is any set, then trivially, $B = \{x \mid x \in B\}$.

As we noted when discussing sets, the same set may be defined by different criteria of membership. When this happens it reflects the logical equivalence of the defining criteria. The equality

$$\{x \mid P(x)\} = \{x \mid Q(x)\}$$

is effectively the same as the proposition

$$(\forall x)\quad P(x) \Leftrightarrow Q(x).$$

Similarly, the inclusion

$$\{x \mid P(x)\} \subseteq \{x \mid Q(x)\}$$

has the force of

$$(\forall x)\quad P(x) \Rightarrow Q(x).$$

The value of the theory of sets as a vehicle for mathematical reasoning lies largely in the free interchangeability of sets and propositional schemes. A scheme may contain a very complicated notion, but the corresponding set is, at least intuitively, very simple. This conceptual simplification serves to increase our span of attention.

In discussing the notion of set, we insisted that each set be so clearly defined that all questions of membership are clearly questions of fact. In pursuance of this idea, we enunciate the principle that (with an exception to be considered in Chapter 11) we shall admit no sets for mathematical discussion which cannot be defined by a propositional scheme. From time to time, and particularly in the next chapter, we shall introduce various abbreviations which will make it possible to define sets without the explicit use of propositional schemes, but this does not abrogate our principle. Thus the list notation introduced in Section 1–6 is to be regarded as an abbreviation: $\{a, b, c\}$ for $\{x \mid x = a \text{ or } x = b \text{ or } x = c\}$.

EXERCISE

Which of the following sets are the same? What inclusion relations hold among these sets? (All variables have domain **R**.)

$A_1 = \{x \mid 0 < x < 2\}$ $A_2 = \{x \mid |x| < 1\}$

$A_3 = \{x \mid x < x^2\}$ $A_4 = \{x \mid (\exists y)\, x = y^2\}$

$A_5 = \{x \mid (\exists y)\, x = 1/y\}$ $A_6 = \{x \mid (\exists y)\, x = 3y\}$

$A_7 = \{x \mid (\exists y, z)\, x = y^2 \text{ and } y + z^2 < 1\}$ $A_8 = \{x \mid (\forall y)\, |x - y| = x - y\}$

$A_9 = \{x \mid (\forall y)\, |x - y| = |y| - |x|\}$ $A_{10} = \{x \mid (\exists y)\, y^2 + xy + x = 1\}$

2-7. THE SET-THEORETIC PARADOXES

In Section 2–6 we established a notation which enables us to replace complicated logical statements by sets. Unfortunately, if one does this too freely, a formal contradiction can arise. There are a number of these so-called paradoxes of set theory. We shall give here what is known as Russell's paradox.

Let us consider the set A of all sets which are not members of themselves; in symbols,

$$A = \{x \mid x \notin x\}. \tag{1}$$

Familiar sets like **I** or **R** are not members of themselves, for **I** is certainly not an integer, nor is **R** a real number. But the set of all sets appears to be a member of itself. Admittedly there is a certain circularity in defining this set, because we ought to have the members of a set clearly defined "before" assembling them into a set. This circularity is the substance of Russell's paradox.

According to 2–6(2),

$$(\forall x) \quad x \in A \Leftrightarrow x \notin x.$$

Since the domain of x is the collection of all possible sets and A itself is a set, we can replace x by A and obtain

$$A \in A \Leftrightarrow A \notin A,$$

a self-contradictory statement.

It is not easy to give precise formal rules for specifying sets by propositional schemes which will be free of this and other paradoxes and nevertheless suffice for set theory as it is actually used in mathematics. Moreover, it is even known to be impossible to give a system of rules which is both adequate for mathematics and demonstrably free of paradoxes. The best we can do is to give rules which are adequate and free of all known paradoxes.*

There are several systems which accomplish this. One of the most convenient of these is based on the idea that some sets are too large to be members of other sets. Russell's paradox is avoided in this system because the set A above is one of these overlarge sets and is not therefore a candidate for membership in itself or, indeed, in any other set. Although we shall not discuss formal rules for set theory in this book, we shall in fact confine ourselves throughout to set-theoretic devices which are admissible under the axiomatic treatment given by Gödel in *The Consistency of the Continuum Hypothesis* (Princeton University Press, 1940).

2–8. DUMMY VARIABLES

We have already remarked that the x appearing in a propositional scheme like $P(x)$ does not refer to anything but serves only to mark the place where an object name is to be put. When the scheme is quantified the role of x becomes even smaller. $(\forall x) P(x)$ is a proposition, and the x which appears no longer marks a place where substitutions are to be made; in a sense all possible substitutions for x have already been made. It is quite unimportant that the letter used is x. In fact, we can replace x by any letter we choose, so long as it remains clear what scheme is being quantified. Thus $(\forall y) P(y)$ and $(\forall z) P(z)$ have the same meaning as $(\forall x) P(x)$, but confusion would ensue if we wrote $(\forall P) P(P)$. Of course, one avoids letters which are ordinarily assigned a fixed significance. For example, it would be all right to use e or π as a variable in an algebraic context, but it would be confusing to do so in an analytic context, because both these letters are usually reserved for numerical constants in analysis.

Starting from a propositional scheme $Q(x, y)$ in two variables, we might quantify one to get the scheme $(\forall y) Q(x, y)$. Here we could use any letter except x or Q

---

* A formal system of mathematics is called *consistent* if and only if it contains no paradoxes; that is, if it is impossible to prove any proposition of the form q *and not-q* by following the rules of the system. Gödel has proved that no consistent system is powerful enough to prove its own consistency. Oddly, any inconsistent system can be used to demonstrate its own consistency. For suppose that it is possible to prove q *and not-q*. Let p be the proposition, "This system is consistent." (We assume, of course, that the system is powerful enough to *state* this proposition.) Since $(q$ *and not-q)* $\Rightarrow p$ is a tautology, it is surely provable in any system which formalizes all of mathematics. Now, p follows by *modus ponens*. Thus the existence within a system of a proof of its own consistency is no guarantee that it is consistent. Gödel went farther. Assuming a system is powerful enough to prove the usual theorems about the arithmetic of the integers, including the unique factorization theorem, he showed that every system that contains a proof of its own consistency is actually inconsistent. To do this, he gave a recipe for converting any proof of consistency into the proof of a self-contradictory statement.

in the place of y, and the resulting scheme would be the same. Suppose that we quantify this new scheme existentially. In so doing we could replace x by any other letter except y or Q:

$$(\exists x)(\forall y) \quad Q(x, y)$$

and

$$(\exists z)(\forall y) \quad Q(z, y) \tag{1}$$

have the same meaning. Since the first quantified variable might have been represented by z instead of y, still another notation for the same proposition is

$$(\exists y)(\forall z) \quad Q(y, z). \tag{2}$$

Compare (1) and (2) carefully. They express the same proposition, but $(\exists y)(\forall z) \, Q(z, y)$ does not.

The same ideas apply to the variable in a scheme used as a criterion of membership. Thus $\{x \mid x^2 > x\}$ might just as well be $\{t \mid t^2 > t\}$.

In any such situation, the variable x is referred to as a *bound* or *dummy* variable. When dummy variables appear it is important to recognize them as such, because it is frequently necessary to change the letter employed to avoid notational confusion. Such changes are often made without comment to the puzzlement of inexperienced readers. A familiar example of a dummy variable is the variable of integration in a definite integral. The x's appearing in $\int_0^1 \sin tx \, dx$ might as well be replaced by y's, z's, or u's, but not by t's.

Some of the difficulties associated with variables are due to the practice of denoting them with letters. We can avoid this at the expense of another kind of complication. We recall that a variable is really a blank space where the name of an object is to be written. Let us mark all such blanks with the same symbol $\square$. If we wish to quantify, we connect the quantifier symbol to the appropriate box by a line. The proposition appearing as (1) or (2) would be

$$(\exists \,)(\forall \,) \quad Q(\square, \square). \tag{3}$$

It is understood that lines which cross are not to be regarded as connected. If several boxes are to be filled with the same name, then they are all connected together. The theorem, "Every real number can be written in the form $x^3 + x$," would appear

$$(\forall \,)(\exists \,) \quad \square^3 + \square = \square.$$

This requires the convention that T-joints *are* connections.

This notation can be used with the set-defining notation of Section 2-6. As an example, the set $\mathbf{Q}$ of rational numbers may be defined as

$$\mathbf{Q} = \{ \, \in \mathbf{R} \mid (\exists \, \in \mathbf{I})(\exists \, \in \mathbf{I}) \, \square = \square / \square \}. \tag{4}$$

For a complicated proposition this notation would involve so many connecting lines that it would be extremely difficult to read. Nevertheless, it can be helpful in understanding the role of dummy variables and the use of quantifiers. We shall illustrate by repeating the proof of 2–5(1) that the product of two rational numbers is rational; we rewrite it as follows:

$$(\forall \cdot \in \mathbf{R})(\forall \cdot \in \mathbf{R}) \ (\square \in \mathbf{Q} \text{ and } \square \in \mathbf{Q}) \Rightarrow \square\square \in \mathbf{Q}. \tag{5}$$

Proof.

(a) Assume $a \in \mathbf{R}$ and $b \in \mathbf{R}$

(b)   assume $a \in \mathbf{Q}$ and $b \in \mathbf{Q}$

(c) $(\exists \cdot \in \mathbf{I})(\exists \cdot \in \mathbf{I}) \ a = \square/\square$ by (4) and (b),

(d)   choose p and q in $\mathbf{I}$ so that $a = p/q$ by (c),

(e) $(\exists \cdot \in \mathbf{I})(\exists \cdot \in \mathbf{I}) \ b = \square/\square$ by (4) and (b),

(f)   choose r and s in $\mathbf{I}$ so that $b = r/s$ by (e),

(g) $ab = (p/q)(r/s) = (pr)/(qs)$ arithmetic,

(h) $pr \in \mathbf{I}, \ qs \in \mathbf{I}$ arithmetic,

(i) $(\exists \cdot \in \mathbf{I})(\exists \cdot \in \mathbf{I}) \ ab = \square/\square$ by (g) and (h),

(j) $ab \in \mathbf{Q}$ by (4) and (i),

(k) $(a \in \mathbf{Q} \text{ and } b \in \mathbf{Q}) \Rightarrow ab \in \mathbf{Q}$ by (b) through (j),

(l) $(\forall \cdot \in \mathbf{R})(\forall \cdot \in \mathbf{R}) \ (\square \in \mathbf{Q} \text{ and } \square \in \mathbf{Q}) \Rightarrow \square\square \in \mathbf{Q}$ by (a) through (k).

Note that every letter appearing in this proof is supposed to represent a specific object; every variable that appears at all is a dummy. This will be true of all the statements we shall encounter except when we are discussing logic.

Once this notation is understood it is clear that the literal variables of the classical notation serve only to indicate the "wiring" of the box notation. The propositions denoted by (1) and (2) are the same, because the letters indicate the same wiring (3); the letters used to indicate the wiring have no relevance to the meaning.

One can have many of the advantages of the wired-box notation without the typographical complications by reserving special letters or symbols for use only as dummy variables. Most mathematicians, however, find that the use of letters similar to those used for objects aids in understanding a proposition or proof. Menger has suggested a compromise: Use corresponding letters in different type faces to denote objects and dummy variables.

CHAPTER 3

THE SET-THEORETIC MACHINERY

Although the set concept can be very helpful in elementary mathematics, its role is in fact only supporting. In advanced mathematics, however, a great deal of the essential logical maneuvering is transferred to the domain of sets. As we noted before, the transfer is accomplished by the correspondence between propositional schemes and sets established in Section 2–6. In this chapter we will define the usual devices for combining sets with one another and for building new sets from old.

First, we will study the simplest set combinations, which mirror the simple logical connectives. Our next goal will be the idea of function in its full generality. Then we return to more sophisticated set combinations, which reflect the logical quantifiers. In the last section we will define the general notion of direct product. Although this definition is important in mathematics, it is not required in the remainder of this book except in a few exercises.

With the exception perhaps of the last section, the ideas in this chapter are quite simple, indeed, deceptively simple. They are, however, absolutely indispensable for the study of modern mathematics.

3–1. BINARY SET COMBINATIONS

3–1.1. Definition. The *union* of two sets A and B is the set obtained by merging A and B. It is denoted by $A \cup B$. In symbols,

$$A \cup B = \{x \mid x \in A \text{ or } x \in B\}.$$

3–1.2. Definition. The *intersection* of two sets A and B is the set of all objects common to A and B. It is denoted $A \cap B$. In symbols,

$$A \cap B = \{x \mid x \in A \text{ and } x \in B\}.$$

The symbols $\cup$ and $\cap$ are read "cup" and "cap," respectively. Whitney, who proposed these names, also suggested the following mnemonic for distinguishing them. The cup sign, since it will hold water, denotes a more inclusive set than the cap sign, which will not.

This terminology and notation are now virtually standard, but some authors use the words *sum* and *product* in place of the words *union* and *intersection*, respectively. The corresponding notation is $A + B$ and $A \cdot B$ or AB, in analogy with ordinary algebra. This analogy goes farther than mere terminology, as we shall see below. Other terms are *join* and *meet* for *union* and *intersection*, respectively.

3–1.3. Proposition. For any sets A, B, and C:

$$A \cup A = A; \tag{1}$$
$$A \cap A = A; \tag{2}$$
$$A \cup B = B \cup A; \tag{3}$$
$$A \cap B = B \cap A; \tag{4}$$
$$A \cup (B \cup C) = (A \cup B) \cup C; \tag{5}$$
$$A \cap (B \cap C) = (A \cap B) \cap C; \tag{6}$$
$$A \cap (B \cup C) = (A \cap B) \cup (A \cap C); \tag{7}$$
$$A \cup (B \cap C) = (A \cup B) \cap (A \cup C); \tag{8}$$
$$A \subseteq B \text{ if and only if } A \cap B = A; \tag{9}$$
$$A \subseteq B \text{ if and only if } A \cup B = B; \tag{10}$$
$$A \subseteq C \text{ and } B \subseteq C \text{ if and only if } A \cup B \subseteq C; \tag{11}$$
$$A \supseteq C \text{ and } B \supseteq C \text{ if and only if } A \cap B \supseteq C. \tag{12}$$

The proofs of these formulas are left as exercises except for the discussion of (8) below.

Formulas (1) and (2) are known as idempotent laws; (3) and (4), as commutative laws; (5) and (6), as associative laws; and (7) and (8), as distributive laws. If we write out (3) through (7) in the $+$ and juxtaposition notation, we get $A + B = B + A$, $AB = BA$, $A + (B + C) = (A + B) + C$, $A(BC) = (AB)C$, and $A(B + C) = AB + AC$. This brings out the remarkable analogy between the union and intersection operations on sets and the sum and product operations on numbers. But (8) becomes $A + BC = (A + B)(A + C)$, which is not usually true for numbers.

3–1.4. Extended binary combinations. We shall often write expressions like $A \cup B \cup C$ or $A \cap B \cap C \cap D$. There are two ways in which we might interpret such expressions. We might imagine that $A \cap B \cap C \cap D$ is an abbreviation for $((A \cap B) \cap C) \cap D$, thus going back to the binary combinations. Altogether there are five ways to interpret $A \cap B \cap C \cap D$ as the result of iterating the binary combinations, but these all denote the same set by virtue of (6). Alternatively we might define

$$A \cap B \cap C \cap D = \{x \mid x \in A, x \in B, x \in C, \text{ and } x \in D\}.$$

If we were trying to write everything in a strictly formal notation, we would have to clarify all such matters in advance. We rely on the meaning, however, and therefore our notation need only be unambiguously interpretable by an intelligent human being who can translate it into a more formal notation if the occasion demands (say, when we are trying to check proofs with a computer). We shall therefore leave out unnecessary parentheses in such expressions without further comment. Furthermore, it is clear that we can rearrange these longer expressions without changing their meaning. Thus $A \cup B \cup C \cup D \cup E = B \cup D \cup C \cup E \cup A$. Extended forms of the distributive laws (7) and (8) are also valid. For example, $A \cap (B \cup C \cup D) = (A \cap B) \cup (A \cap C) \cup (A \cap D)$.

On the other hand, for the present, we must eschew expressions like $A_1 \cup A_2 \cup \ldots \cup A_n$, because it is not at all clear how we could formalize them unless n represents a definite fixed integer. Later, in Chapter 13, we shall prove a theorem in terms of which such expressions can be formalized.

3-1.5. Proposition. Suppose A, B, and C are sets such that $A \subseteq B$. Then

$$A \cup C \subseteq B \cup C \tag{13}$$

and

$$A \cap C \subseteq B \cap C. \tag{14}$$

The proof is obvious. The idea here extends to a general principle. Suppose we have a set defined as a combination of other sets using $\cup$ and $\cap$. If any component of this combination is replaced by a larger set, then the combination is made larger (not strictly larger, but in the sense of $\subseteq$). For example, suppose $A \subseteq B$; then $(A \cup C) \cap (D \cup (A \cap E)) \subseteq (A \cup C) \cap (D \cup (B \cap E))$. Once again we are in no position to formalize this principle as a theorem, but we may apply it in individual instances because the proof in each case can be given using (13) and (14).

3-1.6. Identities and tautologies. As we noted in Chapter 2, set inclusions and equalities reflect logical implications and equivalences. Similarly, the binary set operations, union and intersection, reflect the logical connectives *or* and *and* as shown by their definitions. Each of the identities (1) through (12) is related to a tautology. For example,

$$A \cup (B \cap C) = \{x \mid x \in A \text{ or } (x \in B \text{ and } x \in C)\}$$

and

$$(A \cup B) \cap (A \cup C) = \{x \mid (x \in A \text{ or } x \in B) \text{ and } (x \in A \text{ or } x \in C)\};$$

hence (8) is the same as

$$(\forall x) \quad ((x \in A \text{ or } (x \in B \text{ and } x \in C)) \Leftrightarrow ((x \in A \text{ or } x \in B) \text{ and } (x \in A \text{ or } x \in C))). \tag{15}$$

We prove this latter statement as follows. First we note that

$$(p \text{ or } (q \text{ and } r)) \Leftrightarrow ((p \text{ or } q) \text{ and } (p \text{ or } r)) \tag{16}$$

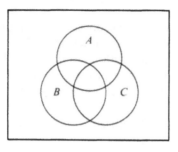

FIG. 3-1. A Venn diagram for three sets.

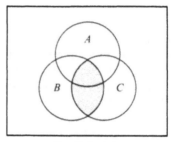

$B \cap C$ shaded

FIGURE 3-2

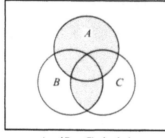

$A \cup (B \cap C)$ shaded

FIGURE 3-3

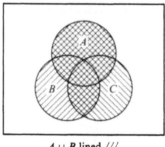

$A \cup B$ lined ///
$A \cup C$ lined \\\

FIGURE 3-4

is a tautology. Choose an arbitrary object a and replace p by $a \in A$, q by $a \in B$, and r by $a \in C$ in (16). This gives

$$(a \in A \text{ or } (a \in B \text{ and } a \in C)) \Leftrightarrow ((a \in A \text{ or } a \in B) \text{ and } (a \in A \text{ or } a \in C)).$$

Since a was chosen arbitrarily, we may deduce (15). Any identity involving $\cup$ and $\cap$ can be proved by a similar argument.

3-1.7. Venn diagrams. A convenient method of verifying set relations is through the use of Venn diagrams. We imagine our sets A, B, and C to be subsets of a rectangle in the plane and draw a figure showing each, making sure to allow for all possible overlappings. Such a figure is called a Venn diagram (Fig. 3-1).

To verify (8) we now shade the portion of Fig. 3-1 corresponding to $B \cap C$ (see Fig. 3-2), and then the portion corresponding to $A \cup (B \cap C)$ (see Fig. 3-3). Then compute $(A \cup B) \cap (A \cup C)$ in the same way (see Figs. 3-4 and 3-5). Since $A \cup (B \cap C)$ and $(A \cup B) \cap (A \cup C)$ represent the same portion of the Venn diagram, we conclude that these sets are the same, not only in this particular case of plane sets, but for any sets A, B, and C!

This leap from the particular to the general can be justified as follows. The eight regions of the Venn diagram correspond to the eight cases to be considered

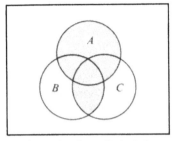
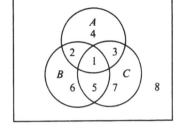

$(A \cup B) \cap (A \cup C)$ shaded

FIGURE 3–5   FIGURE 3–6

TABLE 3–1.   Part of the verification of the tautology (16)

Region	p	q	r	p or q	p or r	(p or q) and (p or r)
1	T	T	T	T	T	T
2	T	T	F	T	T	T
3	T	F	T	T	T	T
4	T	F	F	T	T	T
5	F	T	T	T	T	T
6	F	T	F	T	F	F
7	F	F	T	F	T	F
8	F	F	F	F	F	F

in establishing that (16) is a tautology (see Fig. 3–6). The regions numbered 1 through 8 correspond to the possible assignments of truth values to p, q, and r in Table 3–1. For example, if a is a point in the region numbered 5, then $a \notin A$, $a \in B$, and $a \in C$. Hence, if p, q, and r stand respectively for $a \in A$, $a \in B$, and $a \in C$ as in 3–1.6, they are respectively false, true, and true as in line 5 of Table 3–1. Now $(A \cup B) \cap (A \cup C)$ is represented by the union of regions 1, 2, 3, 4, and 5. This corresponds to the fact that (p or q) and (p or r) is true on the first five lines of Table 3–1 and false on the last three. Because $A \cup (B \cap C)$ is represented by the same subset of the diagram, we know that the truth-value column for p or (q and r) will be the same as that for (p or q) and (p or r). Hence (16) is a tautology.

It is evident that set-theoretic relations involving any finite number of sets can be established through Venn diagrams. The essential thing is that the diagram should show the sets intersecting in the most general possible manner. If there are n sets involved, then the diagram must show 2^n regions. In practice the diagrams become unwieldy for n exceeding 4 or 5.

Venn diagrams can also be used to establish conditional set relations such as: If $A \subseteq C$, then $A \cup (B \cap C) = (A \cup B) \cap C$. We draw the diagram showing A included in C, being careful to take B in "general position" with respect to the other sets. (See Fig. 3–7.)

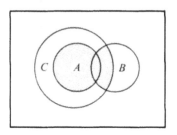

FIG. 3–7. Venn diagram for three sets with the condition that $A \subset C$. The shaded region is either

$$A \cup (B \cap C) \text{ or } (A \cup B) \cap C.$$

3–1.8. Definition. Two sets A and B are said to be *disjoint* if and only if $A \cap B = \emptyset$.

3–1.9. Definition. Let A and B be two sets. The set

$$\{x \mid x \in A \text{ and } x \notin B\}$$

is called the *difference*, *A minus B*, and denoted $A - B$.

The notation using the minus sign is a holdover from the $+$ and juxtaposition notation for union and intersection. Although we can find analogies between set differences and ordinary subtraction which continue those relating set unions and intersections to addition and multiplication (for example, $A \cap (B - C) = (A \cap B) - (A \cap C)$), set differences are more closely related to intersections than to unions because of the logical *and* appearing in the definition. Other notations are in common use, for example, $A \backslash B$, but none is universal.

3–1.10. Definition. When $A \subseteq B$, the set $B - A$ is called the *relative complement of A in B* or the *complement of A relative to B*.

It frequently happens that one considers many subsets of a certain fixed set in the course of an argument. For example, in geometry, one considers subsets of space (i.e., the set of all points); in analysis one often considers subsets of **R**. Such a parent set is often called a *universal* set. If A is a subset of a universal set U, then $U - A$ is called simply the *complement of A*. Evidently one must not refer to the complement of A unless it is perfectly clear what set is to be regarded as universal.

3–1.11. Notation. When it is clear that we have a universal set at hand, we shall denote the complement of a set with $\sim$ written over single characters (e.g., $\tilde{A}$), but to the right of compounds (e.g., $(A \cup B)^{\sim}$).

The word *universal* is really much too strong. It is easy to find arguments involving more than one universal set. For example, in analytic geometry one might have both space and **R** as universal sets. Then, if A is a set of points, $\tilde{A}$ denotes its complement with respect to space; but if A is a set of real numbers, $\tilde{A}$ denotes $\mathbf{R} - A$.

Complements correspond to the logical *not-*; in fact, we might have defined

$$\tilde{A} = \{x \mid x \notin A\}, \tag{17}$$

where the universal set would be the domain of the variable x.

Corresponding to the tautologically equivalent formulas given in 2-2(5) and 2-2(6) we have the set identities

$$(A \cap B)^{\sim} = \tilde{A} \cup \tilde{B} \tag{18}$$

and

$$(A \cup B)^{\sim} = \tilde{A} \cap \tilde{B} \tag{19}$$

whenever a universal set is at hand. These are called DeMorgan's laws. They can be expressed more generally as

$$C - (A \cap B) = (C - A) \cup (C - B) \tag{20}$$

and

$$C - (A \cup B) = (C - A) \cap (C - B). \tag{21}$$

All these identities can easily be established by using Venn diagrams or by going back to the definitions in the fashion of 3-1.6.

3-1.12. Proposition. If A and B are subsets of a universal set, then $A - B = A \cap \tilde{B}$.

The significance of this obvious theorem is that it allows us to carry out manipulations involving set differences with the aid of the identities for $\cup$ and $\cap$ given earlier. We illustrate this by proving: If B and C are disjoint, then $(A - (B \cup C)) \cup C = (A \cup C) - B$. The hypothesis B *and* C *are disjoint* gives $C \subseteq \tilde{B}$ and then $\tilde{B} \cup C = \tilde{B}$. Then

$$
\begin{aligned}
(A - (B \cup C)) \cup C &= (A \cap (B \cup C)^{\sim}) \cup C = (A \cap \tilde{B} \cap \tilde{C}) \cup C \\
&= (A \cup C) \cap (\tilde{B} \cup C) \cap (\tilde{C} \cup C) = (A \cup C) \cap \tilde{B} \cap U \\
&= (A \cup C) \cap \tilde{B} = (A \cup C) - B.
\end{aligned}
$$

There is perhaps some objection to our assumption that there is a universal set available. This assumption is always legitimate, because we can take U to be the union of all the sets involved; in this case, $U = A \cup B \cup C$. One might argue that there is only one truly universal set, namely, the set of all mathematical objects; but if we accept the latter as a bona fide set, we leave the door open to the set-theoretic paradoxes as suggested in Section 2-7.

3-1.13. Duality. DeMorgan's laws (18) and (19) lead to an important principle known as *duality*. Formulas (1) through (12) occur in, pairs, the second member of each pair being obtained from the first (or vice versa) by interchanging $\cup$ and $\cap$ signs, and in the case of (9) through (12) by interchanging $\subseteq$ and $\supseteq$ as well. We shall illustrate how (8) follows from (7) via DeMorgan's laws.

Let X, Y, and Z be any three sets. Since (7) is an identity, we can replace A by $\tilde{X}$, B by $\tilde{Y}$, and C by $\tilde{Z}$ to get

$$\tilde{X} \cap (\tilde{Y} \cup \tilde{Z}) = (\tilde{X} \cap \tilde{Y}) \cup (\tilde{X} \cap \tilde{Z}).$$

Take complements and we have

$$(\tilde{X} \cap (\tilde{Y} \cup \tilde{Z}))^{\sim} = ((\tilde{X} \cap \tilde{Y}) \cup (\tilde{X} \cap \tilde{Z}))^{\sim}.$$

Simplify the two members by using DeMorgan's rules:

$$(\tilde{X} \cap (\tilde{Y} \cup \tilde{Z}))^{\sim} = X \cup (\tilde{Y} \cup \tilde{Z})^{\sim} = X \cup (Y \cap Z)$$

and

$$((\tilde{X} \cap \tilde{Y}) \cup (\tilde{X} \cap Z))^{\sim} = (\tilde{X} \cap \tilde{Y})^{\sim} \cap (\tilde{X} \cap \tilde{Z})^{\sim} = (X \cup Y) \cap (X \cup Z).$$

We have used several times the obvious identity $\tilde{\tilde{A}} = A$. This gives

$$X \cup (Y \cap Z) = (X \cup Y) \cap (X \cup Z).$$

Since X, Y, and Z were chosen arbitrarily, we have the identity (8).

Similar argument will show that any identity in $\cup$ and $\cap$ remains valid when the $\cup$ and $\cap$ signs are interchanged. The identity derived in this way is called the *dual* of the original. If inclusion is involved instead of equality, the $\subseteq$ is reversed at the second step when complements are taken.

3–1.14. It would be pointless to attempt to list all the identities we will use in this book. When set manipulations are involved in a printed proof, it is customary to take fairly long steps with possibly a few hints concerning their justification and to leave the details to the reader. Hence the student must acquire a facility in the formal manipulation of sets comparable to his facility with algebraic manipulation. It is suggested that each student write out the details of each manipulation he meets until he is quite confident of his ability to check them mentally.

EXERCISES

Prove the following set-theoretic identities. Do each one twice, once using Venn diagrams and once by formal manipulation using the identities in the text.

1. $(A \cup B \cup C) \cap (B \cup D) = (A \cap D) \cup B \cup (C \cap D)$.
Write and prove the dual identity also.

2. $(A \cup B) \cap (B \cup C) \cap (C \cup A) = (A \cap B) \cup (B \cap C) \cup (C \cap A)$.
Show that this identity is its own dual.

3. $(A \cup B \cup C)^{\sim} = \tilde{A} \cap \tilde{B} \cap \tilde{C}$.
This is an extended form of (19). To give a formal proof from the identities in the text, one must agree on a definite interpretation of $A \cup B \cup C$ in terms of the binary combination $\cup$, and similarly for $\cap$.

4. $(A - B) \cup C = (A \cup C) - (B - C)$.

5. If $A \cap B \cap C = \emptyset$, then $(A - B) \cup (B - C) \cup (C - A) = A \cup B \cup C$.

6. $(A - B) \cup (B - C) \cup (C - A) = (A - C) \cup (C - B) \cup (B - A)$.

7. $(A \cup (\tilde{B} \cap C))^\sim = (B - A) \cup (\tilde{A} - C)$.

8. The *symmetric difference* of two sets, often denoted by $A \oplus B$, is defined as $(A - B) \cup (B - A)$. Show that for all A, B, C,

$$A \oplus A = \emptyset, \quad A \oplus \emptyset = A, \quad A \oplus B = B \oplus A,$$
$$A \oplus (B \oplus C) = (A \oplus B) \oplus C, \quad A \cap (B \oplus C) = (A \cap B) \oplus (A \cap C).$$

3-2. THE POWER SET

3-2.1. Definition. If A is any set, the power set of A, denoted by $\mathfrak{P}(A)$, is given by

$$\mathfrak{P}(A) = \{B \mid B \subseteq A\}.$$

3-2.2. We shall have frequent occasion to iterate this power set construction. To simplify the notation we shall abbreviate $\mathfrak{P}(\mathfrak{P}(A))$ by $\mathfrak{P}^2(A)$, $\mathfrak{P}(\mathfrak{P}(\mathfrak{P}(A)))$ by $\mathfrak{P}^3(A)$, etc. These abbreviations will involve only definite exponents in the formal part of our work, so it is unnecessary to have any theory of counting to justify them. However, there are some "unofficial" exercises below in which the exponent is used in a more general sense.

EXERCISES

1. Prove: If $A \subseteq B$, then $\mathfrak{P}(A) \subseteq \mathfrak{P}(B)$.

*2. Prove: $(\forall n \in N) \mathfrak{P}^n(\emptyset) \subseteq \mathfrak{P}^{n+1}(\emptyset)$.

*3. How many elements are there in $\mathfrak{P}^n(\emptyset)$?

*4. Suppose that A is a set which can be built up from $\emptyset$ using the list notation repeatedly, as in Section 1-6. Show that for some n, $A \in \mathfrak{P}^n(\emptyset)$.

3-3. ORDERED PAIRS AND DIRECT PRODUCTS

As we stressed in the previous chapter, a set has no internal organization; no member comes before any other. But there is need for new kinds of sets which have some internal organization. The simplest of these new kinds of objects is the ordered pair. This is to be a "set" with two members, one of which comes before the other.

The need for ordered pairs gives us a choice. We might turn to our intuitive sources for the idea of ordered pair in the same way that we did for the idea of set. For the purposes of this book and for almost all purposes of mathematics this would be perfectly satisfactory. But if we should want to formalize mathematics completely, this would cause considerable trouble, because the rules for the manipulation of symbols would have to recognize the distinction between sets and ordered pairs and treat these two kinds of objects differently. The other alternative is to find a set in the sense of Chapter 1 which has the properties of an intuitive ordered pair. This turns out to be possible. The definition of ordered pair given below is the first of a long series of technical devices which impose internal structure on sets without introducing any new intuitive concepts.

3–3.1. Definition. If a and b are any two objects, then the set $\{\{a\}, \{a, b\}\}$ is denoted by $\langle a, b \rangle$ and called an *ordered pair*.

An ordered pair, so defined, has the properties of an intuitive ordered pair. This is the content of the following theorem.

3–3.2. Theorem. If $\langle a, b \rangle = \langle c, d \rangle$, then $a = c$ and $b = d$.

Proof. The hypothesis can be written $\{\{a\}, \{a, b\}\} = \{\{c\}, \{c, d\}\}$. Thus $\{a\} \in \{\{c\}, \{c, d\}\}$. Hence, $\{a\} = \{c\}$ or $\{a\} = \{c, d\}$. In either case $c \in \{a\}$ so $c = a$. Now we may rewrite the hypothesis

$$\{\{a\}, \{a, b\}\} = \{\{a\}, \{a, d\}\}. \tag{1}$$

It remains for us to show that $b = d$. We consider two cases.

CASE 1: $a = b$. Then $\{a, b\} = \{a\}$ and the hypothesis reduces to $\{\{a\}\} = \{\{a\}, \{a, d\}\}$. Then $\{a, d\} \in \{\{a\}\}$ so $\{a, d\} = \{a\}$. Finally $d \in \{a\}$ so $d = a$. This gives $d = b$ in this case.

CASE 2: $a \neq b$. Then $b \notin \{a\}$, hence $\{a, b\} \neq \{a\}$. But $\{a, b\} \in \{\{a\}, \{a, d\}\}$ by (1); hence either $\{a, b\} = \{a\}$ or $\{a, b\} = \{a, d\}$. The first alternative is false. Therefore $\{a, b\} = \{a, d\}$. We have then $b \in \{a, d\}$, so $b = a$ or $b = d$. Again the first alternative is false in this case, so $b = d$.

3–3.3. Definition. Let $\langle a, b \rangle$ be an ordered pair. Then a is called *the first element* of $\langle a, b \rangle$ and b is called *the second element* of $\langle a, b \rangle$.

Theorem 3–3.2 is important in justifying the use of the definite article (in *the first element*) in this definition. The first element of an ordered pair must be an object associated with the ordered pair itself, not with the symbols used to denote the ordered pair. If it were true that $\langle a, b \rangle = \langle c, d \rangle$ but $a \neq c$, say, then both a and c would be first elements of $\langle a, b \rangle$.

We shall occasionally need ordered triples, ordered quadruples, and ordered quintuples. We can define these with the aid of ordered pairs.

3–3.4. Notation. An abbreviation for $\langle \langle a, b \rangle, c \rangle$ is $\langle a, b, c \rangle$; an abbreviation for $\langle \langle a, b, c \rangle, d \rangle$ is $\langle a, b, c, d \rangle$; an abbreviation for $\langle \langle a, b, c, d \rangle, e \rangle$ is $\langle a, b, c, d, e \rangle$. We shall refer to these as ordered triples, quadruples, and quintuples, respectively.

Although we shall not write the formal definition, we will not hesitate to refer, say, to the third element of an ordered quintuple; it is uniquely determined, of course.

We could go on to define ordered sextuples, septuples, etc., but, since we do not yet have a theory of counting, we cannot define an ordered n-tuple. Actually we shall not use n-tuples for $n > 5$.

We might equally well have defined $\langle a, b, c \rangle = \langle a, \langle b, c \rangle \rangle$, etc. There will always be times when one way is preferable to the other, but the difference is rarely, if ever, important.

3–3.5. Definition. Let A and B be any sets. The set of all ordered pairs having first element in A and second element in B is called the *direct product* (or *Cartesian product*) of A and B. It is denoted by $A \times B$. In symbols,

$$A \times B = \{ \langle a, b \rangle \mid a \in A, b \in B \}.$$

A familiar example is $\mathbf{R} \times \mathbf{R}$, the set of all ordered pairs of real numbers, that is, the plane of analytic geometry. By analogy it is quite common to refer to a and b as the first and second *coordinates* of $\langle a, b \rangle$.

3–3.6. Extended direct products. Unlike unions and intersections, direct products are not associative; that is, $(A \times B) \times C \neq A \times (B \times C)$. The distinction, however, is often overlooked. We shall discuss this further in 3–4.17. To be consistent with 3–3.4, we let $A \times B \times C$ be an abbreviation for $(A \times B) \times C$, and similarly for longer direct products.

There are a number of identities involving direct products, unions, and intersections. Some of these are given in the exercises. Like the identities given in Section 3–1, these are usually applied without proof or reference. In checking such identities, one may find it helpful to think of $A \times B$ as a rectangle with A and B as its sides (Fig. 3–8).

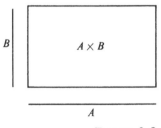

FIGURE 3–8

3–3.7. Definition. The subset $\{ \langle a, a \rangle \mid a \in A \}$ of $A \times A$ is called the *diagonal* of $A \times A$. The subset $\{ \langle a, a, a \rangle \mid a \in A \}$ of $A \times A \times A$ is called the *diagonal* of $A \times A \times A$, and similarly for longer direct products of A with itself.

EXERCISES

1. Why do we not define $\langle a, b \rangle = \{ a, \{ a, b \} \}$ instead of 3–3.1?

2. Would it be appropriate to define $\langle a, b, c \rangle = \{ \{ a \}, \{ a, b \}, \{ a, b, c \} \}$ instead of 3–3.4?

3. Show that $A \times B \subseteq \mathfrak{P}^2(A \cup B)$.

4. Prove the following identities:

$(A \cup B) \times C = (A \times C) \cup (B \times C)$;

$(A \cap B) \times C = (A \times C) \cap (B \times C)$;

$(A \times B) \cap (C \times D) = (A \cap C) \times (B \cap D)$.

5. Prove: $A \times B = \emptyset$ if and only if $A = \emptyset$ or $B = \emptyset$.

6. The following cancellation law is invalid. If $X \times Y = X \times Z$, then $Y = Z$. Prove a corrected statement.

3-4. FUNCTIONS

The idea of function has gradually broadened during the past three hundred years. Once confined strictly to numerical functions which could be expressed with some fairly simple formula, the idea now includes any rule whatever which assigns a definite object to each member of some set.

In one respect the concept has narrowed. Formerly it was accepted that a function might be multiple valued; that is, the rule might sometimes be ambiguous in its assignments. But this usage has almost disappeared. Modern writers use the word *function* only for rules which are unambiguous.

Just as it was undesirable to take ordered pair as a new intuitive concept for formal purposes, so it is undesirable to define a function as a new kind of thing. We shall see that any function, thought of intuitively as a rule, can be recorded faithfully as a set; then we shall take this set as being the function. Hence a function will be a set with certain special properties.

When we deal with a numerical function, say the sine function, we often use a table of values. This is a listing of numbers x with the corresponding values of $\sin x$. Any such table contains only a finite number of entries, and the values of $\sin x$ are usually rounded off. But we can imagine a table which lists each real number x and beside it $\sin x$. We can describe such a table in the language of sets. For each real number x, we form the ordered pair $\langle x, \sin x \rangle$ and then the set of all such ordered pairs,

$$G = \{\langle x, \sin x \rangle \mid x \in \mathbf{R}\}.$$

Clearly, all information about the sine function is recoverable from a knowledge of this set. We shall define the sine function to be G.

The set G is a subset of $\mathbf{R} \times \mathbf{R}$. If we think of $\mathbf{R} \times \mathbf{R}$ as the plane of analytic geometry, G is the set we ordinarily call the *graph* of the sine function. Under the formal definition every function will be its own graph.

As things stand, there is a circularity here, because we have not defined G except in terms of the sine function. The sine function is rather complicated by our present standards, hence it is not surprising that we are unable to define it now. A function closer to our level of technical competence is

$$\{\langle 0, 0 \rangle, \langle 1, 1 \rangle, \langle 2, 4 \rangle, \langle 3, 9 \rangle\}.$$

This is the function which assigns the value 0 to 0, 1 to 1, 4 to 2, and 9 to 3. It is a subset of the function which assigns to each real number its square. All other functions will be sets of ordered pairs of a similar character.

Not every set of ordered pairs will do, however. Consider

$$\{\langle 0, 0\rangle, \langle 1, 1\rangle, \langle 2, 4\rangle, \langle 0, 5\rangle\}.$$

If we try to interpret this as a table of values, there is ambiguity concerning what object is to be assigned to 0. A function must not contain two ordered pairs with the same first element.

3–4.1. Definition. A *function f* is a set of ordered pairs such that distinct members of *f* have distinct first elements; that is,

$$(\forall x, y, z) \quad (\langle x, y\rangle \in f \text{ and } \langle x, z\rangle \in f) \Rightarrow y = z. \tag{1}$$

(See 3–4.11 for a modified definition of function which is in common use.)

A simple but nevertheless important example of a function is the identity function. If *S* is any set, then

$$\{\langle s, s\rangle \mid s \in S\}$$

(i.e., the diagonal of $S \times S$) is the *identity function of S*.

The words *transformation, map, mapping,* and *operator* are often used as synonyms for function. Sometimes these words are reserved for special kinds of functions.

3–4.2. Definitions. Let *f* be a function. The *domain* of *f* is the set of all objects which are the first elements of the members of *f*; that is,

$$\text{domain } f = \{x \mid (\exists y) \langle x, y\rangle \in f\}.$$

A member of the domain of *f* is sometimes called an *argument* of *f*.

The *range* of *f* is the set of all objects which are the second elements of the members of *f*; that is,

$$\text{range } f = \{x \mid (\exists y) \langle y, x\rangle \in f\}.$$

A member of the range of *f* is sometimes called a *value* of *f*.

Requirement (1) tells us that, if $x \in \text{domain } f$, then there is a unique value *y* of *f* such that $\langle x, y\rangle \in f$.

3–4.3. Definition. Let *f* be a function and let $x \in \text{domain } f$. The unique *y* for which $\langle x, y\rangle \in f$ is called the *value of f at x* and is denoted by $f(x)$.

If *f* denotes the identity function of *S*, then $\text{domain } f = \text{range } f = S$ and, for all $s \in S, f(s) = s$.

Suppose X is any set and c any object. Let $g = X \times \{c\}$. Then g is a function with domain $g = X$ and range $g = \{c\}$, and for all $x \in X$, $g(x) = c$. Such a function is often referred to as a *constant* function.

There are many variations in the functional notation used by mathematicians. Of these we single out three as being particularly common. The parentheses may be omitted; that is, fx in place of $f(x)$. The notation may be reversed: xA, where x is the argument and A is the function. The argument may appear as a subscript: φ_y, where φ is the function.

When the domain of a function is a subset of a direct product $A \times B$, one usually writes $f(a, b)$ in place of $f(\langle a, b \rangle)$. Such a function is often referred to as a function of two variables. Thus a function of two real variables is a function whose domain is a subset of $\mathbf{R} \times \mathbf{R}$. A function of three variables is one whose domain is a subset of a triple direct product. Its value at $\langle u, v, w \rangle$ is usually denoted by $f(u, v, w)$.

In classical mathematics one often finds a function f written $f(x)$, where the x plays the role of placeholder, as in the notation for propositional schemes. In modern abstract mathematics there are many contexts in which it is essential to distinguish between the function and its values; consequently it is usually desirable to avoid a notation for a function, like $f(x)$, which does not make the distinction clear.

3–4.4. Let f be a function with domain A. Let B be any set containing range f. In this situation we shall say that f is a function *from A to B*, or that f *carries* or *maps A into B*, or that f is *defined on A with values in B*. This is often expressed in symbols by $f \mid A \to B$.

When $B = \text{range } f$, this fact is often expressed by the word *onto*: f is a function from A *onto* B. The term *surjective* and its variants are also used in place of *onto*; for example, f is a *surjective* function from A to B or f *surjects* A on B. Note that the property of being onto or surjective is not a property of f alone, but a property of f and B together.

3–4.5. Although we have defined a function as a set, we usually think of a function in the more primitive sense as a rule for computing $f(x)$ from x. Correspondingly, when we want to define an individual function, we rarely do so by defining a set of ordered pairs as such; instead we give a formula for finding $f(x)$. A familiar example is, "Let $f(x) = x^3 + 2$," which is intended to define a function f with domain $\mathbf{R}$ (or possibly some other set, such as $\mathbf{Q}$ or $\mathbf{I}$, for which the formula is sensible). Evidently this means the same thing as, "Let

$$f = \{\langle x, y \rangle \mid y = x^3 + 2\}.\text{"}$$

There is of course no requirement that the function be defined by a single expression. The only requirements are those contained in the definition: the rule must be unambiguous in its assignment of a value to each argument, and the associated ordered pairs must form a well-defined set.

When two functions are defined by formulas, one must not assume that, because the formulas are the same, the functions are the same; the domains might be different. On the other hand, functions defined by different formulas may be the same. Since functions are sets of ordered pairs, two functions f and g are the same if and only if domain f = domain g and $(\forall x \in \text{domain } f)\, f(x) = g(x)$. A trivial example is given by $f(x) = x^2 - 1$ and $g(x) = (x - 1)(x + 1)$ for $x \in \mathbf{R}$.

Occasionally a function is defined by a rule which is not obviously unambiguous. We might define $g(a)$ to be the solution of the equation $x^5 + x = a$. Such a definition should be accompanied by a proof that there is indeed only one solution of the equation. The details in a simple case like this one might well be omitted, but one would expect at least some gesture in the direction of showing uniqueness. In this case, we might write, "For each real number a there is exactly one real solution of the equation $x^5 + x = a$. Denote it by $g(a)$."

Whereas elementary mathematics usually deals with single functions from one familiar set to another, there is a growing tendency in advanced mathematics to consider functions having quite sophisticated sets as domains or ranges. One example will suffice here.

Let $\mathfrak{F}$ be the set of all functions from $\mathbf{R}$ to $\mathbf{R}$. For each real number x we define a function h_x from $\mathbf{R}$ to $\mathbf{R}$ by $h_x(y) = x^2 + xy - y^3$. Then $h_x \in \mathfrak{F}$. Now, h itself is a function (with argument written as a subscript) from $\mathbf{R}$ to $\mathfrak{F}$. Detached from any context, this construction is a pointless formality. But as we move up the ladder of abstraction, we find that constructions such as this become commonplace and formal precision in defining them becomes our only guarantee of correctness.

It is a nuisance to introduce a new symbol for a function which is mentioned only a few times. In the classical theory of real functions one refers to the function x^2. This has the disadvantage of failing to distinguish the function from its values. In many situations this disadvantage is compensated by the simplicity of the notation. A common notation which avoids the problem is to write $x \to x^2$ for the function which assigns to each number its square. This notation is particularly useful when one has a function of two variables and wants to keep one of them fixed. For example, if the function h_x in the previous paragraph came up in a context requiring only a passing reference, we would not introduce any new symbol but would simply denote the function by $y \to x^2 + xy - y^3$ (or $t \to x^2 + xt - t^3$, since y is a dummy in this context).

Although the arrow notation is a symbol for a function and is presumably governed by the rules for such symbols, one does not see notations like

$$(x \to x^3 + x^2 + 2x - 1)\left(\frac{3y^2 + y - 2}{2y^3 + 1}\right).$$

3–4.6. Extensions and restrictions. Since functions are sets, it may happen that two functions f and g are related by inclusion. If $f \subseteq g$, then we say that f is a *restriction* of g or that g is an *extension* of f.

Suppose, for example, that f and g are defined by

$$f(x) = \sum_{n=0}^{\infty} x^n \quad \text{for} \quad -1 < x < 1$$

and

$$g(x) = \frac{1}{1 - x} \quad \text{for} \quad x \in \mathbf{R} - \{1\}.$$

Here we have assigned each function its "natural" domain; that is, the set of all real numbers for which the formula is sensible. It is a well-known and important fact that g is an extension of f.

If g is a given function and $A \subseteq$ domain g, then the function f with domain A defined by $f(x) = g(x)$ is called the *restriction of g to A* or *g restricted to A*. Evidently the restrictions of a function are uniquely determined by the original function and the new domain. On the other hand, extensions of functions to strictly larger domains can always be made in many ways, since we can define the extension arbitrarily at new points of the domain. The problem of extending functions is of interest only when we seek an extension with some special property like continuity. In such cases the existence or nonexistence of an extension may be highly significant.

3–4.7. Composition of functions. If f and g are functions, then $h(x) = f(g(x))$ defines a function h with domain

$$\{x \in \text{domain } g \mid g(x) \in \text{domain } f\}.$$

This function h is called the *composition of f and g*; it is denoted by $f \circ g$. (The simpler fg is also very commonly used.)

In practice, one rarely forms the composition unless a substantial part of the range of g is in the domain of f. The following familiar case is typical. Say $f(x) = \sqrt{x}$ and $g(x) = 1 - x^2$, where the domain of f is the set of nonnegative reals and domain g is $\mathbf{R}$. The domain of $f \circ g$ is the interval $[-1, 1]$. The definition is sensible in any event, but we must be prepared for the possibility that $f \circ g$ has the null set for domain. This means that $f \circ g$ is the null set of ordered pairs. Since there is only one empty set, $f \circ g = \emptyset$. One does not ordinarily think of $\emptyset$ as a function, but it satisfies the definition.

3–4.8. Theorem. The composition of functions is associative. That is, for any three functions f, g, and h, $f \circ (g \circ h) = (f \circ g) \circ h$.

Proof. Both $f \circ (g \circ h)$ and $(f \circ g) \circ h$ are given by the formula $f(g(h(x)))$ and in both cases the domain is the set of all x for which the formula is sensible.

3–4.9. Definition. A function f is called *injective* if and only if

$$(\forall x, y \in \text{domain } f) \quad f(x) = f(y) \Rightarrow x = y.$$

The compound word *one-to-one* is sometimes used synonymously. The word *bijective* means *both injective and surjective*. Since surjectivity is not, strictly speaking, a property of a function, neither is bijectivity.

3–4.10. Given any function f, we can form the set

$$\{\langle x, y \rangle \mid \langle y, x \rangle \in f\}$$

consisting of all the ordered pairs of f "turned around." Then f is injective if and only if this new set is itself a function. When this is the case, the new function is denoted by f^{-1} and called the *inverse* of f. Evidently, domain f^{-1} = range f, range f^{-1} = domain f,

$$(\forall x \in \text{domain } f) \quad f^{-1}(f(x)) = x,$$

and

$$(\forall x \in \text{range } f) \quad f(f^{-1}(x)) = x.$$

We can express the last two statements by

$$f^{-1} \circ f \text{ is the identity function on domain } f$$

and

$$f \circ f^{-1} \text{ is the identity function on range } f.$$

Suppose f is any function. It is injective if and only if there exists a function g such that $g \circ f$ is the identity function on domain f. The function g, if it exists, need not be unique, because any extension of f^{-1} will do.

Let f be a function from A to B. It is bijective if and only if there exists a function g from B to A such that $g \circ f$ and $f \circ g$ are the identity functions on A and B, respectively.

3–4.11. Another usage. In many areas of modern mathematics a new and slightly different formalization of the function concept is used. The idea is that a function should not be merely a set of ordered pairs as defined in 3–4.1 but should have attached to it a definite set, called the *codomain*, which includes its range. The word *map* is commonly used in this connection. For this subsection only we shall define a *map* to be an ordered pair $\langle f, B \rangle$ consisting of a function (as defined in 3–4.1) and a set B, the codomain of the map, such that range $f \subseteq B$.

The definitions for functions require a few modifications for their application to maps. Suppose φ is the map $\langle f, \text{codomain } \varphi \rangle$. By definition domain φ = domain f. When $x \in$ domain φ, $\varphi(x)$ denotes $f(x)$. The notation $\varphi \mid A \rightarrow B$, where φ is a map, means that domain φ = A and codomain φ = B. One says "φ is injective" instead of "f is injective." The terms *surjective* and *bijective* now apply quite properly to maps; for example, "φ is surjective" means that "range f = codomain φ."

Suppose $\psi = \langle g, \text{codomain } \psi \rangle$ is a second map. One does not compose the two maps unless the domain of one is the codomain of the other. If codomain ψ =

domain φ, then $\varphi \circ \psi$ (or more commonly $\varphi\psi$) denotes the map $\langle f \circ g,$ codomain $\varphi \rangle$. Thus the domain of $\varphi \circ \psi$ is always the domain of ψ and its codomain is always the codomain of φ.

The notation for restriction can be subsumed into the notation for composition. Suppose that S is a subset of domain φ. Let i be the map $\langle g,$ domain $\varphi \rangle$, where g is the identity function on S. Then $\varphi \circ i$ is the map $\langle f$ restricted to $S,$ codomain $\varphi \rangle$. Any map whose function component is an identity function is called an *inclusion* map. If its domain also coincides with its codomain, it is called an *identity* map.

This modification of the function concept has considerable advantages in many branches of mathematics, but we shall have no need of it in the remainder of this book, and we shall use the noun *map* as a synonym for *function*.

3–4.12. Suppose that f is a function from A to B. If X is any subset of A, then

$$\overline{f}(X) = \{b \in B \mid (\exists x \in X)\, b = f(x)\}$$

and is called the *image of X under f*. This defines a function $\overline{f}$ from $\mathfrak{P}(A)$ to $\mathfrak{P}(B)$ called the map *induced* by f. We recognize $\overline{f}(A)$ as range f and $\overline{f}(X)$ as the range of f restricted to X.

Except in unusual situations, for example if there are subsets of A which are also members of A, one uses the same symbol f to denote the induced map.

3–4.13. Theorem. Suppose that f is a map from A to B and let $\overline{f}$ be the induced map from $\mathfrak{P}(A)$ to $\mathfrak{P}(B)$. If f is injective, then so is $\overline{f}$. If f is surjective, so is $\overline{f}$. If f is bijective, so is $\overline{f}$.

Proof. Suppose that f is injective. Let $X, Y \in \mathfrak{P}(A)$ and suppose that $\overline{f}(X) = \overline{f}(Y)$. We must prove $X = Y$.

Take any $x \in X$. Then $f(x) \in \overline{f}(X)$; so $f(x) \in \overline{f}(Y) = \{f(y) \mid y \in Y\}$. Pick $y \in Y$ so that $f(x) = f(y)$. Since f is injective, $x = y$. Therefore $x \in Y$. This shows that $X \subseteq Y$. The opposite inclusion can be proved in a similar fashion; hence $X = Y$.

Now, suppose that f is surjective. Let $Z \in \mathfrak{P}(B)$. Let $X = \{a \mid f(a) \in Z\}$. Then $(\forall x \in X)\, f(x) \in Z$, hence $\overline{f}(X) \subseteq Z$. Pick any $z \in Z$. We can choose $a \in A$ so that $f(a) = z$, since f is surjective. By the definition of X, $a \in X$, and so $z = f(a) \in \overline{f}(X)$. Therefore $Z \subseteq \overline{f}(X)$, whence $Z = \overline{f}(X)$. This proves that $\overline{f}$ is surjective.

The last statement is an immediate consequence of the other two.

This theorem is very simple and its proof labors the obvious. However, its triviality becomes less apparent if we consider the following fact. Let f be a map from A to B. There may not be an induced map f^* from $A \cup \mathfrak{P}(A)$ to $B \cup \mathfrak{P}(B)$, even if f is bijective. (Give an example.)

We omit the proof of the following similar theorem.

3–4.14. Theorem. Suppose that f is a map from A_1 to A_2 and g is a map from B_1 to B_2. Let h denote the map $\langle a_1, b_1 \rangle \rightarrow \langle f(a_1), g(b_1) \rangle$ from $A_1 \times B_1$ to $A_2 \times B_2$. If both f and g are injective, then so is h. If both f and g are surjective, so is h. If both f and g are bijective, so is h.

The map h is called the map *induced by f and g.*

3–4.15. Suppose that f is a function from A to B. If X is any subset of B, then

$$\{a \in A \mid f(a) \in X\}$$

is called the *inverse image* (or *counterimage*) *of X under f.* It is usually denoted by $f^{-1}(X)$. When X reduces to a one-element set $\{x\}$, it is common to write $f^{-1}(x)$ instead of $f^{-1}(\{x\})$.

The notation f^{-1} is mildly ambiguous. In the present context it represents a function from $\mathfrak{P}(B)$ to $\mathfrak{P}(A)$. Previously, it denoted the function inverse to f (assuming that such a function exists). Suppose, for a moment, that f is bijective. Denote the inverse of f by g. Then for any $X \in \mathfrak{P}(B)$, $f^{-1}(X)$ (in the sense of this subsection) and $g(X)$ (i.e., $\bar{g}(X)$ in the sense of 3–4.12) are the same. There is therefore no ambiguity in most circumstances. If there is any possibility of a misunderstanding, it should be avoided by using a special notation.

3–4.16. Suppose f is a function from A to B. It induces a function f from $\mathfrak{P}(A)$ to $\mathfrak{P}(B)$ ($\bar{f}$ in 3–4.12, but here simply f, as is customary) and a function f^{-1} from $\mathfrak{P}(B)$ to $\mathfrak{P}(A)$. There are many relations between these induced functions and the binary set combinations. Like the set combinations, these are usually applied without comment. We list a few, leaving the proofs to the reader. For all X, X_1, $X_2 \in \mathfrak{P}(A)$ and all Y, Y_1, $Y_2 \in \mathfrak{P}(B)$,

$$f^{-1}(f(X)) \supseteq X; \tag{2}$$
$$f(f^{-1}(Y)) \subseteq Y; \tag{3}$$
$$f(X_1 \cup X_2) = f(X_1) \cup f(X_2); \tag{4}$$
$$f(X_1 \cap X_2) \subseteq f(X_1) \cap f(X_2); \tag{5}$$
$$f^{-1}(Y_1 \cup Y_2) = f^{-1}(Y_1) \cup f^{-1}(Y_2); \tag{6}$$
$$f^{-1}(Y_1 \cap Y_2) = f^{-1}(Y_1) \cap f^{-1}(Y_2). \tag{7}$$

3–4.17. There are a number of useful functions canonically associated with direct products. Most important are

$$\langle a, b \rangle \rightarrow a \quad \text{from} \quad A \times B \text{ to } A$$

and

$$\langle a, b \rangle \rightarrow b \quad \text{from} \quad A \times B \text{ to } B.$$

These are called, respectively, the *first* and *second coordinate projections.* The direct product of three sets is the domain of three coordinate projections as well as three more projections like

$$\langle a, b, c \rangle \rightarrow \langle a, c \rangle.$$

Similarly for longer direct products.

As we remarked before, direct products are not associative. The sets $(A \times B) \times C$ and $A \times (B \times C)$ are not the same (except in trivial cases), but there is evidently a canonical bijection

$$\langle \langle a, b \rangle, c \rangle \rightarrow \langle a, \langle b, c \rangle \rangle$$

from one to the other. One can often afford to "identify" objects mated by this bijection; that is, to ignore the distinction between them. In such cases the direct product is effectively associative.

There is a canonical bijection from $A \times B$ to $B \times A$: $\langle a, b \rangle \rightarrow \langle b, a \rangle$. However, one must be careful about identifying $A \times B$ with $B \times A$ with the aid of this map. When $A = B = \mathbf{R}$, for instance, such an identification would erode the distinction between $\langle 1, 2 \rangle$ and $\langle 2, 1 \rangle$.

3–4.18. Families. Sometimes a function is referred to as a *family*, its domain as the *index set* of the family, and its values as *members* of the family. These names are used when our primary interest is focused on the values of the function, that is, on the members of the family.

The most common kind of family is a sequence, which is a family having $\mathbf{N}$ as its index set. The phrase "a sequence of real numbers" calls up an image of infinitely many real numbers written in a row, such as

$$1, 2, 4, 8, 16, \ldots$$

Technically, however, a sequence is a function; in this case it is presumably the function defined by $f(n) = 2^{n-1}$.

When a function is thought of as a family, it is customary to use a notation like

$$\{A_i \mid i \in I\}.$$

Here A is the function with the argument written as a subscript, and I is the index set. We might refer to a family of sets $\{A_i\}$ *indexed by* the set I.

When the index set of a family is a direct product, say $I \times J$, one typically denotes the members of the family with a double subscript $A_{i,j}$ rather than $A_{\langle i,j \rangle}$.

Families serve as sets in which the members do not all have similar positions. We say that a set B appears twice in the family $\{A_i\}$ if there are two indices i and j for which $B = A_i = A_j$. Thus a family can be regarded as a set permitting multiple memberships. Since the integers have a well-established order, the members of a sequence acquire that order, and the same will be true whenever the index set is ordered. (The general concept of order is discussed in Chapter 6.)

Any set A is the range of a function; for example, the identity function on A. Hence it is always possible to assign indices to the members of A, and thus regard A as a family.

EXERCISES

1. The following sets are all functions:

$$f_1 = \{\langle 1, 1\rangle, \langle 2, 1\rangle, \langle 3, 2\rangle, \langle 4, 0\rangle\}, \qquad f_2 = \{\langle 2, 1\rangle, \langle 4, 1\rangle, \langle 1, 2\rangle, \langle 4, 1\rangle\},$$
$$f_3 = \{\langle 0, 2\rangle, \langle 2, 2\rangle, \langle 1, 4\rangle, \langle 3, 0\rangle\}, \qquad f_4 = \{\langle 4, 1\rangle, \langle 1, 2\rangle, \langle 2, 1\rangle, \langle 0, 5\rangle\}.$$

Write out the domain and range of each of them. Which of them are injective? Is any one a restriction of another? Compute $f_1 \circ f_2$ and $f_2 \circ f_3$. Then compute $(f_1 \circ f_2) \circ f_3$ and $f_1 \circ (f_2 \circ f_3)$.

If f stands for f_3 restricted to $\{0, 1\}$, what is the map $\bar{f}$ induced by f on $\mathfrak{P}(\{0, 1\})$?

2. Are the coordinate projections of a direct product $A \times B$ injective? surjective?

3. Using the notation of 3–4.12, we find that $f \to \bar{f}$ is a map from the set of all functions from A to B to the set of all functions from $\mathfrak{P}(A)$ to $\mathfrak{P}(B)$. Is it injective? surjective?

4. Let f be a map from A to B. Let f^* be the induced map from $A \times A$ to $B \times B$ as in 3–4.14. Let $\bar{f}$ be the induced map from $\mathfrak{P}(A)$ to $\mathfrak{P}(B)$ as in 3–4.12 and let $\bar{\bar{f}}$ be the induced map from $\mathfrak{P}^2(A)$ to $\mathfrak{P}^2(B)$. Show that f^* is $\bar{\bar{f}}$ restricted to $A \times A$. (Recall Exercise 3, p. 39.)

5. Let $\mathfrak{F}$ be the set of all functions from A to B. Let $\mathfrak{G}$ be the set of all functions from $\mathfrak{F}$ to B. Explain fully how the formula $\varphi(a)(f) = f(a)$ defines a function φ from A to $\mathfrak{G}$. Is φ injective? surjective?

6. Let f be a function from A to B. Show that

(a) f is injective if and only if (2) is an equality for every subset X of A;
(b) f is surjective if and only if (3) is an equality for every subset Y of B;
(c) f is injective if and only if (5) is an equality for every two subsets X_1 and X_2 of A.

7. Suppose that f is a function from A to B. If $X \subseteq A$ and $Y \subseteq B$, show that $f(X \cap f^{-1}(Y)) = f(X) \cap Y$.

3–5. RELATIONS

There are many two-place propositional schemes which we think of as asserting a relation between two objects. In the real world we have "x is indebted to y" and "x is the father of y." Among mathematical examples, we find "$x < y$." Often, as in the case of the last example, a symbol expressing the relationship is written between the objects related.

Evidently, we know a relation R completely if we know exactly which pairs of objects are related; that is, if we know

$$\{\langle a, b\rangle \mid a \, R \, b\}.$$

This latter set is sometimes called the *graph* of the relation R. As in the case of functions, we shall define a relation for technical purposes as being its graph.

3–5.1. Definition. A *relation* is a set all of whose members are ordered pairs. A subset of $A \times B$ is sometimes called a *relation from A to B*, and a subset of $A \times A$ is often called a *relation in* the set A.

The diagonal Δ of $A \times A$ is the relation which expresses the conceptual relation of identity, since $\langle a, b \rangle \in \Delta$ if and only if $a = b$.

Obviously every function is a relation, but the converse is not true. In general a relation R can be regarded as a multiple-valued function whose value at x is any y such that $\langle x, y \rangle \in R$.

Sometimes a relation in the sense above is called *binary* to distinguish it from a *ternary* relation, which is a set of ordered triples, or a *quaternary* relation.

3–5.2. Notation. If R is a relation, then $\langle x, y \rangle \in R$ is sometimes written $x \, R \, y$.

We shall see later that the inequality relation $\leq$ for the real numbers is the set

$$\leq \, = \, \{\langle x, y \rangle \mid (\exists z) \, x + z^2 = y\}.$$

Then the fact $\langle 1, 2 \rangle \in \, \leq$ will be expressed $1 \leq 2$, as usual.

3–5.3. Definitions. Let R be a relation. The *domain* of R is the set of all first elements of the members of R. The *range* of R is the set of all second elements of the members of R. In symbols,

$$\text{domain } R \, = \, \{x \mid (\exists y) \, \langle x, y \rangle \in R\},$$
$$\text{range } R \, = \, \{x \mid (\exists y) \, \langle y, x \rangle \in R\}.$$

Evidently these definitions are extensions of the definitions already given for functions. It is clear that $R \subseteq (\text{domain } R) \times (\text{range } R)$.

EXERCISES

1. If R and S are relations, then
$$R \circ S \, = \, \{\langle x, y \rangle \mid (\exists z) \, \langle x, z \rangle \in S \text{ and } \langle z, y \rangle \in R\}$$
is a relation, called the *composition* of R and S. If R and S should be functions, show that this definition coincides with the previous definition of the composition of functions. Show also that the composition of relations is associative; that is,
$$(R \circ S) \circ T \, = \, R \circ (S \circ T).$$

2. If R is a relation and X is any set, define $\overline{R}(X) = \{y \mid (\exists x \in X) \, \langle x, y \rangle \in R\}$. Show that $\overline{R}(\overline{S}(X)) = \overline{(R \circ S)}(X)$ for any relations R and S and any set X. Restricting our attention to sets X which are subsets of the domain of R, we can regard $\overline{R}$ as a function from $\mathfrak{P}(\text{domain } R)$ to $\mathfrak{P}(\text{range } R)$. If $\overline{R} = \overline{S}$, must $R = S$?

3–6. INDEXED UNIONS AND INTERSECTIONS

In Section 3–1 we defined the union and intersection of two sets. By extension we obtained the union and intersection of any finite number of sets. But we need new notations for expressing the union or intersection of infinitely many sets.

These new notations are related to $\cup$ and $\cap$ in the same way that $\exists$ and $\forall$ are related to the binary connectives *or* and *and*.

3-6.1. Definition and notation. Suppose $\{A_i \mid i \in I\}$ is a family of sets and $K \subseteq I$. Then

$$\{x \mid (\exists i \in K)\, x \in A_i\}$$

is denoted by $\bigcup_{i \in K} A_i$. Assuming that K is not empty, then

$$\{x \mid (\forall i \in K)\, x \in A_i\}$$

is denoted by $\bigcap_{i \in K} A_i$. These sets are called the *union* and *intersection* of the family $\{A_i \mid i \in K\}$, respectively.

The condition $K \neq \emptyset$ in the case of intersections is necessary, because

$$\{x \mid (\forall i \in \emptyset)\, x \in A_i\}$$

seems to refer to the inadmissible set of all mathematical objects. But when all of the sets A_i are subsets of a clearly defined universal set U, then $\bigcap_{i \in \emptyset} A_i$ is often interpreted as U; i.e., as $\{x \in U \mid (\forall i \in \emptyset)\, x \in A_i\}$. Usually this interpretation is quite natural and causes no difficulties.

Suppose $B = \bigcup_{i \in I} A_i$. It follows immediately from the definition that $(\forall i \in I)\, A_i \subseteq B$. Moreover, if C is a set such that $(\forall i \in I)\, A_i \subseteq C$, then $B \subseteq C$. Hence we may say that the union of the sets A_i is the smallest set containing them all. In the same way the intersection of the sets A_i is the largest set contained in them all. Reasoning with unions and intersections is often based on these facts.

The index i appearing in the notation is a dummy, since $\bigcup_{j \in K} A_j$ and $\bigcup_{i \in K} A_i$ denote the same set.

The notation for unions and intersections is frequently abbreviated. Thus we might have simply $\bigcup_K A_i$ and $\bigcap_K A_i$. When $K = I$, the union is commonly written $\bigcup_i A_i$ or even $\bigcup A_i$. Similarly, for intersections.

If the set K is finite, then $\bigcup_{i \in K} A_i$ can be written out with the aid of the binary union; for example, if $K = \{1, 2, 3\}$, then $\bigcup_{i \in K} A_i = A_1 \cup A_2 \cup A_3$.

If the index set is $\mathbf{N}$, then the notations

$$\bigcup_{i=1}^{\infty} A_i \qquad \text{or} \qquad \bigcup_{i=1}^{\infty} A_i$$

are commonly used in place of $\bigcup_{i \in \mathbf{N}} A_i$. If K is a string of consecutive integers, say $K = \{m, m+1, \ldots, n\}$, then we use $\bigcup_{i=m}^{n} A_i$.

It is often necessary to denote the union of a set $\mathcal{C}$ of sets which are not provided with indices. Since we can always make $\mathcal{C}$ into a family (the identity function with index set α), such a union is subsumed under our notation. It would be written $\bigcup_{X \in \alpha} X$. A less common notation for this union is $\bigcup \alpha$. The corresponding intersection would be $\bigcap \alpha$. These notations have the advantage of avoiding the sometimes troublesome dummy variable, but the indexed notation causes less trouble in practice.

Among the many formulas analogous to those of Section 3-1 for binary unions and intersections, we formalize just a few:

$$(\bigcup_i A_i) \cap B = \bigcup_i (A_i \cap B), \tag{1}$$

$$(\bigcap_i A_i) \cup B = \bigcap_i (A_i \cup B), \tag{2}$$

$$B - \bigcup_i A_i = \bigcap_i (B - A_i), \tag{3}$$

$$B - \bigcap_i A_i = \bigcup_i (B - A_i), \tag{4}$$

$$\bigcup_i \bigcup_j A_{i,j} = \bigcup_j \bigcup_i A_{i,j} = \bigcup_{\langle i,j \rangle \in I \times J} A_{i,j}, \tag{5}$$

$$\bigcap_i \bigcap_j A_{i,j} = \bigcap_j \bigcap_i A_{i,j} = \bigcap_{\langle i,j \rangle \in I \times J} A_{i,j}. \tag{6}$$

Here (1) and (2) are the distributive laws and (3) and (4) are DeMorgan's laws. These laws are the basis for generalizing the duality principle of 3-1.13 to indexed unions and intersections. Any identity valid for all sets remains valid if all $\cup$ and $\cap$ signs and all $\bigcup$ and $\bigcap$ signs are interchanged. These interchanges reverse all inclusions. Formulas (1) through (6) form three dual pairs. Both (3) and (4) require qualification if the set of indices involved is void, and we use the special interpretation for the null intersection. Then (3) is valid only if B is the universal set and (4) requires that B be a subset of the universal set.

The left-hand member of (5) is intended to represent the union $\bigcup_i B_i$ of the B_i's, where $B_i = \bigcup_j A_{i,j}$ is the union of the family $\{A_{i,j} \mid j \in J\}$ for each fixed i. Similarly for the other double union in (5) and the two double intersections in (6). These formulas may be regarded as combination associative and commutative laws.

Among the many relations connecting set images and inverse images we single out four. Suppose f is a function from A to B. Then for any family $\{X_i\}$ of subsets of A,

$$f(\bigcup_i X_i) = \bigcup_i f(X_i) \tag{7}$$

and

$$f(\bigcap_i X_i) \subseteq \bigcap_i f(X_i). \tag{8}$$

For any family $\{Y_i\}$ of subsets of B,

$$f^{-1}(\bigcup_i Y_i) = \bigcup_i f^{-1}(Y_i) \tag{9}$$

and

$$f^{-1}(\bigcap_i Y_i) = \bigcap_i f^{-1}(Y_i). \tag{10}$$

EXERCISE

Prove the following theorem which concerns a construction of frequent applicability.

Theorem. Suppose $\{f_i \mid i \in I\}$ is a family of functions and $A_i = $ domain f_i. Then $\bigcup_i f_i$ is a function if and only if

$$(\forall i, j \in I)(\forall x \in A_i \cap A_j) \quad f_i(x) = f_j(x). \tag{11}$$

When (11) is satisfied it is appropriate to call $\bigcup_i f_i$ the *least common extension* of the functions f_i. Why? What is its domain? its range?

3-7. INDEXED DIRECT PRODUCTS

We shall define a possibly infinite direct product analogous to the possibly infinite union of Section 3–6. Although this construction will not be used in this book except in a few comments and exercises, it is included for completeness, because it is very commonly used in modern mathematics.

3-7.1. Definition. Let $\{A_i \mid i \in I\}$ be a family of sets. Let $K \subseteq I$. The set of all functions f from K to $\bigcup_K A_i$ such that

$$(\forall i \in K) \quad f(i) \in A_i$$

is called the direct product of the family $\{A_i \mid i \in K\}$. It is denoted by

$$\underset{i \in K}{\text{X}} A_i.$$

The index involved in this notation is a dummy. The notation is abbreviated in the same ways as the notation for unions. We may write $\text{X}_K A_i$. If $K = I$, then we write $\text{X}_i A_i$ or $\text{X}A_i$. When the index set is N, the direct product is often written $\text{X}_{i=1}^{\infty} A_i$.

If we wish to take the direct product of a set $\mathcal{C}$ of sets which is not indexed, we can use $\mathcal{C}$ as the index set and the identity function as a family to convert $\mathcal{C}$ to a family. The direct product of this family is denoted $\text{X}_{A \in \mathcal{C}} A$.

One frequently needs the direct product of a set A with itself many times. Then we consider a constant family on an index set I; that is, $A_i = A$ for all $i \in I$. The direct product of this family is then denoted by $\text{X}_{i \in I} A$ and sometimes A^I.

The direct product so defined is conceptually similar to but not quite the same as the direct product of Section 3–3. Suppose $I = \{1, 2\}$, $A_1 = B$, and $A_2 = C$. Then $\text{X}_{i \in I} A_i$ is the set of functions defined on $\{1, 2\}$ such that $f(1) \in B$ and $f(2) \in C$. Obviously,

$$f \to \langle f(1), f(2) \rangle$$

is a bijection from $\text{X}A_i$ to $B \times C$.

Similarly, if A_1, A_2, A_3 are sets, then $f \to \langle f(1), f(2), f(3) \rangle$ is a bijection from $\mathsf{X}_i \, A_i$ to $A_1 \times A_2 \times A_3$.

With the aid of such bijections we can identify finite direct products in the new sense with direct products in the old sense. Note that if the index set is finite but does not have a natural order, then we must make a choice of how it is to be ordered to make this identification.

3–7.2. Definition. If $j \in I$ and $f \in \mathsf{X}_i \, A_i$, then $f(j)$ is called the jth coordinate of f. The map $f \to f(j)$ from $\mathsf{X} A_i$ to A_j is called the jth *coordinate projection*. The discussion of 3–7.1 should clarify the rationale of these definitions.

3–7.3. Definition. If $K \subseteq I$, then $f \to (f$ restricted to $K)$ is a function from $\mathsf{X}_I \, A_i$ to $\mathsf{X}_K \, A_i$, called the *natural projection* from $\mathsf{X}_I \, A_i$ to $\mathsf{X}_K \, A_i$.

EXAMPLE. Let $I = \{1, 2, 3\}$ and $K = \{2, 3\}$. If we identify $\mathsf{X}_I \, \mathbf{R}$ with $\mathbf{R} \times \mathbf{R} \times \mathbf{R}$ as in 3–7.1 and $\mathsf{X}_K \, \mathbf{R}$ with $\mathbf{R} \times \mathbf{R}$, then the natural projection from $\mathsf{X}_I \, \mathbf{R}$ to $\mathsf{X}_K \, \mathbf{R}$ is $\langle x, y, z \rangle \to \langle y, z \rangle$, which we usually think of as projection on the yz-plane.

EXERCISES

1. Let $\{A_i \mid i \in I\}$ and $\{B_i \mid i \in I\}$ be families of sets with the same index set. Prove $(\mathsf{X}_i A_i) \cap (\mathsf{X}_i B_i) = \mathsf{X}_i (A_i \cap B_i)$.

2. Suppose that $\{A_i\}$, $\{B_i\}$, and $\{C_i\}$ are families with the same index set I. Let $j \in I$ and suppose that $A_i = B_i = C_i$ for all i in I except j. Given $A_j = B_j \cup C_j$, prove $\mathsf{X}_i A_i = (\mathsf{X}_i B_i) \cup (\mathsf{X}_i C_i)$.

3. Suppose that $\{A_{i,j} \mid \langle i, j \rangle \in I \times J\}$ is a family of sets indexed on a direct product. Define a natural bijection from $\mathsf{X}_{I \times J} \, A_{i,j}$ to $\mathsf{X}_i \, (\mathsf{X}_j \, A_{i,j})$.

4. Let $\{A_i \mid i \in I\}$ be a family of sets and suppose that $K \subseteq I$. Show that

$$f \to \langle f \text{ restricted to } K, f \text{ restricted to } I - K \rangle$$

is a bijection from $\mathsf{X}_I \, A_i$ to $(\mathsf{X}_K \, A_i) \times (\mathsf{X}_{I-K} \, A_i)$.

MATHEMATICAL CONFIGURATIONS

Since noneuclidean geometry was discovered early in the last century, there has been a continuous movement toward abstraction in mathematics. As his thought becomes more abstract, the mathematician finds it harder and harder to test his intuition against the physical world. The search for conviction leads to more detailed proofs, more carefully formulated definitions, and a continuous struggle to reach higher levels of precision.

Today's mathematics is based on set theory. Every mathematical concept is described by sets and all mathematical relationships are represented by the interlocking membership relations between the various sets of what we shall call configurations. Mathematics is thus reduced, in a sense, to glorified combinatorial problems. While this approach is decried by some for making mathematics nonintuitive, it does, in fact, lead to a new kind of intuition which is indispensable in modern algebra and valuable in all of mathematics.

In this chapter we shall describe in broad outline how mathematical structures are treated as configurations. In the remainder of the book we shall see a number of detailed examples.

4–1. STRUCTURES AND CONFIGURATIONS

Let A be any set. (What follows is sensible for infinite sets, but the reader is advised to think in terms of finite sets until he has the ideas thoroughly fixed in his mind.) Taking direct products and forming power sets, we can build up many new sets from A; for example,

$$\mathfrak{P}(A), \quad \mathfrak{P}^2(A), \quad A \times A, \quad A \times \mathfrak{P}(A), \quad \mathfrak{P}(A \times \mathfrak{P}(A)). \tag{1}$$

Any member B of any of these sets will be called a *structure* for A, and the ordered pair $\langle A, B \rangle$ will be called a *configuration*.

Let Q be another set and suppose that f is a bijection from A to Q. Now f induces a bijection from each of the sets built up from A to the corresponding set built up from Q. We denote each of these induced maps by the same symbol f. If B is a structure for A, then $f(B)$ is a structure for Q. We shall say that the configuration $\langle Q, f(B) \rangle$ is *isomorphic* to $\langle A, B \rangle$, and that f *effects the isomorphism* or that f is an *isomorphism* of the configuration $\langle A, B \rangle$ onto the configuration $\langle Q, f(B) \rangle$.

Let us consider an example of these ideas. Suppose $A = \{a, b, c, d, e\}$ and $Q = \{p, q, r, s, t\}$ (we assume, of course, that a, b, c, d, and e represent different objects so that A does indeed have five elements, and similarly for Q). Let

$$B = \{\{a, b\}, \{a, c\}, \{a, d, e\}, \{a, b, c, d\}\}.$$

Then $B \in \mathfrak{P}^2(A)$, B is a structure for A, and $\langle A, B \rangle$ is a configuration. Let

$$f = \{\langle a, p \rangle, \langle b, q \rangle, \langle c, r \rangle, \langle d, s \rangle, \langle e, t \rangle\}.$$

Then f is a bijection from A to Q. Evidently,

$$f(B) = \{\{p, q\}, \{p, r\}, \{p, s, t\}, \{p, q, r, s\}\} \in \mathfrak{P}^2(Q),$$

and $\langle Q, f(B) \rangle$ is a configuration isomorphic to $\langle A, B \rangle$.

Now let

$$R = \{\{p, t\}, \{p, q, r, t\}, \{q, s, t\}, \{r, t\}\}.$$

Then $\langle Q, R \rangle$ is isomorphic to $\langle A, B \rangle$, the isomorphism being effected by a different bijection from A to Q. But if

$$S = \{\{p, t\}, \{p, q, r, t\}, \{q, s, t\}, \{r, s\}\},$$

then $\langle Q, S \rangle$ is not isomorphic to $\langle A, B \rangle$.

The assertions of the preceding paragraph could be checked by considering in turn each of the 120 bijections from A to Q. Each of these will convert B into a structure for Q and we could systematically search for R and S among the results. Such a method could theoretically be used to solve any problem concerning the isomorphism of finite configurations, but it would be practically impossible if the basic sets were at all large. The reader no doubt made some observation such as this: The intersection of the members of B is the set $\{a\}$, while the intersection of the members of R is $\{t\}$, and the intersection of the members of S is $\emptyset$. This immediately shows that $\langle Q, S \rangle$ cannot be isomorphic to $\langle A, B \rangle$. Moreover, any isomorphism of $\langle A, B \rangle$ with $\langle Q, R \rangle$ must match a with t. Continuing in this fashion, we can find two bijections from A to Q which effect an isomorphism of the configurations $\langle A, B \rangle$ and $\langle Q, R \rangle$.

Let us analyze the preceding argument. By observation we find that the proposition

$$\bigcap\nolimits_{X \in B} X \ \text{ has exactly one member}$$

is true. It is easy to see that if g is any bijection of A, then

$$\bigcap\nolimits_{X \in g(B)} X \ \text{ has exactly one member}$$

must also be true. Since

$$\bigcap\nolimits_{X \in S} X \ \text{ has exactly one member}$$

is false, we conclude that there is no bijection g such that $g(B) = S$. Facts about a configuration which transfer in this manner to any isomorphic configuration are called *intrinsic* properties of the configuration.

On the contrary, consider the propositions

$$(\exists x, y \in A) \quad x \in y$$

and

$$(\exists x, y \in Q) \quad x \in y.$$

These are corresponding propositions concerning the configurations $\langle A, B \rangle$ and $\langle Q, R \rangle$ above. It could easily happen that the first of these is true and the second false. Thus we might have $d = \{e\}$ without any similar situation arising in Q. This is an example of a nonintrinsic property of a configuration.

When we deal with a configuration we shall consider only intrinsic properties. This amounts to saying that we disregard incidental facts concerning the configuration. Usually it is easy to decide whether or not a given statement concerning a configuration is intrinsic, but occasionally it requires a nontrivial argument. The reader should acquire the habit of checking mentally to see that each mathematical statement he meets is intrinsic.

We should note at this point that there is nothing in the preceding considerations which requires that $Q \neq A$. If $Q = A$, we may still take any bijection f from A to A, and not just the identity. Then there is no reason to expect that $f(B) = B$. Hence two different structures for the same set may determine isomorphic configurations. Continuing the example above, if

$$C = \big\{\{a, b\}, \{b, c\}, \{b, d, e\}, \{a, b, c, d\}\big\},$$

then $\langle A, C \rangle$ is isomorphic to $\langle A, B \rangle$.

In considering configurations it is necessary to keep in mind the basic set as well as the structure set, for the basic set tells us how deeply we are allowed to probe into the relations between the various sets. In the previous example, let $E = \mathfrak{P}(A)$ and $T = \mathfrak{P}(Q)$. Then $B \in \mathfrak{P}(E)$ and $S \in \mathfrak{P}(T)$, hence both $\langle E, B \rangle$ and $\langle T, S \rangle$ are configurations. Furthermore, it is easy to see that they are isomorphic, although $\langle A, B \rangle$ and $\langle Q, S \rangle$ are not.

To be absolutely precise we must remember from which set in the hierarchy (1) the structure is chosen. These sets are not disjoint; for example, $A \times A \subseteq \mathfrak{P}^2(A)$. Such automatic relations cause no difficulty. Suppose, however, that $A = \{x, y, \{x\}, \{x, y\}\}$ and $B = \{\{x\}, \{x, y\}\}$. It is an accident that B is a member of both $A \times A$ and $\mathfrak{P}(A)$. Let f be the bijection

$$\{\langle x, p \rangle, \langle y, q \rangle, \langle \{x\}, r \rangle, \langle \{x, y\}, s \rangle\}$$

from A to the set $Q = \{p, q, r, s\}$. The induced image of B in $Q \times Q$ is $\langle p, q \rangle$, while in $\mathfrak{P}(Q)$ it is $\{r, s\}$. There is no reason to expect these to be the same. If we are concerned with the configuration $\langle A, B \rangle$, we must know whether B appears as a member of $A \times A$ or $\mathfrak{P}(A)$. Such accidents have only a technical interest. It is perhaps worth noting that the reason such a set as $A \cup \mathfrak{P}(A)$ is not admitted to the hierarchy (1) is that a bijection from A to Q need not induce a bijection from $A \cup \mathfrak{P}(A)$ to $Q \cup \mathfrak{P}(Q)$. This should make clear why we were at pains to consider Theorems 3–4.13 and 3–4.14.

The following three theorems are fundamental even though they are very simple.

4–1.2. Theorem. Any configuration is isomorphic to itself.

Proof. The identity function of A is an isomorphism of $\langle A, B \rangle$ onto $\langle A, B \rangle$.

4–1.3. Theorem. If $\langle A, B \rangle$ is isomorphic to $\langle Q, R \rangle$, then $\langle Q, R \rangle$ is isomorphic to $\langle A, B \rangle$.

Proof. If f is an isomorphism of $\langle A, B \rangle$ onto $\langle Q, R \rangle$, then f^{-1} is an isomorphism of $\langle Q, R \rangle$ onto $\langle A, B \rangle$.

4–1.4. Theorem. If $\langle A, B \rangle$ is isomorphic to $\langle Q, R \rangle$ and $\langle Q, R \rangle$ is isomorphic to $\langle X, Y \rangle$, then $\langle A, B \rangle$ is isomorphic to $\langle X, Y \rangle$.

Proof. If f is an isomorphism of $\langle A, B \rangle$ onto $\langle Q, R \rangle$ and g is an isomorphism of $\langle Q, R \rangle$ onto $\langle X, Y \rangle$, then $g \circ f$ is an isomorphism of $\langle A, B \rangle$ onto $\langle X, Y \rangle$.

In many areas of mathematics one considers more general kinds of configurations involving two or more basic sets. There is no difficulty in extending the definitions to this case.

Suppose we have two basic sets A_1 and A_2. The structure would be a member of a set like

$$\mathfrak{P}(A_1 \times \mathfrak{P}(A_2)), \qquad \mathfrak{P}^2(A_1 \times A_2),$$

or   (2)

$$A_2 \times \mathfrak{P}^3(\mathfrak{P}(A_1) \times A_2).$$

The notion of isomorphism now involves two bijections: f_1 from A_1 to Q_1 and f_2 from A_2 to Q_2. We do not care whether A_1 and A_2 are disjoint or not.

Such configurations will appear only by implication in this book.

EXERCISES

In the following exercises find the pairs of configurations which are isomorphic. For each pair, either find an explicit bijection which effects the isomorphism or give reasons why there is none. Assume that distinct symbols in the basic sets represent distinct objects. (Exercise 4 is offered primarily as a puzzle.)

	Basic set	*Structural set*
1.	$\{a, b, c, d, e\}$	$\{\{a\}, \{a, b\}, \{b, c\}, \{b, d, e\}\}$
	$\{f, g, h, i, j\}$	$\{\{f, g\}, \{f, h\}, \{g, i\}, \{h, j\}\}$
	$\{p, q, r, s, t\}$	$\{\{p, q, r\}, \{p, s\}, \{p, t\}, \{s\}\}$

Basic set	Structural set
2. $\{a, b, c, d, e\}$	$\{\{a, b, c\}, \{a, c, e\}, \{b, c\}, \{b, d\}\}$
$\{f, g, h, i, j\}$	$\{\{f, g\}, \{f, h, j\}, \{g, h\}, \{g, h, i\}\}$
$\{p, q, r, s, t\}$	$\{\{p, r\}, \{p, r, s\}, \{p, t\}, \{q, r, t\}\}$
3. $\{a, b, c, d, e\}$	$\{\{a, b, c\}, \{a, c, d\}, \{a, d, e\}, \{b, c, d\}\}$
$\{f, g, h, i, j\}$	$\{\{f, g, j\}, \{f, h, j\}, \{g, h, i\}, \{h, i, j\}\}$
$\{p, q, r, s, t\}$	$\{\{p, q, r\}, \{p, q, t\}, \{p, s, t\}, \{q, r, s\}\}$
4. $\{a, b, c, d, e, f, g, h\}$	$\{\{\{a, b\}, \{c, d\}, \{e, f\}, \{g, h\}\},$
	$\{\{a, c\}, \{b, d\}, \{e, g\}, \{f, h\}\},$
	$\{\{a, d\}, \{b, c\}, \{e, h\}, \{f, g\}\},$
	$\{\{a, e\}, \{b, f\}, \{c, g\}, \{d, h\}\},$
	$\{\{a, f\}, \{b, h\}, \{c, e\}, \{d, g\}\},$
	$\{\{a, g\}, \{b, e\}, \{c, h\}, \{d, f\}\},$
	$\{\{a, h\}, \{b, g\}, \{c, f\}, \{d, e\}\}\}$
$\{s, t, u, v, w, x, y, z\}$	$\{\{\{s, t\}, \{u, w\}, \{v, y\}, \{x, z\}\},$
	$\{\{s, u\}, \{t, w\}, \{v, x\}, \{y, z\}\},$
	$\{\{s, v\}, \{t, y\}, \{u, x\}, \{w, z\}\},$
	$\{\{s, w\}, \{t, x\}, \{u, y\}, \{v, z\}\},$
	$\{\{s, x\}, \{t, z\}, \{u, v\}, \{w, y\}\},$
	$\{\{s, y\}, \{t, v\}, \{u, z\}, \{w, x\}\},$
	$\{\{s, z\}, \{t, u\}, \{v, w\}, \{x, y\}\}\}$

4–2. DEFINITIONS, POSTULATES, AND THEOREMS

Mathematics, when formalized into set theory, is concerned with configurations and classes of configurations. Terms such as *ordered set*, *field*, or *metric space* are applied to the configurations belonging to certain classes of importance. A typical definition is that of an ordered set (we state it here in a form slightly different from that on p. 72):

An *ordered set* is a configuration $\langle A, B \rangle$ such that $B \in \mathfrak{P}(A \times A)$ and

(i) $(\forall x \in A) \langle x, x \rangle \notin B$,
(ii) $(\forall x, y, z \in A)$ if $\langle x, y \rangle \in B$ and $\langle y, z \rangle \in B$, then $\langle x, z \rangle \in B$.

Had we wished to emphasize the fact that we are defining a huge class of configurations, we might have written the definition: Let

$$\mathcal{O} = \{\langle A, B \rangle \mid B \in \mathfrak{P}(A \times A), \text{(i), and (ii)}\}.$$

A member of $\mathcal{O}$ is called an *ordered set*.

The conditions appearing in the definition are called the *axioms* or *postulates* for an ordered set. The word *axiom* has long carried the connotation of being self-evident, but it is hard to find a sense in which these conditions are self-evident. The word *postulate* (from Latin *postulare*, to demand) seems more appropriate here, since we demand that a configuration satisfy our conditions before we apply the term *ordered set* to it.

Examining the postulates in the definition, we can see a qualitative distinction between the unnumbered one, which tells us where in the hierarchy 4–1(1) to look for the structural set, and the remaining two; the former is descriptive, while the latter are restrictive.

It is very important to note that the restrictive postulates (i) and (ii) are intrinsic. Hence, if one of two isomorphic configurations is an ordered set, then so is the other.

Definitions of classes of configurations will always have a similar form; there will be a descriptive postulate followed by one or more restrictive ones. Frequently the descriptive postulate will be run together with some of the restrictive ones. For example, if the structure is a function from A to A, it is clearer to say so directly than to say $f \in \mathfrak{P}(A \times A)$ and then adjoin the definition of a function. Moreover, we shall often describe a configuration as an ordered triple or longer, instead of an ordered pair. When we describe a configuration as $\langle A, B, C \rangle$, we really mean A endowed with the structure $\langle B, C \rangle$.

Often we will define a new class of configurations by simply adding new restrictive postulates to an old definition. Generally this is accomplished by defining an adjective (or other modifier) to be used in conjunction with the old term. For example: An ordered set $\langle A, B \rangle$ is said to be *linearly* ordered if and only if

$$(\forall x, y \in A) \qquad x = y \quad \text{or} \quad \langle x, y \rangle \in B \quad \text{or} \quad \langle y, x \rangle \in B. \tag{1}$$

This serves to distinguish a subclass of the class of all ordered sets. It is, of course, important that all conditions imposed be intrinsic. Although their logical significance is no different, extra conditions imposed on a class of configurations are not likely to be referred to as postulates unless the subclass they define has a certain recognized status.

Once a class of configurations has been defined, we try to find more propositions true of the whole class. These are, of course, theorems. They are obtained by the methods of deduction described in Chapter 2. Any proposition which we can prove from the defining postulates for a class of configurations is automatically true for all members of that class.

There is much variation in the nature of theorems. They often refer to two or more different configurations, sometimes of the same type, sometimes of different types. The configurations may have the same or different basic sets. In the latter case the theorem is by implication concerned with configurations involving several basic sets. One property they invariably possess is that they are always intrinsic; that is, their truth survives, *mutatis mutandis*, when any of the basic sets is replaced by one of its bijective images.

The words *lemma* and *corollary* are common synonyms for the word *theorem*. The former is used to denote theorems which are not regarded as ends in themselves but as steps in the proof of some other theorem. The term *corollary* is applied to those theorems which follow either by a short argument from another theorem or from a slight modification of the proof of another theorem. In recent years there has been a tendency to reserve the word *theorem* to describe results of importance; lesser statements are called *propositions*.

In principle the postulates for a class of configurations may be chosen arbitrarily, except for the proviso that they must be intrinsic. In practice we must take this principle with a grain of salt. Postulate systems are designed to capture the essence of some situation which arises naturally either in mathematics or in the physical world. But unfortunately, concepts and configurations do not always sit well together. Set-theoretic descriptions are often clumsy, oblique, and non-intuitive (recall the definition of ordered pair). Still they are always precise; any vagueness which may exist in an intuitive concept vanishes (or is absorbed into whatever vagueness is inherent in the concept of set) when the definition is written. But with the vagueness may disappear also some essential aspect of the intuitive concept. It takes art, insight, and experience to define a new and useful class of configurations.

Although it may be difficult to find a suitable configurational description of a concept, there may be several descriptions which are equally good. We can discern two reasons for this. There may be two different sets of postulates which describe the same class of configurations, and there may be two classes of configurations which equally well represent the concept.

When two postulate systems determine the same class of configurations, they are said to be *equivalent*. Of two equivalent postulate systems it is a matter of taste and convenience which is to be preferred.

Related to the idea of equivalence is the notion of redundant postulate systems. The reader will easily prove that if $\langle A, B \rangle$ is an ordered set, then

$$(\forall x, y \in A) \quad \langle x, y \rangle \in B \text{ implies } \langle y, x \rangle \notin B. \tag{2}$$

Hence, if we add (2) to the postulates (i) and (ii) for ordered sets, we obtain a new postulate system which defines the same class of configurations. Evidently the same thing will happen whenever we add to a postulate system a theorem deducible from the original postulates. When one postulate in a system is deducible from the remainder, the system is said to be *redundant*.

Redundant systems of postulates are often regarded as inelegant, especially when, as in the above example, the superfluous postulate is an easy consequence of the others. Sometimes, however, it requires an awkward or inconvenient statement of the postulates to avoid redundancy. In such cases a redundant system is usually preferred.

The other possibility, that different classes of configurations are models of the same abstract situation, is one of extraordinary interest in mathematics. Two such postulate systems may be thought of as conceptually equivalent. Unfortunately, the significance of this possibility can hardly be understood apart from substantial examples such as appear in Chapters 5 and 6. Suffice it to say here that the different points of view afforded by conceptually equivalent postulate systems can be a primary source of mathematical power.

EXERCISES

1. Find all ordered sets $\langle A, B \rangle$, where $A = \{1, 2, 3\}$.

2. A *Steiner triple system* is a configuration $\langle A, B \rangle$ such that $B \in \mathfrak{P}^2(A)$ and

(a) every member of B has three members;

(b) every two-element subset of A is a subset of exactly one member of B.

Find a Steiner triple system in which A has seven members.

3. Either of the configurations of Exercise 4, p. 59 might appropriately be called a schedule for a round-robin tournament. Find an appropriate set of postulates for the class of schedules for round-robin tournaments.

4. Using the fundamental relation, "x is a parent of y," try to write a system of postulates for genealogy which reflects the actual biological facts. (This is not easy.)

4–3. CONSISTENCY

Suppose we set up a system of postulates to describe a class of configurations and it turns out that we can prove two contradictory statements from these postulates. Then we conclude that there can be no configuration fulfilling our requirements; i.e., the class of configurations is empty. In this case the postulate system is said to be *inconsistent* or *self-contradictory*. Although it may appear that mathematics should concern itself only with consistent postulate systems (those from which no contradictions can be deduced), we should realize that the nonexistence of configurations satisfying certain conditions may be of great interest. Many important theorems are primarily nonexistence theorems. However, a set of conditions which describes the null set of configurations is not usually presented as a system of postulates.

In general when one introduces a system of postulates, he is at pains to verify that it is consistent. Consistency is usually proved by exhibiting an example of the required kind of configuration. Such an example is called a *model* for the postulate system. The existence of a model for a postulate system should certainly convince us that the system is consistent. But exactly what is meant by the "existence" of a set? After all, sets are an abstract concept.

When there is a finite model for a postulate system (i.e., a configuration in which the basic set has only finitely many members), it can be presented in full, like the example of a configuration in Section 4–1. To verify that the postulates are actually satisfied amounts to checking a finite number of combinatorial facts. Such a model, which one can, so to speak, check personally, is convincing and may be used as the basis for a formal proof that the postulate system is consistent.

Infinite models are a different matter. The use of infinite sets has always raised doubts in the minds of skeptics, and the results of the logicians show that these doubts are by no means ill-founded. Nevertheless, most of mathematics is concerned with configurations like the integers or real numbers for which there is no finite model. G. Peano formalized a postulational description of the integers which involves only the ideas of counting, yet even this system is not known to be consistent. Furthermore, it has been proved by K. Gödel that even if it is consistent, this fact can never be demonstrated. This means that mathematicians must accept the consistency of the integers on faith alone. This they are most happy to do. If anything in mathematics deserves to be called an axiom, it is the consistency of the counting process, one, two, three, . . .

Now, if we grant the existence (whatever it means) of a configuration corresponding to the integers, then we can construct a great number of infinite configurations and among these we can find models for many other postulate systems. The whole of Chapter 9 is devoted to constructing an example of a complete ordered field, the real number system, from a model for the integers. Such a construction is called a proof of consistency *relative* to the integers or simply a relative consistency proof.

4–4. THE CLASSIFICATION PROBLEM

Suppose we have described a class of configurations by means of postulates as in Section 4–2 and granting for the moment that we are able to find at least one such configuration, $\langle A, B \rangle$, the next question is naturally "How many?" If we take this question at its face value, it makes little sense, because, for each bijection f from A to some new set Q we can obtain a new configuration $\langle Q, f(B) \rangle$ of the desired type. Evidently this ought not to be counted as a new example. What we really want is to find all possible different configurations, where different means nonisomorphic. This is called the classification problem.

Given a class $\mathcal{C}$ of configurations defined by postulates, we seek an explicit subclass $\mathcal{D}$ of $\mathcal{C}$ such that

(i) no two members of $\mathcal{D}$ are isomorphic,
(ii) every member of $\mathcal{C}$ is isomorphic to some member of $\mathcal{D}$.

Such a class $\mathcal{D}$ is called a set of *canonical* models for the class $\mathcal{C}$. It is also desirable to have a fairly explicit way to decide to which canonical model a given member of $\mathcal{C}$ is isomorphic.

There are relatively few configuration classes for which this problem has been completely solved. Moreover, since an evaluation of degrees of explicitness is involved, it is not necessarily clear what constitutes a complete solution. But it is fair to say that a large portion of mathematics is devoted to the classification problem for various kinds of configurations.

Classification is trivial for configurations defined by inconsistent postulate systems, for in that case $\mathcal{C} = \emptyset$.

There is another, mathematically important, case in which the classification problem has a simple answer. It may happen that a postulate system is consistent and that all models for it are isomorphic. If so, any model can be taken as the canonical model. In this case the postulate system is said to be *categorical*, and the study of a class of configurations reduces effectively to the study of just one configuration.

In theory at least any intrinsic question concerning the models for a categorical postulate system is answerable by deduction from the postulates. The real numbers are categorically described by a system of postulates that is often quoted in calculus books. Because it is categorical, all of the formulas of analysis can be deduced from this system. On the other hand, the postulates for an ordered set given above are very far from being categorical, and there are many intrinsic questions concerning ordered sets to which the answer varies from one model to the next. The condition 4–2(1) involved in the definition of linearly ordered set is an example. Since 4–2(1) is true for some ordered sets and false for others, we surely cannot prove either it or its denial from the postulates for ordered sets.

In this book we shall study mainly the most basic configurations of analysis, the integers and the real and complex number systems. One of our primary objectives is the proof that our postulational description of these structures is categorical. Although we will meet many other structures, we shall devote little space to their classification.

Until the beginning of this century mathematics was almost entirely concerned with the study of categorically determined configurations, but since then much effort has been expended on classes of configurations, such as fields and topological spaces, determined by noncategorical postulate systems. The objective of these studies is not merely greater generality; it has been found that deep insights into the structure of categorically determined systems, such as the real numbers, can be gained by omitting some postulates and exploring the implications of those which remain.

EXERCISES

1. Show that if we adjoin the postulate,

   A has seven members,

to those for a Steiner triple system (Exercise 2, p. 62), the resulting postulate system is categorical.

2. Show that if we adjoin the postulate,

   A has six members,

to those for a schedule for a round-robin tournament (Exercise 3, p. 62), the resulting postulate system is categorical.

3. Classify configurations $\langle A, R \rangle$, where A has two elements and R is a binary relation in A.

4. Classify configurations $\langle A, B \rangle$, where $B \in \mathfrak{P}^2(A)$ and

(a) A has five members,
(b) B has three members,
(c) every member of B has three members.

CHAPTER 5

EQUIVALENCE

There are many contexts in which two objects, although different, will serve the same purpose equally well, and one frequently says that the two things are equivalent. We shall draw up a precise definition describing the mathematical use of this term. There are two quite different structures which capture the mathematical idea that is involved; our main theorem is a demonstration of this fact. A second theorem concerns a technical device that we shall apply several times in Chapter 9.

5-1. EQUIVALENCE RELATIONS AND PARTITIONS

As we saw in Section 3–5, a relation in the technical sense is just a set of ordered pairs. This means that each intuitive relation is recorded as the set of all ordered pairs which stand in this relation. An equivalence relation is therefore defined as a relation which satisfies certain additional conditions that we attribute to the intuitive notion of equivalence.

5-1.1. Definition. A relation E in the set A is called an *equivalence relation* if and only if

(i) $(\forall a \in A)\, a\, E\, a$;
(ii) $(\forall a, b \in A)$ if $a\, E\, b$, then $b\, E\, a$;
(iii) $(\forall a, b, c \in A)$ if $a\, E\, b$ and $b\, E\, c$, then $a\, E\, c$.

These postulates are known as the *reflexive, symmetric,* and *transitive* laws, respectively.

E is a structure for A in the sense of Section 4–1, but there is no standard term for configurations like $\langle A, E \rangle$.

The relation of equality evidently satisfies this definition, and so does the relation of isomorphism, as we pointed out in Theorems 4–1.2, 4–1.3, and 4–1.4. Both of these are relations in the intuitive sense only. We could, of course, consider the set $\{\langle x, x \rangle\}$ of all ordered pairs having both members the same. This would be the technical relation reflecting equality, but it is one of those overlarge sets which expose us to the set-theoretic paradoxes, so we shall avoid it. For the same reason we shall not represent isomorphism as a technical relation.

It has often been claimed, although incorrectly, that postulates (i), (ii), and (iii) are redundant. The alleged proof of (i) from (ii) and (iii) runs something like this:

$$a \, E \, b \Rightarrow b \, E \, a, \qquad\qquad\qquad\qquad\qquad\qquad \text{by (ii)},$$
$$(a \, E \, b \text{ and } b \, E \, a) \Rightarrow a \, E \, a, \qquad\qquad\qquad\qquad \text{by (iii)},$$
$$a \, E \, b \Rightarrow a \, E \, a.$$

Since this is true for all b, $a \, E \, a$ in any case. What is wrong?

5–1.2. Definitions. Let E be an equivalence relation in A. The function φ from A to $\mathfrak{P}(A)$, defined by

$$\varphi(a) \,=\, \{b \mid a \, E \, b\},$$

is called the *natural* or *quotient* map associated with E. The various sets $\varphi(a)$ are called *equivalence classes* of E. The set of all equivalence classes, that is, the range of φ, is called the *quotient set of A by E*. A member of an equivalence class is often referred to as a *representative* of the class.

5–1.3. Proposition. Let E be an equivalence relation in A and let φ be the associated quotient map. Then $a \, E \, b$ if and only if $\varphi(a) = \varphi(b)$.

Proof. Suppose that $a \, E \, b$. Let x be an arbitrary member of $\varphi(b)$. By the definition of $\varphi(b)$, $b \, E \, x$. Applying (iii), $a \, E \, x$; hence $x \in \varphi(a)$. It follows that $\varphi(b) \subseteq \varphi(a)$. By (ii), $b \, E \, a$, whence $\varphi(a) \subseteq \varphi(b)$ by a similar proof; and thus $\varphi(a) = \varphi(b)$.

Conversely, suppose $\varphi(a) = \varphi(b)$. We know $b \in \varphi(b)$ by (i). Therefore $b \in \varphi(a)$, whence $a \, E \, b$.

5–1.4. Definition. A *partition* $\mathcal{P}$ of the set A is a collection of nonempty subsets of A such that for each $a \in A$, there is exactly one $B \in \mathcal{P}$ with $a \in B$.

This is frequently phrased, "$\mathcal{P}$ is an exhaustive collection of mutually exclusive subsets of A." Here "exhaustive" means that each element of A is in at least one member of $\mathcal{P}$, while "mutually exclusive" implies that no element is in more than one member of $\mathcal{P}$.

Our next theorem can be interpreted as meaning that equivalence relations and partitions are different set-theoretic formalisms for the same underlying idea.

5–1.5. Theorem. Let E be an equivalence relation in A and let $\mathcal{Q}$ be the quotient set of A by E. Then $\mathcal{Q}$ is a partition of A. Conversely, if $\mathcal{P}$ is any partition of A, then there is a unique equivalence relation F in A having $\mathcal{P}$ as its quotient set.

Proof. Let φ be the natural map from A to $\mathcal{Q}$. Given any $a \in A$, there is at least one member of $\mathcal{Q}$ containing a, since $a \in \varphi(a)$ by the reflexive law. Suppose now that $a \in B$ and $B \in \mathcal{Q}$. Since $\mathcal{Q}$ is the range of φ, we can choose x so that $B = \varphi(x)$. Then $a \in \varphi(x)$. By the definition of φ, $x \, E \, a$; by 5–1.3, $\varphi(a) = \varphi(x)$. Thus $B = \varphi(a)$ and $\varphi(a)$ is the only member of $\mathcal{Q}$ containing a. We have proved that $\mathcal{Q}$ is a partition of A.

Conversely, if $\mathcal{P}$ is a partition of A, we may define a relation F in A by

$$a\,F\,b \qquad \text{means} \qquad (\exists X \in \mathcal{P}) \quad a \in X \text{ and } b \in X.$$

There are now three things to verify: (a) F is an equivalence relation; (b) the quotient set of F is $\mathcal{P}$; (c) if E is any equivalence relation in A with quotient set $\mathcal{P}$, then $E = F$. We shall omit the details of these arguments.

The role of reflexivity in this proof should clarify the question raised before 5–1.2.

The following corollary is just a rewording of the theorem.

5–1.6. Corollary. Let A be any set. The function which assigns to each equivalence relation its quotient set is a bijection from the set of all equivalence relations in A to the set of all partitions of A.

EXERCISES

*1. Consider the relation, "is a brother of." Is it reflexive? symmetric? transitive?

2. Give an example of a relation in a set which is symmetric and transitive but not reflexive.

3. Let E be an equivalence relation in the set A and let φ be the corresponding quotient map. Prove that

$$a\,E\,b \Leftrightarrow \varphi(a) \cap \varphi(b) \neq \emptyset.$$

4. List the partitions of the set $\{1, 2, 3\}$. These fall into equivalence classes under the relation of isomorphism (for configurations consisting of a basic set and a partition of it). Exhibit this division into classes.

5. Write out the omitted parts of the proof of Theorem 5–1.5 in the detailed fashion of Section 2–5.

*6. Choose a definite integer m. For integers a and b let "$a \equiv b$" mean "$a - b$ is divisible by m (that is, there is an integer x such that $a - b = mx$)." Prove that $\equiv$ is an equivalence relation. How many equivalence classes are there?

5–2. FACTORING FUNCTIONS

Before it became customary to write mathematics in set-theoretic terms, the particular notion of equality was not clearly distinguished from the general notion of equivalence. Hence in classical texts one often finds equality redefined as some particular equivalence relation. The modern device is to replace the domain A of the equivalence relation by the quotient set $\mathcal{Q}$. Instead of redefining $a = b$ as $a\,E\,b$, we talk of $\varphi(a)$ and $\varphi(b)$. Since $\varphi(a) = \varphi(b)$ if and only if $a\,E\,b$, this replacement serves the same purpose as the old redefinition of equality.

After replacing A by $\mathcal{Q}$ we want to transfer from A to $\mathcal{Q}$ various relationships which may be given between A and other sets. The following theorem tells us when a function with domain A can be transferred to $\mathcal{Q}$.

5-2.1. Theorem. Suppose that E is an equivalence relation in the set A having φ and Q as quotient map and space, respectively. Let f be a function from A to a set S such that

$$(\forall x, y \in A) \quad x\,E\,y \Rightarrow f(x) = f(y). \tag{1}$$

Then there exists a unique function g from Q to S such that $f = g \circ \varphi$.

Proof. Let $g = \{\langle \varphi(x), f(x)\rangle \mid x \in A\}$. Evidently g is a relation with domain Q. We must prove that it is a function. Suppose that two ordered pairs in g, say $\langle \varphi(x), f(x)\rangle$ and $\langle \varphi(y), f(y)\rangle$, have the same first component; that is, $\varphi(x) = \varphi(y)$. By 5-1.3, $x\,E\,y$. By the hypothesis, $f(x) = f(y)$. This shows that g is a function. The formula $f = g \circ \varphi$ and the uniqueness of g are obvious.

Note that the converse of this theorem is clear. If f can be expressed in the form $g \circ \varphi$, then (1) holds.

When a given function f is represented as the composition of other functions, it is often said to be *factored*. When composition is denoted by juxtaposition, the analogy with factorization of numbers is clear.

Theorem 5-2.1 is a special case of a more general theorem. Suppose we have three sets A, X, and S and functions φ from A to X and f from A to S. When can f be factored in the form $g \circ \varphi$?

5-2.2. Theorem. Let φ be a function from A to X and f a function from A to a nonempty set S. For a function g from X to S to exist such that $f = g \circ \varphi$ it is necessary and sufficient that

$$(\forall a, b \in A) \quad \varphi(a) = \varphi(b) \Rightarrow f(a) = f(b). \tag{2}$$

If (2) holds and φ is surjective, then there is only one function g such that $f = g \circ \varphi$.

Proof. Suppose f can be expressed as $f = g \circ \varphi$. Let $a, b \in A$. Suppose that $\varphi(a) = \varphi(b)$. Then

$$f(a) = g(\varphi(a)) = g(\varphi(b)) = f(b).$$

Thus (2) is valid.

Conversely, suppose that (2) holds. Let

$$h = \{\langle \varphi(a), f(a)\rangle \mid a \in A\}.$$

Evidently h is a relation, domain h = range φ, and range h = range f. We shall prove that h is a function. Suppose that $\langle \varphi(a), f(a)\rangle$ and $\langle \varphi(b), f(b)\rangle$ are two members of h with the same first component; that is, $\varphi(a) = \varphi(b)$. By (2) $f(a) = f(b)$. Hence h is a function.

It is now clear that $(\forall a \in A)\, h(\varphi(a)) = f(a)$. But h is defined only on range φ; so we extend it over X. Pick any element $s \in S$. Define $g(x) = h(x)$ if $x \in$ range φ, and $g(x) = s$ if $x \in X -$ range φ. Then g is the required function.

Assume φ is surjective, and let g' be any function from X to S such that $f = g' \circ \varphi$. Suppose $x \in X$. We can choose $a \in A$ so that $\varphi(a) = x$. Then

$$\langle x, g'(x) \rangle = \langle \varphi(a), g'(\varphi(a)) \rangle = \langle \varphi(a), f(a) \rangle \in h,$$

and so $g'(x) = h(x)$. This shows that $g' = h$. Thus in this case, there is only one function g which satisfies $f = g \circ \varphi$, namely h.

EXERCISES

1. Suppose that f is a function with domain A. Prove that

$$\{\langle x, y \rangle \mid f(x) = f(y)\}$$

is an equivalence relation in A. Let φ be the corresponding quotient map. If $f = g \circ \varphi$, as in 5–2.1, prove that g is injective.

2. Let D and E be equivalence relations in the sets A and B, respectively. Let the corresponding quotient maps and sets be φ, ψ, $\mathfrak{Q}$, and $\mathfrak{R}$.

(a) Define an equivalence relation F in $A \times B$ in terms of D and E so that the corresponding quotient set S has a natural bijection to $\mathfrak{Q} \times \mathfrak{R}$.

(b) Suppose that g is a function from $A \times B$ to a set T such that

$$(\forall a_1, a_2 \in A)(\forall b \in B) \quad a_1 \, D \, a_2 \Rightarrow g(a_1, b) = g(a_2, b)$$

and

$$(\forall a \in A)(\forall b_1, b_2 \in B) \quad b_1 \, E \, b_2 \Rightarrow g(a, b_1) = g(a, b_2).$$

Prove that there exists a unique function h from $\mathfrak{Q} \times \mathfrak{R}$ to T such that

$$(\forall a \in A)(\forall b \in B) \quad g(a, b) = h(\varphi(a), \psi(b)).$$

CHAPTER 6

ORDER

The mathematical notion of order is an abstraction of the notion of dominance as expressed in such statements as "$5 > 4$," "$A \subseteq B$," "c comes before d," "x is closer than y," . . .

We shall find two classes of configurations which express this concept. After verifying this fact, we shall give the final definition in a way that recognizes both alternatives. The remainder of the chapter will be devoted to defining a number of terms associated with ordered sets and developing a few of their properties.

6–1. ORDER RELATIONS

Since no member of a set has precedence over any other, order must be described by a configuration. This is accomplished by attaching to the set a catalog detailing the relative precedence of its members.

6–1.1. Definition. A relation S in the set A is called a *strong order relation* if and only if

(i) $(\forall a \in A)\, a \not S a$, and
(ii) $(\forall a, b, c \in A)$ if $a\,S\,b$ and $b\,S\,c$, then $a\,S\,c$.

That is, S is transitive and strictly nonreflexive.

This reflects some but not all of the properties of the relation $<$ as it is used for numbers. In particular, it may happen that $a \neq b$; $a \not S b$; and $b \not S a$. Proper set inclusion is a more typical model for a strong order relation.

Oddly, the null set is a strong order relation in any set A. It describes a trivial order relation which assigns no precedences whatever.

6–1.2. Definition. A relation W in the set A is called a *weak order relation* if and only if

(iii) $(\forall a \in A)\, a\,W\,a$;
(iv) $(\forall a, b \in A)$ if $a\,W\,b$ and $b\,W\,a$, then $a = b$; and
(v) $(\forall a, b, c \in A)$ if $a\,W\,b$ and $b\,W\,c$, then $a\,W\,c$.

We recognize (iii) and (v) as the reflexive and transitive laws. Postulate (iv) is sometimes called the law of *antisymmetry*, because it asserts that W is symmetric as rarely as is consistent with reflexivity.

The relations $\leq$ for numbers and $\subseteq$ for sets are both weak order relations. Now, $\leq$ is related to $<$, for "$x \leq y$" means "$x < y$ or $x = y$." Similarly $\subseteq$ is related to $\subset$. We shall now prove that it is inherent in the postulates that strong and weak order relations are so related.

6-1.3. Proposition. Let A be any set and let D be the diagonal of $A \times A$. If $\varphi(S) = S \cup D$, then φ is a bijection from the set of all strong order relations in A to the set of all weak order relations in A. The inverse map is $\psi : W \rightarrow W - D$.

Proof. Let S be a strong order relation in A and let $W = S \cup D$. We shall prove that W is a weak order relation.

Obviously, "$a \, W \, b$" means "$a \, S \, b$ or $a = b$." Clearly, W is reflexive. If $a \, W \, b$ and $b \, W \, a$ but $a \neq b$, then we should have $a \, S \, b$ and $b \, S \, a$, whereupon (ii) implies $a \, S \, a$, contradicting (i). Thus (iv) holds. To prove that W is transitive suppose that $a \, W \, b$ and $b \, W \, c$. There are four cases:

$$
\begin{aligned}
a \, S \, b \qquad &\text{and} \qquad b \, S \, c, \\
a \, S \, b \qquad &\text{and} \qquad b = c, \\
a = b \qquad &\text{and} \qquad b \, S \, c, \\
a = b \qquad &\text{and} \qquad b = c.
\end{aligned}
$$

The first three lead to $a \, S \, c$ and the last to $a = c$. In any case $a \, W \, c$.

This completes the proof that W is a weak order relation and hence that φ maps the set of strong order relations into the set of weak order relations.

Start again with a weak order relation W and let $S = W - D$. Now, "$a \, S \, b$" means "$a \, W \, b$ but $a \neq b$." Evidently S is nonreflexive. To verify that S is transitive assume $a \, S \, b$ and $b \, S \, c$. Then $a \, W \, b$ and $b \, W \, c$, hence $a \, W \, c$. If $a = c$, our conditions become $a \, W \, b$ and $b \, W \, a$, whence $a = b$, contradicting $a \, S \, b$. Therefore $a \neq c$, giving $a \, S \, c$ as required.

We have proved that S is a strong order relation, and therefore ψ maps the set of weak order relations into the set of strong order relations. Finally, since $\psi \circ \varphi$ and $\varphi \circ \psi$ are both identity maps, φ is a bijection, and the proposition is proved.

A word on the meaning of this proposition. The intuitive notion of order springs from examples provided by $\subset$ or such a relation as "a is an ancestor of b." In an effort to abstract what he regards as relevant common properties of these relations, one mathematician writes 6-1.1 as a set-theoretic description of such situations. Another, who habitually thinks in terms of $\subseteq$ and whose language is such that each person is his own ancestor, might write 6-1.2. There is no *a priori* reason to believe that they will capture the same fundamental idea. Our proposition serves to convince us that the two definitions reflect different set-theoretic descriptions of the same abstract concept.

Since strong and weak order relations are so neatly in correspondence with each other, there is no reason to prefer one to the other in most contexts, although frequently one may prove more convenient in a specific situation. We shall treat them equally, therefore, in our definition of an ordered set.

6–1.4. Definition. An *ordered set* is a triple $\langle A, S, W \rangle$ consisting of a set A, a strong order relation S in A, and the corresponding (in the sense of the preceding proposition) weak order relation W.

6–1.5. Notation. When there is no serious danger of confusion, one refers to an ordered set by its first member. Thus "the ordered set A" actually refers to a triple $\langle A, S, W \rangle$, where S and W are supposed to be well defined from the context.

It is conventional throughout mathematics to refer to a configuration by the name of its basic set. Under this convention the sentence, "x is a member of the ordered set A," means "x is a member of the set A in which corresponding strong and weak order relations have been singled out,"

The literal interpretation of this phrase, using the fact that a configuration is an ordered pair and therefore a set with two members, is ridiculous. It is a penalty for our insistence on defining concepts as configurations. Certainly one should think of an ordered set as a set with its elements arranged internally.

6–1.6. To *order a set* A means to specify an order relation in A. We shall give two examples of the customary phraseology.

"The relation $\subseteq$ weakly orders the subsets of X." Here the set ordered is $\mathfrak{P}(X)$. Since the symbols $\subset$ and $\subseteq$ have permanent significance and do not denote sets, $\langle \mathfrak{P}(X), \subset, \subseteq \rangle$ is not a meaningful notation. The ordered set intended is really $\langle \mathfrak{P}(X), S, W \rangle$, where

$$S = \{\langle Y, Z \rangle \mid Y \subset Z \subseteq X\}$$

and

$$W = \{\langle Y, Z \rangle \mid Y \subseteq Z \subseteq X\}.$$

Of course the order relation would always be written with the symbols $\subset$ and $\subseteq$ and the sets S and W would never appear at all.

"If '$a < b$' means '$a \leq \frac{1}{2}b$,' then $<$ is a strong order relation in the set of positive numbers."

Emphasis on the distinction between weak and strong order relations is uncommon; so when an order relation is defined it is frequently not stated whether it is the weak or the strong form.

6–1.7. Subsets. Let B be any subset of an ordered set A. Then B is itself an ordered set using the same ordering. Technically, if $\langle A, S, W \rangle$ is an ordered set, then

$$\langle B, S \cap (B \times B), W \cap (B \times B) \rangle$$

is an ordered set. It is customary to refer to any subset of an ordered set as an ordered set without stating that it is this inherited order relation that is meant. This convention is unambiguous because of the following obvious fact. If C is a subset of B, then the order structure inherited by C directly from A is the same as that inherited by C as a subset of B.

EXERCISES

1. Prove that the diagonal of $A \times A$ is the only relation in A which is both an equivalence relation and an order relation.

2. Suppose that S and W are corresponding strong and weak order relations in a set A. Prove that if $a\,S\,b$ and $b\,W\,c$, then $a\,S\,c$.

3. Show that in the notation of Exercise 1, p. 50 the transitive law can be expressed $S \circ S \subseteq S$. If S^{-1} denotes the relation obtained by reversing all the ordered pairs of S, then 6–1.1(i) can be expressed as $S \cap S^{-1} = \emptyset$. What is the corresponding expression of 6–1.2(iii) and (iv)?

4. Suppose that R is a relation in the set A which satisfies (iv) and (v). Prove that there is a unique pair of corresponding order relations S and W for which $S \subseteq R \subseteq W$.

6–2. MAPS OF ORDERED SETS

6–2.1. Definitions. Let $\langle A, S, W \rangle$ and $\langle B, T, X \rangle$ be ordered sets and let φ be a function from A to B. Then φ is said to be *strongly order-preserving* if and only if

(i) $(\forall a_1, a_2 \in A)$ if $a_1\,S\,a_2$, then $\varphi(a_1)\,T\,\varphi(a_2)$.

It is said to be *weakly order-preserving* if and only if

(ii) $(\forall a_1, a_2 \in A)$ if $a_1\,W\,a_2$, then $\varphi(a_1)\,X\,\varphi(a_2)$.

It is easy to see that (i) implies (ii), but not vice versa; hence, if a function is described as order-preserving without either of the adverbs, it is usually intended that it is weakly order-preserving.

We shall omit the obvious definitions of strongly and weakly *order-reversing* maps. The word *isotone* is a synonym for *order-preserving*, while *monotone* means either *order-preserving* or *order-reversing*.

6–2.2. Proposition. Let $\langle A, S, W \rangle$ and $\langle B, T, X \rangle$ be ordered sets. A bijection φ from A to B is an isomorphism of the configuration $\langle A, S, W \rangle$ onto $\langle B, T, X \rangle$ if and only if both φ and its inverse are order-preserving, either strongly or weakly.

We leave the proof of this proposition as an exercise.

In less formal terms the proposition says that, for ordered sets a bijection φ is an isomorphism if and only if

$$a_1\,S\,a_2 \Leftrightarrow \varphi(a_1)\,T\,\varphi(a_2).$$

This criterion is written directly in terms expressing the idea of order. It does not befog the issue in terms of the artificial set-theoretic formalism that we have adopted for purposes of mathematical precision. The validity of the proposition is therefore evidence that we have not inadvertently introduced extraneous notions into our set-theoretic formulation of the idea of ordered set.

Propositions of this general type are often stated in connection with the definition of a new class of configurations. Sometimes such a proposition is stated as the definition of isomorphism. This reflects a less completely organized outlook on mathematics than we are attempting to maintain.

The following result, occasionally useful in itself, is more important as an illustration of the use of equivalence relations and quotient sets. Constructions similar to this one are common; the reader should familiarize himself with this technique.

The object is to take a relation which is not quite a weak order relation and make it into one in a natural way by "identifying" certain sets of elements.

6–2.3. Proposition. Let X be a reflexive and transitive relation in the set A. If "$b \cong c$" means "$b \, X \, c$ and $c \, X \, b$," then $\cong$ is an equivalence relation. Let φ and Q be the corresponding quotient map and set. There is a unique weak order relation W in Q such that

$$\varphi(b) \, W \, \varphi(c) \qquad \text{if and only if} \qquad b \, X \, c. \tag{1}$$

Moreover, suppose R is a weak order relation in the set E and f is a function from A to E such that

$$b \, X \, c \qquad \text{implies} \qquad f(b) \, R \, f(c). \tag{2}$$

There exists a unique map g of Q into E such that $f = g \circ \varphi$ and this map is weakly order-preserving.

Proof. We omit the proof that $\cong$ is reflexive, symmetric, and transitive.

Define $W = \{\langle \varphi(b), \varphi(c) \rangle \mid b \, X \, c\}$. Evidently $W \subseteq Q \times Q$. We shall show that W is a weak order relation.

Suppose that $Q \in Q$. Choose $b \in Q$. Since $b \, X \, b$, $\varphi(b) \, W \, \varphi(b)$; that is, $Q \, W \, Q$. Thus W is reflexive.

Suppose that $Q_1, Q_2 \in Q$, $Q_1 \, W \, Q_2$, and $Q_2 \, W \, Q_1$. Choose representatives $a \in Q_1$, $b \in Q_2$, $c \in Q_2$, and $d \in Q_1$ so that $a \, X \, b$ and $c \, X \, d$. We have $b \, X \, c$ (because $b \cong c$), $c \, X \, d$, and $d \, X \, a$ (because $a \cong d$); hence $b \, X \, a$. Therefore $a \cong b$ and $Q_1 = \varphi(a) = \varphi(b) = Q_2$. Thus W is antisymmetric.

Suppose that $Q_1, Q_2, Q_3 \in Q$, $Q_1 \, W \, Q_2$, and $Q_2 \, W \, Q_3$. Choose representatives $a \in Q_1$, $b \in Q_2$, $c \in Q_2$, and $d \in Q_3$ so that $a \, X \, b$ and $c \, X \, d$. Since $b \cong c$, $b \, X \, c$ and therefore $a \, X \, d$. This shows that $\varphi(a) \, W \, \varphi(d)$; that is, $Q_1 \, W \, Q_3$. Thus W is transitive.

The "if" part of (1) follows immediately from the definition of W. Conversely, suppose that $\varphi(b) \, W \, \varphi(c)$. Choose $a \in \varphi(b)$ and $d \in \varphi(c)$ so that $a \, X \, d$. Since $a \cong b$ and $c \cong d$, we have $b \, X \, a$ and $d \, X \, c$. This gives $b \, X \, c$, which proves the rest of (1).

If W' is another relation in Q (not even necessarily an order relation) such that

$$\varphi(b) \, W' \, \varphi(c) \qquad \text{if and only if} \qquad b \, X \, c,$$

then (1) gives

$$\varphi(b) \, W' \, \varphi(c) \qquad \text{if and only if} \qquad \varphi(b) \, W \, \varphi(c),$$

whence $W' = W$. This proves uniqueness.

Consider a function f from A to E such that (2) holds. If $b \cong c$, then $b\, X\, c$ and $c\, X\, b$; so $f(b)\, R\, f(c)$ and $f(c)\, R\, f(b)$. Since R is antisymmetric, $f(b) = f(c)$. By Theorem 5–2.1 there is a unique function g from $\mathcal{Q}$ to E such that $f = g \circ \varphi$. We must prove that g is weakly order-preserving. Suppose that $Q_1\, W\, Q_2$. Choose $b \in Q_1$ and $c \in Q_2$ so that $b\, X\, c$. Then $f(b)\, R\, f(c)$. But $g(Q_1) = g(\varphi(b)) = f(b)$ and $g(Q_2) = g(\varphi(c)) = f(c)$; hence $g(Q_1)\, R\, g(Q_2)$. This concludes the proof.

It is worth noting that if X were a weak order relation to begin with, φ would turn out to be an isomorphism of the ordered set A onto the ordered set $\mathcal{Q}$.

EXERCISES

1. Let φ be a weakly order-preserving, injective map of one ordered set into another. Prove that φ is strongly order-preserving.

2. Let φ be a bijective order-preserving map from one ordered set to another. Must φ be an isomorphism?

*3. Let $\leq$ be the usual weak order relation in the set N of positive integers. Let $X = \leq \cup \{\langle 6, 5 \rangle\}$. Show that X satisfies the hypotheses of 6–2.3 and describe explicitly the quotient set and its order relation W. Show that the resulting ordered set is isomorphic to N.

6–3. LINEAR ORDER

6–3.1. Definitions. A weak order relation W in the set A will be called *linear* if and only if

$$(\forall a, b \in A) \qquad a\, W\, b \quad \text{or} \quad b\, W\, a.$$

A strong order relation S in the set A will be called *linear* if and only if

$$(\forall a, b \in A) \qquad a\, S\, b, \quad a = b, \quad \text{or} \quad b\, S\, a.$$

The latter condition is known as the rule of *trichotomy*. It is easily verified that its three alternatives are exclusive.

6–3.2. Proposition. Let S and W be corresponding strong and weak order relations in the set A. Then S is linear if and only if W is linear.

The point of this proposition, whose proof we omit, is that the linearity of an ordering is independent of which representation of the ordering is chosen.

6–3.3. Terminology. When describing an order relation we may use the word *partial* to emphasize the fact that trichotomy is not demanded or not yet proved. Some writers, however, use the word *ordering* to mean *linear ordering* unless the word *partial* is explicitly used. The words *total* and *simple* are often used as synonyms for *linear* in the context of order relations.

Suppose that a and b are in the domain of a strong order relation S. Then a and b are said to be *comparable* if and only if one of the following statements is true: $a = b$, $a\,S\,b$, $b\,S\,a$. If all of these statements are false, they are said to be *incomparable*. The trichotomy postulate can therefore be phrased, "Any two elements are comparable."

Since ordinary English seems to contain few words which directly refer to partial orderings, we are obliged to borrow words which carry intuitive implications of linear ordering, and the mathematician must beware lest these implications improperly influence his thinking. Words of size, height, temporal lapse, or spatial extent are frequently used to describe order relations in mathematical works. Thus "$a < b$" might be written "a comes before b," and the term *least upper bound*, which we shall define in the next section, exhibits side-by-side two different interpretations of mathematical order as intuitive order.

Order relations are almost invariably denoted by a symbol more or less similar to $<$, such as $\subset$ or $\prec$. The symbol $<$ seems to be reserved for strong order relations, while $\subset$, as we remarked before, is often used for weak set inclusion. Once a symbol for a strong order relation is introduced, it is customary to hybridize it with an equality sign to denote the corresponding weak order relation, e.g., $\le$, $\leqq$, or $\subseteq$. It is quite common to use the same symbol for the order relations of two different sets. This rarely causes confusion; indeed it often facilitates understanding.

If a relation in a set is an order relation, either strong or weak, partial or total, then the reversed relation is easily seen to be an order relation of the same species, and it is very commonly denoted by simply reversing the symbol for the original relation when the latter is $<$ or one of its modifications. This practice would cause considerable confusion if the definitions were taken too literally. The words, *less* and *more*, *lower* and *upper*, or *before* and *after*, always refer, respectively, to the small and large sides of the symbol regardless of which way it is written. This must be borne in mind when applying the definitions in the next section.

Superlative order words and the words *maximum* and *minimum* are used under a strict convention. Suppose $\langle A, <, \le \rangle$ is an ordered set. The phrases, "a is the least element of A" and "a is the minimum element of A," mean "for all $b \in A$, $a \le b$." The weaker assertion, "for all $b \in A$, $b \not< a$," would be written, "a is a *minimal* element of A." Let A be the set of nonempty subsets of a set E. Order A by inclusion. Then each one-element subset of E is a minimal element of A, but A has no least element (unless E has only one element). The word *maximal* is used similarly. There can be at most one maximum element of A, but there can be any number of maximal elements. The distinction disappears if A is linearly ordered.

EXERCISES

1. Let φ be an injective order-preserving map from an ordered set A to a linearly ordered set B. Must A be linearly ordered?

2. Let φ be a bijective order-preserving map from one linearly ordered set to another. Must φ be an isomorphism?

3. Let φ be a surjective weakly order-preserving map from a linearly ordered set A to an ordered set B. Must B be linearly ordered?

6–4. BOUNDS

6–4.1. Definitions. Let $<$ be a weak order relation in the set A and let B be a subset of A. B is said to be *bounded above* if and only if

$$(\exists c \in A)(\forall b \in B) \quad b < c.$$

An element c with this property is called an *upper bound* of B. B is said to be *bounded below* if and only if

$$(\exists d \in A)(\forall b \in B) \quad d < b.$$

An element d with this property is called a *lower bound* of B. B is said to be *bounded* if and only if it is bounded both above and below.

The upper and lower bounds of a set are always considered with respect to the weak order relation to allow for the possibility that the bound might fall in the set.

If the order relation in question were denoted by $\geq$, then an element c such that $(\forall b \in B)\ c \geq b$ would be called an upper bound, although it is a lower bound if the definition is applied literally. (See the remarks in 6–3.3.)

Assuming that A itself is not empty, the null set is always a bounded subset of A.

6–4.2. Definitions. Let f be a function with values in an ordered set. Then f is said to be *bounded, bounded above,* or *bounded below* if and only if range f is bounded, bounded above, or bounded below, respectively.

6–4.3. Proposition. Let $<$ be a weak order relation in the set A and let B be a subset of A. There exists at most one upper bound c_0 of B such that

$$\text{for all upper bounds } c \text{ of } B, c_0 < c, \tag{1}$$

and there exists at most one lower bound d_0 of B such that

$$\text{for all lower bounds } d \text{ of } B, d < d_0. \tag{2}$$

Proof. Suppose c_0 and c_1 are upper bounds for B such that (1) holds and also

$$\text{for all upper bounds } c \text{ of } B, c_1 < c.$$

Since c_0 is an upper bound, $c_1 < c_0$. Since c_1 is an upper bound, we obtain $c_0 < c_1$ from (1). Hence $c_0 = c_1$. The proof for lower bounds is similar.

6–4.4. Definitions. Let B be a subset of an ordered set A. If an upper bound c_0 of B satisfying (1) exists, it is called the *least upper bound* or *supremum* of B. If a lower bound d_0 satisfying (2) exists, it is called the *greatest lower bound* or *infimum* of B. Note that nothing has been said as to whether or not the least upper bound is in B.

6-4.5. Notation. We shall use the notations sup B and inf B for the least upper bound and greatest lower bound. The abbreviations lub and glb are also commonly used. Sometimes max B and min B are used, but usually this notation implies that the bound is actually a member of B. If f is a function with values in an ordered set, then a notation such as $\sup_x f(x)$ or $\sup f(x)$ means the supremum of the range of f.

If B is a subset of an ordered set, there is certainly no guarantee that its supremum exists; hence when one writes sup B, he is by implication asserting its existence.

EXERCISES

1. Let A be an ordered set with weak order relation $<$. Let B be a subset of A for which both inf B and sup B exist. At first glance, it appears that necessarily

$$\text{inf } B < \text{sup } B. \tag{3}$$

But this is not true. Give a counterexample and state the additional hypothesis that is necessary to ensure (3).

2. Let X be any set and consider $\mathfrak{P}(X)$ ordered by inclusion. If $B \subseteq \mathfrak{P}(X)$, show that sup B and inf B exist.

*3. Consider the set of positive integers $\mathbf{N}$ with the usual order. For which subsets A of $\mathbf{N}$ does inf A exist? For which subsets A does sup A exist?

4. Prove the following theorem which is frequently invoked tacitly:

Let A be an ordered set with weak order relation $<$. Let B and C be subsets of A having least upper bounds. Then

(i) if $B \subseteq C$, then sup $B <$ sup C;
(ii) if $(\forall b \in B)(\exists c \in C) b < c$, then sup $B <$ sup C.

5. An ordered set $\langle A, <, \leq \rangle$ is called a *lattice* if and only if each two-element subset of A has both a supremum and an infimum. In lattices, sup $\{a, b\}$ is often denoted by $a \vee b$ while inf $\{a, b\}$ is denoted by $a \wedge b$. (The author reads these symbols as "jug" and "jag," respectively.) Prove that the associative and commutative laws hold in any lattice A; that is, $(\forall a, b, c \in A)$

$$a \vee (b \vee c) = (a \vee b) \vee c, \qquad a \wedge (b \wedge c) = (a \wedge b) \wedge c,$$
$$a \vee b = b \vee a, \qquad\qquad a \wedge b = b \wedge a.$$

The distributive laws

$$(\forall a, b, c) \quad a \wedge (b \vee c) = (a \wedge b) \vee (a \wedge c), \tag{4}$$
$$(\forall a, b, c) \quad a \vee (b \wedge c) = (a \vee b) \wedge (a \vee c) \tag{5}$$

may fail, but prove the weaker laws:

$$(\forall a, b, c) \quad a \wedge (b \vee c) \geq (a \wedge b) \vee (a \wedge c),$$
$$(\forall a, b, c) \quad a \vee (b \wedge c) \leq (a \vee b) \wedge (a \vee c).$$

Assuming that the distributive law (4) holds, prove that the other distributive law (5) also holds, and vice versa. A lattice in which these laws hold is called a *distributive* lattice.

6–5. COMPLETE ORDERED SETS

6–5.1. Proposition. Let A be an ordered set. If A has either of the following properties, it has both:

(i) every nonempty subset of A which is bounded above has a supremum;
(ii) every nonempty subset of A which is bounded below has an infimum.

Proof. We denote the weak order relation in A by $<$, and suppose that (i) holds. Let C be a nonempty subset of A which is bounded below. Let B be the set of all lower bounds for C. By hypothesis, B is not empty and it is bounded above by every member of C. Now, (i) implies that B has a supremum, say b_0. If $c \in C$, then c is an upper bound of B, so $b_0 < c$. Thus b_0 itself is a lower bound for C. For any lower bound b of C we have $b \in B$ and therefore $b < b_0$. This shows that b_0 is the infimum of C.

We have thus checked that (i) implies (ii), and obviously the same argument, *mutatis mutandis*, will prove that (ii) implies (i).

6–5.2. Definition. An ordered set is called *complete* if and only if it has properties (i) and (ii).

Sometimes the word *complete* is used to mean that every subset of A, including the null set, has a supremum, without requiring the existence of an upper bound. With slight changes, the argument of 6–5.1 will prove that, if every subset of A has a supremum, then every subset of A has an infimum. Note that the supremum of the empty subset of A must be the least element of A. An author who used *complete* in this new sense would probably call an ordered set satisfying (i) and (ii) *relatively* complete.

Completeness is an extraordinarily powerful property, as we shall see subsequently. Its role in the next two proofs is worthy of study.

6–5.3. Theorem. Let $\langle A, <, \leq \rangle$ and $\langle P, <, \leq \rangle$ be complete ordered sets neither of which contains either a maximal or a minimal element. Suppose that B is a subset of A which meets every interval in A; that is,

$$(\forall a_1, a_2 \in A) \quad a_1 < a_2 \quad \text{implies} \quad (\exists b \in B)\, a_1 < b < a_2. \tag{1}$$

Let Q be a subset of P which has the same property relative to P. Let φ be an isomorphism of B onto Q. There is a unique weakly order-preserving map from A to P which extends φ. This map is an isomorphism of A onto P.

To understand the meaning of this theorem, let A be the real numbers and let B be the rational numbers. Then B has property (1). Let P and Q also be the real and rational numbers respectively. The theorem says that every order-preserving isomorphism of the rationals onto themselves has a unique extension as an order-preserving map of the reals onto the reals. This is essentially the only application we shall make of this theorem, but it is not any more difficult to prove it in full generality. The proof will use the following lemma.

6–5.4. Lemma. Let $\langle A, <, \leq \rangle$ be a complete ordered set which contains no minimal elements. Let B be a subset of A such that (1) holds. If θ is a weakly order-preserving map of A into itself such that $(\forall b \in B) \; \theta(b) = b$, then θ is the identity.

Proof. Choose any $a \in A$. Since it is not minimal, we can choose $a_1 < a$. Then (1) shows that $S(a) = \{b \in B \mid b < a\}$ is not empty. Evidently a is an upper bound for this set, hence $a \geq \sup S(a)$. If $\sup S(a) < a$, then we can choose $b_0 \in B$ so that $\sup S(a) < b_0 < a$. Now, $b_0 \in S(a)$; hence $b_0 \leq \sup S(a)$, which is a contradiction. This proves that $a = \sup S(a)$.

If $b \in S(a)$, then $a > b$. Since θ is weakly order-preserving,

$$\theta(a) \geq \theta(b) = b.$$

Therefore $\theta(a) \geq \sup S(a) = a$. If $\theta(a) > a$, we can choose $b_1 \in B$ so that $a < b_1 < \theta(a)$. Applying θ to the first of these inequalities, we obtain $\theta(a) \leq \theta(b_1) = b_1$, contradicting the second. Therefore $\theta(a) = a$. Thus θ is the identity.

Proof of Theorem 6–5.3. Let $a \in A$. Since it is neither maximal nor minimal, we can choose a_1 and a_2 so that $a_1 < a < a_2$. By (1) we can choose b_1 and b_2 so that $a_1 < b_1 < a < b_2 < a_2$. This shows that b_2 is an upper bound for the set $S(a) = \{b \in B \mid b < a\}$, and that the latter is not empty. Since φ is order-preserving, $\varphi(b_2)$ is an upper bound for the nonempty set $\varphi(S(a))$. Hence we may define $\bar\varphi$ by

$$\bar\varphi(a) = \sup \varphi(S(a)).$$

We shall prove that $\bar\varphi$ is weakly order-preserving and an extension of φ.

If $a_1 \leq a_2$, it follows immediately from the transitive law that $S(a_1) \subseteq S(a_2)$. Therefore

$$\varphi(S(a_1)) \subseteq \varphi(S(a_2))$$

and

$$\bar\varphi(a_1) = \sup \varphi(S(a_1)) \leq \sup \varphi(S(a_2)) = \bar\varphi(a_2).$$

Thus $\bar\varphi$ is weakly order-preserving.

Let $b_0 \in B$. It is an upper bound for $S(b_0)$, hence $\varphi(b_0)$ is an upper bound for $\varphi(S(b_0))$. Therefore

$$\varphi(b_0) \geq \sup \varphi(S(b_0)) = \bar\varphi(b_0).$$

If $\varphi(b_0) > \bar\varphi(b_0)$, we can choose $q \in Q$ so that $\varphi(b_0) > q > \bar\varphi(b_0)$. Since φ is surjective, there is a $b_1 \in B$ such that $\varphi(b_1) = q$. Since φ is an isomorphism, $b_0 > b_1$; so $b_1 \in S(b_0)$. Hence $q \in \varphi(S(b_0))$, while $q > \bar\varphi(b_0) = \sup \varphi(S(b_0))$, which is a contradiction. This proves that $\varphi(b_0) = \bar\varphi(b_0)$. Hence $\bar\varphi$ is an extension of φ.

Let ψ be the inverse of φ. By similar arguments we can show that

$$\bar{\psi}(p) = \sup \{\psi(q) \mid q \in Q \text{ and } q < p\}$$

defines a weakly order-preserving map of P into A which extends ψ. Now, consider $\bar{\psi} \circ \bar{\varphi}$; it is a weakly order-preserving map of A into itself, which is the identity on B. By Lemma 6–5.4, it is the identity on A. Similarly, $\bar{\varphi} \circ \bar{\psi}$ is the identity on P. Now, by 6–2.2 we see that $\bar{\varphi}$ is an isomorphism of A onto P.

Suppose that φ^* is a weakly order-preserving map of A into P which extends φ. According to the lemma, $\bar{\psi} \circ \varphi^*$ is the identity function on A. Since $\bar{\varphi} \circ \bar{\psi}$ is the identity function on P, by a purely algebraic argument worthy of special note we see that $\varphi^* = \bar{\varphi}$; for

$$\varphi^* = (\bar{\varphi} \circ \bar{\psi}) \circ \varphi^* = \bar{\varphi} \circ (\bar{\psi} \circ \varphi^*) = \bar{\varphi}.$$

This establishes the uniqueness of $\bar{\varphi}$ and completes the proof.

6–5.5. The Knaster fixed-point theorem. Let $\langle A, <, \leq \rangle$ be a complete ordered set with a largest element and a smallest element. Let φ be a weakly order-preserving map of A into itself. Then there is an element $a \in A$ such that $\varphi(a) = a$. (Such an element is called a *fixed point* of φ.)

Proof. Let $B = \{x \in A \mid x \leq \varphi(x)\}$; it is not empty, since it contains the least element of A, and it is bounded above by the largest element of A. Hence we may let $a = \sup B$.

Suppose $b \in B$. Then $b \leq \varphi(b)$ and $b \leq a$. Since φ is order-preserving, $\varphi(b) \leq \varphi(a)$ and therefore $\varphi(a) \geq b$. This shows that $\varphi(a)$ is an upper bound for B, so $\varphi(a) \geq \sup B = a$. Again using the fact that φ is order-preserving, we see that $\varphi(\varphi(a)) \geq \varphi(a)$, whence $\varphi(a) \in B$. Therefore $\varphi(a) \leq \sup B = a$. Thus $\varphi(a) = a$, as required.

Let A be the closed interval $[0, 1]$ of the real line. The theorem asserts that any increasing function φ from $[0, 1]$ to itself has a fixed point. This means that the graph of φ meets the diagonal. (See Fig. 6–1.)

6–5.6. Corollary. Under the hypotheses of the theorem, φ has a largest fixed point and a smallest fixed point.

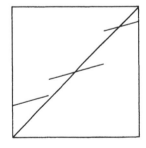

FIGURE 6-1

Proof. It is evident from the definition of B in the previous proof that a is the largest fixed point. A dual argument establishes the existence of a smallest.

EXERCISES

1. Is every subset of a complete ordered set complete?

2. Show that the set of all subsets of some fixed set, ordered by inclusion, is complete.

3. Show that the set of all equivalence relations in some fixed set, ordered by inclusion, is complete.

4. Let A and B satisfy the hypothesis of 6–5.3. Prove that if B is linear, so is A. Show, however, that this is false if A is not assumed to be complete.

5. Let A be any set and let $\mathcal{W}$ be the set of all weak order relations in A. Inclusion defines an order relation in $\mathcal{W}$. Prove the following facts:

   (a) The intersection of any nonvoid set of weak order relations is a weak order relation.
   (b) $\mathcal{W}$ is a complete ordered set.
   (c) An element $W \in \mathcal{W}$ is maximal (in the ordering of $\mathcal{W}$) if and only if it is a linear ordering of A.
   (d) Suppose $\mathfrak{X}$ is a nonvoid subset of $\mathcal{W}$ which is linearly ordered by inclusion. Then $(\bigcup_{X \in \mathfrak{X}} X) \in \mathcal{W}$.

6–6. WELL-ORDERING

6–6.1. Definition. Let $<$ denote the weak order relation in a linearly ordered set A. A is called *well-ordered* if and only if every nonempty subset of A contains a least element; i.e., if B is a nonempty subset of A, then

$$(\exists b_0 \in B)(\forall b \in B) \quad b_0 < b.$$

(This means that every nonempty subset must have an infimum *within itself*.)

Assuming that there are sufficiently many elements in a well-ordered set A, A must contain a first element, say f, then a second element (the least element of $A - \{f\}$), then a third, and so on. Thus the definition describes a kind of order significantly related to the counting process; indeed, the most important well-ordered set is the natural numbers in their usual order. The following example illustrates a more complicated possibility.

Let $\mathbf{N}$ be the set of all positive integers. Let "$a < b$" mean "a is odd and b is even, or $b - a$ is a nonnegative even integer." Then $<$ is a weak order relation in $\mathbf{N}$ which puts the integers in the order

$$1, 3, 5, \ldots, 2, 4, 6, \ldots$$

It is easily checked that $<$ well-orders $\mathbf{N}$. Note that 2 has infinitely many predecessors but no immediate predecessor.

6–6.2. Transfinite Induction. Propositions concerning the elements of a well-ordered set can sometimes be proved by a generalization of mathematical induction as follows.

Let A be a well-ordered set and let $P(x)$ be a propositional scheme having A as the domain of the variable. Suppose that, for any $a \in A$, we can prove $P(a)$ if we assume $P(b)$ for all predecessors b of a. Then $P(a)$ is true for all $a \in A$. If $<$ denotes the strong order relation in A, we can state this more formally as follows.

From
$$(\forall a \in A) \;\; ((\forall b < a)\, P(b)) \Rightarrow P(a) \tag{1}$$
we may infer
$$(\forall a \in A) \;\; P(a).$$

For, consider $S = \{a \in A \mid \text{not-}P(a)\}$. We must prove that S is empty. If not, S has a least member. Call this s_0. By the definition of S and s_0, $(\forall b < s_0)\, P(b)$. Then from (1) we get $P(s_0)$, contradicting the fact that $s_0 \in S$.

This technique of inference, called the *second principle of finite induction* when applied to the positive integers in the usual order, is known as *transfinite induction* when applied to more complicated well-ordered sets.

Ordinary induction, which we shall discuss in Chapter 7, reasons from each case to the next with a special argument at the start. Note that we should never reach 2 in the example of 6–6.1 by such a line of reasoning. On the other hand, transfinite induction does not formally require a special argument at the start. If a_0 is the first element of A, then $(\forall b < a_0)\, P(b)$ is surely true since the quantification is over the null set. Therefore (1) implies $P(a_0)$ automatically.

A celebrated theorem of advanced set theory asserts that every set can be well-ordered; that is, if E is any set, there exist order relations S and W for E such that $\langle E, S, W \rangle$ is a well-ordered set. The proof is nonconstructive, but it opens the way to inductive proofs and constructions in very large sets.

6–6.3. Considerable use is made in algebra of the *minimum condition*, the analogue of well-ordering for partially ordered sets. A partially ordered set is said to satisfy the minimum condition if and only if every nonempty subset contains a minimal element. The *maximum condition* is defined dually.

EXERCISES

1. Suppose that A is an ordered set such that every nonempty subset of A contains a least element. Prove that A is well-ordered.

2. Is every subset of a well-ordered set well-ordered?

3. Let $\langle X, S, W \rangle$ be a well-ordered set. Let F be a function defined on X such that, for each $x \in X$, $F(x)$ is a linearly ordered set. We denote the strong order relations in each of the sets $F(x)$ by $<$. Consider $P = \bigtimes_{x \in X} F(x)$. If $g, h \in P$, let $g < h$ if and only if
$$(\exists x \in X) \;\; (g(x) < h(x) \text{ and } (\forall y \in X)\, y\, S\, x \Rightarrow g(y) = h(y)).$$
Prove that $<$ is a strong linear order relation for P. This is called the *lexicographic ordering* of P.

By considering the special case where $X = \{1, 2, 3, 4, 5\}$, with the usual order, and for each $x \in X$, $F(x)$ is the alphabet in its usual order, explain the term *lexicographic*.

CHAPTER 7

MATHEMATICAL INDUCTION

Mathematical induction is a basic technique of all branches of mathematics; yet it does not fall directly under our logical rules. In this chapter we shall develop the ideas of inductive proof and inductive construction. To do this we require a description of the integers as a configuration. Our description, and therefore our results, will of course apply equally well to any configuration isomorphic to the integers.

To emphasize the configurational point of view, we shall study a class of configurations which has not received enough mathematical attention to have a standard name. Following Dedekind, who analyzed this class in his fundamental paper "Was sind und was sollen die Zahlen?",* we shall call them *chains*.

7-1. CHAINS

7-1.1. Definition. A *chain* is a configuration $\langle S, f \rangle$ consisting of a set S and a function f from S to S. The function f will be called the *successor* function.

Such a function is often called a *unary operation* in the set S.

7-1.2. Definition. A *subchain* of the chain $\langle S, f \rangle$ is a chain $\langle T, g \rangle$ such that $T \subseteq S$ and $g \subseteq f$.

We note that the definition is worded to admit the trivial cases: $\langle \emptyset, \emptyset \rangle$ and $\langle S, f \rangle$ are both subchains of $\langle S, f \rangle$.

7-1.3. Proposition. Let $\langle T, g \rangle$ be a subchain of the chain $\langle S, f \rangle$ and let $\langle U, h \rangle$ be a subchain of $\langle T, g \rangle$. Then $\langle U, h \rangle$ is a subchain of $\langle S, f \rangle$.

We omit the proof of this obvious fact. The definitions of subconfigurations are invariably chosen so that transitivity holds as above.

In the case of ordered sets any subset of the basic set can be chosen as the basic set of a subconfiguration, but this is not the case for chains.

---

* "What Are Numbers and What Should They Be?" The paper was originally published as a pamphlet and is available in English translation under the title, "The Nature and Meaning of Numbers," one of two papers in a book, *Essays on the Theory of Numbers* by Richard Dedekind, translated by W. W. Beman, Open Court, La Salle, Illinois, 1948.

7–1.4. Proposition. Let $\langle S, f \rangle$ be a chain. For a subset T of S to be the basic set of a subchain of $\langle S, f \rangle$ it is necessary and sufficient that

$$(\forall t \in T) \quad f(t) \in T. \tag{1}$$

If T satisfies this condition there is a unique function g such that $\langle T, g \rangle$ is a subchain of $\langle S, f \rangle$.

Proof. To prove necessity, suppose that $\langle T, g \rangle$ is a subchain of $\langle S, f \rangle$. Choose $t \in T$. Then $g(t) \in T$ by the definition of chain. Now, $g \subseteq f$ means that f is an extension of g, so $f(t) = g(t)$ and thus $f(t) \in T$. This proves (1).

To show sufficiency, we suppose that (1) holds and let g be the restriction of f to T. Now, (1) asserts that range $g \subseteq T$, and therefore $\langle T, g \rangle$ is a chain. It is a subchain of $\langle S, f \rangle$, since $T \subseteq S$ and $g \subseteq f$.

Let h be any function such that $\langle T, h \rangle$ is a subchain of $\langle S, f \rangle$. Since $h \subseteq f$, h is a restriction of f. Since domain $h = T$, h is f restricted to T; that is, $h = g$. This proves the uniqueness of the successor function for T.

We shall often follow the usual practice of referring to a chain by the name of its basic set when there is no doubt about what function is involved. Following this convention, we shall call any subset satisfying the criterion of 7–1.4 a subchain. The uniqueness of the successor function for T established in 7–1.4 is essential for the validity of this convention.

The following proposition, whose proof we shall omit, translates the general definition of isomorphism into convenient terms for chains.

7–1.5. Proposition. Let $\langle S, f \rangle$ and $\langle T, g \rangle$ be chains and let φ be a bijection from S to T. A necessary and sufficient condition that φ be an isomorphism of the chains is that $\varphi \circ f = g \circ \varphi$.

EXERCISES

1. Suppose that $S = \{1, 2, 3, 4, 5, 6\}$ and $f = \{\langle 1, 2 \rangle, \langle 2, 3 \rangle, \langle 3, 4 \rangle, \langle 4, 5 \rangle, \langle 5, 6 \rangle, \langle 6, 4 \rangle\}$. Then $\langle S, f \rangle$ is a chain. Find all the subchains of S.

2. Let $T = \{1, 2, 3, 4, 5\}$ and $g = \{\langle 1, 3 \rangle, \langle 2, 5 \rangle, \langle 3, 4 \rangle, \langle 4, 1 \rangle, \langle 5, 2 \rangle\}$. Then $\langle T, g \rangle$ is a chain. Find all subchains of T which are isomorphic to a subchain of S (cf. Exercise 1).

3. Show that the chain $\langle S, f \rangle$, defined by $S = \{1, 2, 3, 4\}$, $f = \{\langle 1, 2 \rangle, \langle 2, 3 \rangle, \langle 3, 2 \rangle, \langle 4, 3 \rangle\}$, is nontrivially isomorphic to itself; i.e., there is an isomorphism which is not the identity function of S.

7–2. INDUCTIVE PROOF

The intuitive natural numbers form a chain when we consider the successor function $n \rightarrow n + 1$. The essential idea of mathematical induction is that, starting from 1 and repeatedly applying the successor function, we can eventually reach any natural number. We shall formalize this in the general setting of chains.

Aside from its immediate significance for this section, the next proposition is important as an example of a method of finding least subsystems which applies in many contexts.

7–2.1. Proposition. Let $\langle S, f \rangle$ be a chain and let E be any subset of S. Among the subchains of S which contain E there is a least subchain (in the ordering by inclusion). This least subchain U has the following property:

$$(\forall u \in U) \qquad u \in E \quad \text{or} \quad u \in f(U). \tag{1}$$

Proof. Let T be the set of all subchains of S which contain E. Since $S \in T$, the latter is not void. Let

$$U = \bigcap_{T \in T} T.$$

It is clear that $E \subseteq U$. We shall prove that U is a subchain of S, whereupon it becomes obvious that U is the least member of T. Let $u \in U$. For any $T \in T$ we have $u \in T$. Since T is a subchain, 7–1.4 implies $f(u) \in T$. This shows that $(\forall T \in T) \, f(u) \in T$, or $f(u) \in \bigcap_{T \in T} T = U$. Thus we have proved

$$(\forall u \in U) \; f(u) \in U.$$

By 7–1.4, U is a subchain.

The proof of the last statement is indirect. Let us suppose that there is an element u_0 such that $u_0 \in U - E$ and $u_0 \notin f(U)$. Consider $U' = U - \{u_0\}$. Since $u_0 \notin E$, $E \subseteq U'$. And since $u_0 \notin f(U)$, $f(U') \subseteq f(U) \subseteq U'$ and U' is a subchain by 7–1.4. But this contradicts the fact that U is the least subchain containing E.

7–2.2. Definition. In the notation of Proposition 7–2.1, E will be said to *generate* the subchain U. If E consists of a single element s of S, then s will be said to generate U.

The latter sentence is really a terminological inaccuracy; in accordance with the first definition we should say "$\{s\}$ generates U." Such inexactness is common in mathematics. It rarely causes any confusion and often affords notational simplification.

If E is a subset of S and U is the least subchain of S containing E, then U is also the least subchain of U containing E. This depends on 7–1.3. It explains why we can say that E generates U without referring to the original chain S except implicitly via the successor function f, which is understood to determine the successor function for all the chains involved.

Intuitively, it should be clear that the chain generated by E is

$$E \cup f(E) \cup f(f(E)) \cup \ldots$$

But the latter expression does not formally define a set, because the three dots have only intuitive meaning. The point of Proposition 7-2.1 is that we can define this set formally by the intersection procedure.

7-2.3. We are now in a position to examine the structure of proofs by mathematical induction. Suppose that $\langle S, f \rangle$ is a chain and that $P(x)$ is a propositional scheme in which the variable x has domain S. Let E be a subset of S and let U be the subchain of S generated by E.

Suppose that we have proved

$$(\forall s \in S) \qquad P(s) \quad \text{implies} \quad P(f(s)) \tag{2}$$

and

$$(\forall s \in E) \quad P(s). \tag{3}$$

Then we may infer

$$(\forall s \in U) \quad P(s). \tag{4}$$

In the usual case, S is the set of natural numbers, f is $s \to s + 1$, $E = \{1\}$, and $U = S$.

It should be clear that this is not one of the inferences of Section 2–5. However, we can prove that (4) follows from (2) and (3) using only the kinds of inference discussed in Chapter 2. Let

$$T = \{s \mid P(s)\}. \tag{5}$$

Statement (2) implies that T is a subchain of S, and (3) implies that $E \subseteq T$. Then, since U is the least subchain of S containing E, we conclude that $U \subseteq T$. Now (4) follows directly from (5) and the definition of $U \subseteq T$.

The argument has two essential steps. First, by forming the set T we in effect convert the propositional scheme into a set. Second, we apply the defining property of U, which involves by implication a universal quantifier with domain $\mathfrak{P}(S)$ rather than S. The interchange of sets and propositional schemes together with the quantification over $\mathfrak{P}(S)$ means that, in effect, we have allowed quantification over propositional schemes.

This inference, when applied to the natural numbers, is known as the *first principle of mathematical induction*. As we have just seen, it depends on the fact that the natural numbers form a chain generated by 1. The second principle of induction, which we discussed in 6–6.2, depends on the fact that the natural numbers form a well-ordered set. It happens that the order structure of the natural numbers and the chain structure are closely related (see Exercise 5, p. 88 and Exercises 1 and 2, p. 93), and consequently the two forms of induction are also closely related. But when seen in their general settings, the two forms are much farther apart. The circular chain of implications given in Section 2–5 can be regarded as an application of the generalized form of the first principle of induction (where $S = \{a, b, c, d\}$ and $f = \{\langle a, b \rangle, \langle b, c \rangle, \langle c, d \rangle, \langle d, a \rangle\}$) although there is no related order structure. The second principle, as we noted in 6–6.2, is applicable to well-ordered sets of a more complicated character than the natural numbers, but the first principle is not.

EXERCISES

1. Let $\langle S, f \rangle$ be a chain. Show that the union of any family of subchains of S is itself a subchain. Let E be any subset of S. Show that among the subchains of S disjoint from E there is a largest. Suppose that E is a subchain of S, and let T be the largest subchain of S disjoint from E. Prove that $S - T$ is a subchain.

2. Let A be any set and let R be a relation in A. Show that among the transitive relations in A which contain R there is a smallest.

3. Let R be a symmetric relation in the set A. Let S be the least transitive relation in A which contains R. Prove that S is symmetric.

4. Let A be any set and let R be a relation in A. Show that among the equivalence relations in A which contain R there is a smallest.

5. Let $\langle S, <, \leq \rangle$ be a well-ordered set with no greatest element. For each $s \in S$, let $f(s)$ be the least element greater than s. Show that the chain $\langle S, f \rangle$ is generated by the set $E = S - \text{range } f$. Show also that E is the least subset of S which generates S. Write out the propositional scheme that defines the function f.

7-3. THE NATURAL NUMBERS AND INDUCTIVE DEFINITIONS

The intuitive natural numbers arise from the counting process. We count one, two, three, four, . . . , and no matter how far we go, we can always count one more. Counting one more is a unary operation in the set of numbers. Two facts about this unary operation characterize the natural numbers as a configuration: starting from 1, we can eventually reach any natural number, and each step produces a new natural number. We shall formalize these properties as the defining postulates for a simple chain.

7-3.1. Definition. A chain $\langle S, f \rangle$ will be called *simple* if and only if

(a) S is generated by an element not in the range of f, and
(b) f is injective.

Our study of the class of simple chains has two immediate objectives. First, we shall prove that it is legitimate to define functions by induction. Then we shall prove that the postulates for a simple chain are categorical. A third objective, the construction of the real number system starting from a simple chain, will be the subject of Chapter 9.

7-3.2. Proposition. If $\langle S, f \rangle$ is a simple chain, then there is exactly one element of S not in $f(S)$. Moreover, this element is the only generator of S.

Proof. By 7-3.1(a) there is an element, say a, which generates S and is not in $f(S)$. According to 7-2(1), $S = \{a\} \cup f(S)$. This shows that there is only one element of S not in $f(S)$. Suppose b is a generator of S. Then 7-2(1) gives $S = \{b\} \cup f(S)$. Since $a \notin f(S)$, $a \in \{b\}$; that is, $a = b$, and the generator of S is unique.

7-3.3. Definition. The generator of a simple chain S will be called the *first element* of S.

7-3.4. Although the terms *chain* and *simple chain* are no longer current in mathematics, the postulates for a simple chain coincide in effect with what are called Peano's axioms for the natural numbers. These are usually stated as follows:

Axiom 1. 1 is a natural number.

Axiom 2. To every natural number x there is assigned a uniquely determined natural number x', called its successor.

Axiom 3. 1 is not the successor of any natural number.

Axiom 4. If $x' = y'$, then $x = y$.

Axiom 5. Suppose a subset T of the natural numbers has the following properties:

  (a) $1 \in T$,
  (b) if $t \in T$, then $t' \in T$.

Then T contains all natural numbers.

Comparing these axioms with our postulates, we see that Axiom 2 amounts to the assertion that the natural numbers are a chain with the successor function $x \to x'$. Axiom 4 asserts that the successor function is injective, while Axioms 3 and 5 coincide with 7–3.1(a). Axiom 1 is purely terminological.

The first natural number is often taken to be 0. This makes no difference from the configurational point of view.

7-3.5. Our next theorem is concerned with defining a function by induction. Written informally in terms of the set $\mathbf{N}$ of positive integers, it says that, if we prescribe $\varphi(1)$ and a method of calculating $\varphi(n + 1)$ from $\varphi(n)$, a unique function φ with domain $\mathbf{N}$ is thereby determined. If we prescribe that $\varphi(1) = b$ and that $\varphi(n + 1)$ should be obtained from $\varphi(n)$ by applying the function g (that is, $\varphi(n + 1) = g(\varphi(n))$), then $\varphi(2) = g(b)$, $\varphi(3) = g(\varphi(2)) = g(g(b))$, $\varphi(4) = g(\varphi(3))$, etc. The prescription gives a rule for calculating the value of φ at any integer. This is obvious; the theorem is concerned with a subtler point.

While we may think of a function as being a rule for calculation, our definition says that it is a set of ordered pairs. Furthermore, we have agreed not to consider any sets except those definable by the set-builder notation of Section 2–6 using propositional schemes and quantifiers. The point of the theorem is that φ can be defined in this manner.

Evidently, φ ought to be the set

$$\{\langle 1, b\rangle, \langle 2, g(b)\rangle, \langle 3, g(g(b))\rangle, \ldots\}. \tag{1}$$

Our problem is to replace this informal description by a formal definition.

We have already faced the problem of defining formally a set which is informally described as

$$\{c, h(c), h(h(c)), \ldots\},$$

where each member of the list is calculated from its predecessor by applying a function h. This is just Proposition 7-2.1 if we replace E by $\{c\}$ and f by h. Hence the proof will consist of an application of 7-2.1 to obtain the set φ (that is, (1)) followed by a detailed verification that φ is indeed a function with the required properties.

Since we do not yet have the set $\mathbf{N}$ as a formal entity, the proposition is stated in terms of a simple chain $\langle S, f \rangle$. To make sure that the values, b, $g(b)$, $g(g(b))$, etc., of the hoped-for function φ are all defined, we shall insist that range $g \subseteq$ domain g, which amounts to requiring that $\langle \text{domain } g, g \rangle$ should be a chain.

7-3.6. Theorem. Let $\langle S, f \rangle$ be a simple chain with first element a. Let $\langle T, g \rangle$ be any chain, and let $b \in T$. There exists a unique function φ from S to T such that

$$\varphi(a) = b, \tag{2}$$

and

$$(\forall s \in S) \quad \varphi(f(s)) = g(\varphi(s)). \tag{3}$$

Proof. There is no difficulty in proving by induction that there is at most one such function φ. We shall omit this step and turn to the question of existence.

Define a function h from $S \times T$ to itself by

$$h(s, t) = \langle f(s), g(t) \rangle.$$

Then $\langle S \times T, h \rangle$ is a chain. Let φ be the subchain of $S \times T$ generated by $\langle a, b \rangle$. We shall prove that φ is the required function.

Evidently, φ is a relation, domain $\varphi \subseteq S$, and range $\varphi \subseteq T$. We shall prove next that domain $\varphi = S$. Since $\langle a, b \rangle \in \varphi$, $a \in$ domain φ. If $s \in$ domain φ, pick t so that $\langle s, t \rangle \in \varphi$. Then $h(s, t) = \langle f(s), g(t) \rangle \in \varphi$, so $f(s) \in$ domain φ. Now it follows by induction that $S \subseteq$ domain φ, whence domain $\varphi = S$.

To prove that φ is actually a function, let R be the set of all members of S to which φ assigns a unique value; that is,

$$R = \{s \in S \mid (\forall t, u) \, ((\langle s, t \rangle \in \varphi \text{ and } \langle s, u \rangle \in \varphi) \Rightarrow t = u)\}.$$

Now suppose $\langle a, t \rangle \in \varphi$. By 7-2.1, either $\langle a, t \rangle = \langle a, b \rangle$ or $\langle a, t \rangle \in h(\varphi)$. The latter statement means that we can choose $\langle s_1, t_1 \rangle$ so that $\langle a, t \rangle = h(s_1, t_1) = \langle f(s_1), g(t_1) \rangle$; but then $a = f(s_1)$, contrary to $a \notin$ range f. Therefore $\langle a, t \rangle = \langle a, b \rangle$ and $t = b$. This proves $a \in R$.

We shall prove the inductive step in the contrapositive form. Suppose $s \in S$ and that $f(s) \notin R$. Then choose t and u in T so that $t \neq u$, $\langle f(s), t \rangle \in \varphi$, and $\langle f(s), u \rangle \in \varphi$. By 7-2.1, since $f(s) \neq a$, we can choose $\langle s_1, t_1 \rangle$ and $\langle s_2, t_2 \rangle$ in φ so that $h(s_1, t_1) = \langle f(s), t \rangle$ and $h(s_2, t_2) = \langle f(s), u \rangle$. Applying the definition of

h, we get $f(s_1) = f(s)$, $f(s_2) = f(s)$, $g(t_1) = t$, $g(t_2) = u$. Since f is injective, $s_1 = s_2$; since $t \neq u$, $t_1 \neq t_2$. Now, $\langle s, t_1 \rangle \in \varphi$ and $\langle s, t_2 \rangle \in \varphi$, so $s \notin R$.

Induction gives $R = S$, and therefore φ is a function. Since $\langle a, b \rangle \in \varphi$, we have (2). Since $\langle s, \varphi(s) \rangle \in \varphi$ and φ is a chain, $h(s, \varphi(s)) = \langle f(s), g(\varphi(s)) \rangle \in \varphi$ or $\varphi(f(s)) = g(\varphi(s))$.

7–3.7. Corollary. Let $\langle S, f \rangle$ be a simple chain having first element a. Let $\langle T, g \rangle$ be any chain and let ψ map some set E into T. There exists a unique function φ from $E \times S$ to T such that

$$(\forall e \in E) \quad \varphi(e, a) = \psi(e), \tag{4}$$
and
$$(\forall e \in E)(\forall s \in S) \quad \varphi(e, f(s)) = g(\varphi(e, s)). \tag{5}$$

Although the existence of such a function would appear to be a direct application of the previous theorem, our insistence that φ should be definable by a propositional scheme makes this a trifle more subtle. For each fixed $e \in E$ the theorem tells us that there is a function φ_e such that $\varphi_e(a) = \psi(e)$ and $\varphi_e \circ f = g \circ \varphi_e$. The existence of such a function for each e does not directly tell us that there is a function of two variables $\langle e, s \rangle \rightarrow \varphi_e(s)$. We must show that such a function can be defined as a set of ordered pairs under our rule for defining sets. The uniqueness of the individual functions φ_e makes this possible.

Proof of 7–3.7. Let F be the set of all functions from S to T, and let

$$\varphi = \{ \langle \langle e, s \rangle, t \rangle \mid (\exists h \in F) \ h(a) = \psi(e), h \circ f = g \circ h, \text{ and } h(s) = t \}.$$

Obviously φ is a relation with domain $E \times S$. We must show that it is a function. Suppose that $\langle \langle e, s \rangle, t_1 \rangle \in \varphi$ and $\langle \langle e, s \rangle, t_2 \rangle \in \varphi$. Choose h_1 and h_2 in F so that

$$h_1(a) = \psi(e), \quad h_1 \circ f = g \circ h_1, \quad h_1(s) = t_1,$$
$$h_2(a) = \psi(e), \quad h_2 \circ f = g \circ h_2, \quad h_2(s) = t_2.$$

The uniqueness assertion of Theorem 7–3.6 shows that $h_1 = h_2$, whence $t_1 = t_2$. It is now a routine matter to verify (4) and (5).

The arguments used to prove Theorem 7–3.6 can be extended slightly to prove the corollary directly. Define a successor function h in $(E \times S) \times T$ by

$$h(\langle e, s \rangle, t) = \langle \langle e, f(s) \rangle, g(t) \rangle$$

and consider the subchain φ generated by

$$\{ \langle \langle e, a \rangle, \psi(e) \rangle \mid e \in E \}.$$

Then φ turns out to be the required function.

7–3.8. Theorem. Any two simple chains are isomorphic; in fact, if $\langle S, f \rangle$ and $\langle T, g \rangle$ are simple chains, there is a unique isomorphism from S to T.

The technical details of the proof will be simplified by the following lemma, which should be compared with Lemma 6–5.4.

7-3.9. Lemma. Let $\langle S, f \rangle$ be a simple chain and let a be its first element. If θ is a function from S to S such that $\theta \circ f = f \circ \theta$ and $a \in$ range θ, then θ is the identity.

Proof. Suppose $a = \theta(b)$. If $b \neq a$, we can choose c so that $b = f(c)$. Then $\theta(f(c)) = f(\theta(c))$, contrary to the fact that $a \notin$ range f. Therefore $b = a$; hence $\theta(a) = a$. Now it follows immediately by induction, using $\theta \circ f = f \circ \theta$, that $(\forall s \in S)\ \theta(s) = s$; thus θ is the identity.

Proof of Theorem 7–3.8. Let a and b be the first elements of S and T, respectively. According to 7–3.6 there is a function φ from S to T such that

$$\varphi(a) = b \quad \text{and} \quad \varphi \circ f = g \circ \varphi. \tag{6}$$

There is also a function ψ from T to S such that

$$\psi(b) = a \quad \text{and} \quad \psi \circ g = f \circ \psi. \tag{7}$$

Now, $\psi(\varphi(a)) = a$ and

$$\psi \circ \varphi \circ f = \psi \circ g \circ \varphi = f \circ \psi \circ \varphi$$

from (6) and (7). By Lemma 7–3.9, $\psi \circ \varphi$ is the identity on S. By similar argument $\varphi \circ \psi$ is the identity of T. This proves that φ is a bijection from S to T, and (6) with 7–1.5 shows that φ is an isomorphism of the chains.

Suppose now that φ' is any isomorphism from the chain $\langle S, f \rangle$ to the chain $\langle T, g \rangle$. Then $\psi \circ \varphi'$ is the identity on S by the lemma. Then

$$\varphi' = (\varphi \circ \psi) \circ \varphi' = \varphi \circ (\psi \circ \varphi') = \varphi.$$

Thus there is only one isomorphism from S to T.

We have proved that simple chains are categorically determined. As we remarked above, the intuitive natural numbers apparently form a simple chain by using the counting operation $n \rightarrow n + 1$. Hence the postulates for a simple chain should serve as an adequate foundation for the study of the natural numbers. In Chapter 9 we shall show how the arithmetic and order properties of the natural numbers can be derived from the postulates.

On the other hand, we must realize that the postulates do not describe a unique configuration, only a class of isomorphic configurations. In particular, the natural numbers from 2 on, with the same successor function, form a simple chain; and many other simple chains can be constructed. In Chapter 10 we shall introduce one particular simple chain which we shall call *the* natural numbers. Until then we shall consider all simple chains on an equal footing.

EXERCISES

1. Let $\langle S, f \rangle$ be a chain such that f is injective and $f(S) \subset S$. Prove that S contains a simple subchain.

2. Let $\langle S, <, \leq \rangle$ be a nonvoid, well-ordered set with no greatest element. Suppose also that every nonempty subset of S bounded above contains its supremum. For each $s \in S$ let $f(s)$ be the least element greater than s. Prove that $\langle S, f \rangle$ is a simple chain.

CHAPTER 8

FIELDS

When the principal properties of addition and multiplication are abstracted we are led to a class of configurations called fields. After a brief study of binary operations, we shall define fields and prove that many familiar formulas of arithmetic are valid in all fields.

8–1. BINARY OPERATIONS

8–1.1. Definition. A *binary* operation in the set S is a function from $S \times S$ to S. A *ternary* operation in the set S is a function from $S \times S \times S$ to S.

Operations of higher order may be defined similarly. The subject known to mathematicians as algebra is concerned with the theory of sets endowed with one or more operations. Binary operations have received by far the most study. Addition and multiplication in the real numbers are the best known examples of binary operations.

The function sign for binary operations is customarily written between its arguments, as in $3 + 4$, $2 \cdot 6$, etc. Symbols like $+$ or $\cdot$ are frequently used for abstract operations which share some of the properties of ordinary addition and multiplication. We shall adopt this practice starting in the next section.

8–1.2. Definition. The binary operation b in the set S is said to be *associative* if and only if

$$(\forall x, y, z \in S) \quad b(x, b(y, z)) = b(b(x, y), z).$$

If we imagine computing the right- and left-hand sides of the above equation by using a table of the function b, it is clear that there is no *a priori* reason that they should come out the same; hence the associative law is a stringent condition. It is well known that addition and multiplication are associative operations in the ordinary numbers; on the other hand, we are equally well acquainted with the nonassociative operation of subtraction, $b(x, y) = x - y$.

8–1.3. Definition. A binary operation b in the set S is said to be *commutative* if and only if

$$(\forall x, y \in S) \quad b(x, y) = b(y, x).$$

94

Addition and multiplication are commutative, but subtraction is not.

Among ordinary numbers, zero and one possess extraordinary properties with respect to multiplication. We shall abstract these properties in the following definitions.

8–1.4. Definitions. Let b be a binary operation in the set S. An element $u \in S$ is called an *identity* (sometimes *unit* or *neutral*) element for b if and only if

$$(\forall x \in S) \quad b(u, x) = b(x, u) = x.$$

An element $z \in S$ is called a *zero* element for b if and only if

$$(\forall x \in S) \quad b(z, x) = b(x, z) = z.$$

An element $u \in S$ is called a *left-identity* element for b if and only if

$$(\forall x \in S) \quad b(u, x) = x.$$

Right-identity elements and *left-* and *right-zero* elements are defined in an analogous manner.

It is obvious that, if b is a commutative operation, then any left-identity element is actually an identity element and similarly a left-zero element is actually a zero element.

8–1.5. Proposition. Let b be a binary operation in the set S. Then there is at most one identity element for b and at most one zero element for b.

Proof. Let u and u' be identity elements for b. Then

$$u' = b(u, u') = u.$$

Similarly for zero elements.

EXERCISES

1. Restate the definition of isomorphism for configurations consisting of a set and a binary operation in that set, using terms adapted to this class of configurations. Let the binary operation in both configurations be denoted by some medial symbol.

2. Suppose that b is a binary operation in the set S. Prove that if b has both a left-identity and a right-identity then it has just one of each kind and these are the same. Give examples to prove that it is possible that b has

    (a) more than one left-identity and more than one left-zero,
    (b) more than one left-identity and more than one right-zero,
    (c) more than one left-identity and a zero,
    (d) an identity and more than one left-zero,
    (e) a left-identity which is also a right-zero (S having more than one element).

*3. Which of the following binary operations defined on the ordinary numbers are associative? commutative? have a left-identity? a right-identity? a left-zero? a right-zero?

(a) $(x, y) \rightarrow x$

(b) $(x, y) \rightarrow |x - y|$

(c) $(x, y) \rightarrow x + y + xy$

(d) $(x, y) \rightarrow xy - 2x - y + 4$

(e) $(x, y) \rightarrow xy + y - 1$

(f) $(x, y) \rightarrow \sqrt{x^2 + y^2}$ (let S be the nonnegative numbers)

(g) $(x, y) \rightarrow \max(x, y)$

(h) $(x, y) \rightarrow \max(x^2 - y, y^2 - x)$

8-2. FIELDS

8-2.1. Definition. A *field* is a configuration $\langle F, a, m \rangle$ consisting of a set F and two binary operations a and m in F such that, writing $x + y$ for $a(x, y)$ and xy for $m(x, y)$, and with quantifiers ranging over F,

(i_a) $(\forall w, x, y) (w + x) + y = w + (x + y)$, (i_m) $(\forall w, x, y) (wx)y = w(xy)$,

(ii_a) $(\forall x, y) x + y = y + x$, (ii_m) $(\forall x, y) xy = yx$,

(iii) $(\forall w, x, y) w(x + y) = (wx) + (wy)$, and

(iv) $(\exists z, u)$ $\begin{cases} (\forall x)\, x + z = x \text{ and } xu = x, \\ (\forall b)(\exists x)\, b + x = z, \\ (\forall b \neq z)(\exists x)\, bx = u, \\ u \neq z. \end{cases}$

We are demanding two associative, commutative binary operations. Postulate (iii) is the familiar *distributive law;* it is the only postulate which effectively interlocks the two operations. Without such an interlocking postulate the study of configurations involving two operations would be pointless, for one might as well study the operations separately. Postulate (iv) demands first the existence of identity elements for both operations and then the solvability of certain equations; the last part is simply to ensure that F has more than one element.

The definition is intended to describe most of the principal algebraic properties of the rational, real, and complex number systems. However, one must be careful not to ascribe too many of the properties of conventional numbers to a general field. The following example will illustrate this point and also convince us that the requirements for a field are consistent.

EXAMPLE. Let $F = \{p, q\}$, where $p \neq q$. The binary operations in F are most conveniently presented in square tables:

$a:$	p	q
p	p	q
q	q	p

$m:$	p	q
p	p	p
q	p	q

Then $\langle F, a, m \rangle$ is a field.

Because F has only two elements, the postulates reduce to a finite list of statements: 8–2.1 (i), (ii), (iii) comprise 32 statements (counting in the most generous manner; there are many duplications and tautologies), and (iv) becomes 8 statements when the choice $z = p, u = q$ is made.

It is, of course, not necessary to prove the existence of a field, but it is rather comforting if we can do so. When one writes down a complicated set of postulates there is always the distinct possibility that they are self-contradictory, and the contradiction may be quite subtle. It is known, for example, that if F is finite, it must have a number of elements equal to a prime power; hence, the postulates would become self-contradictory if we were to adjoin a new one:

F has six elements.

We shall introduce further requirements shortly which will, among other things, guarantee that F is infinite. It will then be impossible to give a complete explicit example and we shall be obliged to continue, as mathematicians always have, without any absolute assurance that our postulates are consistent.

8–2.2. Notation. When there is no question about what the operations are, a field is usually referred to by the name of its basic set. The operations in a field are almost invariably designated by $+$ and juxtaposition (or medial dot), as was done in the definition, and are called addition and multiplication. From Proposition 8–1.5 it follows that the unit elements of addition and multiplication are uniquely determined; hence it is appropriate to assign them permanent names. By analogy with ordinary numbers they are designated 0 and 1, respectively. Even when several fields are under consideration at once, these symbols are ordinarily used for the units of each. Moreover, they are also used with their customary significance as counting numbers. This multiplicity of connotations rarely causes any confusion.

We shall immediately adopt the usual conventions of algebra to simplify our notation. We will write 2 for $1 + 1$, a^2 for aa, etc., and leave out as many parentheses as we can. We agree that an exponent affects only the element immediately to its left, that multiplication is done before division (to be introduced shortly) and both before addition or subtraction. Expressions like $(ab)c$ will be written abc, since there is no ambiguity of meaning by virtue of 8–2.1(i_m), and we will constantly apply the commutative and distributive laws without explicit reference.

It is well known that continued sums or products can be rearranged freely by virtue of the associative and commutative laws, and various theorems can be proved to this effect (see Chapter 13). At the moment we would be hard pressed even to state such a theorem, so we must postpone such considerations. But we are adopting conventions which suggest that these theorems are already available for use. Does this mean that we are abandoning our standards? Not at all! Before we investigate these questions, we will apply the conventions only in contexts in which the full justification could be given by using the postulates;

in other words, we are only abbreviating. If we were to write all these steps out, we would gain neither rigor nor clarity.

Consider the equations

$$(ab)^2 = a^2 b^2 \quad \text{and} \quad (ab)^n = a^n b^n.$$

The first is an abbreviation for the statement $(ab)(ab) = (aa)(bb)$, which follows easily from 8–2.1(i_m) and (ii$_m$). The abbreviation is unambiguous even though the exponents have not yet been defined. But the second statement is not an abbreviation, for one cannot write down n (ab)'s without knowing what n is. There might still be no objection if it were known that only one value of n was involved, for we could then imagine the whole thing written down and justified step by step. But if we prefix the universal quantifier, then certainly

$$(\forall n) \quad (ab)^n = a^n b^n$$

cannot be completely justified by direct application of 8–2.1(i_m) and (ii$_m$); formal definition of exponents and induction will be necessary to obtain this theorem. We shall defer this matter until we have a theory of finite sets and counting.

8-2.3. Definition. A *subfield* of the field $\langle F, a, m \rangle$ is a field $\langle G, a', m' \rangle$ for which $G \subseteq F$, $a' \subseteq a$ and $m' \subseteq m$.

The significance of this definition is that the operations of G must coincide with the operations of F. As in the case of chains, not every subset of F is the basic set for a subfield, but each subfield is determined by its basic set. We need not, therefore, distinguish between a subfield and its basic set. As in 7–1.3, the relation of being a subfield of a field is transitive.

EXERCISES

*1. Determine which of the following systems endowed with the usual addition and multiplication operations are fields: the positive integers, the nonnegative integers, the integers, the positive rationals, the rationals, the nonnegative real numbers, the real numbers. In each case, find which postulates are valid and which fail.

2. Verify that $\langle F, a, m \rangle$ of the Example on p. 96 is indeed a field.

3. Show that the last part of postulate (iv) for fields can be replaced by

F contains at least two elements;

that is, prove that this replacement leads to an equivalent system of postulates.

4. Rephrase the definition of isomorphism for fields using terms appropriate to this class of configurations.

5. Let G be a subfield of the field F. Show that the additive identity of G must coincide with the additive identity of F and the multiplicative identity of G must coincide with the multiplicative identity of F. (If this were not true, we could not afford to denote the additive identity in any field by 0 and the multiplicative identity by 1. Observe that the second statement would be false if we allowed a field to have only one element.)

8-3. THE ELEMENTARY ARITHMETIC OF FIELDS

8-3.1. Proposition. Let F be a field. Then

(a) $(\forall a, b)(\exists x)\, a + x = b$;
(b) $(\forall a, x, y)$ if $a + x = a + y$, then $x = y$;
(c) $(\forall a \neq 0)(\forall b)(\exists x)\, ax = b$;
(d) $(\forall a \neq 0)(\forall x, y)$ if $ax = ay$, then $x = y$;
(e) 0 is a zero element for multiplication.

(All quantifiers have domain F.)

Proof. (a) Let a and b be given. By 8-2.1(iv) we can choose c so that $a + c = 0$. Now,

$$a + (c + b) = (a + c) + b = 0 + b = b;$$

hence we may choose $x = c + b$.

(b) Let a, x, and y be given such that $a + x = a + y$. Choose c so that $a + c = 0$. Then

$$x = 0 + x = (c + a) + x = c + (a + x) = c + (a + y)$$
$$= (c + a) + y = 0 + y = y.$$

The proofs of (c) and (d) are similar.

(e) Let $x \in F$. Then

$$x \cdot 0 + x \cdot 0 = x(0 + 0) = x \cdot 0 = x \cdot 0 + 0,$$

and (e) follows from (b).

8-3.2. Notation. The unique solution of the equation $a + x = b$ is denoted by $b - a$, and if $b = 0$, simply by $-a$. The unique solution of the equation $ax = b$ ($a \neq 0$) is denoted

$$b/a \qquad \text{or} \qquad \frac{b}{a}.$$

The expression $b/0$ has no connotation.

8-3.3. Proposition. Let F be a field. Then for all $a, b, c, d \in F$,

(a) $a + (b - c) = (a + b) - c$;   (b) $a - (b + c) = (a - b) - c$;
(c) $-(-a) = a$;   (d) $a(b - c) = ab - ac$;
(e) $(-a)b = -(ab)$;   (f) $(-a)(-b) = ab$;
(g) if $ab = 0$, then either $a = 0$ or $b = 0$;
(h) $(a/b) \cdot (c/d) = ac/bd$ provided $b, d \neq 0$;
(i) $(a/b)/(c/d) = ad/bc$, provided $b, c, d \neq 0$;
(j) $(a/b) + (c/d) = (ad + bc)/bd$ provided $b, d \neq 0$;
(k) $(a/b) - (c/d) = (ad - bc)/bd$ provided $b, d \neq 0$.

Proof. We prove only (d) and (h).

(d) $ab = a((b - c) + c) = a(b - c) + ac$, and the result follows from the definition of $ab - ac$.

(h) Let $a/b = x$ and $c/d = y$. Then

$$a = bx, \qquad c = dy, \qquad ac = (bx)(dy) = (bd)(xy).$$

Since (g) shows that $bd \neq 0$, (h) follows from the definition of ac/bd.

Proposition 8–3.3 illustrates the fact that the basic rules for addition, subtraction, multiplication, and division are valid in all fields. We can hardly expect to formalize all of the arithmetic identities we may subsequently use; the above list is only a sample.

EXERCISES

1. Supply the proofs not given in the text.

2. Let S be a subset of a field F. Prove that the following conditions are necessary and sufficient for S to be a subfield:

(a) if $s, t \in S$, then $s - t \in S$;
(b) if $s, t \in S$, $t \neq 0$, then $s/t \in S$; and
(c) S has at least two elements.

On the other hand, show that if (b) is replaced by

(b') if $s, t \in S$, then $st \in S$,

then (a), (b'), and (c) are not sufficient for S to be a subfield.

3. Suppose that F and G are fields and that ϕ is a function from F to G such that

$$\phi(a + b) = \phi(a) + \phi(b) \qquad \text{and} \qquad \phi(ab) = \phi(a)\phi(b).$$

Prove:

(a) $\phi(a - b) = \phi(a) - \phi(b)$ and $\phi(0) = 0$;
(b) either range $\phi = \{0\}$ or ϕ is injective and range ϕ is a subfield of G;
(c) in the latter case, $\phi(1) = 1$ and $\phi(a/b) = \phi(a)/\phi(b)$ if $b \neq 0$.

(Note that the operation signs on the left refer to F and those on the right to G. Universal quantification over F is understood.)

8–4. WHOLE NUMBERS AND RATIONAL NUMBERS

8–4.1. Proposition. Let F be a field. Among subsets S of F having the properties

(a) $1 \in S$, and   (b) if $s \in S$, then $s + 1 \in S$,

there is a smallest.

Proof. There is at least one such set, namely F, and the intersection of all such sets is the required smallest one.

If we note that F, endowed with the function $x \to x + 1$, is a chain, we will recognize this as part of Proposition 7–2.1.

8–4.2. Notation. In any field F the smallest subset having Properties 8–4.1(a) and (b) will be denoted $N(F)$.

If F is the real field, then $N(F)$ consists of the natural numbers one, two, three, etc. In general $N(F)$ consists of those members of F which one can get by adding up 1's; that is, 1, $1 + 1$, $(1 + 1) + 1$, etc. We can make this description more precise by defining a function g from the natural numbers (i.e., a simple chain) to F as follows: Let $g(1) = 1$ and continue inductively by $g(n + 1) = g(n) + 1$. We see that range g has Properties 8–4.1(a) and (b); hence $N(F) \subseteq$ range g. Conversely, range $g \subseteq N(F)$ by induction; therefore range $g = N(F)$.

8–4.3. Proposition. In any field F, if $x \in N(F)$ and $y \in N(F)$, then $x + y \in N(F)$ and $xy \in N(F)$.

Proof. Let $x \in N(F)$ and define $B = \{w \mid x + w \in N(F)\}$. Then B has Properties 8–4.1(a) and (b). Therefore $N(F) \subseteq B$. Hence for any $y \in N(F)$, $x + y \in N(F)$, which is the first conclusion.

Let $x \in N(F)$ and define $C = \{w \mid xw \in N(F)\}$. Obviously $1 \in C$. If $s \in C$, then $xs \in N(F)$ and $x(s + 1) = xs + x \in C$ by the first part of the proposition. Thus C has property (b), whence $N(F) \subseteq C$, and the second conclusion follows.

8–4.4. Notation. In any field F we shall write $I(F)$ for

$$N(F) \cup \{0\} \cup \{x \mid -x \in N(F)\}.$$

If F is the real number field, then $I(F)$ consists of the integers, positive, negative, and zero. There is a temptation to call the members of $I(F)$ *integers* of F, but the term *integer* is used in a more inclusive sense in algebraic number theory.

8–4.5. Lemma. An element of a field F is in $I(F)$ if and only if it can be expressed as the difference of two members of $N(F)$.

Proof. Note that the set $\{x \mid x - 1 = 0 \text{ or } x - 1 \in N(F)\}$ has Properties 8–4.1(a) and (b); hence it includes $N(F)$. Therefore, if $u \in N(F)$, then either $u - 1 = 0$ or $u - 1 \in N(F)$; in either case, $u - 1 \in I(F)$.

Now fix $u \in N(F)$. The set $E = \{x \mid u - x \in I(F)\}$ has Property 8–4.1(a). If $x \in E$, then $u - x \in I(F)$. There are three cases. If $u - x \in N(F)$, then

$$u - (x + 1) = (u - x) - 1 \in I(F)$$

as above. If $u - x = 0$, then

$$u - (x + 1) = -1 \in I(F).$$

If $-(u - x) \in N(F)$, then

$$-(u - (x + 1)) = -(u - x) + 1 \in N(F).$$

Thus $u - (x + 1) \in I(F)$ in any case, and E has Property 8–4.1(b). Therefore, $N(F) \subseteq E$. This proves that

$$(\forall u, v \in N(F)) \quad u - v \in I(F).$$

Hence every element which can be expressed as the difference of two members of $N(F)$ is in $I(F)$. Conversely, every element of $I(F)$ can be so expressed. For, if $a \in N(F)$, then $a = (a + 1) - 1$; if $-a \in N(F)$, then $a = 1 - (-a + 1)$; and $0 = 1 - 1$.

8–4.6. Proposition. In any field F, if $x \in I(F)$ and $y \in I(F)$, then $x + y \in I(F)$, $x - y \in I(F)$, and $xy \in I(F)$.

Proof. Let x and y be given in $I(F)$. By Lemma 8–4.5 we can write

$$x = s - t, \qquad y = u - v,$$

where $s, t, u, v \in N(F)$. Hence,

$$x + y = (s + u) - (t + v), \qquad x - y = (s + v) - (t + u),$$

and

$$xy = (su + tv) - (sv + tu).$$

All of these are in $I(F)$ by 8–4.3 and 8–4.5.

8–4.7. Notation. For any field F we shall write

$$Q(F) = \{z \in F \mid (\exists x, y \in I(F)) \, z = x/y\}.$$

When F is the real field, $Q(F)$ is the set of rational numbers.

8–4.8. Proposition. $Q(F)$ is a subfield of F and, in fact, the smallest subfield of F.

Proof. That $Q(F)$ is a subfield of F follows immediately from 8–3.3(h) through (k) and Proposition 8–4.6.

Let S be any subfield of F. We know that $1 \in S$ (Exercise 5, p. 98) and therefore S has Properties 8–4.1(a) and (b). Hence $N(F) \subseteq S$. It follows from 8–4.5 that $I(F) \subseteq S$. Then $Q(F) \subseteq S$ from 8–4.7. Therefore $Q(F)$ is the smallest subfield of F.

The smallest subfield of a given field is frequently called the *prime* subfield.

8–4.9. Definition. A field F is said to have *characteristic* 0 (*characteristic* ∞ is also in common use) if and only if $0 \notin N(F)$. It is said to be *modular* if and only if $0 \in N(F)$.

The real, rational, and complex number fields have characteristic 0 while any finite field such as the example of Section 8–2 is modular. There are also infinite modular fields.

Recall the function g defined after 8–4.2 which maps the natural numbers onto $N(F)$. F has or does not have characteristic 0 according as g is or is not injective. For suppose g is not injective. Let n be the least integer for which

$$(\exists m < n)\ g(m) = g(n),$$

and select such an m. If $m > 1$, then

$$g(m - 1) + 1 = g(m) = g(n) = g(n - 1) + 1,$$

whence $g(m - 1) = g(n - 1)$, contrary to the choice of n. Therefore $m = 1$ and $g(n) = g(1) = 1$; so $g(n - 1) = 0$ and $0 \in N(F)$. Conversely, if $0 \in N(F)$, then there is a least integer p such that $g(p) = 0$. We have $g(p + 1) = g(1)$, so g is not injective. The integer p appearing in the last argument is called the *characteristic* of F. It is easily shown to be a prime.

8–4.10. Proposition. Let F be a field of characteristic 0. Then $\langle N(F),\ f \rangle$, where f is the function $x \to x + 1$, is a simple chain having 1 as its first element.

Proof. The function f certainly maps $N(F)$ into itself, so $\langle N(F), f \rangle$ is a chain. By 8–3.1(b) f is injective. If T is a subchain of $N(F)$ with $1 \in T$, then T has Properties 8–4.1(a) and (b). Therefore $N(F) \subseteq T$, whence $T = N(F)$. Thus $N(F)$ is generated by 1. If $1 = f(x)$, where $x \in N(F)$, then $x = 0$, contrary to the assumption $0 \notin N(F)$. This shows that $N(F)$ is a simple chain.

8–4.11. Theorem. Let F and G be fields of characteristic 0. Then $Q(F)$ and $Q(G)$ are isomorphic fields and the isomorphism of $Q(F)$ onto $Q(G)$ is unique.

Proof. See Exercise 4 below.

EXERCISES

1. If F is a modular field, then $I(F) = N(F)$.

*2. If F is a modular field, then $Q(F) = N(F)$.

3. Prove the existence of a least subfield in any field by an argument analogous to the proof of 7–2.1. Be sure to note how the result of Exercise 5, p. 98, is involved.

4. Suppose F and G are fields of characteristic 0.

(a) Show that there exists a unique function ϕ from $N(F)$ to $N(G)$ such that
   (i) $\phi(1) = 1$,
   (ii) $(\forall x)\ \phi(x + 1) = \phi(x) + 1$.
Furthermore, show that ϕ is a bijection,
   (iii) $(\forall x, y)\ \phi(x + y) = \phi(x) + \phi(y)$, and
   (iv) $(\forall x, y)\ \phi(xy) = \phi(x)\phi(y)$.

(b) Show that ϕ can be extended to be a bijection from $I(F)$ to $I(G)$ so that (iii) and (iv) remain valid. (We keep the same symbol for the extended function. Since the domain of the quantifiers is understood to be the domain of ϕ, when ϕ is extended, (iii) and (iv) become more inclusive.)

(c) Show that ϕ can be further extended to be a bijection from $Q(F)$ to $Q(G)$ so that (iii) and (iv) remain valid.

(d) Finish the proof of Theorem 8–4.11.

8–5. ORDERED FIELDS

8–5.1. Definition. An *ordered field* is a configuration $\langle F, +, \cdot, <, \leq \rangle$ such that

(a) $\langle F, +, \cdot \rangle$ is a field;
(b) $\langle F, <, \leq \rangle$ is a linearly ordered set;
(c) if $x < y$, then $x + z < y + z$; and
(d) if $0 < x$ and $0 < y$, then $0 < xy$.

Postulates (c) and (d) serve to bind the order structure to the field structure of F. Universal quantification over F is, of course, understood. When order and field structures are separately imposed on a set it is customary to abbreviate reference to (c) and (d) by describing the structures as *compatible* if these postulates are satisfied. The same word is used whenever different kinds of structure are imposed on a set, provided the intertwining postulates are well known.

We shall refer to an ordered field by its underlying set and usually designate its order relations by $<$ and $\leq$. An element x will be called *positive, negative, nonnegative,* or *nonpositive* according as $x > 0$, $x < 0$, $x \geq 0$, $x \leq 0$.

The real and rational number fields with their usual order are examples of ordered fields. The complex number field is not ordered, nor is it possible to impose a compatible order structure upon it, because -1 is a square in the complex numbers but -1 is never a square in any ordered field as we shall see below. On the other hand, it is possible to impose order on certain fields in more than one way. Consider, for example, the field F of all rational functions of one real variable; that is, functions defined on the set of real numbers $\mathbf{R}$ (with a finite number of exceptions due to zeros of the denominator) by rational formulas like

$$f(x) = \frac{a_0 + a_1 x + \cdots + a_m x^m}{b_0 + b_1 x + \cdots + b_n x^n},$$

where the a's and b's are real and not all the b's are zero. The arithmetic is to be performed by formal manipulation of the formulas, so we agree that x and x^2/x represent the same field element (although they don't represent the same function, since one has 0 in its domain and the other does not). A compatible order relation is defined in F by

$$f < g \Leftrightarrow (\exists r \in \mathbf{R})(\forall x > r)\ f(x) < g(x)$$

(i.e., the graph of f ultimately lies below that of g if we look far to the right.) A different compatible ordering is defined by

$$f < g \Leftrightarrow (\exists r \in \mathbf{R})(\forall x < r) \; f(x) < g(x)$$

(i.e., we compare graphs to the left).

8–5.2. Let F be an ordered field and let S be a subfield of F. Then S also inherits an order structure from F which is clearly compatible with its field structure. Hence any subfield of an ordered field is itself an ordered field.

An isomorphism of one ordered field onto another is, of course, a map which is both an isomorphism of the field structures and an isomorphism of the order structures. It is possible that two ordered fields are isomorphic as fields and isomorphic as ordered sets (via different maps) but not isomorphic as ordered fields.

8–5.3. Proposition. The set P of all positive elements in an ordered field F satisfies the following conditions:

(a) If $x \in P$ and $y \in P$, then $x + y \in P$ and $xy \in P$.

(b) For any $x \in F$, exactly one of the following statements is true:

$$x \in P, \qquad x = 0, \qquad \text{or} \qquad -x \in P.$$

Conversely, if P is a subset of a field F (no order specified) satisfying these conditions, then there is a unique ordering of F compatible with its field structure with respect to which P is the set of all positive elements.

We omit the proof which is straightforward.

If we had not made a prior study of order relations, it would have been easier to define an ordered field in terms of the set P. Hence the definition of an ordered field is frequently given in a form similar to the following. An ordered field is a quadruple $\langle F, P, +, \cdot \rangle$, where $\langle F, +, \cdot \rangle$ is a field, $P \subseteq F$, and for which 8–5.3(a) and (b) hold. The order relation is then introduced in terms of P. This is another example of a concept which can be described by two different classes of configurations.

8–5.4. Notation. If F is an ordered field, we shall denote by $P(F)$ the set of its positive elements; that is,

$$P(F) = \{x \in F \mid x > 0\}.$$

We now state a few elementary inequalities which we shall need later. These represent only a sample of the inequalities we may use in what follows.

8–5.5. Proposition. In any ordered field

(a) $a > 0$ if and only if $-a < 0$;

(b) if $a > b$ and $c > d$, then $a + c > b + d$;

(c) if $a > 0$ and $b < 0$, then $ab < 0$;
(d) if $a < 0$ and $b < 0$, then $ab > 0$;
(e) if $a > b$ and $c > 0$, then $ac > bc$;
(f) if $a > b$ and $c < 0$, then $ac < bc$;
(g) if $a > b \geq 0$ and $c > d \geq 0$, then $ac > bd$;
(h) $a^2 \geq 0$;
(i) either $a = b = 0$ or $a^2 + b^2 > 0$;
(j) $1 > 0$;
(k) if $a > 0$, then $1/a > 0$.

Proof. We shall prove three of these statements. Suppose $a < 0$ and $b < 0$, then $-a > 0$ and $-b > 0$; hence $ab = (-a)(-b) > 0$, and (d) is proved.

If $a > 0$, then (h) follows from 8–5.1(d); if $a = 0$, it is obvious; and if $a < 0$, it is a special case of (d).

Since $1 = 1^2$ and $1 \neq 0$, (j) immediately follows from (h).

Among other things, we see that -1 cannot be a square in any ordered field, nor can it even be a sum of squares. We have already remarked that the complex numbers cannot be ordered in a manner compatible with their field structure, and we can immediately extend this remark: A necessary condition for a field F to be orderable is that -1 is not a sum of squares in F. This condition is also sufficient, but the proof requires advanced set-theoretic techniques.

8–5.6. Lemma. In any ordered field,

(a) $N(F) = P(F) \cap I(F)$,
(b) 1 is the least member of $N(F)$,
(c) if $x, y \in I(F)$ and $x < y + 1$, then $x \leq y$.

Proof. The set $P(F)$ has Properties 8–4.1(a) and (b); therefore $N(F) \subseteq P(F)$, whence $N(F) \subseteq P(F) \cap I(F)$. The opposite inclusion follows from the definition of $I(F)$ and 8–5.3(b).

The set $S = \{x \mid x \geq 1\}$ has Properties 8–4.1(a) and (b). Hence $N(F) \subseteq S$, which establishes (b).

The hypothesis $x < y + 1$ gives $y - x + 1 \in I(F) \cap P(F) = N(F)$. Hence $1 \leq y - x + 1$, and $x \leq y$ follows.

8–5.7. Proposition. An ordered field has characteristic 0.

Proof. $0 \notin P(F)$; hence $0 \notin N(F)$ by 8–5.6(a).

8–5.8. Theorem. If F is an ordered field, $N(F)$ is well-ordered.

Proof. Let S be a subset of $N(F)$. We must prove that either S is empty or S contains a least element. We suppose therefore that S contains no least element and prove that S is empty.

Let $M = \{x \in N(F) \mid (\forall s \in S) x < s\}$. Obviously $S \cap M$ is empty; hence the proof will be completed if we prove $M = N(F)$. By virtue of 8–5.7 and 8–4.10

we may use induction in $N(F)$. If $1 \in S$, it would be the least element of S by 8–5.6(b); hence $1 \notin S$, and therefore $1 \in M$. If $m \in M$ but $m + 1 \notin M$, then we can choose $s_1 \in S$ so that $s_1 \leq m + 1$. Since s_1 is not the least element of S, we can choose $s_2 \in S$ so that $s_2 < m + 1$. By 8–5.6(c), $s_2 \leq m$, contradicting $m \in M$. Thus $m \in M$ implies $m + 1 \in M$. Induction gives $M = N(F)$, and the theorem is proved.

EXERCISE

Let F be any field of characteristic 0. Prove that $Q(F)$ can be ordered in exactly one way. Deduce that if F and G are ordered fields and ϕ is the isomorphism of $Q(F)$ onto $Q(G)$ (Theorem 8–4.11), then ϕ is order-preserving. (*Note.* Although the prime subfield of a field of characteristic 0 can always be ordered in exactly one way, the field itself may not be orderable or it may be orderable in several ways.)

8–6. ARCHIMEDEAN ORDERED FIELDS

8–6.1. Definition. An ordered field F is *Archimedean* if and only if

$$(\forall x \in F)(\exists n \in N(F)) \quad n > x.$$

We can paraphrase this definition by saying, "the natural numbers of F are arbitrarily large."

8–6.2. Proposition. If F is an Archimedean ordered field, then

(a) $(\forall x > 0)(\forall y)(\exists n \in N(F)) \, nx > y$;
(b) $(\forall x > 0)(\exists n \in N(F)) \, 1/n < x$;
(c) if $x < y$, then $(\exists r \in Q(F)) \, x < r < y$.

Proof. For (a) and (b) we need only to choose $n > y/x$ and $n > 1/x$, respectively. To prove (c), assume first that $x > 0$; then choose $n \in N(F)$ so that $n(y - x) > 1$. Multiplying this out, we have $ny > 1 + nx$. There are elements of $N(F)$ greater than nx; so let m be the least such element. Now $m - 1$ is either 0 or in $N(F)$; in any case $m - 1 \leq nx$. Combining, we have

$$nx < m \leq nx + 1 < ny,$$

whence $x < m/n < y$, and (c) is established for $x > 0$. If $x \leq 0$, choose $k \in N(F)$ so that $k > -x$. Then $0 < x + k < y + k$ and we can choose an $r \in Q(F)$ such that $x + k < r < y + k$. Then $r - k \in Q(F)$ and $x < r - k < y$ as required.

EXERCISES

1. Prove that if an ordered field has any one of the properties (a), (b), (c) of 8–6.2, then it is Archimedean.

2. Let F be any field of characteristic 0. Show that the unique order which can be imposed on $Q(F)$ (Exercise, p. 107) is Archimedean.

*3. Show that the ordered field of rational functions discussed in 8–5.1 is not Archimedean.

8–7. COMPLETE ORDERED FIELDS

8–7.1. Definition. An ordered field $\langle F, +, \cdot, <, \le \rangle$ is called *complete* if and only if $\langle F, <, \le \rangle$ is a complete linearly ordered set.

8–7.2. Proposition. A complete ordered field is Archimedean.

Proof. Let F be a complete ordered field and x be any element of F. If there is no member of $N(F)$ greater than x, then x is an upper bound for $N(F)$. Since F is complete, we may set $y = \sup N(F)$. Now, $y - 1$ is not an upper bound for $N(F)$, so we can choose $n \in N(F)$ so that $y - 1 < n$. Then $y < n + 1$, contradicting the fact that y is an upper bound for $N(F)$.

8–7.3. Lemma. Let x and y be nonnegative elements of an ordered field. Then $x^2 < y^2$ if and only if $x < y$.

This lemma follows immediately from elementary inequalities.

8–7.4. Proposition. In a complete ordered field every nonnegative element has a unique nonnegative square root.

Proof. Let F be a complete ordered field and let a be a nonnegative element of F. Lemma 8–7.3 shows that a has at most one nonnegative square root. The meat of the proposition is the existence.

We first choose a convenient interval in which we are sure the square root must fall. Let
$$B = \{x \in F \mid 0 \le x \le 1 + a\}.$$

(Since $(1 + a)^2 > a$, the square root of a must be in B, if it exists.) With the inherited order structure B is a complete ordered set with a largest element and a smallest element.

Define
$$\phi(x) = x + \frac{1}{2(1 + a)}(a - x^2) \qquad \text{for} \qquad x \in B.$$

If $0 \le x < y \le 1 + a$, then
$$1 - \frac{x + y}{2(1 + a)} > 0;$$
hence
$$\phi(y) - \phi(x) = (y - x)\left(1 - \frac{x + y}{2(1 + a)}\right) > 0.$$

Therefore ϕ is order-preserving. Moreover,

$$0 \leq \phi(0) < \phi(1 + a) = 1 + a - \frac{1 + a + a^2}{2(1 + a)} < 1 + a,$$

so ϕ maps B into itself.

According to the Knaster fixed-point theorem (6–5.5), there is a number $b \in B$ such that $\phi(b) = b$; that is,

$$b = b + \frac{1}{2(1 + a)}(a - b^2).$$

This gives $b^2 = a$ and proves that the square root exists.

This proof typifies the use of fixed-point theorems. The function ϕ could have been defined by $\phi(x) = x + m(a - x^2)$ with any m small enough to guarantee that ϕ is increasing. The choice $m = 1/2(1 + a)$ is the least possible choice.

We shall use the symbol $\sqrt{}$ to designate square roots as usual: If a is a non-negative member of an ordered field, then $\sqrt{a}$ designates its unique nonnegative square root (if it exists). The following facts about square roots are useful.

8–7.5. Proposition. In a complete ordered field

(a) $a \leq \sqrt{a^2}$;

(b) if a and b are nonnegative, $\sqrt{ab} = \sqrt{a}\sqrt{b}$;

(c) $\sqrt{(a + c)^2 + (b + d)^2} \leq \sqrt{a^2 + b^2} + \sqrt{c^2 + d^2}$.

Proof. If $a \geq 0$, then $a = \sqrt{a^2}$. If $a < 0$, then $a < \sqrt{a^2}$ (in fact, $a = -\sqrt{a^2}$). Since $\sqrt{a} \geq 0$ and $\sqrt{b} \geq 0$, $\sqrt{a}\sqrt{b} \geq 0$. Now, $(\sqrt{a}\sqrt{b})^2 = ab$ and (b) follows.

Since all squares are nonnegative,

$$(ac + bd)^2 \leq (ac + bd)^2 + (ad - bc)^2 = (a^2 + b^2)(c^2 + d^2);$$

therefore, $ac + bd \leq \sqrt{a^2 + b^2}\sqrt{c^2 + d^2}$. Hence

$$\begin{aligned}
(a + c)^2 + (b + d)^2 &= a^2 + b^2 + 2(ac + bd) + c^2 + d^2 \\
&\leq a^2 + b^2 + 2\sqrt{a^2 + b^2}\sqrt{c^2 + d^2} + c^2 + d^2 \\
&= (\sqrt{a^2 + b^2} + \sqrt{c^2 + d^2})^2.
\end{aligned}$$

Now (c) follows because its right-hand member is surely nonnegative.

8–7.6. Proposition. Suppose that $\langle F, +, \cdot, <, \leq \rangle$ is a complete ordered field. Then $\langle F, +, \cdot \rangle$ has only one compatible ordering.

Proof. Let $P(F)$ denote the set of positive elements of F in the given ordering. Let S be the set of positive elements of F in some ordering compatible with its field structure. The proposition will be established if we prove that $S = P(F)$.

Every element of $P(F)$ is a square, and all squares except 0 are positive in every compatible ordering of F. Therefore $P(F) \subseteq S$.

Suppose that $s \in S - P(F)$. Then $s \neq 0$ and $s \notin P(F)$, so $-s \in P(F)$ and $-s$ is a square. But this shows that $-s \in S$, contradicting $s \in S$. We therefore conclude that $S - P(F) = \emptyset$ or $S \subseteq P(F)$. Thus we have $S = P(F)$ as required.

As a result of this proposition, we may refer to a field in which no ordering is specified, as a complete ordered field, meaning that the field in question has a compatible ordering in which it is complete. Since the ordering in question is unique, no ambiguity is introduced. From the proof above we see that the ordering is most easily described by specifying its positive elements as the set of all nonzero squares, using 8–5.3.

The structure we shall call the real number system is to be an idealization of the numbers we use to measure quantity in the physical world. Our experience with physical measurement suggests that it should be an ordered field, and the identification of numbers with the points of an ideal straight line suggests that it should be complete. So we demand that the real numbers should form a complete ordered field. This immediately raises two questions. First, does this describe our structure adequately; that is, are the postulates for a complete ordered field categorical? If so, we can expect to answer all questions about "the real numbers" on the basis of our postulates. Second, is it perhaps overdescribed to the point of inconsistency? If so, we should have to reexamine our intuitive reasons for believing that "the real numbers" are a complete ordered field. The first of these questions we answer in the following theorem; the second we will consider in the next chapter.

8–7.7. Theorem. If F and G are complete ordered fields, then there exists a unique bijection from F to G which is an isomorphism of the field structures. This bijection is also an isomorphism of the order structures.

Proof. According to Theorem 8–4.11 and the exercise on page 107, there is a bijection ϕ from $Q(F)$ to $Q(G)$ which is an isomorphism of both the field and order structures.

Consider the complete ordered sets $\langle F, <, \leq \rangle$ and $\langle G, <, \leq \rangle$. By 8–6.2(c), $Q(F)$ meets every interval in F and $Q(G)$ meets every interval in G. Therefore, Theorem 6–5.3 tells us that there is an isomorphism ψ of $\langle F, <, \leq \rangle$ onto $\langle G, <, \leq \rangle$ which extends ϕ. We will prove that ψ is an isomorphism of $\langle F, +, \cdot \rangle$ onto $\langle G, +, \cdot \rangle$ as well. In other words we will prove

$$\psi(x + y) = \psi(x) + \psi(y) \tag{1}$$

and

$$\psi(xy) = \psi(x)\psi(y) \tag{2}$$

for all $x, y \in F$. The proof will be indirect.

Suppose $\psi(x) + \psi(y) < \psi(x + y)$. We can choose r' and s' in $Q(G)$ so that

$$\psi(x) < r', \qquad \psi(y) < s', \qquad \text{and} \qquad \psi(x + y) > r' + s'. \tag{3}$$

(Using 8–6.2(c), choose r' so that $\psi(x) < r' < \psi(x + y) - \psi(y)$. Then choose s' so that $\psi(y) < s' < \psi(x + y) - r'$.)

Since ϕ is a bijection from $Q(F)$ to $Q(G)$, we can choose r and s in $Q(F)$ so that $\phi(r) = r'$ and $\phi(s) = s'$. Because ϕ is a field isomorphism $\phi(r + s) = r' + s'$. Since ψ is an extension of ϕ, (3) becomes

$$\psi(x) < \psi(r), \qquad \psi(y) < \psi(s), \qquad \text{and} \qquad \psi(x + y) > \psi(r + s).$$

Using the fact that ψ is an order isomorphism, we obtain

$$x < r, \qquad y < s, \qquad \text{and} \qquad x + y > r + s.$$

But these inequalities are inconsistent, so we conclude that $\psi(x) + \psi(y) \not< \psi(x + y)$.

Similarly, the assumption $\psi(x) + \psi(y) > \psi(x + y)$ leads to a contradiction. This establishes (1).

From (1) we deduce $\psi(-x) = -\psi(x)$, and it follows easily that we need only prove (2) for positive x and y. With this additional assumption, either

$$\psi(x)\psi(y) < \psi(xy) \quad \text{or} \quad \psi(x)\psi(y) > \psi(xy)$$

leads to a contradiction by an argument analogous to the one above. For example, if $\psi(x)\psi(y) < \psi(xy)$, choose r' and s' in $Q(G)$ so that

$$\psi(x) < r', \qquad \psi(y) < s', \qquad \text{and} \qquad \psi(xy) > r's'; \text{ etc.}$$

Thus there exists a bijection from F to G which is an isomorphism of both the field and order structures.

Now suppose that θ is any bijection from F to G which is an isomorphism of the field structures. The restriction of θ must be an isomorphism of $Q(F)$ onto $Q(G)$. Since there is only one such isomorphism, θ must be an extension of ϕ. Now, the nonnegative elements of both F and G can be characterized in terms of the field structures: they are the squares. Hence θ must carry nonnegative elements of F into nonnegative elements of G. Therefore θ is order-preserving. But, by 6–5.3, there is only one order-preserving extension of ϕ over F, and we know that it is ψ. Thus $\theta = \psi$ and the theorem is proved.

In the next chapter we shall need the following lemma.

8–7.8. Lemma. Let F be an ordered field and let P be the set of positive elements of F. If P is complete in the ordering inherited from F, then F is complete.

Proof. Let X be any nonvoid subset of F bounded above by b. Choose $x_0 \in X$ and let

$$Y = \{x - x_0 + 1 \mid x \in X\} \cap P.$$

(We "slide" X up to make sure it meets P.) It is clear that Y is a nonempty subset of P bounded above by $b - x_0 + 1$. Then $x_0 - 1 + \sup Y$ is easily seen to be the least upper bound of X.

CHAPTER 9

THE CONSTRUCTION OF THE
REAL NUMBERS

In this chapter we shall show how to construct a complete ordered field from a simple chain. This construction is important for two reasons. It proves that any contradiction inherent in the postulates for a complete ordered field, that is, the real number system, is latent in the postulates for a simple chain, which is a far less complicated structure whose consistency is almost guaranteed by our intuition. Perhaps more important to the novice in abstract mathematics is that it provides an excellent demonstration of set-theoretical techniques.

By studying successively the arithmetic of the rational numbers, the positive rationals, the positive real numbers, and finally the whole real number system, we will retrace both the historical development of the number system and the development most people learn in elementary school. Putting everything in formal terms, however, exposes a number of difficulties and forces us to make rather sophisticated constructions which throw the emphasis on a different aspect of the theory.

The work of this chapter is quite similar to that of Chapter 8, but there is an essential difference. In Chapter 8 we considered a *given* field in which the familiar algebraic laws were true by assumption. In this chapter we will *define* operations in various sets. Most of our effort will be devoted to proving that these new operations do indeed obey the usual rules. Although this chapter is logically independent of Chapter 8, conceptually it is entirely dependent on field theory or, more fundamentally, on what we know to be true of "the real numbers."

To reduce the notation we shall reserve, throughout this chapter but only in this chapter, various type faces for the members of the different systems. The domain of a quantifier is automatically determined by the font of the letter quantified.

9–1. THE ARITHMETIC OF THE NATURAL NUMBERS

We fix our attention on a simple chain $\langle \mathrm{S}, f \rangle$ whose first element we denote by e (see Section 7–3). Lower-case roman letters will be reserved to designate members of S. We should think of S as being the set of natural numbers ($\mathrm{e} = 1, f(\mathrm{e}) = 2$, etc.), but to help avoid any appeal to our preconceptions concerning numbers we shall not use this suggestive notation.

Note that we do not discuss the existence of the simple chain S. In spite of its intuitive simplicity a simple chain carries within itself the germs of all the difficulties in logic and mathematics; we are obliged to take its existence as axiomatic.

The chain $\mathcal{S}$ comes to us endowed with only one structure, its successor function. We shall define two binary operations, which we may call addition and multiplication, and a linear order structure in $\mathcal{S}$. Of course the definitions are so chosen that, if $\mathcal{S}$ is identified with the set of natural numbers, the arithmetic is the usual arithmetic.

9–1.1. Proposition. There exists a unique function $+$ (written medially) from $\mathcal{S} \times \mathcal{S}$ to $\mathcal{S}$ such that for all p, q,

$$p + e = f(p), \tag{1}$$

$$p + f(q) = f(p + q). \tag{2}$$

This function has the following properties: For all p, q, r,

(i) $(p + q) + r = p + (q + r)$,
(ii) $p + q = q + p$.

Proof. The existence and uniqueness of a function satisfying conditions (1) and (2) follow directly from the general theorem on inductive definition of functions (7–3.7).

To prove (i) let p and q be arbitrary. We find directly that

$$(p + q) + e = f(p + q) = p + f(q) = p + (q + e).$$

Suppose that $(p + q) + n = p + (q + n)$. Then

$$(p + q) + f(n) = f((p + q) + n) = f(p + (q + n))$$
$$= p + f(q + n) = p + (q + f(n)).$$

Therefore $(\forall r)\,(p + q) + r = p + (q + r)$. Since p and q were arbitrarily chosen, (i) holds universally.

To prove (ii) we need two inductions. First, $e + e = e + e$, clearly; and if $n + e = e + n$, then

$$f(n) + e = f(f(n)) = f(n + e) = f(e + n) = e + f(n).$$

Therefore,

$$(\forall r) \quad r + e = e + r.$$

This is our subsidiary goal which gives us the first step in the second induction. Pick an arbitrary p. We know that $p + e = e + p$. If $p + n = n + p$, then

$$p + f(n) = f(p + n) = f(n + p) = n + f(p) = n + (p + e)$$
$$= n + (e + p) = (n + e) + p = f(n) + p,$$

where we used (i) at the sixth step. Therefore $(\forall q)\, p + q = q + p$. Since p was arbitrary, (ii) holds.

Next we introduce a linear order relation in S. It is the usual ordering: "$p < q$" means "p comes before q in the sequence $e, f(e), f(f(e)), \ldots$" Although we could define order by formalizing this idea (see Exercise 5), it is more convenient to use the addition structure.

9–1.2. Proposition. Let $<$ be the relation in S defined by

$$p < q \quad \text{if and only if} \quad (\exists x) \ p + x = q.$$

Then $<$ is a strong linear order relation in S.

Proof. We first prove that $<$ is transitive. Suppose that $p < q$ and $q < r$. Choose x and y so that $p + x = q$ and $q + y = r$. Then $p + (x + y) = (p + x) + y = q + y = r$. Therefore $p < r$.

Next we prove that $<$ is strictly nonreflexive; that is, $(\forall p) \ p \not< p$. This is accomplished by induction with both steps indirect. Suppose that $e < e$. Then choose x so that $e = x + e = f(x)$; this contradicts the fact that e is the first element of S. Therefore $e \not< e$. Now suppose $f(n) < f(n)$. Choose y so that $f(n) + y = f(n)$. Then

$$f(n) = y + f(n) = f(y + n) = f(n + y).$$

Since f is injective,

$$n = n + y, \quad \text{or} \quad n < n.$$

This proves that $n \not< n$ implies $f(n) \not< f(n)$, which completes the induction.

Having proved that $<$ is a strong order relation in S, we must prove that it is linear; that is,

$$(\forall p) \ ((\forall r) \ p \text{ is comparable with } r). \tag{3}$$

This we do by induction on p. If $r \neq e$, then there is an s such that $r = f(s) = s + e = e + s$, so $e < r$. Hence $(\forall r) \ e$ is comparable with r.

Now assume that

$$(\forall r) \ n \text{ is comparable with } r.$$

Choose any r. If $r = e$, then $f(n)$ is comparable with r as we have just proved. If $r \neq e$, then say $r = f(s)$. Now n is comparable with s. We readily see that $f(n) < r, f(n) = r$, or $r < f(n)$ according as $n < s, n = s$, or $s < n$. Therefore,

$$(\forall r) \ f(n) \text{ is comparable with } r.$$

This completes the inductive proof of (3).

9–1.3. Proposition. There exists a unique function (written by juxtaposition) from $S \times S$ to S such that for all p, q,

$$pe = p, \tag{4}$$

$$pf(q) = (pq) + p. \tag{5}$$

This function has the following properties: For all p, q, r,

(i) $p(q + r) = pq + pr$,
(ii) $(pq)r = p(qr)$,
(iii) $pq = qp$,
(iv) if $pq = pr$, then $q = r$.

Proof. The existence and uniqueness of a function satisfying (4) and (5) follow from the general theorem on inductive definition (7–3.7).

To reduce the number of parentheses required, we shall adopt the usual conventions about the interpretation of mixed sums and products.

Choose p and q arbitrarily. Then

$$p(q + e) = pf(q) = pq + p = pq + pe.$$

If $p(q + n) = pq + pn$, then

$$p(q + f(n)) = pf(q + n) = p(q + n) + p = (pq + pn) + p$$
$$= pq + (pn + p) = pq + pf(n).$$

Therefore, $(\forall r) \, p(q + r) = pq + pr$. Since p and q were chosen arbitrarily, (i) is proved.

Again choose p and q arbitrarily. We have

$$(pq)e = pq = p(qe).$$

Suppose that $(pq)n = p(qn)$. Then

$$(pq)f(n) = (pq)n + pq = p(qn) + pq = p(qn + q) = p(qf(n)).$$

Thus (ii) is proved by induction on r.

To prove that multiplication is commutative we need two intermediate results:

$$(\forall p, q) \quad f(p)q = pq + q, \tag{6}$$

$$(\forall r) \quad er = r. \tag{7}$$

We have $f(p)e = f(p) = p + e = pe + e$. Assume that $f(p)n = pn + n$. Then

$$f(p)f(n) = f(p)n + f(p) = (pn + n) + (p + e)$$
$$= (pn + p) + (n + e) = pf(n) + f(n).$$

This establishes (6) by induction on q. We know that $ee = e$. Suppose that $en = n$. Then

$$ef(n) = en + e = n + e = f(n).$$

This proves (7).

We return to the proof of (iii). Choose q arbitrarily. By (7), eq = q = qe. Assume nq = qn. Then

$$f(n)q = nq + q = qn + q = qf(n).$$

This proves (iii) by induction on p.

Next we prove

if s < t, then ps < pt.   (8)

If s < t, we can choose u so that s + u = t; then

$$ps + pu = p(s + u) = pt,$$

whence ps < pt.

Suppose that pq = pr. Then either q < r or r < q leads to a contradiction with (8). Since < is a linear order relation, q = r is the only remaining possibility.

EXERCISES

1. Prove that (∀p, q, r) if p + r = q + r, then p = q.

2. Prove that (∀p, q, r) p < q if and only if p + r < q + r.

3. Show that the ordering of 9–1.2 is the only ordering of S for which (∀s) s < f(s).

4. Show that S is well-ordered by the ordering of 9–1.2.

5. For each p let S_p denote the subchain of S generated by p (see 7–2.2). Let < be the relation in S defined by

p < q   if and only if   q ∈ S_p.

Using only the properties of chains, prove that < is a weak linear order relation in S. Prove that < coincides with ≤.

6. Show that there is a unique binary operation ∗ in S such that

(a) e ∗ e = e,

(b) (∀p) f(p) ∗ e = f(f(p ∗ e)), and

(c) (∀p, q) p ∗ f(q) = f(f(p ∗ q)).

Show that ∗ is commutative.

9–2. FRACTIONS AND RATIONAL NUMBERS

A rational number is ordinarily defined as a number which can be expressed as the quotient of two integers. Since we cannot actually divide any two integers within S, a fraction must be interpreted as a new kind of entity. A formal fraction is just an ordered pair of elements of S. We shall think of the first element of a pair as the numerator and the second element as the denominator. Now, many different formal fractions represent the same rational number, for example ⟨2, 3⟩ and ⟨4, 6⟩; but ⟨2, 3⟩ ≠ ⟨4, 6⟩. Since we cannot redefine equality, we introduce an equivalence relation in the set of formal fractions and define a rational number as an equivalence class. The set Q of rational numbers is therefore a quotient set of the set $s \times s$ of formal fractions.

The next step is to endow Q with addition, multiplication, and order. We do this by using some of the well-known rules for the arithmetic of fractions as definitions. The remainder of the section is devoted to proving that many other facts about rational numbers are consequences of the definitions.

9–2.1. Proposition. Let $\sim$ be the relation in $s \times s$ defined by

$$\langle p, q \rangle \sim \langle r, s \rangle \qquad \text{if and only if} \qquad ps = qr. \tag{1}$$

Then $\sim$ is an equivalence relation.

Proof. Both the reflexivity and symmetry of $\sim$ follow from the commutative law for multiplication. To prove transitivity suppose that $\langle p, q \rangle \sim \langle r, s \rangle$ and $\langle r, s \rangle \sim \langle t, u \rangle$, which means that $ps = qr$ and $ru = st$. Then

$$s(pu) = (sp)u = (ps)u = (qr)u = q(ru) = q(st) = (qs)t = s(qt).$$

By the cancellation law 9–1.3(iv) $pu = qt$ or $\langle p, q \rangle \sim \langle t, u \rangle$.

9–2.2. The quotient set of $\sim$ will be denoted by Q and the natural map from $s \times s$ to Q will be denoted by $\overline{}$ (written over its argument). Lower case Greek letters will be reserved for members of Q, and in particular we let $\epsilon = \overline{\langle e, e \rangle}$.

9–2.3. Proposition. There exists a unique function $\oplus$ from $Q \times Q$ to Q such that

$$(\forall p, q, r, s) \quad \overline{\langle p, q \rangle} \oplus \overline{\langle r, s \rangle} = \overline{\langle ps + qr, qs \rangle}. \tag{2}$$

This function has the following properties: For all $\alpha, \beta, \gamma,$

(i) $$(\alpha \oplus \beta) \oplus \gamma = \alpha \oplus (\beta \oplus \gamma),$$

(ii) $$\alpha \oplus \beta = \beta \oplus \alpha.$$

Proof. It is clear that there is at most one function from $Q \times Q$ to Q, such that (2) holds. If we let

$$\oplus = \{\langle\langle \alpha, \beta \rangle, \gamma \rangle \mid (\exists p, q, r, s) \langle p, q \rangle \in \alpha, \langle r, s \rangle \in \beta, \text{ and } \langle ps + qr, qs \rangle \in \gamma\},$$

then we have defined a relation $\oplus$ having domain $Q \times Q$ and range in Q. The problem is to show that $\oplus$ is a function; in other words, to prove that if α and β are given, then there is exactly one γ such that $\langle\langle \alpha, \beta \rangle, \gamma \rangle \in \oplus$. It is obvious that there is at least one such γ.

Suppose then $\langle\langle \alpha, \beta \rangle, \gamma \rangle \in \oplus$ and $\langle\langle \alpha, \beta \rangle, \delta \rangle \in \oplus$. We can select p, q, r, s, t, u, v, w so that

$$\langle p, q \rangle \in \alpha, \quad \langle r, s \rangle \in \beta, \quad \langle ps + qr, qs \rangle \in \gamma,$$
$$\langle t, u \rangle \in \alpha, \quad \langle v, w \rangle \in \beta, \quad \langle tw + uv, uw \rangle \in \delta.$$

Then $\langle p, q \rangle \sim \langle t, u \rangle$; hence $pu = qt$. Now,

$$(ts + ur)(qs) = (qt)(ss) + (qr)(us) = (pu)(ss) + (qr)(us) = (us)(ps + qr),$$

and therefore

$$\langle ts + ur, us \rangle \sim \langle ps + qr, qs \rangle \quad \text{or} \quad \overline{\langle ts + ur, us \rangle} = \gamma.$$

Similarly, from $\langle r, s \rangle \sim \langle v, w \rangle$ we derive

$$\overline{\langle ts + ur, us \rangle} = \delta.$$

Thus $\gamma = \delta$ and we have proved that $\oplus$ is a function.

We shall interrupt the proof to comment on the foregoing. To define a function having the quotient set for some equivalence relation as its domain it is often necessary to make the values of the function depend formally on representatives of its arguments rather than on the arguments themselves. In such cases, one must prove that the values of the function in fact depend only on the arguments and not on the particular choice of representatives. This is the content of the last paragraph. Essentially the same idea is involved in Theorem 5–2.1.

We shall now repeat the proof of the existence of the addition function in the phraseology customarily used in such contexts.

Define $\alpha \oplus \beta$ as follows: For any representatives $\langle p, q \rangle$ of α and $\langle r, s \rangle$ of β,

$$\alpha \oplus \beta = \overline{\langle ps + qr, qs \rangle}.$$

We must prove that $\alpha \oplus \beta$ is independent of the choice of representatives. Suppose then that $\langle t, u \rangle$ is another representative of α. By calculation we find that

$$\langle ts + ur, us \rangle \sim \langle ps + qr, qs \rangle,$$

and therefore $\alpha \oplus \beta$ is independent of the choice of the representative of α. Suppose that $\langle v, w \rangle$ is another representative of β. Then

$$\langle pw + qv, qw \rangle \sim \langle ps + qr, qs \rangle,$$

and therefore $\alpha \oplus \beta$ is independent of the choice of the representative of β. Hence $\alpha \oplus \beta$ is well defined. (We are, in effect, applying the result of Exercise 2(b), p. 69.

We now return to prove (i). Choose representatives $\langle p, q \rangle$ of α, $\langle r, s \rangle$ of β, and $\langle t, u \rangle$ of γ. According to (2) one representative of $\alpha \oplus \beta$ is $\langle ps + qr, qs \rangle$, and therefore $\langle (ps + qr)u + (qs)t, (qs)u \rangle$ is a representative of $(\alpha \oplus \beta) \oplus \gamma$. Also,

$$\langle ru + st, su \rangle \in \beta \oplus \gamma$$

and

$$\langle p(su) + q(ru + st), q(su) \rangle \in \alpha \oplus (\beta \oplus \gamma).$$

But

$$\langle (ps + qr)u + (qs)t, (qs)u \rangle = \langle p(su) + q(ru + st), q(su) \rangle$$

by direct calculation in S; hence $(\alpha \oplus \beta) \oplus \gamma = \alpha \oplus (\beta \oplus \gamma)$.

Using the same notation, we find that $\langle rq + sp, sq \rangle$ is a representative of $\beta \oplus \alpha$. Since $\langle rq + sp, sq \rangle = \langle ps + qr, qs \rangle$, we conclude that $\alpha \oplus \beta = \beta \oplus \alpha$. This completes the proof.

9–2.4. Proposition. There exists a unique function $\cdot$ from $Q \times Q$ to Q such that

$$(\forall p, q, r, s) \quad \overline{\langle p, q \rangle} \cdot \overline{\langle r, s \rangle} = \overline{\langle pr, qs \rangle}. \tag{3}$$

This function has the following properties. For all α, β, γ

(i) $(\alpha \cdot \beta) \cdot \gamma = \alpha \cdot (\beta \cdot \gamma)$,
(ii) $\alpha \cdot \beta = \beta \cdot \alpha$,
(iii) $\alpha \cdot (\beta \oplus \gamma) = \alpha \cdot \beta \oplus \alpha \cdot \gamma$,
(iv) $\alpha \cdot \epsilon = \epsilon \cdot \alpha = \alpha$,
(v) for every α, there is a unique δ such that $\alpha \cdot \delta = \epsilon$.

Proof. The first statement presents the same problem as the corresponding statement concerning addition. We must prove that (3) can be used to define a function; that is, the function $\cdot$ is well defined if we set $\alpha \cdot \beta = \overline{\langle pr, qs \rangle}$ for any choice of representatives $\langle p, q \rangle$ of α and $\langle r, s \rangle$ of β. Let $\langle t, u \rangle$ be another representative of α. Then $pu = qt$ and therefore

$$(tr)(qs) = (qt)(rs) = (pu)(rs) = (us)(pr)$$

so

$$\langle tr, us \rangle \sim \langle pr, qs \rangle.$$

This shows that $\alpha \cdot \beta$ is independent of the choice of the representative of α. Similarly, it is independent of the choice of the representative of β.

Statements (i), (ii), and (iii) follow in just the same way as 9–2.3(i) and (ii) for addition while (iv) is a trivial computation. To prove the existence part of (v) let $\langle p, q \rangle \in \alpha$ and let $\delta = \overline{\langle q, p \rangle}$. Then $\alpha \cdot \delta = \overline{\langle pq, qp \rangle} = \epsilon$. The uniqueness follows by an argument applicable to any associative, commutative binary operation with identity: If $\alpha \cdot \eta = \epsilon$ then

$$\eta = \eta \cdot \epsilon = \eta \cdot (\alpha \cdot \delta) = (\alpha \cdot \eta) \cdot \delta = \epsilon \cdot \delta = \delta.$$

9–2.5. Since the element δ of 9–2.4(v) is uniquely determined by α, we shall denote it by α^*. Then for any element α we shall have $\alpha \cdot \alpha^* = \alpha^* \cdot \alpha = \epsilon$. The uniqueness shows that $\alpha^{**} = \alpha$.

9–2.6. Proposition. Let $<$ be the relation in Q defined by

$$\alpha < \beta \qquad \text{if and only if} \qquad (\exists \gamma) \quad \alpha \oplus \gamma = \beta.$$

Then $\prec$ is a strong linear order relation in $\mathbb{Q}$. Furthermore,

(i) if $\alpha \prec \beta$, then $(\exists\delta)\ \alpha \prec \delta,\ \delta \prec \beta$;
(ii) if $\alpha \prec \beta$, then $(\forall\delta)\ \alpha \oplus \delta \prec \beta \oplus \delta$;
(iii) if $\alpha \prec \beta$, then $(\forall\delta)\ \alpha \cdot \delta \prec \beta \cdot \delta$;
(iv) if $\alpha \prec \beta$, then $\beta^* \prec \alpha^*$.

Proof. The relation $\prec$ is certainly transitive by virtue of 9–2.3(i). We must prove that it is strictly nonreflexive; that is, $\alpha \oplus \gamma = \alpha$ is impossible. Supposing such an equation were possible, let $\langle p, q \rangle$ and $\langle t, u \rangle$ be representatives of α and γ respectively. Then $\langle \overline{pu + qt, qu} \rangle \sim \langle p, q \rangle$ or $(pu + qt)q = qup$. Rewriting this as $puq + qtq = puq$, we have a contradiction of 9–1.2. This proves that $\prec$ is a strong order relation.

Next we consider linearity. If α and β are two members of $\mathbb{Q}$ represented by $\langle p, q \rangle$ and $\langle r, s \rangle$ respectively, then either $ps = qr$, $ps + t = qr$, or $ps = qr + u$ by appropriate choice of t or u according to 9–1.2. Corresponding to these three cases we find that $\alpha = \beta$, $\alpha + \gamma = \beta$, or $\alpha = \beta + \delta$, where $\gamma = \langle \overline{t, qs} \rangle$ or $\delta = \langle \overline{u, qs} \rangle$; that is, $\alpha = \beta$, $\alpha \prec \beta$, or $\beta \prec \alpha$. This proves that $\prec$ is linear.

To prove (i) we first note that, for any $\gamma \in \mathbb{Q}$, there is a ξ such that $\xi \oplus \xi = \gamma$; indeed, if $\gamma = \langle \overline{t, u} \rangle$, we take $\xi = \langle \overline{t, u + u} \rangle$. Now, if $\alpha \prec \beta$, let γ satisfy $\alpha \oplus \gamma = \beta$. We choose ξ as above, and set $\delta = \alpha \oplus \xi$.

The commutative and distributive laws make (ii) and (iii) obvious. Finally (iv) follows by an indirect argument. Suppose $\alpha \prec \beta$. If $\alpha^* = \beta^*$, then

$$\epsilon = \alpha\alpha^* \prec \beta\alpha^* = \beta\beta^* = \epsilon,$$

which is a contradiction. If $\alpha^* \prec \beta^*$, then

$$\epsilon = \alpha\alpha^* \prec \beta\alpha^* \prec \beta\beta^* = \epsilon,$$

again a contradiction. Now $\beta^* \prec \alpha^*$ is the only remaining possibility.

EXERCISES

1. Suppose that $\alpha \cdot \alpha = \alpha$. Prove that $\alpha = \epsilon$.

2. Prove that there exists a unique map Ω from $\mathbb{S}$ to $\mathbb{Q}$ such that
$$\Omega(x + y) = \Omega(x) \oplus \Omega(y),$$
$$\Omega(xy) = \Omega(x) \cdot \Omega(y).$$
Prove that this map is injective and order-preserving.

3. Let F be any ordered field. Show that there is a unique isomorphism of $\langle \mathbb{Q}, \oplus, \cdot \rangle$ onto $\langle P(F) \cap Q(F), +, \cdot \rangle$, where the operations in the latter configuration are the restrictions of those in F. Show also that this isomorphism is order-preserving.

4. Show by argument with representatives that there is a unique relation L in $\mathbb{Q}$ such that $\langle \overline{p, q} \rangle L \langle \overline{r, s} \rangle$ if and only if $ps < qr$. Prove that L is a strong linear order relation. Then prove that L coincides with the order relation of 9–2.6.

5. Define binary operations $+$ and $\times$ in $S \times S$ as follows:

$$\langle p, q \rangle + \langle r, s \rangle = \langle ps + qr, qs \rangle,$$
$$\langle p, q \rangle \times \langle r, s \rangle = \langle pr, qs \rangle.$$

Show that these operations are commutative and associative, but that the distributive law does not hold. Show that the operations $\oplus$ and $\cdot$ in Q can be obtained from the operations $+$ and $\times$ by the general method for the construction of functions on quotient spaces (Theorem 5–2.1 and Exercise 2(b), p. 69).

9–3. THE POSITIVE REAL NUMBERS

The discovery by the Pythagoreans in the sixth century B.C. that the ratio of the side of a square to its diagonal cannot be expressed as a ratio of whole numbers, is undoubtedly one of the greatest landmarks in the history of pure mathematics. The proof itself was a triumph, but more important perhaps was the problem it posed in the geometrical theory of proportion, for the solution of that problem required a profound analysis of the whole concept of ratio. The Greek solution to this problem can be found in the fifth book of Euclid. It is attributed to Eudoxus of Cnidos, a mathematician and astronomer of the fourth century B.C.

Although the Greeks were aware that the ratios of geometric magnitudes could be treated as numbers and were able to carry out numerical calculations, they never formalized this fact, nor did they ever produce a theory of real numbers for its own sake. It was not until after the development of the calculus that mathematicians turned their attention to the foundations of the real number system. The first purely set-theoretic attempt to build a foundation for analysis was made by Dedekind in the latter part of the nineteenth century. His theory is essentially the same as that given in Chapter 7 and in this chapter.

In his work on the foundations of the real number system Dedekind returned to the old Greek theory, adding a simple but essential new idea. The Greek criterion for the equality of two ratios asserts that a ratio is characterized by two sets of rational numbers, those that are greater than the given ratio and those that are less. Dedekind defines a real number as *being* a partition of the rational numbers into two classes such that all the numbers of one class are less than all those of the other. Such a partition is often called a Dedekind cut. The Greek criterion assumes that the concept of ratio exists *a priori*. By turning the criterion into a definition Dedekind freed the concept of real numbers from its dependence on *a priori* notions, such as geometrical ratio, and based it squarely on set theory.

The definition given below is really the same as Dedekind's; for purely technical reasons it seems somewhat easier to concentrate on just one of the two classes.

9–3.1. Definition. Let $\mathcal{P}$ be the set of all subsets A of Q having the following properties:

(i) $\emptyset \subset A \subset Q$;
(ii) if $\alpha \in A$ and $\beta < \alpha$, then $\beta \in A$; and
(iii) if $\alpha \in A$, then $(\exists \gamma \in A) \, \alpha < \gamma$.

It is easily seen that (ii) is equivalent to the following:

$$\text{if } \beta \notin A, \text{ then } \beta \text{ is an upper bound for A,} \tag{1}$$

while property (iii) asserts that A has no greatest member.

The nontriviality of the definition is guaranteed by

$$E = \{\alpha \mid \alpha < \epsilon\} \in \mathcal{P}. \tag{2}$$

This, of course, requires proof. We prove (i) by noting that $(\epsilon + \epsilon)^* \in E$ by 9–2.6(iv), but $\epsilon \notin E$. Since $<$ is transitive, E satisfies (ii), while (iii) follows directly from 9–2.6(i).

9–3.2. We shall reserve upper-case Roman letters for members of $\mathcal{P}$ and keep the letter E for the number defined by (2).

If we imagine for a moment that we have all real numbers already at our disposal, then $\mathbb{Q}$ should be thought of as the set of all positive rational numbers. If $A \in \mathcal{P}$, then $A \subset \mathbb{Q}$, and we should identify A with the real number sup A, which exists because A is nonvoid and bounded above. Set inclusion reflects numerical inequality, since $A \subseteq B$ implies sup $A \leq$ sup B. Hence we define the order structure of $\mathcal{P}$ directly in terms of set inclusion.

9–3.3. Proposition. Set inclusion is a linear order relation in $\mathcal{P}$ and $\mathcal{P}$ is complete in this order.

Proof. Set inclusion defines an order relation among the subsets of any set. We must prove it is linear when confined to $\mathcal{P}$; that is, if $A \not\subseteq B$ then $B \subseteq A$.

Suppose then $A \not\subseteq B$. Choose $\alpha \in A$ so that $\alpha \notin B$. Let $\beta \in B$. If $\alpha < \beta$, then $\alpha \in B$ by 9–3.1(ii), which is a contradiction; surely $\beta \neq \alpha$; so $\beta < \alpha$. Therefore $\beta \in A$. Hence $B \subseteq A$.

Completeness remains to be shown. Let T be any nonempty subset of $\mathcal{P}$ bounded above, say, by B. Let C be the union of the members of T. If we can prove that $C \in \mathcal{P}$, then it is obvious that $C = $ sup T.

First $\emptyset \subset C \subseteq B \subset \mathbb{Q}$, so 9–3.1(i) holds. Suppose that $\alpha \in C$. Choose $T \in T$ so that $\alpha \in T$. If $\beta < \alpha$, then $\beta \in T \subseteq C$; thus C satisfies 9–3.1(ii). Moreover, we can choose $\gamma \in T$ so that $\alpha < \gamma$; then $\gamma \in C$, so C has no greatest element.

9–3.4. Lemma. Suppose that $A \in \mathcal{P}$ and $\xi \in \mathbb{Q}$. Then there exists $\alpha \in A$ such that $\alpha \oplus \xi \notin A$.

Proof. The statement contrary to the conclusion is

$$(\forall \alpha \in A) \quad \alpha \oplus \xi \in A. \tag{3}$$

We shall deduce a contradiction from this. Since A is not empty, we choose any member β of A. Then $\beta \oplus \overline{\langle e, e \rangle} \cdot \xi = \beta \oplus \xi \in A$. Moreover, from

$$\beta \oplus \overline{\langle n, e \rangle} \cdot \xi \in A$$

and (3) follows

$$\beta \oplus \overline{\langle n + e, e \rangle} \cdot \xi = \beta \oplus \overline{\langle n, e \rangle} \cdot \xi \oplus \xi \in A$$

and therefore

$$(\forall p) \quad \beta \oplus \overline{\langle p, e \rangle} \cdot \xi \in A. \tag{4}$$

Now choose $\eta \notin A$ and representatives $\langle w, x \rangle \in \xi$ and $\langle y, z \rangle \in \eta$. We compute

$$\beta \oplus \overline{\langle xy, e \rangle} \cdot \xi > \overline{\langle xy, e \rangle} \cdot \overline{\langle w, x \rangle} = \overline{\langle wy, e \rangle} \geq \overline{\langle y, z \rangle} = \eta;$$

therefore $\beta \oplus \overline{\langle xy, e \rangle} \cdot \xi \notin A$ by (1), contradicting (4).

9–3.5. Proposition. The function $\oplus$, defined by

$$A \oplus B = \{ \alpha \oplus \beta \mid \alpha \in A, \beta \in B \}, \tag{5}$$

maps $\mathcal{P} \times \mathcal{P}$ into $\mathcal{P}$. Furthermore, for all A, B, C,

(i) $(A \oplus B) \oplus C = A \oplus (B \oplus C)$,
(ii) $A \oplus B = B \oplus A$,
(iii) $A \subset A \oplus B$,
(iv) if $A \subset B$, then $(\exists D) \, A \oplus D = B$,
(v) $A \subset B$ if and only if $A \oplus C \subset B \oplus C$,
(vi) if $A \oplus C = B \oplus C$, then $A = B$.

Proof. The first point at issue is whether $A \oplus B$, as defined by (5), is a member of $\mathcal{P}$. It is certainly a nonempty subset of $\mathcal{Q}$. Choose $\xi \notin A$ and $\eta \notin B$. If $\xi \oplus \eta \in A \oplus B$, then pick $\alpha \in A$ and $\beta \in B$ so that $\xi \oplus \eta = \alpha \oplus \beta$. We have $\alpha < \xi$ and $\beta < \eta$ by (1) and therefore

$$\alpha \oplus \beta < \xi \oplus \beta < \xi \oplus \eta,$$

which is a contradiction. Thus $\xi \oplus \eta \notin A \oplus B$ and $A \oplus B \subset \mathcal{Q}$ is established.

Suppose that $\gamma \in A \oplus B$ and $\delta < \gamma$. Choose $\alpha \in A$ and $\beta \in B$ so that $\gamma = \alpha \oplus \beta$. Since $\delta \gamma^* < \epsilon$,

$$\delta \gamma^* \alpha < \alpha \quad \text{and} \quad \delta \gamma^* \alpha \in A.$$

Similarly,

$$\delta \gamma^* \beta \in B.$$

Finally,

$$\delta = \delta \gamma^* (\alpha \oplus \beta) = \delta \gamma^* \alpha \oplus \delta \gamma^* \beta \in A \oplus B,$$

verifying 9–3.1(ii) for $A \oplus B$.

To check 9–3.1(iii) let $\gamma = \alpha \oplus \beta \in A \oplus B$, where $\alpha \in A, \beta \in B$. Take $\varsigma \in A$ so that $\alpha < \varsigma$; then $\gamma < \varsigma \oplus \beta \in A \oplus B$. This finishes the proof of the first statement.

Any member of $(A \oplus B) \oplus C$ has the form $(\alpha \oplus \beta) \oplus \gamma$, where $\alpha \in A$, $\beta \in B$, and $\gamma \in C$. By the associative law this is $\alpha \oplus (\beta \oplus \gamma)$, a member of

$A \oplus (B \oplus C)$. Therefore,

$$(A \oplus B) \oplus C \subseteq A \oplus (B \oplus C).$$

The opposite inclusion is proved in the same way. Hence (i) is true. Formula (ii) is proved similarly. It is immediate from 9–3.1(ii) that $A \subseteq A \oplus B$. Then (iii) follows easily from 9–3.4.

The proof of (iv) is longer. Suppose that $A \subset B$. Let

$$D = \{\delta \mid (\exists \eta)\, \eta \notin A, \eta + \delta \in B\}.$$

It is easy to check that $D \in \mathcal{P}$ and $A \oplus D \subseteq B$, so we shall only concern ourselves with the opposite inclusion. Let β be any member of B; we shall prove $\beta \in A \oplus D$. This is trivial if $\beta \in A$, since $A \subseteq A \oplus D$; so we assume $\beta \notin A$. We can find a ξ such that $\beta \oplus \xi \in B$ by 9–3.1(iii) and 9–2.6. By 9–3.4, we can pick $\alpha \in A$ so that $\alpha \oplus \xi \notin A$. Now, $\alpha < \beta$ by (1), so we can choose δ so that $\beta = \alpha \oplus \delta$. Since $\alpha \oplus \xi \notin A$ and $\alpha \oplus \xi \oplus \delta = \beta \oplus \xi \in B$, we conclude that $\delta \in D$. Now, $\beta = \alpha \oplus \delta$ shows that $\beta \in A \oplus D$. This gives $B \subseteq A \oplus D$, which proves (iv).

If $A \subset B$, choose D so that $B = A \oplus D$. Then $A \oplus C \subset A \oplus C \oplus D = B \oplus C$ by (iii) and the associative and commutative laws. This proves one half of (v). The other half of (v) and (vi) follow from this because $\mathcal{P}$ is linearly ordered by inclusion.

9–3.6. Lemma. Suppose that $A \in \mathcal{P}$ and $\epsilon < \zeta$. Then there exists $\alpha \in A$ such that $\alpha \cdot \zeta \notin A$.

Proof. Pick any element β of A. If $\beta \cdot \zeta \notin A$, then set $\alpha = \beta$ and we are done. Assume therefore that $\beta \cdot \zeta \in A$. Now, $\beta < \beta \cdot \zeta$ by 9–2.4(iv) and 9–2.6(iii); hence we can find ξ such that $\beta \cdot \zeta = \beta \oplus \xi$. Pick $\alpha \in A$ by Lemma 9–3.4 so that $\alpha \oplus \xi \notin A$. Since $\beta \oplus \xi \in A$, $\beta \oplus \xi < \alpha \oplus \xi$, and therefore $\beta < \alpha$. Hence

$$\xi = \epsilon \cdot \xi = \beta \cdot \beta^* \cdot \xi < \alpha \cdot \beta^* \cdot \xi$$

and

$$\alpha \oplus \xi < \alpha \oplus \alpha \cdot \beta^* \cdot \xi = \alpha \cdot \beta^* \cdot (\beta \oplus \xi) = \alpha \cdot \beta^* \cdot \beta \cdot \zeta = \alpha \cdot \zeta.$$

Therefore $\alpha \cdot \zeta \notin A$ by (1).

9–3.7. Proposition. The function $\cdot$, defined by

$$A \cdot B = \{\alpha \cdot \beta \mid \alpha \in A, \beta \in B\},$$

maps $\mathcal{P} \times \mathcal{P}$ into $\mathcal{P}$. Furthermore, for all A, B, C,

(i) $(A \cdot B) \cdot C = A \cdot (B \cdot C)$,   (ii) $A \cdot B = B \cdot A$,
(iii) $A \cdot (B \oplus C) = A \cdot B \oplus A \cdot C$,   (iv) $A \cdot E = E \cdot A = A$,
(v) for every A, there is a unique D such that $A \cdot D = E$.

Proof. The first statement and formulas (i) through (iv) work out routinely, except for a slight wrinkle in (iii). To prove

$$A \cdot B \oplus A \cdot C \subseteq A \cdot (B \oplus C),$$

select a typical member $\alpha_1 \cdot \beta \oplus \alpha_2 \cdot \gamma$ of the first set, where $\alpha_1, \alpha_2 \in A$, $\beta \in B$, and $\gamma \in C$. If $\alpha_1 = \alpha_2$, the distributive law in $\mathfrak{Q}$ applies directly. If not, let α be the larger of α_1 and α_2; then

$$\alpha_1 \cdot \beta \oplus \alpha_2 \cdot \gamma \prec \alpha \cdot (\beta \oplus \gamma),$$

which proves $\alpha_1 \cdot \beta \oplus \alpha_2 \cdot \gamma \in A \cdot (B \oplus C)$ by 9–3.1(ii). The opposite inclusion follows immediately from the distributive law for $\mathfrak{Q}$.

We now turn our attention to (v). Let A be fixed and set

$$D = \{\delta \mid (\exists \eta) \ \delta \prec \eta \text{ and } \eta^* \notin A\}.$$

We shall omit verifying that $D \in \mathcal{P}$ and proceed to the proof that $A \cdot D = E$.

Choose any member of $A \cdot D$, say $\alpha \cdot \delta$, where $\alpha \in A$ and $\delta \in D$. Pick η so that $\delta \prec \eta$ and $\eta^* \notin A$. Then

$$\alpha \prec \eta^* \qquad \text{and} \qquad \alpha \cdot \delta \prec \alpha \cdot \eta \prec \eta^* \eta = \epsilon;$$

therefore $\alpha \cdot \delta \in E$. This proves $A \cdot D \subseteq E$.

Now let σ be any element of E. Then $\epsilon = \epsilon^* \prec \sigma^*$, hence by 9–3.6 we can choose $\alpha_0 \in A$ so that $\alpha_0 \cdot \sigma^* \notin A$. Then choose $\alpha_1 \in A$ so that $\alpha_0 \prec \alpha_1$; from this it follows that $\alpha_1^* \prec \alpha_0^*$. Let $\delta = \alpha_1^* \cdot \sigma$ and $\eta = \alpha_0^* \cdot \sigma$. We find by easy computation that $\eta^* = \alpha_0 \cdot \sigma^* \notin A$ and $\delta \prec \eta$; hence $\delta \in D$. Now, $\sigma = \alpha_1 \cdot \alpha_1^* \cdot \sigma = \alpha_1 \cdot \delta \in A \cdot D$. Therefore $E \subseteq A \cdot B$, which completes the proof of existence. Uniqueness follows as in the proof of 9–2.4(v).

If we had hastily defined $D = \{\delta \mid \delta^* \notin A\}$, the proof would have failed. Why?

9–3.8. Since the element D of 9–3.7(v) is uniquely determined by A, we shall denote it by A*. Then for any element A of $\mathcal{P}$ we shall have $A \cdot A^* = A^* \cdot A = E$ and $A^{**} = A$.

EXERCISES

1. Prove that there exists a unique map Ψ of $\mathfrak{Q}$ into $\mathcal{P}$ such that

$$\Psi(\alpha \oplus \beta) = \Psi(\alpha) \oplus \Psi(\beta),$$

and

$$\Psi(\alpha \cdot \beta) = \Psi(\alpha) \cdot \Psi(\beta).$$

Prove that this map is injective and order-preserving.

2. Let F be a complete ordered field. Show that there is a unique isomorphism from $\langle \mathcal{P}, \oplus, \cdot \rangle$ to $\langle P(F), +, \cdot \rangle$, where the operations of the latter configuration are the restrictions of those of F.

9–4. REAL NUMBERS

It is perhaps surprising that negative numbers were introduced into mathematics so late. The theory of the positive real numbers, although not completely formulated except in geometric terms, was in the minds of mathematicians from before the days of Euclid, but even in the seventeenth century negative numbers were referred to as *false* numbers. No doubt this was due to the difficulty in associating, in any direct way, a physical image with a negative number.

We shall introduce negative numbers by considering formal subtraction problems. Since there are many different subtraction problems with the same answer, we must once again pass to equivalence classes. The construction is entirely analogous to that for fractions.

9–4.1. Proposition. Let $\cong$ be the relation in $\mathcal{P} \times \mathcal{P}$ defined by

$$\langle A, B \rangle \cong \langle C, D \rangle \quad \text{if and only if} \quad A \oplus D = B \oplus C.$$

Then $\cong$ is an equivalence relation.

We shall omit the proof.

9–4.2. We will denote the quotient set of $\cong$ by $\mathcal{R}$ and the natural map from $\mathcal{P} \times \mathcal{P}$ to $\mathcal{R}$ by ‾ (written over its argument). Bold-face lower-case letters will be reserved for members of $\mathcal{R}$. In addition, we shall use the special notations z for $\overline{\langle E, E \rangle}$ and $\mathbf{u}$ for $\overline{\langle E \oplus E, E \rangle}$.

9–4.3. Proposition. There exist functions $+$ and $\cdot$ from $\mathcal{R} \times \mathcal{R}$ to $\mathcal{R}$ such that, for all A, B, C, D,

$$\overline{\langle A, B \rangle} + \overline{\langle C, D \rangle} = \overline{\langle A \oplus C, B \oplus D \rangle},$$
$$\overline{\langle A, B \rangle} \cdot \overline{\langle C, D \rangle} = \overline{\langle A \cdot C \oplus B \cdot D, A \cdot D \oplus B \cdot C \rangle}.$$

Furthermore, $\langle \mathcal{R}, +, \cdot \rangle$ is a field.

Proof. Our first conclusion depends on showing that if

$$\langle A, B \rangle \cong \langle W, X \rangle \quad \text{and} \quad \langle C, D \rangle \cong \langle Y, Z \rangle,$$

then

$$\langle A \oplus C, B \oplus D \rangle \cong \langle W \oplus Y, X \oplus Z \rangle.$$

This follows easily from the results of the preceding section.

The second conclusion is similar, but the calculation will be simplified if one shows the independence of the choice of representatives for one argument at a time.

The associative, commutative, and distributive laws now follow by direct but tedious calculations from the corresponding laws for $\mathcal{P}$. To verify 8–2.1(iv) we take z and u of that section as $\mathbf{z}$ and $\mathbf{u}$, respectively. Then we must prove that

$$(\forall \mathbf{x}) \quad \mathbf{x} + \mathbf{z} = \mathbf{x}, \quad \text{and} \quad \mathbf{x} \cdot \mathbf{u} = \mathbf{x},$$
$$(\forall \mathbf{x})(\exists \mathbf{y}) \quad \mathbf{x} + \mathbf{y} = \mathbf{z},$$
$$(\forall \mathbf{x} \neq \mathbf{z})(\exists \mathbf{y}) \quad \mathbf{x} \cdot \mathbf{y} = \mathbf{u},$$
$$\mathbf{z} \neq \mathbf{u}.$$

The last of these statements reduces to $E \oplus E \neq E \oplus E \oplus E$, which follows from 9–3.5(iii). The first is completely straightforward calculation. The second follows from the observation that $\langle \overline{A, B} \rangle + \langle \overline{B, A} \rangle = \mathbf{z}$.

To prove the third statement, suppose that $\mathbf{x} \neq \mathbf{z}$ and let $\langle A, B \rangle \in \mathbf{x}$. Then $A \neq B$, since $\langle \overline{A, A} \rangle = \mathbf{z}$. Since $\mathcal{P}$ is linearly ordered, we have either $A = B \oplus C$ or $A \oplus D = B$ by an appropriate choice of C or D (see 9–3.5(iv)). If $A = B \oplus C$, we set $\mathbf{y} = \langle \overline{C^* \oplus E, E} \rangle$ and calculate that $\mathbf{x} \cdot \mathbf{y} = \mathbf{u}$. If $A \oplus D = B$, then $\langle \overline{A, B} \rangle \cdot \langle \overline{E, D^* \oplus E} \rangle = \mathbf{u}$; so we take $\mathbf{y} = \langle \overline{E, D^* \oplus E} \rangle$. This completes the proof.

We note that the proof shows that

$$-\langle \overline{A, B} \rangle = \langle \overline{B, A} \rangle. \tag{1}$$

9–4.4. Proposition. There is a unique strong order relation $<$ in $\mathcal{R}$ such that

$$\langle A, B \rangle < \langle C, D \rangle \quad \text{if and only if} \quad A \oplus D \subset B \oplus C. \tag{2}$$

Moreover, $<$ is a complete linear order relation compatible with the field structure of $\mathcal{R}$.

Proof. It is obvious that there is at most one relation in $\mathcal{R}$ satisfying (2). That there is one can be verified by the usual computation to show that the validity of the second half of (2) is independent of the choice of representatives. In view of the remainder of the theorem, the following argument is shorter.

Let Φ be the map of $\mathcal{P}$ into $\mathcal{R}$ defined by

$$\Phi(A) = \langle \overline{A \oplus E, E} \rangle.$$

It is easy to verify that

$$\langle \overline{A, B} \rangle \in \text{range } \Phi \quad \text{if and only if} \quad B \subset A,$$
$$\Phi(A \oplus B) = \Phi(A) + \Phi(B),$$
$$\Phi(A \cdot B) = \Phi(A) \cdot \Phi(B).$$

Furthermore, for any $\mathbf{x}$ exactly one of the statements

$$\mathbf{x} = \mathbf{z}, \quad \mathbf{x} \in \text{range } \Phi, \quad \text{or} \quad -\mathbf{x} \in \text{range } \Phi$$

is valid. According to 8–5.3, there is a unique linear ordering of $\mathcal{R}$ compatible with its field structure such that range Φ is the set of positive elements. If the strong form of this ordering is denoted by $<$, then the necessary and sufficient condition that $\langle \overline{A, B} \rangle < \langle \overline{C, D} \rangle$ is that

$$\langle \overline{C, D} \rangle - \langle \overline{A, B} \rangle = \langle \overline{C, D} \rangle + \langle \overline{B, A} \rangle = \langle \overline{B \oplus C, A \oplus D} \rangle \in \text{range } \Phi$$

or

$$A \oplus D \subset B \oplus C.$$

Another elementary calculation now shows that Φ is an injective order-preserving map of $\mathcal{P}$ into $\mathcal{R}$. Therefore the positive elements of $\mathcal{R}$ are complete in the ordering of $\mathcal{R}$ and by Lemma 8–7.8, $\mathcal{R}$ is complete.

This finishes the proof of the proposition and also achieves the goal of this chapter. We have constructed a complete ordered field from a simple chain.

It is worth reviewing how the field $\mathcal{R}$ is related to the original chain $\mathcal{S}$. Formal fractions are members of $\mathcal{S} \times \mathcal{S}$ and rational numbers are sets of formal fractions. Hence a rational number is a member of $\mathfrak{P}(\mathcal{S} \times \mathcal{S})$ and $\mathcal{Q} \subseteq \mathfrak{P}(\mathcal{S} \times \mathcal{S})$. The members of $\mathcal{P}$ are sets of rationals; hence $\mathcal{P} \subseteq \mathfrak{P}^2(\mathcal{S} \times \mathcal{S})$. A member of $\mathcal{R}$ is an equivalence class of ordered pairs of members of $\mathcal{P}$; that is, a member of $\mathfrak{P}(\mathcal{P} \times \mathcal{P})$. Hence, finally,

$$\mathcal{R} \subseteq \mathfrak{P}(\mathfrak{P}^2(\mathcal{S} \times \mathcal{S}) \times \mathfrak{P}^2(\mathcal{S} \times \mathcal{S})).$$

The way has indeed been long. In the preface to "Was sind und was sollen die Zahlen?" Dedekind wrote "This memoir can be understood by anyone possessing what is usually called good common sense; no technical philosophic, or mathematical, knowledge is in the least degree required. But I feel conscious that many a reader will scarcely recognize in the shadowy forms which I bring before him his numbers which all his life long have accompanied him as faithful and familiar friends; he will be frightened by the long series of simple inferences corresponding to our step-by-step understanding, by the matter-of-fact dissection of the chains of reasoning on which the laws of numbers depend, and will become impatient at being compelled to follow out proofs for truths which to his supposed inner consciousness seem at once evident and certain."*

---

* From the translation by W. W. Beman cited on page 84.

CHAPTER 10

COMPLEX NUMBERS

The last great extension of the number system is to the complex numbers. Whereas negative numbers have been generally accepted as legitimate, complex numbers are still regarded with great suspicion by most nonmathematicians. Even the standard terminology of the subject involving the adjectives *imaginary* and *complex* suggests that the concept is both mystical and difficult.

While there are many reasonable ways to use negative numbers in the physical world, there are only a few areas in which the direct use of complex numbers is appropriate (alternating current theory is one). This lack of everyday applications unquestionably serves to make the complex numbers less intuitive and more abstract than the real numbers, even to professional mathematicians. From the set-theoretic point of view, however, all numbers are abstract, and there is little point in distinguishing degrees of abstraction after the complicated construction of Chapter 9.

The need for complex numbers must have been felt from the time that the formula for solving quadratic equations was discovered, but the importance of complex numbers to mathematics far transcends the mere existence of square roots of negative numbers. The remarkable formula of DeMoivre,

$$(\cos x + i \sin x)^n = \cos nx + i \sin nx,$$

and the related formula of Euler,

$$e^{ix} = \cos x + i \sin x,$$

were just the first of many insights into the nature of functions which depend on complex numbers. It is one of the truly amazing facts of mathematics that the use of complex numbers simplifies many problems, from the convergence of series to the evaluation of definite integrals, which on their face seem to belong strictly to the real domain.

This chapter is devoted to defining and constructing a complex number system and developing the most fundamental tools for complex analysis. In Chapter 15 we shall return to the subject and prove Euler's formula and the so-called fundamental theorem of algebra.

10–1. COMPLEX NUMBER SYSTEMS

10–1.1. Definition. A *complex number system* is a configuration $\langle \mathbf{C}, +, \cdot, \mathbf{R}, i \rangle$ such that

   (i) $\langle \mathbf{C}, +, \cdot \rangle$ is a field,
   (ii) $\mathbf{R}$ is a subfield of $\mathbf{C}$,
   (iii) $\mathbf{R}$ is a complete ordered field (see 8–7.6 for the conventional use of this term),
   (iv) $i \in \mathbf{C}$ and $i^2 = -1$, and
   (v) $(\forall x \in \mathbf{C})(\exists a, b \in \mathbf{R})\, x = a + bi$.

10–1.2. Proposition. Let $\mathbf{R}_o$ be a complete ordered field and let

$$\mathbf{C} = \mathbf{R}_o \times \mathbf{R}_o.$$

Define binary operations $+$ and $\cdot$ in $\mathbf{C}$ by

$$\langle a, b \rangle + \langle c, d \rangle = \langle a + c, b + d \rangle,$$
$$\langle a, b \rangle \cdot \langle c, d \rangle = \langle ac - bd, bc + ad \rangle.$$

Let $\mathbf{R} = \{\langle r, 0 \rangle\}$ and $i = \langle 0, 1 \rangle$. Then $\langle \mathbf{C}, +, \cdot, \mathbf{R}, i \rangle$ is a complex number system.

Proof. The verification of the postulates is a routine computation. The zero element is $\langle 0, 0 \rangle$ and the unit element is $\langle 1, 0 \rangle$. If $\langle a, b \rangle \neq \langle 0, 0 \rangle$, its multiplicative inverse is given by $\langle a/(a^2 + b^2), -b/(a^2 + b^2) \rangle$.

10–1.3. Proposition. If $\langle \mathbf{C}, +, \cdot, \mathbf{R}, i \rangle$ is a complex number system, then every element of $\mathbf{C}$ has a unique representation in the form $a + bi$, where a and b are in $\mathbf{R}$.

Proof. Postulate 10–1.1(v) asserts that every element has at least one such representation. Suppose that $a + bi = c + di$, where $a, b, c, d \in \mathbf{R}$. If $b \neq d$, then $i = (a - c)/(d - b) \in \mathbf{R}$, which contradicts the fact that -1 is not the square of any element of $\mathbf{R}$ (see 8–5.5(h) and (j)). Hence $b = d$ and $a = c$.

10–1.4. Proposition. Any two complex number systems are isomorphic; in fact, if $\langle \mathbf{C}_1, +, \cdot, \mathbf{R}_1, i_1 \rangle$ and $\langle \mathbf{C}_2, +, \cdot, \mathbf{R}_2, i_2 \rangle$ are complex number systems, then there is a unique isomorphism of $\mathbf{C}_1$ onto $\mathbf{C}_2$.

Proof. According to 8–7.7 there is a unique isomorphism φ of $\mathbf{R}_1$ onto $\mathbf{R}_2$ for $\mathbf{R}_1$ and $\mathbf{R}_2$ regarded as fields. Any element x of $\mathbf{C}_1$ has a unique representation in the form $x = a + bi_1$, where $a, b \in \mathbf{R}_1$. Hence we may define

$$\psi(x) = \varphi(a) + \varphi(b)i_2$$

and readily find that ψ is an isomorphism of $\mathbf{C}_1$ onto $\mathbf{C}_2$.

   If ψ' is any isomorphism of $\mathbf{C}_1$ onto $\mathbf{C}_2$ (as complex number systems), then ψ' maps $\mathbf{R}_1$ onto $\mathbf{R}_2$ and i_1 onto i_2. As an isomorphism of the field $\mathbf{R}_1$ onto the field

$\mathbf{R}_2$, ψ' must agree with φ on $\mathbf{R}_1$, by 8–7.7. Hence if $x = a + bi$, where $a, b \in \mathbf{R}_1$, then

$$\psi'(x) = \psi'(a) + \psi'(b)\psi'(i_1) = \varphi(a) + \varphi(b)i_2 = \psi(x).$$

Thus $\psi' = \psi$ and the uniqueness is established.

Why do we not consider the complex numbers as simply a field? For the purpose of algebra this would be satisfactory, but for analysis it is not. In analysis we must have a definite idea of which numbers are real, because it is in terms of the real subfield that we define the notion of absolute value and through the notion of absolute value that we handle the notion of limit. The geometrical picture of the complex numbers, with which the reader is no doubt familiar, suggests that we should be able to define $\mathbf{R}$ in terms of the field structure of $\mathbf{C}$; but here our intuition is misguided, for it can be shown, albeit by nonconstructive methods, that $\mathbf{C}$ has many subfields which satisfy the requirements imposed on $\mathbf{R}$. It is necessary therefore to single out an appropriate subfield to be *the* real subfield. There are two square roots of -1 in $\mathbf{C}$, and in the definition of a complex number system we single out one of them. The theory of complex numbers could be developed without a definite choice of which of the square roots of -1 is to be called i, but this would cause some difficulties. For example, the imaginary part of a complex number (as defined in 10–3.1) would not then be intrinsic.

EXERCISES

1. Let x be any complex number which is not real. Show that every complex number z can be expressed uniquely in the form $z = a + bx$, where a and b are real.

A *division algebra* (or skew field) is, in effect, a field with noncommutative multiplication. More precisely it is a configuration $\langle D, +, \cdot \rangle$ satisfying the field postulates 8–2.1(i_a), (i_m), (ii_a), (iii), and (iv) as well as

(iii') $(\forall w, x, y)$ $(x + y)w = xw + yw$,

which no longer follows from 8–2.1(iii). It is possible to prove that the right identity for multiplication and the right multiplicative inverses demanded in 8–2.1(iv) are, in fact, two-sided.

2. Let D be a division algebra and let

$$E = \{e \mid (\forall d \in D)\, de = ed\}.$$

Show that $\langle E, +, \cdot \rangle$ is a field. It is called the *center* of D.

3. Show that there exists a division algebra Q with the following properties: the center of Q is a complete ordered field R, and Q contains elements i, j, and k such that

$$(\forall q \in Q)(\exists a, b, c, d \in R) \quad q = a + bi + cj + dk,$$

and

$$i^2 = j^2 = k^2 = ijk = -1.$$

Show that any two such division algebras are isomorphic. This division algebra is known as the *quaternion algebra*.

10–2. PERMANENT NOTATION

The fundamental number systems of analysis are the natural numbers and the real and complex number systems. We have described all these as configurations and shown that they are categorically determined. This means, of course, that it does not make the slightest difference which simple chain, complete ordered field, or complex number system we consider. If, however, a reference to *the real number 1*, say, is to make sense, we must make a definite choice. A convenient choice is one which makes a real number just a special complex number.

Hence we shall now imagine that a particular complex number system $\langle C, +, \cdot, R, i \rangle$ is singled out which we shall hereafter refer to simply as **C**. Members of **C** will be called *complex numbers* and members of **R**, *real numbers*. We shall use the notation **N**, **I**, and **Q** instead of $N(C)$, $I(C)$, and $Q(C)$, respectively (see Section 8–4). Members of these sets are called *positive integers*, *integers*, and *rational numbers*, respectively. We shall also use **N*** for the set of nonnegative integers; that is, $N^* = N \cup \{0\}$. Both **N** and **N*** are often referred to as the set of *natural numbers;* that is, some authors regard zero as a natural number and some do not. We note the inclusions $N \subset N^* \subset I \subset Q \subset R \subset C$.

We know that **R** has exactly one order structure compatible with its field structure. We shall denote this order relation as usual by $<$ and $\leq$. The field **C** has no order structure, nor can it be assigned one compatible with its field structure, because -1 is a square in **C**. Therefore the relations $<$ and $\leq$ are defined only for real numbers and have no meaning at all for nonreal complex numbers. By extension, whenever an order comparison is made between two complex numbers, it is also asserted that the numbers are real; hence the phrase "let ϵ be a positive number" means "let ϵ be a positive real number." On the other hand, the term *bounded* in conjunction with complex numbers is used under a different convention (see 10–3.8).

10–3. CONJUGATES AND ABSOLUTE VALUES

10–3.1. Definition. If $z \in C$, then the unique real numbers a and b for which $z = a + bi$ (see 10–1.3) are called the *real* and *imaginary parts* of z and are denoted by $\text{Re}(z)$ and $\text{Im}(z)$, respectively.

When a complex number is written in the form $a + bi$, it is usually understood that a and b are real.

10–3.2. Definition. A complex number z for which $\text{Re}(z) = 0$ is called *purely imaginary*.

10–3.3. Definition. The *conjugate* of a complex number z is the number

$$\bar{z} = \text{Re}(z) - \text{Im}(z)i.$$

The bar notation is widely used without introduction.

10–3.4. Proposition. For any complex numbers w and z,

(a) $\overline{w + z} = \overline{w} + \overline{z}$,   (b) $\overline{w - z} = \overline{w} - \overline{z}$,

(c) $\overline{wz} = \overline{w} \cdot \overline{z}$,   (d) $\overline{w/z} = \overline{w}/\overline{z}$,

(e) $\overline{\overline{z}} = z$,   (f) z is real if and only if $z = \overline{z}$,

(g) z is purely imaginary if and only if $z = -\overline{z}$,

(h) $\operatorname{Re}(z) = \frac{1}{2}(z + \overline{z})$,   (i) $\operatorname{Im}(z) = (1/2i)(z - \overline{z})$.

We shall omit the proof of this proposition.

10–3.5. Formulas (a) and (c) mean that the map $z \to \overline{z}$ is an automorphism of **C** when **C** is regarded as a field only; that is, an isomorphism of **C** onto itself. (It is not an automorphism of **C**, however, when **C** is regarded as a complex number system. Why not?) From this it follows that a similar formula holds for the conjugate of any arithmetic combination of complex numbers; for example,

$$\overline{(x + y^2)/z} = (\overline{x} + \overline{y}^2)/\overline{z}.$$

While we shall not state this formally, we note that each individual instance can easily be justified by sufficiently many applications of (a), (b), (c), and (d).

10–3.6. Definition. If $z \in$ **C**, then

$$|z| = \sqrt{\operatorname{Re}(z)^2 + \operatorname{Im}(z)^2}.$$

(Here the radical refers to the unique nonnegative square root which exists in any complete ordered field as shown in 8–7.4.) This real number is called the *absolute value* of z.

10–3.7. Proposition. For all complex numbers z, w,

(a) $|z| = \sqrt{z\overline{z}}$,   (b) $|z| = |\overline{z}|$,

(c) $|z| = 0$ if and only if $z = 0$,

(d) $- |z| \leq \operatorname{Re}(z) \leq |z|$ (the first equality holds if and only if $z \leq 0$; the second, if and only if $z \geq 0$),

(e) $|\operatorname{Re}(z)| \leq |z|$, $|\operatorname{Im}(z)| \leq |z|$,   (f) $|wz| = |w|\,|z|$,

(g) $|w + z|^2 = |w|^2 + |z|^2 + 2\operatorname{Re}(w\overline{z})$,

(h) $|w + z| \leq |w| + |z|$ (equality holds if and only if $z = 0$ or $w/z \geq 0$),

(i) $\big||w| - |z|\big| \leq |w - z|$ (equality holds if and only if $z = 0$ or $w/z \geq 0$).

Proof. We shall prove only the last two of these formulas. Using (g), (d), (f), and (b) in that order, we have

$$|w + z|^2 \leq |w|^2 + |z|^2 + 2|w|\,|z| = (|w| + |z|)^2.$$

Hence (h) is true. In applying (d), we see that equality holds if and only if $w\overline{z} \geq 0$. Since either $z = 0$ or $w\overline{z} = |z|^2(w/z)$, the criterion for equality claimed in (h) is established.

To prove (i), we note that

$$|w| = |(w - z) + z| \leq |w - z| + |z|;$$

hence $|w| - |z| \leq |w - z|$. Similarly, $|z| - |w| \leq |w - z|$. Since $||w| - |z||$ is either $|w| - |z|$ or $|z| - |w|$, (i) follows. If $z = 0$, then equality in (i) is obvious. If $w/z = \lambda \geq 0$, then

$$||w| - |z|| = |\lambda|z| - |z|| = |\lambda - 1| |z| = |\lambda z - z| = |w - z|.$$

Conversely, suppose that equality holds in (i). Say,

$$|w| - |z| = |w - z|;$$

that is,

$$|w| = |w - z| + |z|.$$

By (h) either $z = 0$ or $(w - z)/z \geq 0$, which implies that either $z = 0$ or $w/z \geq 0$. The case in which $|z| - |w| = |w - z|$ leads by the same argument to $w = 0$ or $z/w \geq 0$. From this it follows that either $z = 0$ or $w/z \geq 0$.

10-3.8. Definition. A set S of complex numbers is said to be *bounded* if and only if the set

$$\{|s| \mid s \in S\}$$

is bounded. A function f taking values in $\mathbf{C}$ is said to be *bounded* if and only if range f is bounded.

EXERCISES

1. Derive the relation $|w + z|^2 + |w - z|^2 = 2|w|^2 + 2|z|^2$.

2. Suppose that x, y, and z are complex numbers and $x \neq 0$. Show that if

$$|x + y + z| = |x| + |y| + |z|,$$

then there are nonnegative real numbers r and s such that $y = rx$ and $z = sx$.

10-4. EXPONENTS

It is desirable at this point to present a formal theory of integral exponents. All the results of this section are familiar from high school algebra. Proofs are omitted and it is suggested that the reader supply them, although they are in some cases rather longwinded.

10-4.1. Definition. If $a \in \mathbf{C}$ and $n \in \mathbf{N}^*$, then a^n is defined recursively as follows: $a^0 = 1$, $a^{n+1} = a^n a$ for $n \in \mathbf{N}^*$.

10–4.2. Proposition. For all $a, b \in \mathbf{C}$ and all $m, n \in \mathbf{N}^*$,

(a) $a^m a^n = a^{m+n}$,
(b) $(a^m)^n = a^{mn}$,
(c) $(ab)^n = a^n b^n$.

10–4.3. Definition. If $a \in \mathbf{C} - \{0\}$ and $n \in \mathbf{N}$, then $a^{-n} = (1/a)^n$.

10–4.4. Proposition. The laws of exponents (a), (b), and (c) are valid for all $a, b \in \mathbf{C} - \{0\}$ and all $m, n \in \mathbf{I}$.

EXERCISES

1. Suppose that S is a set endowed with an associative binary operation (written multiplicatively). Define a^n for $a \in S$ and $n \in \mathbf{N}$. Prove that 10–4.2(a) and (b) hold. Suppose furthermore that S has a two-sided unit element e and that

$$(\forall s \in S)(\exists t \in S) \quad st = e.$$

(These assumptions make S a *group*.) Prove that there is a unique extension of the function just defined over $S \times \mathbf{I}$ such that 10–4.2(a) holds. Prove that the extension satisfies (b).

2. Let a be a positive real number and let n be a positive integer. Show that there exists a unique positive number x such that $x^n = a$.

CHAPTER 11

COUNTING AND THE SIZE OF SETS

Since prehistoric times men have compared physical sets by pairing off elements until one is exhausted. The fact that the result of such a comparison is independent of how the pairing is done was probably the first discovered theorem of mathematics.

When we leave the domain of physical sets for the domain of mathematical sets, we can carry along the idea of pairing off. We pair the elements of A with those of B by finding a suitable function. Suppose that we can pair the elements of A with those of B until one set at least is exhausted. If A is exhausted, the pairs define an injection from A to B. If B is exhausted, the pairs define an injection from B to A. Alternatively, after exhausting B we can continue making pairs, using some elements of B over again, and obtain a surjection from A to B. We compare two sets therefore by investigating the existence of injective, surjective, or bijective functions from one set to the other.

In this chapter we shall develop the formal theory of set comparison just far enough to prove the facts most commonly used in analysis. An informal treatment of the general theory of cardinal numbers follows. In the course of our study we shall encounter a new set-theoretic principle, the axiom of choice. This occasions an informal discussion of axiomatic set theory which is later extended in connection with the theory of cardinals.

11–1. SIMILARITY AND DOMINANCE

11–1.1. Definition. Let A and B be sets. We shall say that A and B are *similar* and write $A \sim B$ if and only if there exists a bijection from A to B.

11–1.2. Definition. Let A and B be sets. We shall say that A *dominates B weakly* and write $A \gtrsim B$ or $B \lesssim A$ if and only if there exists an injection from B to A. We shall say that A *dominates B* and write $A \succ B$ or $B \prec A$ if and only if $A \gtrsim B$ and $A \nsim B$.

The term *similar* is fairly standard in this connection. The term *dominates* is less so. Mathematicians often write or speak as if "$\sim$" means "has the same number of elements as" and "$\prec$" means "has fewer elements than." The concepts of similarity and dominance are certainly intended to model the ideas expressed by these phrases, but except in the domain of finite sets their properties are significantly different from what we should expect from the ordinary meanings of these terms. Using ordinary language, we should expect to find that there are more integers than even integers, but the bijection $n \to 2n$ shows that the set of even integers is, in fact, similar to the set of all integers. Until he is thoroughly familiar with the properties of similarity and dominance, therefore, the student should avoid these locutions.

It is easy to verify that, as the name and notation suggest, similarity is an equivalence relation. In fact, it is just isomorphism for the class of configurations consisting of a set with no structure.

The notation also suggests that $<$ is a strong order relation, but this is not so easy to prove. The following facts are obvious:

(a) $\lesssim$ is transitive and reflexive,
(b) if $A \sim B$, $B < C$, and $C \sim D$, then $A < D$.

The transitivity of $<$ is less obvious. Given $A < B$ and $B < C$, it follows immediately that $A \lesssim C$. We must eliminate the possibility that $A \sim C$. This is accomplished by the Schröder-Bernstein theorem.

11-1.3. The Schröder-Bernstein Theorem. Let A and B be sets. If there exists an injection from A to B and an injection from B to A, then there exists a bijection from A to B. In symbols, if $A \lesssim B$ and $B \lesssim A$, then $A \sim B$.

Proof. Suppose that f is an injection from A to B and g is an injection from B to A. Consider the set $\mathfrak{P}(A)$ ordered by inclusion. It is a complete ordered set with a largest element A and a smallest element $\emptyset$. Define a function θ from $\mathfrak{P}(A)$ to itself by

$$\theta(X) = A - g(B - f(X)).$$

Since θ is order-preserving, the Knaster fixed-point theorem (6–5.5) tells us that there is a set Z such that $\theta(Z) = Z$. This means that

$$Z = A - g(B - f(Z)),$$

which shows that $g(B - f(Z)) = A - Z$. Because g is injective, g^{-1} is a bijection from $A - Z$ to $B - f(Z)$. Also, f is a bijection from Z to $f(Z)$. Hence, if we define

$$h(x) = f(x) \qquad \text{for} \quad x \in Z$$

and

$$h(x) = g^{-1}(x) \qquad \text{for} \quad x \in A - Z,$$

then h is a bijection from A to B.

Although short and easy to check, this proof has a serious demerit. It is not easy to see what the bijection h actually is, because a fixed point Z of θ is not easily found. Hence we shall give another proof which is completely explicit.

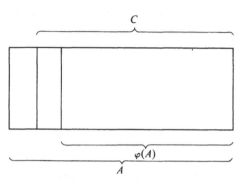

FIGURE 11-1

11-1.4. Lemma. Let φ be an injection from a set A to itself. If C is any set such that $A \supseteq C \supseteq \varphi(A)$, then $A \sim C$.

Before giving the formal proof, let us look at two diagrams. Figure 11-1 represents the situation described in the hypothesis. Since φ is injective, it carries the complex $A \supseteq C \supseteq \varphi(A)$ into a smaller version of the same thing. If φ is iterated, the successive images of A and C alternate:

$$A \supseteq C \supseteq \varphi(A) \supseteq \varphi(C) \supseteq \varphi(\varphi(A)) \supseteq \varphi(\varphi(C)) \supseteq \cdots$$

This is schematically represented in Fig. 11-2. Here φ maps each of the vertical strips at the left bijectively onto its right-hand neighbor once removed. Hence a bijection from A to C is obtained if we allow φ to act on the shaded strips and keep everything else fixed. Now we formalize this argument.

Proof of 11-1.4. Let $X_1 = A - C$ and define inductively $X_{n+1} = \varphi(X_n)$. Let $Y = \bigcup_{n \in N} X_n$. Define a function ψ from A to itself by

$$\psi(p) = \varphi(p) \quad \text{if} \quad p \in Y,$$
$$\psi(p) = p \quad \text{if} \quad p \notin Y.$$

We shall prove that ψ is in fact a bijection from A to C.

To begin with we note that ψ maps Y into Y and $A - Y$ into $A - Y$. Hence, if $\psi(p_1) = \psi(p_2)$, either both p_1 and p_2 are in Y or both are in $A - Y$. In the first case, we conclude $p_1 = p_2$, because φ is injective; in the second case, the same conclusion is trivial. Thus ψ is injective. For any $p \in A$, either

$$\psi(p) = p \in A - Y \subseteq C \quad \text{or} \quad \psi(p) = \varphi(p) \in \varphi(A) \subseteq C,$$

thus ψ maps A into C. Finally, choose any element $q \in C$. If $q \in A - Y$, then $\psi(q) = q$. If $q \in Y$, choose n so that $q \in X_n$; here $n \neq 1$, since $X_1 \cap C = \emptyset$, and therefore $X_n = \varphi(X_{n-1})$. We can choose an element $x \in X_{n-1}$ so that $\varphi(x) = q$. Since $x \in Y$,

$$\psi(x) = q,$$

which shows that range $\psi = C$.

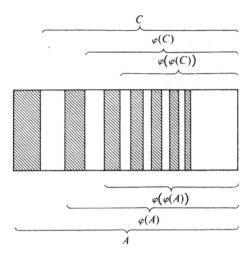

$$\overbrace{\hspace{4cm}}^{C}$$

$$\varphi(C)$$

$$\varphi(\varphi(C))$$

$$\varphi(\varphi(A))$$

$$\varphi(A)$$

$$A$$

FIGURE 11-2

Second proof of the Schröder-Bernstein Theorem. Suppose that f is an injection from A to B and g is an injection from B to A. Set $\varphi = g \circ f$. Then φ is an injection from A to itself and $A \supseteq g(B) \supseteq \varphi(A)$. By the lemma there is a bijection ψ from A to $g(B)$. Then $g^{-1} \circ \psi$ is an explicit bijection from A to B.

11-1.5. Corollary. $A \prec B$ and $B \precsim A$ if and only if $A \sim B$.

11-1.6. Corollary. The relation $\prec$ is transitive.

The Schröder-Bernstein theorem reduces the problem of proving the similarity of two sets to that of proving that each dominates the other weakly. This is often a good deal easier than proving similarity directly, because the injections involved do not have to be related to each other. The following proposition is a good illustration.

11-1.7. Proposition. $\mathbf{N} \times \mathbf{N} \sim \mathbf{N}$.

Proof. The mapping $n \rightarrow \langle n, 1 \rangle$ is an injection from $\mathbf{N}$ to $\mathbf{N} \times \mathbf{N}$; hence we need only construct an injection from $\mathbf{N} \times \mathbf{N}$ to $\mathbf{N}$.

Define a function g from $\mathbf{N} \times \mathbf{N}$ to $\mathbf{N}$ by $g(r, s) = (r + s)^2 + r$. Suppose that $g(r, s) = g(t, u)$; that is,

$$(r + s)^2 + r = (t + u)^2 + t. \tag{1}$$

If $r + s > t + u$, then

$$t - r = (r + s)^2 - (t + u)^2$$
$$\geq (t + u + 1)^2 - (t + u)^2 = 2t + 2u + 1 > t - r,$$

which is a contradiction.

Similarly, $r + s < t + u$ is impossible. Therefore $r + s = t + u$. Now, (1) gives $r = t$ and finally $s = u$. Hence g is injective.

A bijection from $\mathbf{N}$ to $\mathbf{N} \times \mathbf{N}$ can be thought of as an arrangement of the elements of $\mathbf{N} \times \mathbf{N}$ in a simple sequence. Undoubtedly the easiest way to see that this can be done is to inspect Fig. 11–3.

If we had already established theorems on the factorization of integers, we could prove that

$$\langle k, m \rangle \rightarrow 2^{k-1}(2m - 1)$$

is a bijection from $\mathbf{N} \times \mathbf{N}$ to $\mathbf{N}$.

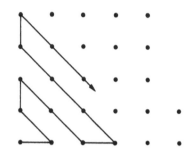

FIG. 11–3. $\mathbf{N} \times \mathbf{N}$ is represented as the set of all lattice points (points with integral coordinates) in the first quadrant. The broken arrow shows how to arrange these points in a simple sequence.

To prove that one set actually dominates another we must both construct an injection in one direction and prove that none exists in the other. The following theorem is a fine example of such a nonexistence proof. The argument bears a striking resemblance to Russell's paradox (see Section 2–7).

11–1.8. Theorem. Let A be any set and let $\mathfrak{P}(A)$ be its power set. There is no surjection from A to $\mathfrak{P}(A)$.

Proof. Let f be any function from A to $\mathfrak{P}(A)$ and let $X = \{a \in A \mid a \notin f(a)\}$. We shall prove that f is not surjective by showing that $X \notin$ range f.

Suppose that $b \in A$. If $b \notin f(b)$, then $b \in X$; hence $X \neq f(b)$. If $b \in f(b)$, then $b \notin X$, and again $X \neq f(b)$.

It is worthwhile to work through this proof by taking A to be a set with two elements and several explicit functions f. The case $A = \emptyset$ is also interesting.

11–1.9. Corollary. For any set A, $A < \mathfrak{P}(A)$.

Proof. Evidently, $a \rightarrow \{a\}$ is an injection from A to $\mathfrak{P}(A)$.

EXERCISES

1. Prove that $\mathbf{I} \sim \mathbf{N}$.

2. Prove that $\mathbf{N} \times \mathbf{N} \times \mathbf{N} \sim \mathbf{N}$.

3. How is the explicit bijection from A to B found in the second proof of the Schröder-Bernstein theorem related to the bijection found by using the Knaster fixed-point theorem?

4. Prove that $\mathbf{R} \sim \{x \in \mathbf{R} \mid 0 < x < 1\}$.

5. Prove that $\mathbf{R} \sim \mathbf{R} - \mathbf{N}$.

6. Prove that $\mathfrak{P}(\mathbf{N}) \sim \mathfrak{P}(\mathbf{N}) \times \mathfrak{P}(\mathbf{N})$.

11-2. FINITE SETS

Everyone is acquainted with the fundamental facts about finite sets, so the theorems of this section are bound to seem obvious. We must remember, however, that they refer to abstract sets and configurations. If they tell us nothing new about our intuitive ideas of counting, they may at least serve to convince us that our abstract description of the counting process satisfactorily represents the intuitive facts.

11-2.1. Definition. For each $k \in N^*$, let $N_k = \{x \in N \mid x \leq k\}$. In particular, N_0 is the null set. The sets N_k so defined will be called *segments* of the integers.

11-2.2. Proposition. Let p and q be nonnegative integers. Then $N_p \precsim N_q$ if and only if $p \leq q$.

Proof. If $p \leq q$, it is clear that $N_p \precsim N_q$. Hence we only need to prove that, if $N_p \precsim N_q$, then $p \leq q$.

Suppose this is false. Let p be the least nonnegative integer such that, for some $q < p$, an injection exists from N_p to N_q, and let f be such an injection. Since $p > 0$, N_p is not void. Therefore N_q is not void and $q > 0$. Thus $p - 1$ and $q - 1$ are both nonnegative integers.

Define a function g from $N_q - \{f(p)\}$ to N_{q-1} by

$$g(i) = i \qquad \text{if} \qquad i < f(p),$$
$$g(i) = i - 1 \qquad \text{if} \qquad i > f(p).$$

Then g is injective and $g \circ f$ is an injection from N_{p-1} to N_{q-1}. Since $q - 1 < p - 1$, this contradicts the original choice of p, and the proposition is proved.

11-2.3. Proposition. Let p and q be nonnegative integers. If there exists a surjection from N_p to N_q, then $p \geq q$.

Proof. Since f is surjective, for each $i \in N_q$, the set $f^{-1}(i)$ is not void. Define a function g from N_q to N_p by letting $g(i)$ be the least integer in $f^{-1}(i)$. Then g is injective. Therefore $q \leq p$ by 11-2.2.

11-2.4. Proposition. Let p and q be nonnegative integers. Then $N_p \sim N_q$ if and only if $p = q$, and $N_p \prec N_q$ if and only if $p < q$.

Proof. This proposition follows immediately from 11-2.2 and 11-2.3.

11-2.5. Definition. A set E is called *finite* if and only if it is similar to some segment of the integers. Otherwise it is *infinite*.

Since similarity is an equivalence relation, it is clear that a set similar to a finite set is itself finite.

11–2.6. Theorem. If E is a finite set, then there is a unique nonnegative integer p such that $E \sim \mathbf{N}_p$.

Proof. This follows immediately from 11–2.4 and 11–2.5.

11–2.7. Definition. If E is a finite set, then the unique integer p of 11–2.6 is called the *cardinal* of E. We shall denote it by card E. Evidently, similar finite sets have the same cardinal.

11–2.8. Proposition. $\mathbf{N}$ is infinite.

Proof. Suppose on the contrary that $\mathbf{N}$ is finite. Then there exists a bijection f from $\mathbf{N}$ to some segment of the integers $\mathbf{N}_p$. The restriction of f to $\mathbf{N}_{p+1}$ is then an injection from $\mathbf{N}_{p+1}$ to $\mathbf{N}_p$. This contradicts 11–2.2.

The proofs of the following propositions are left as exercises for the reader.

11–2.9. Proposition. A set E is finite if and only if there exists an injection from E to some segment of the integers. If so, card E is the least integer p such that there exists an injection from E to $\mathbf{N}_p$.

11–2.10. Proposition. A set E is finite if and only if there exists a surjection from some segment of the integers to E. If so, card E is the least integer p such that there exists a surjection from $\mathbf{N}_p$ to E.

11–2.11. Proposition. If $A \subset B$ and B is finite, then A is finite and card $A <$ card B.

11–2.12. Theorem. If $A \subseteq B$, B is finite, and card $A =$ card B, then $A = B$.

11–2.13. Proposition. If A and B are finite sets, then $A \cup B$ is finite and

$$\text{card } (A \cup B) + \text{card } (A \cap B) = \text{card } A + \text{card } B.$$

11–2.14. Proposition. If $\mathcal{B}$ is a finite set of finite sets, then $\bigcup_{A \in \mathcal{B}} A$ is finite.

11–2.15. Proposition. If A and B are finite sets, then $A \times B$ is finite and

$$\text{card } (A \times B) = (\text{card } A)(\text{card } B).$$

11–2.16. Proposition. A nonvoid finite subset of a linearly ordered set has a largest element and a smallest element.

EXERCISES

1. Prove that a nonvoid, finite ordered set contains a minimal element.

2. Prove that a finite linearly ordered set is isomorphic (as an ordered set) to some segment of the integers.

3. Prove that if A is a finite set and E is infinite, then $A < E$.

4. Let R be a relation in the set A. Let S be the least transitive relation in A which contains R (see Exercise 2, p. 88). Let $a, b \in A$. Prove that a necessary and sufficient condition that $a\, S\, b$ is that there exist an integer $k \geq 2$ and a function g from $\mathbf{N}_k$ to A such that (i) $g(1) = a$, (ii) $g(k) = b$, and (iii) $(\forall n < k)\, g(n)\, R\, g(n+1)$.

11-3. COUNTABLE SETS

When one passes from finite sets to infinite sets, the concept of similarity loses much of its value, largely because the analogue of the powerful theorem 11–2.12 is false. Among infinite sets those similar to $\mathbf{N}$ are the most tractable, for a bijection from $\mathbf{N}$ to a set E can be regarded as an arrangement of the elements of E in a sequence.

11–3.1. Definition. A set E is called *countable* if and only if it is finite or similar to $\mathbf{N}$. Otherwise it is *uncountable*.

Since similarity is transitive, any set similar to a countable set is itself countable. Therefore it follows immediately from 11–1.7 that $\mathbf{N} \times \mathbf{N}$ is countable. Corollary 11–1.9 tells us that $\mathfrak{P}(\mathbf{N})$ is uncountable. We shall prove in Chapter 13 that $\mathbf{R}$ is similar to $\mathfrak{P}(\mathbf{N})$, hence $\mathbf{R}$ is also uncountable.

The words *denumerable* and *enumerable* are often used as synonyms for *countable*, but some authors reserve one or the other of these words for the narrower meaning, *similar to* $\mathbf{N}$; i.e., they exclude the finite sets.

A family of sets is said to be countable if and only if its index set is countable. The adjectives *finite*, *infinite*, and *uncountable* are used in the same way.

11–3.2. Proposition. If S is an infinite subset of $\mathbf{N}$, then there exists a unique order-preserving bijection from $\mathbf{N}$ to S.

Proof. If $k \in \mathbf{N}^*$, then $S \cap \mathbf{N}_k$ is a finite set. Since S is infinite, $S - \mathbf{N}_k$ is not empty; that is, S contains an integer greater than k.

We define a function from $\mathbf{N}$ to S by induction. Let $f(1)$ be the least member of S. For all $n \in \mathbf{N}$, let $f(n+1)$ be the least member of S greater than $f(n)$. It follows from the definition of f that $f(n+1) > f(n)$. By induction on p we find that $(\forall p)\, f(n+p) > f(n)$. Hence f is injective and order-preserving.

We must still prove that range $f = S$. Suppose the contrary. Then there is a least element s in $S -$ range f. Surely s is not itself the least element of S, since in that case

$$s = f(1) \in \text{range}\, f.$$

Hence $\mathbf{N}_{s-1} \cap S$ is not void. As a finite nonvoid subset of a linearly ordered set, $\mathbf{N}_{s-1} \cap S$ contains a largest element, say t. Now $t < s$, so $t \in$ range f. Let $t = f(p)$. Then the definition of f gives $f(p+1) = s$; thus $s \in$ range f. This establishes the existence. The uniqueness is left to the reader.

The following theorem can be interpreted as meaning that there are no infinite sets of a size between all finite sets and $\mathbf{N}$ itself; that is, there is no infinite set E such that $E < \mathbf{N}$.

11–3.3. Proposition. A set E is countable if and only if $E \lesssim \mathbf{N}$.

Proof. If E is countable, an immediate consequence of the definitions is that $E \lesssim \mathbf{N}$. The converse is trivial if E is finite. We assume therefore that E is infinite and that f is an injection from E to $\mathbf{N}$. Since E is similar to range f, the latter is infinite. Hence there exists a bijection g from $\mathbf{N}$ to range f. Now, $g^{-1} \circ f$ is a bijection from E to $\mathbf{N}$; thus $E \sim \mathbf{N}$, and E is countable.

11–3.4. Corollary. Any subset of a countable set is countable.

11–3.5. Proposition. A nonvoid set E is countable if and only if there exists a surjection from $\mathbf{N}$ to E.

Proof. Suppose that E is nonvoid and countable. Let f be an injection from E to $\mathbf{N}$. Pick any element $e \in E$. Define g from $\mathbf{N}$ to E by

$$g(n) = f^{-1}(n) \quad \text{if} \quad n \in \text{range } f$$

and

$$g(n) = e \quad \text{if} \quad n \notin \text{range } f.$$

Then g is surjective.

Now suppose that φ is a surjection from $\mathbf{N}$ to E. For each $x \in E$, the set $\varphi^{-1}(x)$ is not empty, so we can define $\psi(x)$ as the least integer in $\varphi^{-1}(x)$. Then ψ is an injection from E to $\mathbf{N}$.

11–3.6. Corollary. A function with a countable domain has a countable range.

EXERCISES

1. Prove that the set $\mathbf{Q}$ of rational numbers is countable.

2. Prove that the union of a finite family of countable sets is countable.

3. Show that the set $\mathfrak{F}$ of all finite subsets of $\mathbf{N}$ is countable.

4. Construct an injection from $\mathbf{R}$ to $\mathfrak{P}(\mathbf{N})$.

5. Let S be a set which is well-ordered by a weak order relation W and also by the reverse order relation $R = \{\langle s, t \rangle \mid \langle t, s \rangle \in W\}$. Prove that S is finite.

6. Let L be a linearly ordered set. Suppose there exists a weakly order-preserving surjection φ from $\mathbf{N}$ to L. Show that L is either finite or isomorphic to $\mathbf{N}$. Show also that there is at most one element x in L such that $\varphi^{-1}(x)$ is infinite.

11–4. ANOTHER FORM OF INDUCTIVE DEFINITION

In the next section and many times later in the book we shall define functions on $\mathbf{N}$ by a form of induction not directly covered by the fundamental theorem 7–3.6. That theorem tells us that we can define a function φ by giving $\varphi(1)$ together with a rule for calculating $\varphi(k + 1)$ from $\varphi(k)$. It is often necessary to make $\varphi(k + 1)$ depend on all of the values $\varphi(1), \varphi(2), \ldots, \varphi(k)$, and possibly k itself. Even the simple recursive definition of the factorial function,

$$\varphi(1) = 1,$$
$$\varphi(k + 1) = (k + 1)\varphi(k),$$

is not covered directly by Theorem 7–3.6. In this section we shall prove that these more elaborate methods of inductive definition do indeed lead to a function. As in the case of the previous theorem on this subject, the issue is not whether the new kind of rule determines a unique value of $\varphi(n)$ for each integer n. The problem is to show that the set of ordered pairs which is the function φ can be defined by a propositional scheme in keeping with the convention of Section 2–6.

Actually this more powerful form of inductive construction can be deduced rather simply from the older form. The trick is to construct, not the sequence of values $\varphi(1)$, $\varphi(2)$, $\ldots$, but the sequence $\varphi_1, \varphi_2, \ldots$ of partial functions, where φ_k is the restriction of φ to N_k. To define φ_{k+1} in terms of φ_k amounts to defining $\varphi(k + 1)$ in terms of φ_k, which in turn amounts to defining $\varphi(k + 1)$ in terms of $\varphi(1)$, $\varphi(2)$, $\ldots$, $\varphi(k)$, and k (since we can recover k from φ_k: $k = \operatorname{card} \varphi_k$).

11–4.1. Theorem. Let E be any set and let $\mathfrak{F}$ be the set of all functions from a segment of the integers to E. Let G be any function from $\mathfrak{F}$ to E. Then there exists a unique function φ from N to E such that

$$(\forall k \in N^*) \quad \varphi(k + 1) = G(\varphi_k), \tag{1}$$

where φ_k stands for the restriction of φ to N_k.

The possibility that $k = 0$ in (1) is explained by the fact that $\emptyset$ is a function from N_0 to E; that is, $\emptyset \in \mathfrak{F}$. There is no need therefore to have a special prescription for $\varphi(1)$; it is given by (1) as $G(\emptyset)$. The function G is the rule by which we extend the partially defined φ to the next larger integer.

Proof. Since the uniqueness is obvious, we shall prove only the existence of such a function.

Define a function H from $\mathfrak{F}$ to $\mathfrak{F}$ as follows: Suppose $f \in \mathfrak{F}$ and f has domain N_k. Let $H(f)$ be the function from N_{k+1} to E defined by

$$H(f)(i) = f(i) \quad \text{if} \quad i \in N_k,$$
$$H(f)(k + 1) = G(f). \tag{2}$$

(This definition of $H(f)$ could be stated more succinctly but perhaps less clearly: $H(f) = f \cup \{\langle 1 + \operatorname{card} f, G(f)\rangle\}$.)

According to Theorem 7–3.6 there is a function ψ from N^* to $\mathfrak{F}$ such that

$$\psi(0) = \emptyset$$

and

$$(\forall k \in N^*) \quad \psi(k + 1) = H(\psi(k)).$$

It is easy to prove by induction that the domain of $\psi(k)$ is $\mathbf{N}_k$ and that $\psi(p)$ is an extension of $\psi(k)$ if $p > k$. It follows that $\varphi = \bigcup_{k\in\mathbf{N}} \psi(k)$ is a function from $\mathbf{N}$ to E. Furthermore, φ_k, the restriction of φ to $\mathbf{N}_k$, is just $\psi(k)$.

If $k \in \mathbf{N}^*$, then

$$\varphi(k + 1) = \psi(k + 1)(k + 1) = H(\psi(k))(k + 1) = G(\psi(k)) = G(\varphi_k),$$

where the third equality is from (2). This proves the existence.

This theorem is almost always applied tacitly. To write out the formal details that justify an inductive construction under this theorem usually involves a good bit of notation which can and should be dispensed with. However, to elucidate what is involved, we shall look at two examples.

First, consider the factorial function. We let $E = \mathbf{R}$, $G(\emptyset) = 1$, and for any function f from $\mathbf{N}_k$ ($k > 0$) to $\mathbf{R}$, let $G(f) = (k + 1)f(k)$. Then the function φ satisfying (1) is indeed the factorial function.

Now, for a more complicated case, suppose that we seek a real-valued function φ such that $\varphi(1) = 3$, $\varphi(2) = 5$, and

$$\varphi(k + 1) = \varphi(k) + \sqrt{\varphi(k - 1) - k} \tag{3}$$

for $k > 1$. Note that it is not obvious that the recursion will actually go on computing values of φ in this case, because the radicand in (3) might turn out to be negative at some stage.

Take $E = \mathbf{R}$. Then $\mathfrak{F}$ is the set of all real-valued functions defined on segments of the integers. We must define the function G from $\mathfrak{F}$ to $\mathbf{R}$. First, we arrange the initial values of φ by defining

$$G(\emptyset) = 3, \tag{4}$$

$$G(f_0) = 5, \tag{5}$$

where f_0 is the function from $\mathbf{N}_1$ to $\mathbf{R}$ defined by $f_0(1) = 3$. Now the difficulty posed by the radical in (3) appears: there is no obvious way to define G on the rest of $\mathfrak{F}$. We do the best we can, however. For each integer $k > 1$, let

$$\mathfrak{F}_k = \{f \in \mathfrak{F} \mid \mathrm{domain}\, f = \mathbf{N}_k \text{ and } f(k - 1) \geq k\}$$

and set

$$G(f) = f(k) + \sqrt{f(k - 1) - k} \tag{6}$$

if $f \in \mathfrak{F}_k$. Finally, we define G on the rest of $\mathfrak{F}$ in any convenient manner, say,

$$G(f) = 0 \tag{7}$$

if $f \in \mathfrak{F} - \bigcup_k \mathfrak{F}_k - \{\emptyset\} - \{f_0\}$.

Now, the theorem guarantees the existence of a function φ satisfying (1), but it is not clear that φ satisfies (3). (Indeed, it would not if, instead of (5), we had defined $G(f_0) = 3$. For then $\varphi_5 \notin \mathfrak{F}_5$, whence $\varphi(6) = G(\varphi_5) = 0$ according to (7). Thus (3) fails for $k = 5$.)

Instead of proving just (3), we shall prove the two statements

$$\varphi(n) = \varphi(n-1) + \sqrt{\varphi(n-2)} - n + 1 \qquad \text{for} \quad n > 2 \qquad (8)$$

and

$$\varphi(n) \geq n + 2 \qquad \text{for all } n. \qquad (9)$$

We can verify directly that (8) is true for $n = 3$ and that (9) is true for $n = 1, 2$, and 3. We continue by induction. Assume that (8) and (9) hold for $n = k \geq 3$. First we find that

$$\varphi_k(k-1) = \varphi(k-1) \geq (k-1) + 2 > k,$$

and therefore $\varphi_k \in \mathfrak{F}_k$. Hence

$$\varphi(k+1) = G(\varphi_k) = \varphi_k(k) + \sqrt{\varphi_k(k-1)} - k$$
$$= \varphi(k) + \sqrt{\varphi(k-1)} - k \geq (k+2) + 1 = k + 3.$$

Thus (8) and (9) hold for $n = k + 1$. This proves (8) and (9). But (8) is equivalent to (3).

The situation encountered here is common. In the notation of Theorem 11–4.1 we have a function G^* from a subset $\mathfrak{F}^*$ of $\mathfrak{F}$ to E and we want to find a function φ from $\mathbf{N}$ to E such that

$$(\forall k \in \mathbf{N}^*) \quad \varphi(k+1) = G^*(\varphi_k). \qquad (10)$$

The theorem does not apply unless $\mathfrak{F}^* = \mathfrak{F}$. Hence we extend G^* arbitrarily to be a function G from $\mathfrak{F}$ to E and apply the theorem using G. Then we establish (10) by showing that the partial functions φ_k are always in $\mathfrak{F}^*$. This amounts to proving that successive applications of G^* as suggested by (10) are actually possible. The theorem says that, if a recursive computation will produce new values indefinitely, then the ordered pairs can be assembled to form a function; that is, the appropriate set of ordered pairs can be defined with a complicated propositional scheme.

This second point is generally omitted altogether. The omission is natural, because the question at issue is important only for the formalization of mathematics. Once we understand how the situation can be managed, there is no point in worrying about it. On the other hand, if it is not obvious that a recursion will produce a value for every integer, then the proof must be given since this is a point of substance.

If the recursion (3) came up in a paper, the argument just given would probably be condensed somewhat as follows.

We shall define a function from $\mathbf{N}$ to $\mathbf{R}$ by induction. Let $\varphi(1) = 3$ and $\varphi(2) = 5$. Suppose that $k \geq 2$ and $\varphi(1), \varphi(2), \ldots, \varphi(k)$ have been defined so that

$$\varphi(n) \geq n + 2 \qquad \text{for} \quad 1 \leq n \leq k.$$

Then $\varphi(k - 1) - k \geq 1$; so we can put

$$\varphi(k + 1) = \varphi(k) + \sqrt{\varphi(k - 1) - k}$$

and then obtain $\varphi(k + 1) \geq (k + 2) + 1 = k + 3$.

EXERCISE

Prove that there exists a function φ from $\mathbf{N}$ to $\mathbf{R}$ such that $\varphi(1) = 2$ and

$$(\forall k \in \mathbf{N}) \quad \varphi(k + 1) = \sqrt{\varphi(k) - 1/k}.$$

Give the constructions necessary to bring in Theorem 11–4.1.

11–5. THE AXIOM OF CHOICE

We shall begin by proving an "obvious" theorem. Unfortunately, a careful examination of the proof uncovers a difficulty which we cannot overcome by using only elementary set theory. We are thus led to a new set-theoretic convention, usually known as the axiom of choice. A brief discussion of the nature of abstract or axiomatic set theory follows.

11–5.1. Proposition. If E is an infinite set, then $\mathbf{N} \precsim E$.

Since the proof will take us far afield, we shall state two easy corollaries before starting the proof.

11–5.2. Corollary. Every infinite set contains an infinite countable subset.

11–5.3. Corollary. A set is infinite if and only if it is similar to one of its proper subsets.

Proof. We already know (11–2.11) that no finite set is similar to one of its proper subsets. Suppose then that E is an infinite set; we shall prove that E is similar to one of its proper subsets.

Let f be an injection from $\mathbf{N}$ to E. Define g from E to E by

$$g(x) = x \qquad\qquad \text{if} \quad x \notin \text{range } f,$$
$$g(x) = f(1 + f^{-1}(x)) \qquad \text{if} \quad x \in \text{range } f.$$

Now, g is easily seen to be a bijection from E to $E - \{f(1)\}$.

11–5.4. Proof of 11–5.1. We shall define the required injection from $\mathbf{N}$ to E by induction. Since E is certainly not empty, we choose any element $e \in E$ and let $\varphi(1) = e$. If $\varphi(1), \varphi(2), \ldots, \varphi(k)$ have already been defined, then

$$E - \{\varphi(1), \varphi(2), \ldots, \varphi(k)\} = E - \text{range } \varphi_k \neq \emptyset,$$

since E is infinite; so choose any member of this set and call it $\varphi(k + 1)$. The construction makes it clear that φ is injective, and the proposition is proved. *But Theorem 11-4.1 does not justify this construction.* Before reading further the reader should try to write out the details for this proof similar to those given in the last section to establish the existence of the function satisfying the recursion 11-4(3). He will find that he cannot justify the construction either with 11-4.1 or with the more primitive Theorem 7-3.6.

The easiest way to see that Theorem 11-4.1 does not apply is to note that it asserts the uniqueness of the function φ, whereas there is nothing unique about the function φ constructed in our pseudoproof of 11-5.1.

To avoid the complexity of Theorem 11-4.1, let us look at a proposition which involves the same difficulty but requires only Theorem 7-3.6.

11-5.5. Proposition. Let E be a nonvoid ordered set with no maximal elements. Then there exists an order-preserving injection from $\mathbf{N}$ to E.

Proof. We construct the required injection ψ by induction. Since E is not empty, we can choose $e \in E$ and let $\psi(1) = e$. Suppose that $\psi(k)$ has been defined. Choose $\psi(k + 1)$ so that $\psi(k + 1) > \psi(k)$ in the order of E. This is possible because $\psi(k)$ is not a maximal element of E.

If these instructions do indeed define a function, then the function is obviously an order-preserving injection from $\mathbf{N}$ to E.

Again, it is clear that this cannot be justified by Theorem 7-3.6, because ψ is certainly not unique. The hypothesis of 7-3.6 demands a function g to serve as the rule for calculating $\varphi(k + 1)$ from $\varphi(k)$, and the instruction *choose* is not sufficiently explicit to be defined in terms of a function. Without a definite function g the proof of Theorem 7-3.6 collapses completely.

There is no question that the rule given will enable us to find values of ψ for as many integers as we please. But we are claiming to define a function; a function is a set and, according to the convention of Section 2-6, the only legitimate method of defining a set is by means of a propositional scheme.

Let us imagine all the values of ψ being calculated, one after another, each by means of a separate act of choosing an element from a nonvoid set. The legitimacy of choosing an element from a nonvoid set stems from the meaning of the word *nonvoid*. If we have an explicit finite number of nonvoid sets, we can make a choice from each, one at a time, and justify each step of the argument in terms of the meaning of *nonvoid*. But when we claim that functions are defined by the constructions of 11-5.4 and 11-5.5, we are, in effect, requiring that infinitely many choices be made at once. This we cannot justify by an appeal to the meaning of *nonvoid*.

Not only is there no known method for defining the function required by Propositions 11-5.1 and 11-5.5, but it is known that no method exists within the austere convention of Section 2-6. We must therefore either abandon these two propositions and many others like them or liberalize our criteria for deciding when a set has been properly defined. We shall follow the latter course.

It would be unwise, however, to give up all the rules of set formation. An uncontrolled use of set theory leads all too rapidly to logical contradictions. It is better therefore to introduce a particular exception to our old convention. From now on we shall agree on the following:

11–5.6. Set-theoretic convention. If $\{H(i) \mid i \in I\}$ is any family of nonvoid sets, then it is legitimate to infer the existence of a function h with domain I such that

$$(\forall i \in I) \quad h(i) \in H(i).$$

In other words, we arbitrarily legitimize the act of choosing simultaneously one member from each of a presumably infinite number of nonvoid sets.

Recall the definition of the direct product of a family of sets (3–7.1): $\times_{i \in I} H(i)$ is the set of all functions h with domain I such that $(\forall i \in I) \, h(i) \in H(i)$. Therefore, our new convention amounts to this: If each of the sets $H(i)$ is not void, then we accept without proof the statement

$$\times_{i \in I} H(i) \text{ is not void.}$$

Since it would probably not occur to anyone unacquainted with the subtleties of set theory to even suspect that the direct product of a family of nonvoid sets might be empty, it is fair to say that our new convention is reasonable.

With the aid of this new convention we can readily justify the construction of 11–5.4. Let $I = \mathfrak{P}(E) - \{\emptyset\}$, and let H be the identity function on I. We infer the existence of a function h from $\mathfrak{P}(E) - \{\emptyset\}$ to E such that $h(X) \in X$ for every nonvoid subset X of E. Now Theorem 11–4.1 can be applied: There exists a function φ from $\mathbf{N}$ to E such that

$$(\forall k \in \mathbf{N}^*) \quad \varphi(k + 1) = h(E - \text{range } \varphi_k).$$

This function is the required injection from $\mathbf{N}$ to E.

In effect, the single choice of the function h replaces the infinite sequence of choices required in the inductive construction. Note that the choice of h determines the function φ uniquely.

11–5.7. Proposition. Let S and T be any two sets. If f is a surjection from S to T, then there exists an injection g from T to S such that $f \circ g$ is the identity map of T.

Proof. If f is surjective, $\{f^{-1}(t) \mid t \in T\}$ is a family of nonvoid sets, so there exists a function g with domain T such that

$$(\forall t \in T) \quad g(t) \in f^{-1}(t).$$

This is the required injection.

11–5.8. Corollary. Suppose S is a set and T is any nonvoid set. A necessary and sufficient condition that $S \gtrsim T$ is that there exist a surjection from S to T.

Proof. If f is a surjection from S to T, then the proposition asserts the existence of an injection from T to S; that is, $T \lesssim S$. Conversely, suppose h is an injection from T to S. Since T is not void, pick $t \in T$. Define $\varphi(s) = h^{-1}(s)$ if $s \in$ range h, and $\varphi(s) = t$ if not. Then φ is the required surjection.

It is worth noting that the necessity for the convention 11–5.6 arises in part at least from our preference for stating theorems in abstract form. When an explicit family $\{H(i) \mid i \in I\}$ of nonvoid sets is involved, it often happens that we can define a choice function h explicitly. If each of the sets $H(i)$ were a subset of N, for example, we could let $h(i)$ be the least member of $H(i)$. We have already used this method of choice in the proof of 11–3.5, which is just a special case of 11–5.8.

11–5.9. Proposition. The union of a countable family of countable sets is countable.

Proof. Let $\{E_i \mid i \in I\}$ be a countable family of countable sets. We must prove that $\bigcup_{i \in I} E_i$ is countable. If there are indices i for which $E_i = \emptyset$, we may suppress them without affecting the union. Moreover, if $I = \emptyset$, the result is trivial. Hence we shall assume that $I \neq \emptyset$ and that $(\forall i) \, E_i \neq \emptyset$.

Let g be a surjection from N to I. Then

$$\bigcup_I E_i = \bigcup_N E_{g(n)}.$$

Thus it is sufficient to prove that $\bigcup_{n \in N} F_n$ is countable, where for each n, F_n is nonvoid and countable.

Let S_n be the set of surjective functions from N to F_n. By 11–3.5, $\{S_n \mid n \in N\}$ is a family of nonvoid sets. Let φ be a function with domain N such that

$$(\forall n) \quad \varphi(n) \in S_n.$$

Then $\langle n, m \rangle \rightarrow \varphi(n)(m)$ is a surjection from $N \times N$ to $\bigcup_N F_n$. Hence $\bigcup_N F_n$ is countable by 11–3.6.

Our new convention is usually called the axiom of choice. To understand why it is called an axiom, to explain its relation to the rest of set theory, and to clarify further our insistence on defining sets with the aid of propositional schemes, let us look briefly at the idea of abstract or axiomatic set theory. We shall do so within the framework of the naive set theory we have been using.

We have agreed that for technical purposes the natural number system is a configuration. This configuration is only a set-theoretic model of our intuitive ideas of the natural numbers, which have a conceptual existence quite independent of the ideas of set theory. We adopt this model because it seems to offer great precision in expressing our ideas about the "true" natural numbers.

Similarly, we can form a set-theoretic model of set theory itself. We define a class of configurations by means of postulates chosen to reflect the properties we attribute to intuitive sets. Intuitive set theory is concerned with objects, which we call sets, and a relation, membership, which holds between certain pairs of sets. The configuration model of this situation is a set S together with a binary relation m in S. Members of S will be called *a-sets*. The binary relation m is the abstract representation of membership; it should not be confused with membership itself, which remains as an intuitive notion.

A model for abstract set theory is therefore a configuration $\langle S, m \rangle$, where $m \in \mathfrak{P}(S \times S)$, such that some suitable list of restrictive postulates is satisfied. Many different postulate systems seem to be acceptable, but they are all quite complicated. We cannot discuss them in any detail here. Instead we shall examine a few possible postulates to illustrate the ideas involved.

First, let us write a postulate which expresses the fact that an intuitive set is determined by its members and not, for example, by the actual criterion of membership:

$$(\forall s, t) \quad ((\forall u) \; u \, m \, s \Leftrightarrow u \, m \, t) \Rightarrow s = t. \tag{1}$$

(If two a-sets s and t have the same members (in the m-sense), they are the same.)

In our work with intuitive sets we frequently form new sets from one or more old sets. We have a choice in our intuitive imagery. We can think of actually forming a new set which did not previously exist, or we can think of turning our attention toward some already existing set. The former does not make sense in the abstract context, because all a-sets exist *a priori*; they are just the members of S. Instead of forming a new a-set, we prove that there exists in S an a-set having the desired properties. Of course we need postulates on which to base such proofs, and among these we might have

$$(\forall s, t)(\exists u)(\forall v) \quad v \, m \, u \Leftrightarrow (v = s \text{ or } v = t). \tag{2}$$

Since the element u of this postulate is uniquely determined by the elements s and t according to (1), we may appropriately denote it by $\{s, t\}_m$. The notation is intended to remind us that the a-set $\{s, t\}_m$ is related to the a-sets s and t by m in the same way that the intuitive set $\{x, y\}$ is related to the sets x and y by $\in$. We shall abbreviate $\{s, s\}_m$ by writing $\{s\}_m$.

Postulate (2) also guarantees the existence of the abstract ordered pair $\langle s, t \rangle_m$, which is defined to be $\{\{s\}_m, \{s, t\}_m\}_m$ by analogy with 3–3.1. An a-set f is an *a-relation* if and only if

$$(\forall u) \quad u \, m \, f \Rightarrow (\exists s, t) \; u = \langle s, t \rangle_m, \tag{3}$$

and it is an *a-function* if and only if furthermore

$$(\forall s, t, u) \quad (\langle s, t \rangle_m \, m \, f \text{ and } \langle s, u \rangle_m \, m \, f) \Rightarrow t = u. \tag{4}$$

Other postulates similar to (2) provide for the existence of the abstract analogues of the domain and range of every a-relation, unions and intersections, the power a-set of an a-set, etc. The postulates are designed expressly so that all of the set-theoretic constructions used in mathematics can be imitated abstractly. At the same time the postulates must not permit an argument analogous to Russell's paradox or any of the other known paradoxes of set theory, since this would mean that the postulate system is inconsistent.

Once we have achieved a postulate system which appears to be satisfactory, we can check through various mathematical proofs to see whether they can in fact be carried out in abstract set theory, using only steps which can be justified by the postulates. This gives us a firm objective standard for deciding when a proof is correct.

If we adopt some particular postulate system for abstract set theory and agree that the criterion for accepting an intuitive set-theoretic argument is that its analogue can be justified by the postulates, then we are in effect agreeing that the set of all intuitive sets endowed with the membership relation is a configuration satisfying the postulates. We can have at best intuitive reasons for believing this. It is really an act of faith.

The situation is the same when we agree to accept only those statements about "the" natural numbers which are provable for simple chains. Evidently, "the" natural numbers have a conceptual existence which is quite independent of set theory, and it is an act of faith on our part to agree that the set of "the" natural numbers endowed with the "count-one-more" function is a configuration satisfying the postulates for a simple chain.

When we leave the domain of abstract configurations, what we call postulates take on a different significance. In the abstract domain we are in charge; we can frame postulates as we please and simply exclude from consideration configurations which fail to satisfy them. But we cannot take this attitude toward concepts which have any sort of independent existence. What are appropriately called postulates in the context of configurations become axioms when we deal with independent conceptions. They are axioms because they are accepted as true, or at least granted for purposes of argument, for intuitive reasons. Thus when we describe "the" natural numbers as a simple chain, the postulates for a simple chain become axioms about "the" natural numbers. Similarly, when we agree that the set of all intuitive sets is a model for abstract set theory, the postulates for the abstract theory become axioms about the set of intuitive sets. We are thus led to the idea of axioms for set theory.

Since it embraces all of mathematics, the set of all intuitive sets has a very complicated structure. Consequently, it takes a very complicated system of axioms to describe it. In fact, one of Gödel's famous theorems says, in effect, that it cannot be described completely by any system of axioms. In other words, given any list of axioms for set theory, there will always be questions about sets which cannot be settled by deduction from this list.

This being the case, it is not surprising that there is more than one accepted system of axioms for set theory. All of the usual systems will justify all of the reasoning used in this book up to and including Section 11–4. Indeed, our careful treatment of the existence questions associated with Theorems 7-3.6 and 11-4.1 was designed to simplify the task for anyone who wants to check that the theorems of this book follow from the axioms for elementary set theory. In an axiomatic set theory, Convention 11-5.6 becomes a new axiom, called the axiom of choice.

11-5.10. The axiom of choice. The direct product of a family of nonvoid sets is not void.

This axiom is not always included among the axioms for set theory. It is the first of the axioms which belong to the domain of "advanced" set theory. We shall use it throughout the remainder of this book.

The axiom of choice has some bizarre consequences which for a long time persuaded many mathematicians that it would lead to a new set-theoretic paradox. From the abstract point of view this would mean that a postulate system including the axiom of choice (rewritten in terms of the abstract membership relation m, of course) along with the noncontroversial postulates would be inconsistent. However, it was proved by Gödel in 1938 that the axiom of choice is consistent with a plausible postulate system for elementary set theory. This means that any paradox obtained from using the axiom of choice can be modified to give a paradox which can be derived without using the axiom of choice. (This awkward statement is necessitated by the fact discussed in Section 2-7, that one cannot have a postulate system for set theory which is demonstrably free of paradoxes.) Furthermore, it was proved by Cohen in 1963 that the denial of the axiom of choice is also consistent with the rest of set theory.

The results of Gödel and Cohen show that it is a matter of personal taste whether a mathematician should or should not use the axiom of choice. There are many areas of pure mathematics in which some form of the axiom of choice seems essential. On the other hand, the advent of computing machinery has reawakened interest in problems of effective computation. For such problems the axiom of choice is worthless, since one invokes it precisely when he has no way to construct a suitable function.

At one time there was considerable argument among mathematicians concerning the axiom of choice. Many mathematicians took great pains to avoid its use and carefully pointed out any places where they had found it indispensable. After the appearance of Gödel's consistency proof the controversy died down. Most mathematicians in recent years have used the axiom without comment even when it could be avoided. Perhaps Cohen's proof of the complete independence of this axiom, together with a host of other new results in abstract set theory which have stemmed from his methods, heralds a new era of careful attention to the set-theoretic basis of mathematical proofs. It is too early to tell.

Among the consequences of the axiom of choice is the theorem that every set can be well-ordered (see Section 6-6). This result is actually equivalent to the axiom of choice; that is, if we add it in place of the axiom of choice to the axioms for elementary set theory, then the axiom of choice becomes a theorem.

EXERCISES

1. Show that every infinite set is the union of two disjoint infinite sets.

2. Show that every infinite set is the union of a countable family of mutually disjoint infinite sets.

In the following exercises arrange your proofs to show explicitly how the axiom of choice enters the picture.

3. Prove Proposition 11–5.5.

4. Let S be any ordered set. Prove that S satisfies the maximum condition (6–6.3) if and only if it satisfies the following condition, known as the ascending-chain condition: There exists no order-preserving injection from N to S.

5. Assume that every set can be well-ordered. Prove the axiom of choice.

6. Replace the axiom of choice by the following weaker axiom: The direct product of a countable family of nonvoid sets is not void. Prove Proposition 11–5.1.

11–6. CARDINAL NUMBERS

The concept of cardinal number, which we have already defined for finite sets, can be extended to infinite sets. In this section we shall discuss the theory informally. Aside from a few elementary propositions which we leave for the reader to prove, the proofs of the theorems are quite complicated and do not lie along the way toward the foundations of analysis.

We have seen that similarity is an equivalence relation in the class of all sets. Let us agree to call the equivalence classes of this relation *cardinal numbers* and to denote the quotient map by card. Thus card A, the cardinal of A, is the equivalence class containing A. There are two objections to be raised against this definition. First, the new definition of the function card disagrees with the one given in Section 11–2. Second, because this definition involves the class of all sets, it is very close to bringing in the logical paradoxes; indeed, with the aid of the unstated axioms for set theory which govern this book, one can prove that such a function card does not exist. One can avoid the difficulty by letting a cardinal number be a canonically chosen member of an equivalence class; for example, we might take N_k itself as the canonical member of the class of sets similar to N_k. In any event, neither objection is important for an informal discussion.

In Proposition 6–2.3 we proved that a transitive and reflexive relation in a set induces an equivalence relation in that set and a weak order relation in the corresponding quotient set. Let us apply this result to the relation $\precsim$. The induced equivalence relation is "$A \precsim B$ and $B \precsim A$," which, by a corollary of the Schröder-Bernstein theorem, is similarity. Hence the quotient set is precisely the set of all cardinal numbers. We shall denote the weak form of the induced order relation by $\leq$ and the strong form by $<$. Then we have

$$\text{card } A < \text{card } B \quad \text{if and only if} \quad A \prec B,$$
$$\text{card } A \leq \text{card } B \quad \text{if and only if} \quad A \precsim B.$$

Let us look at the structure of the ordered set of cardinals. It is clear that the finite cardinals are order-isomorphic to the set $\mathbf{N}^*$ and that all finite cardinals precede all infinite cardinals. Proposition 11–5.1 shows that there is a least infinite cardinal, namely, card $\mathbf{N}$. This cardinal is traditionally denoted $\aleph_0$. (Here $\aleph$ is the first letter, aleph, of the Hebrew alphabet.) These facts suggest that the ordering of the cardinals is linear. This suggestion is supported by the intuitive idea that given any two sets P and Q, one can pair off elements from P with elements from Q until one set is exhausted, thereby proving either $P \precsim Q$ or $Q \precsim P$. This idea can easily be formalized into a proof using the well-ordering theorem.

Moreover, it turns out that the cardinal numbers are not merely linearly ordered; they are well-ordered. There is no largest cardinal number, because if A is any set, $A < \mathfrak{P}(A)$, and therefore card $A <$ card $\mathfrak{P}(A)$. Hence for any cardinal there is always a next larger cardinal. The cardinal immediately following $\aleph_0$ is denoted $\aleph_1$, the next after that $\aleph_2$, etc. Even this infinite list does not exhaust the possibilities. For suppose we choose sets A_n, $n \in \mathbf{N}^*$, so that card $A_n = \aleph_n$, and let $B = \bigcup_n A_n$. Then $A_{n+1} \precsim B$, so

$$\aleph_n < \aleph_{n+1} = \text{card } A_{n+1} \leq \text{card } B$$

for all $n \in \mathbf{N}^*$. Thus card B is a new cardinal larger than any in the previous list. It is denoted by $\aleph_\omega$. Furthermore, there are cardinals still larger than this.

Pursuit of this line of reasoning leads to one of the most important of the logical paradoxes, known as Cantor's paradox. Suppose we consider a set $\mathfrak{B}$ whose members include at least one set having every possible cardinal number. Let $C = \bigcup_{A \in \mathfrak{B}} A$. Just as above, it follows that card $A \leq$ card C for every $A \in \mathfrak{B}$. Apparently, then, card C is the largest cardinal, contrary to the fact that card $C <$ card $\mathfrak{P}(C)$. This paradox, like Russell's, disappears in a properly formulated axiomatic set theory.

Since cardinal numbers are a generalization of the integers used in counting, they are often used grammatically in the same way as integers. Thus the statement "A is similar to $\mathbf{N}$" might be phrased, "A has $\aleph_0$ members."

Binary operations generalizing addition and multiplication of integers can be introduced into the set of all cardinals. Let $\mathbf{p}$ and $\mathbf{q}$ be cardinals and choose representatives $P \in \mathbf{p}$ and $Q \in \mathbf{q}$. We define multiplication by

$$\mathbf{p} \cdot \mathbf{q} = \text{card } (P \times Q),$$

which is independent of the choice of representatives. Addition is defined by

$$\mathbf{p} + \mathbf{q} = \text{card } (P \cup Q)$$

with the proviso that P and Q be disjoint (it is clearly possible to select disjoint representatives). The associative, commutative, and distributive laws follow easily from these definitions. For finite cardinals these operations agree with the usual arithmetic of $\mathbf{N}^*$; for infinite cardinals they are almost trivial. Unless both $\mathbf{p}$ and $\mathbf{q}$ are finite,

$$\mathbf{p} + \mathbf{q} = \max\{\mathbf{p}, \mathbf{q}\}.$$

If, furthermore, neither $\mathbf{p}$ nor $\mathbf{q}$ is zero, then

$$\mathbf{p} \cdot \mathbf{q} = \max\{\mathbf{p}, \mathbf{q}\}.$$

The proof for general cardinals is rather complicated, but special proofs can be given for the cardinals of direct interest to us. Thus 11–1.7 implies that

$$\aleph_0 \cdot \aleph_0 = \aleph_0,$$

and

$$\aleph_0 + \aleph_0 = \aleph_0$$

results from the representation of $\mathbf{N}$ as the union of the even and odd integers.

Cardinals may be used as exponents under the following definition. With the same notation as above, let $\mathbf{p}^{\mathbf{q}}$ be the cardinal of the set of all functions from Q to P. This is, of course, independent of the choice of the representatives P and Q. If Q is finite and we take $\mathbf{q} = \operatorname{card} Q$ as defined in 11–2.7, then this new use of exponents agrees with the usual one. The standard laws of exponents are valid:

$$\mathbf{p}^{\mathbf{q}+\mathbf{r}} = \mathbf{p}^{\mathbf{q}} \cdot \mathbf{p}^{\mathbf{r}}, \tag{1}$$

$$(\mathbf{p} \cdot \mathbf{q})^{\mathbf{r}} = \mathbf{p}^{\mathbf{r}} \cdot \mathbf{q}^{\mathbf{r}}, \tag{2}$$

$$(\mathbf{p}^{\mathbf{q}})^{\mathbf{r}} = \mathbf{p}^{\mathbf{q}\cdot\mathbf{r}}. \tag{3}$$

Much of the interest in cardinal exponentiation stems from the fact that $\operatorname{card} \mathfrak{P}(A) = 2^{\operatorname{card} A}$. To prove this we shall exhibit a bijection from $\mathfrak{P}(A)$ to the set of all functions from A to the two-element set $\{0, 1\}$. If X is a subset of A, we define its so-called *characteristic function* f_X by

$$f_X(a) = 1 \quad \text{if} \quad a \in X,$$
$$f_X(a) = 0 \quad \text{if} \quad a \notin X.$$

Then $X \to f_X$ is the required bijection. Thus Theorem 11–1.9 becomes $\mathbf{p} < 2^{\mathbf{p}}$ for every cardinal $\mathbf{p}$. This is, of course, a well-known inequality for finite cardinals.

Our previous remark that $\mathbf{R}$ is similar to $\mathfrak{P}(\mathbf{N})$ can now be expressed by

$$\operatorname{card} \mathbf{R} = 2^{\aleph_0}.$$

By virtue of the usual correspondence between $\mathbf{R}$ and the points of a geometric line, $2^{\aleph_0}$ is also the cardinal of the line or continuum. Hence it is frequently referred to as the *cardinal of the continuum* and denoted by $\mathbf{c}$. It happens that most of the uncountable sets which arise in analysis have cardinal $\mathbf{c}$. Thus $\mathbf{C}$, which is clearly similar to $\mathbf{R} \times \mathbf{R}$, has cardinal

$$\mathbf{c} \cdot \mathbf{c} = \mathbf{c},$$

a formula which can be derived easily from the rules for exponents without relying on the more sophisticated general theory. Similarly, the direct product of any finite or countable family of lines has cardinal $\mathbf{c}$. In particular, euclidean three-space has cardinal $\mathbf{c}$.

Unlike the simple operations, exponentiation raises extremely difficult problems of calculation which have not yet been fully clarified, although great strides have recently been made. Shortly after infinite cardinals were introduced into mathematics, the question arose as to the value of $2^{\aleph_0}$ and it was conjectured that

$$2^{\aleph_0} = \aleph_1.$$

This conjecture is known as the *hypothesis of the continuum*. The history of the hypothesis is remarkably similar to that of the axiom of choice. It has some startling consequences which at first led mathematicians to search for a disproof. Then Gödel proved that the hypothesis of the continuum is consistent with the rest of set theory. Finally Cohen showed that its denial is also consistent.

At first glance it seems impossible that the hypothesis of the continuum can be independent. For independence means that there is doubt about the cardinal number of $\mathfrak{P}(\mathbf{N})$. Since $\mathbf{N}$ is categorically determined by the postulates for a simple chain, how can this be?

Let us consider a configuration $\langle S, m \rangle$ which satisfies some postulate system for abstract set theory. For any element t in S there is a set $T = \{s \in S \mid s \, m \, t\}$ (note that we are here using the intuitive concept of set). In a sense t represents the intuitive set T. If u is an a-subset of t, that is, if

$$(\forall s \in S) \quad s \, m \, u \Rightarrow s \, m \, t,$$

then $U = \{s \in S \mid s \, m \, u\}$ is an intuitive subset of T. However if V is an intuitive subset of T, there is no reason to expect that there is a $v \in S$ such that

$$V = \{s \in S \mid s \, m \, v\}.$$

The postulates for the configuration S will provide for the existence of v whenever V can be defined by a propositional scheme built up from schemes of the form $x \, m \, y$, but there may be subsets of T which cannot be so described.

Once it is understood that the concept of subset in an abstract model of set theory does not necessarily agree with the intuitive concept of subset, it is easy to see that the cardinal of $\mathfrak{P}(\mathbf{N})$ is not necessarily determined by the postulates for set theory, for one model may recognize more subsets of $\mathbf{N}$ than does another.

Now one might argue as follows: In the realm of intuitive sets, subsets are formed in the freest possible manner; hence the cardinal number of $\mathfrak{P}(\mathbf{N})$ in intuitive set theory must be as large as it is in any model of set theory. But this is incorrect. The cardinal number of $\mathfrak{P}(\mathbf{N})$ might appear to be large in some model of set theory not because many intuitive subsets of N are recognized by the model, but because of the paucity of bijective a-functions within the model. Returning to the a-sets t and u and the corresponding intuitive sets T and U defined above, suppose that S contains no bijective a-function from t to u. Then t and u have different cardinal numbers in the abstract sense. Nevertheless, there might exist in the intuitive realm a bijection from T to U; if so, t and u represent intuitive sets having the same cardinal.

Much of the theory of cardinal numbers is independent of the axioms for elementary set theory. Most of the statements made in this discussion concerning the addition, multiplication, and comparison of cardinals are proved by using the well-ordering theorem or some other variant of the axiom of choice. Many of these statements are false in models for set theory which do not satisfy the axiom of choice. For example, it can be shown that the cardinal numbers are not linearly ordered in such a model.

EXERCISE

State the definitions of addition, multiplication, and exponentiation for cardinal numbers, giving full attention to the question of independence of the choice of representatives. Prove that addition and multiplication are associative and commutative and that multiplication distributes over addition. Prove the formulas (1), (2), and (3) for exponentiation.

CHAPTER 12

LIMITS

Analysis is distinguished as a branch of mathematics by its frequent appeals to the notion of limit. Differentiation, integration, and infinite summation are all applications of the limit concept. So far we have used this concept only twice, in showing the existence of square roots of positive numbers and in proving that any two complete ordered fields are isomorphic. Both of these proofs appealed directly to the completeness of the real numbers. We shall now develop a theory of limits for both real and complex numbers which will serve as a prototype for a more general theory in Chapter 14.

12–1. CONVERGENT SEQUENCES

12–1.1. Definition. Let X be any set. A *sequence* of members of X is a function from **N** to X. Elements in the range of a sequence are usually called *terms* of the sequence. Sometimes they are called members or elements of the sequence.

A sequence is a special kind of family. We shall apply the term *sequence* to a function with domain **N** only when the emphasis is strongly on the values taken by the function and the order in which they appear. The intuitive picture is of infinitely many elements written down one after another, possibly with repetitions. This idea is carried into the notation, for one often writes something like, "Let $u_1, u_2, u_3, \ldots$ be a sequence of $\ldots$," which is really a substitute for the formal "Let u be a function (argument written as a subscript) from **N** to $\ldots$" We shall follow the common practice of writing a dummy argument for sequences. Thus we shall write $\{x_n\}$ for the sequence informally denoted by $x_1, x_2, x_3, \ldots$ This is an abbreviation of the general notation $\{x_n \mid n \in \mathbf{N}\}$ described in 3–4.18.

The term *sequence* is also applied more generally to denote a function whose domain is any set of consecutive integers. There are, therefore, finite sequences, which have finite domains, and sequences with all of **I** as domain. From an abstract point of view, we shall be concerned in this chapter with functions from a set S to **R** or **C**, where S is an ordered set isomorphic to **N**. We shall confine ourselves to sequences with domain **N**, but, after minor changes, the results apply equally well to sequences with domains of the form $\{n \in \mathbf{I} \mid n \geq k\}$.

When we refer to the number of terms of a sequence having a certain property, we always mean the number of indices for which the corresponding term of the sequence has the property. Thus the number of terms of $\{x_n\}$ which equal 1 is the cardinal of the set $\{n \in \mathbf{N} \mid x_n = 1\}$.

In accordance with the convention for functions, a sequence $\{x_n\}$ of real numbers is *bounded above* if and only if its range is bounded above; i.e., if and only if

$$(\exists M \in \mathbf{R})(\forall n) \quad x_n \le M.$$

Similarly, it is *bounded below* if and only if its range is bounded below. A sequence $\{z_n\}$ of complex numbers is *bounded* if and only if the sequence $\{|z_n|\}$ is bounded above. A sequence $\{x_n\}$ of real numbers is also a sequence of complex numbers, of course. It is bounded if and only if it is both bounded above and bounded below.

12–1.2. Definition. A sequence $\{x_n\}$ of real numbers is said to be *increasing* if and only if

$$(\forall m, n) \quad m > n \Rightarrow x_m \ge x_n.$$

It is said to be *decreasing* if and only if

$$(\forall m, n) \quad m > n \Rightarrow x_m \le x_n.$$

A sequence is said to be *monotone* if and only if it is either increasing or decreasing. The word *strictly* is used in conjunction with any of these words to mean that the inequalities are all strong; thus "$\{x_n\}$ is strictly increasing" means

$$(\forall m, n) \quad m > n \Rightarrow x_m > x_n.$$

The word *weakly* is used to emphasize that weak inequalities are possible. Regarded as a function from the ordered set $\mathbf{N}$ to the ordered set $\mathbf{R}$, a sequence is increasing if it is weakly order-preserving and strictly increasing if it is strongly order-preserving.

12–1.3. Definition. Let ζ be a complex number and let $\{z_n\}$ be a sequence of complex numbers. The sequence $\{z_n\}$ is said to *converge to* ζ, in symbols, $z_n \to \zeta$, if and only if

$$(\forall \epsilon > 0)(\exists M \in \mathbf{R})(\forall n > M) \quad |\zeta - z_n| < \epsilon. \tag{1}$$

A sequence of complex numbers is said to *converge* if and only if there exists a complex number to which it converges; in symbols,

$$(\exists \xi \in \mathbf{C})(\forall \epsilon > 0)(\exists M \in \mathbf{R})(\forall n > M) \quad |\xi - z_n| < \epsilon. \tag{2}$$

Otherwise, it is said to *diverge*.

There is some ellipsis in indicating the domains of the variables involved in this definition. The variable indicated by the symbol ϵ is intended to have as its domain the set of positive real numbers, and the variable indicated by n is to have domain $\{n \in \mathbf{N} \mid n > M\}$, as suggested by its appearance as a subscript.

It is easy to see that (1) is equivalent to

$$(\forall \epsilon > 0)(\exists m \in \mathbf{N})(\forall n \geq m) \quad |\zeta - z_n| \leq \epsilon. \tag{3}$$

In fact, the three changes which have been made in transforming (1) into (3) can all be made independently to give six more propositions equivalent to (1).

12–1.4. EXAMPLE

Consider the sequence defined by

$$z_n = \frac{n}{n+1} + \frac{1}{n^2} i.$$

This sequence converges to 1. Since we have not studied any proofs involving so many quantifiers directly since Chapter 2, we shall prove this fact in the fashion of Section 2–5:

(a) Let δ be any positive real number.
(b) Let p be any integer such that $p > 2/\delta$.

(c) $\quad |1 - z_p| = \left| \frac{1}{p+1} - \frac{1}{p^2} i \right| \leq \left| \frac{1}{p+1} \right| + \left| \frac{i}{p^2} \right| = \frac{1}{p+1} + \frac{1}{p^2}$

$$< \frac{1}{p} + \frac{1}{p} = \frac{2}{p} < \delta,$$

(d) $(\forall n > 2/\delta) |1 - z_n| < \delta$ $\qquad\qquad$ by (b) and (c),
(e) $(\exists M \in \mathbf{R})(\forall n > M) |1 - z_n| < \delta$ $\qquad$ by (d),
(f) $(\forall \epsilon > 0)(\exists M \in \mathbf{R})(\forall n > M) |1 - z_n| < \epsilon$ $\qquad$ by (a) through (e).

This last proposition is by definition the statement that $z_n \to 1$. Step (c) is really a long combination of steps which can be justified without difficulty by propositions appearing in Chapters 8 and 10.

In this proof we have used letters δ and p as the temporary names of real numbers and different letters ϵ and n to denote the variables in the quantified propositions (d), (e), and (f). As we remarked in Chapter 2, this is not usually done. Most authors would write only steps (a), (b), and (c) and omit the introduction of formal quantifiers. If the quantifiers were actually written, however, the variables appearing would undoubtedly be denoted by the same letters as the numbers appearing in the calculation (c). This practice helps to elucidate the dependence of the steps in a proof on the temporary assumptions. This dependence is not usually shown by indentation, since proofs are customarily written in a continuous text.

The choice of letters in propositions of this kind is subject to some well-established conventions, which are extremely useful, although they have no logical status. The letters δ and ϵ are usually reserved for positive real numbers which are presumed to be small. When they denote variables, it is presumed that they will be replaced by the name of a small number. The use of the letter ϵ in the definition of convergence indicates that interest is attached to the possibility that it may be replaced by (the name of) a small number. Similarly, the use of the capital letter

M to denote the next variable suggests that the choice will often have to be a large positive number. When the quantification is expressed in words, these conventions are often emphasized by parenthetical phrases. Thus the definition (1) of convergence might begin, "For every positive ϵ, no matter how small, there exists a number M so large that . . ."

From a formal point of view it makes absolutely no difference what letters we use to denote the variables in (1), provided only that they are different. However, if we were to write

$$(\forall M > 0)(\exists \epsilon \in \mathbf{R})(\forall n > \epsilon) \quad |\zeta - z_n| < M$$

in place of (1), most mathematicians would be temporarily confused, because the code has been violated.

Proving a statement involving a string of quantifiers may be likened to a chess problem. We interpret the universal quantifier as indicating our opponent's move (our opponent is the malevolent genie mentioned on p. 23) and the existential quantifier as indicating our move.

Suppose we wish to prove that the sequence $\{z_n\}$ converges. We first express the desired conclusion in symbols:

$$(\exists \xi)(\forall \epsilon > 0)(\exists M \in \mathbf{R})(\forall n > M) \quad |\xi - z_n| < \epsilon. \tag{4}$$

Since this begins with an existential quantifier, we must move first by choosing a number ξ. Since the next quantifier is universal, our opponent moves next by choosing a positive number ϵ. He will presumably make the best move he can, which is to say he will choose ϵ so that

$$(\exists M \in \mathbf{R})(\forall n > M) \quad |\xi - z_n| < \epsilon$$

is false, if possible. Now it is our move and we must choose a number M. We make our move with a knowledge of the previous moves; that is, we may let M depend on the numbers ξ and ϵ. Finally, our opponent chooses an integer n exceeding M and the burden is on us to prove the inequality $|\xi - z_n| < \epsilon$.

The directions for the usual mate-in-two chess problem can be written somewhat imprecisely as

$$(\exists \text{ white move})(\forall \text{ black moves})(\exists \text{ white move}) \quad \text{black is checkmated.}$$

If we expand the phrase "black is checkmated," we get

$$(\exists \text{ white move})(\forall \text{ black moves})(\exists \text{ white move})(\forall \text{ black moves})$$
$$\text{white can capture black's king,}$$

and the analogy with (4) is now clear. The solution to such a problem consists of an explicit choice of white's first move and a list of white's answers to each of the possible black moves. It is usually left to the reader to verify that black is checkmated at the end of each of these sequences of moves, which means that the reader must examine each of the possible second black moves to find out whether black's king can be captured.

Because only a finite number of moves are possible in any chess position, the complete solution to a chess problem can be exhibited by listing all the possibilities. In the convergence proof the quantified variables have infinite domains, so that the proof cannot be given by listing all the possibilities. Sometimes, as in Example 12–1.4, we shall describe by explicit formulas how our moves depend on the preceding moves, but equally often we will invoke some existentially quantified proposition already known or assumed.

An existentially quantified proposition is a rather subtle thing, in fact, much too subtle for a full discussion here. A brief example will show what the difficulties are. Convergent sequences are often used practically to compute numbers like π, e, or log 2. Suppose we know that a certain sequence $\{x_n\}$ of easily computed numbers converges to π. If we want to compute π to five decimals, we might specify ϵ in the definition of convergence to be 10^{-6} to obtain the existentially quantified proposition

$$(\exists M \in \mathbf{R})(\forall n > M) \quad |\pi - x_n| < 10^{-6}.$$

Now, unless we know or can find out what value of M is needed, the above proposition is no help at all in evaluating π. On the other hand, if, say, we know that the choice $M = 8$ will do, then we can get a good estimate of π by calculating x_9. Assuming that x_9 turns out to be between 3.141593 and 3.141594, we can conclude that $3.141592 < \pi < 3.141595$ and thus obtain the first five decimals of the expansion of π. The practical mathematician must distinguish between an existentially quantified proposition which, like 12–1.4(e) above, merely fails to record the actual value that makes the proposition true and one which is established by methods which cannot be reduced to actual computation. As we remarked in Chapter 2, the abstract mathematician usually ignores this distinction.

The definition of convergence has a nice geometrical interpretation. Let us identify the complex number $z = x + iy$ $(x, y \in \mathbf{R})$ with the point $\langle x, y \rangle$ of the Cartesian plane of analytic geometry. Then $|z|$ is the distance of the point z from the origin and $|w - z|$ is the distance from the point z to the point w. The complex numbers $\zeta, z_1, z_2, \ldots$ in the definition of convergence are now points of the plane. We shall see (Fig. 12–1) that $z_n \to \zeta$ if and only if

(i) every circle with center ζ encloses all but a finite number of the points z_n.

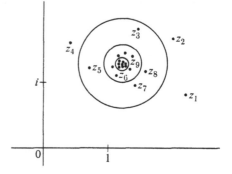

FIG. 12–1. Convergent sequence in the complex plane. Every circle centered at the limit point encloses all but a finite number of the terms of the sequence.

To prove this, assume first that (i) holds. Let $\epsilon > 0$ be given. The circle of radius ϵ about ζ encloses all but a finite number of the points z_n. This means that $|\zeta - z_n| < \epsilon$ except for a finite set F of indices n. If F is empty, let $M = 0$; otherwise, let M be the largest member of F. Then

$$(\forall n > M) \quad |\zeta - z_n| < \epsilon$$

and we have proved that $z_n \to \zeta$.

Next, assume that $z_n \to \zeta$; that is, assume (1). Given any circle C with center ζ, let its radius be ρ. Specialize (1) to get

$$(\exists M \in \mathbf{R})(\forall n > M) \quad |\zeta - z_n| < \rho,$$

and choose a number K such that

$$(\forall n > K) \quad |\zeta - z_n| < \rho.$$

Then z_n lies within the circle C, except possibly for some of the indices in the set $\{n \mid 1 \leq n \leq K\}$; that is, except for a finite set of indices.

We have just seen that the double quantifier $(\exists M \in \mathbf{R})(\forall n > M)$ may appropriately be rendered, "for all but a finite number of integers n." Another suitable translation is "for all sufficiently large n." These are helpfully suggestive phrases for thinking about convergence. Thus the criterion (1) for the convergence of a sequence may be expressed as:

For all positive ϵ, for all sufficiently large n, $|\zeta - z_n| < \epsilon$.

When two or more quantifiers are written out, there is always some danger of misunderstanding the order in which they are to be applied. In the last example we must remember that the meaning of the words *sufficiently large* may (and most probably will) depend on ϵ, just as the choice of M in the more formal expression (1) may depend on ϵ.

The definition of $z_n \to \zeta$ means that the far-out terms of the sequence $\{z_n\}$ are better and better approximations to ζ. But there is no requirement that each new term be a better approximation than its predecessor. If we set $x_n = 1/n^2$ for even n and $x_n = 1/n$ for odd n, the sequence $\{x_n\}$ converges to 0, but the even terms are much better approximations to 0 than the adjacent odd ones. The only requirement is that *eventually* the terms become arbitrarily good approximations to 0.

There is no objection to the possibility that one or more of the terms of a sequence should be exactly ζ. Indeed, if we define $z_n = 1$ for all n, then $\{z_n\}$ is a perfectly good example of a sequence which converges to 1.

12-1.5. Proposition. If a sequence of complex numbers converges, then there is only one number to which it converges.

Proof. Suppose $\{z_n\}$ is a sequence of complex numbers which converges to both ζ and η. We must prove that $\zeta = \eta$. We proceed indirectly.

Suppose $\zeta \neq \eta$. Then $\delta = \frac{1}{3}|\zeta - \eta|$ is a positive number. By the definition of $z_n \to \zeta$, we can choose a number M_1 such that

$$(\forall n > M_1) \quad |\zeta - z_n| < \delta.$$

By the definition of $z_n \to \eta$, we can choose a number M_2 such that

$$(\forall n > M_2) \quad |\eta - z_n| < \delta.$$

Now pick an integer k greater than both M_1 and M_2 (this is certainly possible). Then $|\zeta - z_k| < \delta$ and $|\eta - z_k| < \delta$. Therefore,

$$|\zeta - \eta| = |(\zeta - z_k) - (\eta - z_k)| \le |\zeta - z_k| + |\eta - z_k| < \delta + \delta = \frac{2}{3}|\zeta - \eta|,$$

which is false. Hence $\zeta \neq \eta$ is impossible and the proposition is proved.

This being established, the following definition becomes sensible.

12-1.6. Definition. If $\{z_n\}$ is a convergent sequence of complex numbers, then the unique number ζ to which it converges is called the *limit* of the sequence $\{z_n\}$. This number is often denoted by

$$\lim_{n \to \infty} z_n \qquad \text{or} \qquad \lim_{n \to \infty} z_n.$$

For formal purposes the symbols $\to$ and ∞ should be regarded as parts of an indivisible symbol $\lim_{\to \infty}$. The n which appears is a dummy index, since $\lim_{p \to \infty} z_p$ means exactly the same thing as $\lim_{n \to \infty} z_n$.

The expression $\lim_{n \to \infty} z_n$ is usually read "the limit of z_n as n tends to (or approaches) infinity." Both the notation and the terminology are holdovers from an interpretation of sequences, and of functions in general, different from what is current today. The expression z_n was once conceived as denoting a variable number; that is, a number that changes as n changes. In this interpretation $\lim_{n \to \infty} z_n$ referred to the behavior of z_n as n "approaches infinity," that is, gets larger and larger. The interpretation of a function as a variable number is suggestive and therefore useful, but it is hard to interpret the more sophisticated set-theoretic constructs in a manner consistent with this idea, so it has been generally abandoned.

The abbreviated notations $\lim_n z_n$ and $\lim z_n$ are also standard.

The expression $\lim z_n$ is frequently used with a slightly different meaning. For example, "$\lim z_n$ does not exist" means "the sequence $\{z_n\}$ diverges," and "does $\lim z_n$ exist?" means "does the sequence $\{z_n\}$ converge?" There is a convention, however, that the use of the expression $\lim z_n$ without context to the contrary includes the assertion that the sequence $\{z_n\}$ converges.

12-1.7. Proposition. A convergent sequence of complex numbers is bounded.

Proof. Suppose $z_n \to \zeta$. In the definition of convergence, set $\epsilon = 1$ and pick a number M such that

$$(\forall n > M) \quad |\zeta - z_n| < 1.$$

If $k > M$, we have

$$|z_k| \le |\zeta| + |\zeta - z_k| < |\zeta| + 1.$$

Therefore, a bound for the sequence $\{|z_n|\}$ is given by the largest member of the finite set

$$\{|\zeta| + 1\} \cup \{|z_n| \mid 1 \le n \le M\}.$$

It must be admitted that, strictly speaking, the instruction "set $\epsilon = 1$" in the last paragraph is not sensible, for ϵ in (1) is a variable and 1 is a number. Nevertheless, the intent is clear. Such locutions are commonly used in mathematics.

From the discussion of convergence following the definition it should be clear that the convergence of a sequence depends only on its "infinite part" and is unaffected by modifying, adjoining, or omitting any finite number of terms. We shall state a proposition to this effect but leave the proof to the reader.

12–1.8. Proposition. Suppose that $\{w_n\}$ and $\{z_n\}$ are sequences of complex numbers. Let k be an integer. Suppose that $w_n = z_{n+k}$ for all but a finite number of indices n. If either $\lim w_n$ or $\lim z_n$ exists, then so does the other limit and $\lim w_n = \lim z_n$.

EXERCISES

1. Prove that a constant sequence of complex numbers is convergent.

2. Demonstrate the equivalence of the criteria (1) and (3) for convergence; that is, ζ and $\{z_n\}$ satisfy (1) if and only if they satisfy (3).

3. Suppose that (1) were replaced by

$$(\forall \epsilon \ge 0)(\exists M \in \mathbf{R})(\forall n > M) \;\; |\zeta - z_n| \le \epsilon.$$

Describe what $z_n \to \zeta$ would mean under this definition.

4. Prove that if $\{x_n\}$ is a convergent sequence of real numbers, then $\lim x_n$ is real. Rewrite the definition of convergence for real sequences without using absolute-value signs.

5. Examine the real sequences defined by the following formulas. Which ones are monotone? bounded? convergent? For those that converge, give explicit formulas for choosing the M corresponding to a given ϵ in the notation of (1). For those that do not converge, prove this fact in full.

(a) $x_n = 1 - (-\tfrac{1}{2})^n$ (b) $x_n = (-1)^n + 1/n$

(c) $x_n = n^2/(n + 1)$ (d) $x_n = 1/n - (-\tfrac{1}{2})^{n+2}$

(e) Define $\{x_n\}$ recursively by $x_0 = 0$, $x_1 = 1$, and

$$x_{n+1} = \tfrac{2}{3}x_n + \tfrac{1}{3}x_{n-1} \qquad \text{for} \qquad n > 1.$$

6. Which of the following sequences of complex numbers are convergent?

(a) $z_n = \dfrac{n^2 + in + 1}{n + i + in^2}$ (b) $z_n = \dfrac{n^3 + 3i}{n^2 + i^n}$

(c) $z_n = (-\tfrac{1}{2} + \tfrac{1}{2}\sqrt{3}\,i)^n$ (d) $z_n = (-\tfrac{1}{2} + \tfrac{1}{2}\sqrt{2}\,i)^n$

7. For what complex values of ρ is the sequence $\{\rho^n\}$ bounded? convergent?

12–2. LIMITS AND ARITHMETIC

Suppose $\{w_n\}$ and $\{z_n\}$ are convergent sequences of complex numbers. We can form a new sequence by adding corresponding terms. This new sequence will converge to $\lim w_n + \lim z_n$. In this section we shall prove this and a number of similar facts relating limits to the arithmetic operations.

12–2.1. Proposition. Suppose $\{w_n\}$ and $\{z_n\}$ are convergent sequences of complex numbers. Then the sequences $\{w_n + z_n\}$, $\{w_n - z_n\}$, and $\{w_n z_n\}$ are all convergent and

(a) $\lim (w_n + z_n) = \lim w_n + \lim z_n$,
(b) $\lim (w_n - z_n) = \lim w_n - \lim z_n$,
(c) $\lim w_n z_n = (\lim w_n)(\lim z_n)$.

Moreover, if $\lim z_n \neq 0$, then $\{w_n/z_n\}$ is convergent and
(d) $\lim w_n/z_n = (\lim w_n)/(\lim z_n)$.

(This last statement requires some special interpretation. If $z_n = 0$ for even one index n, the sequence $\{w_n/z_n\}$ is not defined. The hypothesis $\lim z_n \neq 0$ implies that there are at most a finite number of indices n for which $z_n = 0$ and therefore w_n/z_n is undefined. Since the convergence of a sequence is unaffected by any finite number of terms, it is customary to accept $\{w_n/z_n\}$ as a bona fide sequence in any context in which only convergence is at stake.)

Proof. Let $\eta = \lim w_n$ and $\varsigma = \lim z_n$. Then

$$(\forall \epsilon > 0)(\exists M \in \mathbf{R})(\forall n > M) \qquad |\eta - w_n| < \epsilon, \tag{1}$$

$$(\forall \epsilon > 0)(\exists M \in \mathbf{R})(\forall n > M) \qquad |\varsigma - z_n| < \epsilon. \tag{2}$$

Let $\delta > 0$ be given. By (1) we can choose a real number M_1 such that

$$(\forall n > M_1) \quad |\eta - w_n| < \delta/2. \tag{3}$$

By (2) we can choose a real number M_2 such that

$$(\forall n > M_2) \quad |\varsigma - z_n| < \delta/2. \tag{4}$$

Let P be the larger of M_1 and M_2. Let k be any integer greater than P. By (3) and (4) we have

$$|\eta - w_k| < \delta/2 \qquad \text{and} \qquad |\varsigma - z_k| < \delta/2.$$

Therefore,

$$|(\eta + \varsigma) - (w_k + z_k)| = |(\eta - w_k) + (\varsigma - z_k)| \leq |\eta - w_k| + |\varsigma - z_k| < \delta. \tag{5}$$

Hence

$$(\forall n > P) \quad |(\eta + \varsigma) - (w_n + z_n)| < \delta.$$

Thus

$$(\exists M \in \mathbf{R})(\forall n > M) \quad |(\eta + \varsigma) - (w_n + z_n)| < \delta.$$

Finally,

$$(\forall \epsilon > 0)(\exists M \in \mathbf{R})(\forall n > M) \quad |(\eta + \zeta) - (w_n + z_n)| < \epsilon;$$

that is,

$$\lim (w_n + z_n) = \eta + \zeta = \lim w_n + \lim z_n.$$

This proves (a).

Before going on to the other parts of the proposition, a number of comments on this proof are in order. The last three steps, which serve to introduce the quantifiers formally, are usually omitted. The essential steps are the calculation (5) and the specialization of (1) and (2) to obtain (3) and (4) by setting $\epsilon = \delta/2$.

Why can we set $\epsilon = \delta/2$? Simply because $\delta/2$ is a positive number. Why do we set $\epsilon = \delta/2$? This is the interesting question. Most of the skill in finding mathematical proofs consists of choosing appropriate specializations of universally quantified propositions. No great skill was required in this case. We simply looked ahead to see how the calculation (5) would work out. Once we had found the inequality

$$|(\eta + \zeta) - (w_k + z_k)| \leq |\eta - w_k| + |\zeta - z_k|,$$

it was obvious that we had only to make both terms on the right smaller than $\delta/2$. Other choices would do equally well. We might make both terms less than $\delta/3$, or the first term less than $2\,\delta/3$ and the second less than $\delta/3$. These other choices would presumably lead to different choices of M_1 and M_2 and therefore to different values of P. Fortunately only the existence of an appropriate P matters. We are under no obligation to find the least value of P.

The proof of (b) is almost identical with the proof of (a). The calculation (5) must be replaced by

$$|(\eta - \zeta) - (w_k - z_k)| = |(\eta - w_k) - (\zeta - z_k)| \leq |\eta - w_k| + |\zeta - z_k| < \delta,$$

and some signs must be changed in the subsequent steps.

Instead of going ahead with a formally organized proof of (c), we shall show how such a proof might actually be discovered.

The desired conclusion is

$$(\forall \epsilon > 0)(\exists M \in \mathbf{R})(\forall n > M) \quad |\eta\zeta - w_n z_n| < \epsilon.$$

Hence the problem is to estimate $|\eta\zeta - w_k z_k|$ for large values of k. The first step is to express this number in terms of the numbers $|\zeta - z_k|$ and $|\eta - w_k|$. Writing $w_k = \eta - (\eta - w_k)$ and $z_k = \zeta - (\zeta - z_k)$, we find that

$$\eta\zeta - w_k z_k = \zeta(\eta - w_k) + \eta(\zeta - z_k) - (\eta - w_k)(\zeta - z_k).$$

Therefore

$$|\eta\zeta - w_k z_k| \leq |\zeta|\,|\eta - w_k| + |\eta|\,|\zeta - z_k| + |\eta - w_k|\,|\zeta - z_k|. \tag{6}$$

To make the latter sum less than δ, let us try to make each of its terms less than $\delta/3$. This suggests that we try to make

$$|\eta - w_k| < \delta/3|\zeta|, \tag{7}$$

$$|\zeta - z_k| < \delta/3|\eta|, \tag{8}$$

$$|\eta - w_k||\zeta - z_k| < \delta/3. \tag{9}$$

Since large enough k's will make both $|\eta - w_k|$ and $|\zeta - z_k|$ as small as we please, we are essentially through. There are some details which require further attention, however.

It could be that $\zeta = 0$, in which case (7) is not sensible. This would not really matter, because if $\zeta = 0$, the first term on the right of (6) would surely be 0. It is easier, however, to avoid discussing this special case, so we shall demand

$$|\eta - w_k| < \frac{\delta}{3(1 + |\zeta|)}, \tag{10}$$

and similarly,

$$|\zeta - z_k| < \frac{\delta}{3(1 + |\eta|)}. \tag{11}$$

If δ were a large number (which is not likely, to be sure, but we want to avoid any special discussion of the size of δ in our proof), then (10) and (11) would not imply (9). We can take care of this by imposing another condition on $|\eta - w_k|$, namely, $|\eta - w_k| < 1$. Then (6) becomes

$$|\eta\zeta - w_k z_k| \leq |\zeta||\eta - w_k| + (|\eta| + |\eta - w_k|)|\zeta - z_k|$$

$$< |\zeta|\frac{\delta}{3(1 + |\zeta|)} + (|\eta| + 1)\frac{\delta}{3(1 + |\eta|)} < \tfrac{2}{3}\delta.$$

Now it appears that this calculation would be tidier if we had 2's instead of 3's in (10) and (11).

We shall now give the proof in formal style. In this final form the essential steps appear in an order almost completely reversed from the order in which we discovered them. It is an unfortunate fact that mathematical proofs, as usually written, often obscure the route taken by their discoverers in finding them.

Let δ be any positive number. Since $w_n \to \eta$, we can choose M_1 so that

$$|\eta - w_k| < \frac{\delta}{2(1 + |\zeta|)} \quad \text{and} \quad |\eta - w_k| < 1$$

are both true for all $k > M_1$. (We could, of course, write this as one inequality: $|\eta - w_k| < \inf\{1, \delta/2(1 + |\zeta|)\}$.) Since $z_n \to \zeta$, we can choose M_2 so that

$$|\zeta - z_k| < \frac{\delta}{2(1 + |\eta|)}$$

for all $k > M_2$. Let P be the larger of M_1 and M_2. If $k > P$, then

$$|\eta\zeta - w_k z_k| \leq |\zeta| |\eta - w_k| + (|\eta| + |\eta - w_k|)|\zeta - z_k|$$

$$< |\zeta| \frac{\delta}{2(1 + |\zeta|)} + (|\eta| + 1) \frac{\delta}{2(1 + |\eta|)} < \delta.$$

Therefore,

$$(\forall n > P) \quad |\eta\zeta - w_n z_n| < \delta.$$

Hence

$$(\exists M \in \mathbf{R})(\forall n > M) \quad |\eta\zeta - w_n z_n| < \delta,$$

and finally,

$$(\forall \epsilon > 0)(\exists M \in \mathbf{R})(\forall n > M) \quad |\eta\zeta - w_n z_n| < \epsilon.$$

This is conclusion (c) of the proposition.

To prove (d) we begin with the special case in which $w_n = 1$ for all n. This means that we must prove $\lim 1/z_n = 1/\zeta$.

Let $\delta > 0$ be given. Both $|\zeta|/2$ and $|\zeta|^2 \delta/2$ are positive, so we can choose M so that

$$(\forall n > M) \quad |\zeta - z_n| < \inf \{|\zeta|/2, |\zeta|^2 \delta/2\}.$$

Now, if $k > M$, then

$$|\zeta| \leq |\zeta - z_k| + |z_k| < \frac{|\zeta|}{2} + |z_k|,$$

and therefore $|z_k| > |\zeta|/2$. Furthermore,

$$\left| \frac{1}{\zeta} - \frac{1}{z_k} \right| = \left| \frac{z_k - \zeta}{\zeta z_k} \right| = \frac{|z_k - \zeta|}{|\zeta| |z_k|} < \frac{|\zeta|^2 \delta/2}{|\zeta| |\zeta|/2} = \delta.$$

We shall omit the formal introduction of the quantifiers. This finishes the proof for the special case. Note that any terms of the sequence $\{1/z_n\}$ which are undefined because $z_n = 0$ do not figure at all in the proof.

The general case of (d) follows by combining the special case with (c).

Taking $\{w_n\}$ to be a constant sequence in (c) leads to a special case which is so important that we state it formally.

12-2.2. Corollary. If λ is a complex number and $\{z_n\}$ is a convergent sequence of complex numbers, then $\{\lambda z_n\}$ is convergent and

$$\lim \lambda z_n = \lambda \lim z_n.$$

12-2.3. Proposition. Suppose $\{w_n\}$ and $\{z_n\}$ are sequences of complex numbers. If $\{w_n\}$ is bounded and $z_n \to 0$, then $w_n z_n \to 0$.

We shall omit the proof, which the reader should be able to supply without difficulty.

12–2.4. Proposition. A sequence $\{z_n\}$ of complex numbers converges if and only if both the sequences $\{\text{Re } z_n\}$ and $\{\text{Im } z_n\}$ converge. If $\{z_n\}$ converges, then $\{\bar{z}_n\}$ and $\{|z_n|\}$ converge and we have

(a) $\lim \text{Re } z_n = \text{Re}(\lim z_n)$,   (b) $\lim \text{Im } z_n = \text{Im}(\lim z_n)$,

(c) $\lim \bar{z}_n = \overline{\lim z_n}$,   (d) $\lim |z_n| = |\lim z_n|$.

Proof. Suppose, to begin with, that $\{\text{Re } z_n\}$ and $\{\text{Im } z_n\}$ converge. Then, by 12–2.2 and 12–2.1, $\{z_n\} = \{\text{Re } z_n + i \text{ Im } z_n\}$ converges.

Conversely, suppose that $\{z_n\}$ converges; say, $\zeta = \lim z_n$. Let $\delta > 0$ be given. Choose M so that

$$(\forall n > M) \quad |\zeta - z_n| < \delta.$$

If $k > M$, then

$$|\text{Re } \zeta - \text{Re } z_k| = |\text{Re}(\zeta - z_k)| \leq |\zeta - z_k| < \delta,$$

by 10–3.7(e). This proves (a). Formulas (b), (c), and (d) follow in the same way from 10–3.7(e), (b), and (i).

12–2.5. Corollary. If a sequence of real numbers converges, its limit is real.

12–2.6. Proposition. Let $\{x_n\}$ and $\{y_n\}$ be convergent sequences of real numbers. If $x_n \leq y_n$ for infinitely many indices n, then $\lim x_n \leq \lim y_n$.

Note, however, that the similar statement for strong inequalities is false. Even from $(\forall n) \, x_n < y_n$ we cannot conclude that $\lim x_n < \lim y_n$.

Proof. Let $\lim x_n = \xi$ and $\lim y_n = \eta$. We must prove that $\xi \leq \eta$. Suppose, on the contrary, that $\xi > \eta$. Then $\delta = (\xi - \eta)/2$ is positive and we can choose M so that

$$(\forall n > M) \qquad |\xi - x_n| < \delta \quad \text{and} \quad |\eta - y_n| < \delta.$$

The hypothesis implies that we can choose an integer $k > M$ such that $x_k \leq y_k$. Then we have

$$\xi \leq x_k + |\xi - x_k| < y_k + \delta \leq \eta + |\eta - y_k| + \delta < \eta + 2\delta = \xi,$$

which is false. Therefore $\xi > \eta$ is impossible.

From the results of this section we can derive similar results for arithmetic combinations of any finite number of sequences. For example, if $x_n \rightarrow \xi$, $y_n \rightarrow \eta$, and $z_n \rightarrow \zeta$, then

$$\lim \frac{x_n^3 - y_n \bar{z}_n}{|x_n| + \text{Re } y_n + \text{Im } z_n} = \frac{\xi^3 - \eta \bar{\zeta}}{|\xi| + \text{Re } \eta + \text{Im } \zeta},$$

provided $|\xi| + \text{Re } \eta + \text{Im } \zeta \neq 0$. At the moment we lack the vocabulary to state a formal theorem which covers all such possibilities. But each particular case can be established by reference to the theorems of this section.

EXERCISES

1. Prove that if $\{x_n\}$ is a convergent sequence of nonnegative real numbers, then $\{\sqrt{x_n}\}$ is convergent and $\lim \sqrt{x_n} = \sqrt{\lim x_n}$.

2. Prove that the following limits exist:

$$\lim_{n \to \infty} \frac{n^2 + 2n - 1}{3n^2 - 5n + 3} \qquad\qquad \lim_{n \to \infty} \frac{n^2 - 6n + 7}{n^3 + 2n - 5}$$

$$\lim_{n \to \infty} \sqrt{n + 1} - \sqrt{n} \qquad\qquad \lim_{n \to \infty} n(\sqrt{n^2 + 1} - n)$$

3. Suppose that $\{x_n\}$ is a convergent sequence of complex numbers. Prove that

$$\lim_{n \to \infty} \frac{n^2 x_n + 4n - 1}{n^2 x_n - 5n + 2}$$

exists, provided $\lim x_n \neq 0$. Show by examples that if $\lim x_n = 0$, this limit may not exist, or it may exist and take any prescribed value.

4. Suppose $\{x_n\}$ is a sequence of real numbers such that for all n, $x_n^3 + \frac{1}{2}x_n = 1/n$. Prove that $\{x_n\}$ converges.

12–3. INFINITY AND THE EXTENDED REAL NUMBER SYSTEM

If $A \subseteq \mathbf{R}$, then under the definition of supremum (6–4.4) sup A may or may not exist. It is convenient to supply a supremum, denoted by $+\infty$, to subsets of $\mathbf{R}$ which are unbounded above. It is similarly desirable to have an infimum, $-\infty$, for subsets of $\mathbf{R}$ which are unbounded below. This brief section is devoted to introducing the quasi-numbers $+\infty$ and $-\infty$.

12–3.1. Definition. Let $\mathbf{R}^*$ be a set consisting of $\mathbf{R}$ and two new elements, $+\infty$ and $-\infty$; that is, $\mathbf{R}^* = \mathbf{R} \cup \{+\infty, -\infty\}$. Extend the order relation $<$ from $\mathbf{R}$ to $\mathbf{R}^*$ by defining $-\infty < +\infty$ and

$$(\forall r \in \mathbf{R}) \quad -\infty < r < +\infty.$$

So extended, $<$ is a strong linear order relation in $\mathbf{R}^*$. Denote the corresponding weak order relation by $\leq$.

Elements of $\mathbf{R}$ are often referred to as *finite* members of $\mathbf{R}^*$, while $+\infty$ and $-\infty$ are said to be *infinite*. The infinite members of $\mathbf{R}^*$ are not usually referred to as numbers, the latter term being reserved for members of $\mathbf{R}$ itself.

Although the symbols $+\infty$ and $-\infty$ are read "plus infinity" and "minus infinity" respectively, they should be regarded as indivisible symbols, because we do not wish to suggest any connection with the operations of addition and subtraction. Indeed, we shall make no effort to extend the field operations of $\mathbf{R}$ over $\mathbf{R}^*$. While partial extensions with reasonable properties are possible, we shall have no need of them.

The plain symbol ∞ is often used as an abbreviation for $+\infty$, as in the notation for limits. On the other hand, if $\{x_n\}$ is a sequence defined for all integers (positive and negative), then the usual notation might not be sufficiently clear. One would write $\lim_{n \to -\infty} x_n$ for the limit of the sequence $\{x_{-k} \mid k \in \mathbf{N}\}$ and $\lim_{n \to +\infty} x_n$ for the limit of $\{x_n \mid n \in \mathbf{N}\}$. These notations are self-explanatory if we conceive of n as moving along the domain of the sequence (in this case $\mathbf{I}$) toward $-\infty$ or $+\infty$. While this conception is inappropriate for modern mathematics, we cannot afford to forget it completely, since it is the source of many standard notations.

The set $\mathbf{R}^*$ is called the *extended real number system*. It is endowed with an order structure and partially defined binary operations. To follow the postulational path, we should define *an* extended real number system as a configuration $\langle \mathbf{R}^*, +, \cdot, <, \le \rangle$ satisfying the following conditions.

(i) $\langle \mathbf{R}^*, <, \le \rangle$ is a complete linearly ordered set with a largest element and a smallest element.

(ii) If $\mathbf{R} = \mathbf{R}^* - \{\sup \mathbf{R}^*, \inf \mathbf{R}^*\}$ and $<_{\mathbf{R}}$ and $\le_{\mathbf{R}}$ are the restrictions to $\mathbf{R}$ of $<$ and $\le$, then $\langle \mathbf{R}, +, \cdot, <_{\mathbf{R}}, \le_{\mathbf{R}} \rangle$ is a complete ordered field.

It follows easily from the corresponding facts for complete ordered fields that an extended real number system exists and is categorically determined. This being so, we select a particular extended real number system $\mathbf{R}^*$. For convenience we arrange that the complete ordered field $\mathbf{R}^* - \{\sup \mathbf{R}^*, \inf \mathbf{R}^*\}$ is $\mathbf{R}$ itself.

If $A \subseteq \mathbf{R}$, then $A \subset \mathbf{R}^*$. Consequently we can (attempt to) calculate $\sup A$ in the ordered set $\mathbf{R}$ or in the ordered set $\mathbf{R}^*$. We shall temporarily distinguish these possibilities by writing $\sup A$ and $\sup^* A$, respectively.

12–3.2. Proposition. Suppose $A \subseteq \mathbf{R}$. Then $\sup^* A$ exists. Furthermore, $\sup A$ exists if and only if $\sup^* A \in \mathbf{R}$, and, if so, $\sup A = \sup^* A$.

Proof. Suppose first that $\sup A$ exists; say, $\sup A = \lambda$. It follows immediately from the definitions that $\sup^* A$ exists and, in fact, $\sup^* A = \lambda$. Hence $\sup^* A \in \mathbf{R}$.

Now suppose that $\sup A$ does not exist. Then either $A = \emptyset$ or A is not bounded above in $\mathbf{R}$. In the first case, $\sup^* A = -\infty$; in the second, $\sup^* A = +\infty$. In either case, $\sup^* A$ exists, but $\sup^* A \notin \mathbf{R}$.

As a consequence of this proposition, we shall no longer distinguish the supremum of a subset of $\mathbf{R}$ calculated in $\mathbf{R}$ from that calculated in $\mathbf{R}^*$. From now on when we write $\sup A$, where $A \subseteq \mathbf{R}$, the calculation is to be made in $\mathbf{R}^*$, so we admit the possibilities $\sup A = +\infty$ and $\sup A = -\infty$. When $\sup A \in \mathbf{R}$, the result is the same as if we had made the calculation in $\mathbf{R}$.

An analogous proposition holds for infima, so we shall always allow infima to be calculated in $\mathbf{R}^*$. Hence, if $A \subseteq \mathbf{R}$, $\inf A$ always exists. But $\inf A = +\infty$ if A is empty, and $\inf A = -\infty$ if A is unbounded below.

Thus we have provided a supremum and an infimum for every subset of $\mathbf{R}$. This is convenient in analysis, but we have only transferred to a slightly more manageable place the difficulties associated with an empty or unbounded set. We must constantly bear in mind that $\sup A$ and $\inf A$ need not be finite.

12–3.3. Proposition. Every subset of $\mathbf{R}^*$ has both a supremum and an infimum.

Proof. Suppose $B \subseteq \mathbf{R}^*$. If $+\infty \in B$, then $\sup B = +\infty$. If $+\infty \notin B$, it is easy to see that $\sup B = \sup (B \cap \mathbf{R})$ and the latter exists by 12–3.2.

A similar argument is valid for infima, or we may note, as in the discussion following 6–5.2, that the existence of a supremum for every subset of an ordered set implies the existence of an infimum for every subset as well.

EXERCISE

Show that, as an ordered set, $\mathbf{R}^*$ is isomorphic to $\{x \in \mathbf{R} \mid 0 \leq x \leq 1\}$.

12–4. SUPERIOR AND INFERIOR LIMITS

There may be limit numbers associated with divergent sequences. We shall not take up the general theory of these limit numbers until Chapter 14, but here we shall study two of them, the superior limit and the inferior limit, which pertain only to sequences of real numbers. We shall allow them to take values in $\mathbf{R}^*$ and thus ensure that they exist for every sequence.

12–4.1. Definition. Let $\{x_n\}$ be a sequence of real numbers. The *superior limit* of this sequence, denoted by

$$\limsup_{n \to \infty} x_n,$$

is $\inf_k \sup \{x_n \mid n \geq k\}$. The *inferior limit*, denoted by

$$\liminf_{n \to \infty} x_n,$$

is $\sup_k \inf \{x_n \mid n \geq k\}$.

The indicated bounds are to be calculated in $\mathbf{R}^*$; hence there is no question of existence. Abbreviated notations such as $\limsup_n$ or $\liminf$ are commonly used. The classical notations were $\overline{\lim}$ for $\limsup$ and $\underline{\lim}$ for $\liminf$; they are still used fairly often.

Occasionally one forms the superior or inferior limit of a sequence in $\mathbf{R}^*$. The definition is sensible in any ordered set, but there will be problems of existence unless the ordered set is complete and has a largest and a smallest element.

12–4.2. Proposition. If $\{x_n\}$ is a sequence of real numbers, then

$$\liminf x_n \leq \limsup x_n.$$

Proof. Let $y_k = \sup \{x_n \mid n \geq k\}$ and $z_k = \inf \{x_n \mid n \geq k\}$. The sequence $\{y_n\}$ is (weakly) decreasing, since the y's are the suprema of successively smaller sets. Similarly, $\{z_n\}$ is increasing. Then for any integers p and q,

$$z_p \leq z_{p+q} \leq y_{p+q} \leq y_q.$$

Therefore for each integer p,

$$z_p \leq \inf_q y_q = \limsup x_n.$$

Finally,

$$\liminf x_n = \sup_p z_p \leq \limsup x_n.$$

12–4.3. Proposition. Let $\{x_n\}$ be a sequence of real numbers. Then

(a) $\limsup x_n < +\infty$ if and only if $\{x_n\}$ is bounded above;

(b) $\liminf x_n > -\infty$ if and only if $\{x_n\}$ is bounded below.

Proof. If $\{x_n\}$ is bounded above, then $\sup \{x_n \mid n \geq 1\}$ is finite, and therefore

$$\limsup x_n = \inf_k \sup \{x_n \mid n \geq k\} < +\infty.$$

On the other hand, if $\limsup x_n < +\infty$, we can choose k so that

$$\sup \{x_n \mid n \geq k\}$$

is finite. Then $\{x_n\}$ is bounded by the largest member of the finite set

$$\{x_1, x_2, \ldots, x_{k-1}, \sup \{x_n \mid n \geq k\}\}.$$

The proof of (b) is similar.

It can happen, of course, that $\limsup x_n = -\infty$ or $\liminf x_n = +\infty$; for example, let $x_n = -n$.

12–4.4. Proposition. Let $\{x_n\}$ be a sequence of real numbers and let $a \in R^*$:

(a) If $a < \limsup x_n$, then $a < x_n$ for infinitely many indices n.

(b) If $a \leq x_n$ for infinitely many indices n, then $a \leq \limsup x_n$.

(c) If $a > \limsup x_n$, then $a > x_n$ for all but finitely many indices n.

(d) If $a \geq x_n$ for all but finitely many indices n, then $a \geq \limsup x_n$.

(e) If $a > \liminf x_n$, then $a > x_n$ for infinitely many indices n.

(f) If $a \geq x_n$ for infinitely many indices n, then $a \geq \liminf x_n$.

(g) If $a < \liminf x_n$, then $a < x_n$ for all but finitely many indices n.

(h) If $a \leq x_n$ for all but finitely many indices n, then $a \leq \liminf x_n$.

Proof. We shall prove only (a) through (d), since the other statements are dual to these. Let $y_k = \sup \{x_n \mid n \geq k\}$.

Suppose $a < \limsup x_n$. Choose any integer k. We have $a < y_k$. By the definition of y_k there is an integer $n \geq k$ such that $a < x_n$. Thus we have proved that $(\forall k)(\exists n \geq k) \, a < x_n$, which is to say, $a < x_n$ for infinitely many indices n.

Suppose $a \leq x_n$ for infinitely many indices n. Then $a \leq y_k$ for every integer k, and $a \leq \inf y_k = \limsup x_n$.

Suppose $a > \limsup x_n$. Then $a > \inf y_k$, and so we can choose p so that $a > y_p$. Then $a > x_n$ for all indices n except possibly some of those in the set $\{1, 2, \ldots, p-1\}$, that is, for all but a finite set of indices.

Suppose $a \geq x_n$ for all but finitely many indices n. Then we can choose k so that $a \geq x_n$ for all $n \geq k$. Now, $a \geq y_k$; so

$$a \geq \inf y_k = \limsup x_n.$$

Let $\{x_n\}$ be a real sequence and consider the sets

$$L = \{r \in \mathbf{R} \mid r < x_n \text{ for infinitely many indices } n\},$$
$$U = \{r \in \mathbf{R} \mid r < x_n \text{ for only finitely many indices } n\}.$$

Obviously, every real number is either in L or in U. If $r \in L$ and $s < r$, then $s \in L$. If $r \in U$ and $s > r$, then $s \in U$. It can happen that one of these two sets is empty. If neither is empty, then L and U are rays with a common endpoint when $\mathbf{R}$ is identified as usual with a geometric line. This endpoint is $\limsup x_n$; it may belong to L or U. If L is empty, then $\limsup x_n = -\infty$, and if U is empty, then $\limsup x_n = +\infty$. This characterization of the superior limit as the division point between numbers exceeded finitely often and those exceeded infinitely often is extremely useful. A similar characterization can, of course, be given for the inferior limit. It is the division point between numbers which exceed only finitely many terms of the sequence and those which exceed infinitely many. (See Fig. 12-2.)

FIG. 12-2. The sequence $(-1)^{n(n-1)/2} - (-1)^n/n$. The superior limit is 1, the inferior limit -1.

Note how the strong and weak inequalities occur in the parts of Proposition 12-4.4. The pairs (a) and (b) and (c) and (d) are not exactly converse pairs simply because $\limsup x_n$ may belong to either of the sets L or U.

12-4.5. Proposition. Let $\{x_n\}$ be a sequence of real numbers. In order for $\{x_n\}$ to converge, it is necessary and sufficient that both $\liminf x_n = \limsup x_n$ and the latter be finite. If so, then

$$\lim x_n = \liminf x_n = \limsup x_n.$$

Proof. Suppose first that

$$\liminf x_n = \limsup x_n = \xi \in \mathbf{R}.$$

We will prove that $\lim x_n = \xi$. Let $\epsilon > 0$ be given. Since $\xi - \epsilon < \liminf x_n$, by 12-4.4(g) we can choose an integer k_1 such that

$$(\forall n > k_1) \quad \xi - \epsilon < x_n.$$

Since $\xi + \epsilon > \limsup x_n$, by 12–4.4(c) we can choose an integer k_2 such that

$$(\forall n > k_2) \quad x_n < \xi + \epsilon.$$

If k is the larger of k_1 and k_2, then

$$(\forall n > k) \quad |\xi - x_n| < \epsilon.$$

This proves that $\lim x_n = \xi$.

Conversely, suppose $\lim x_n = \eta$. Given $\epsilon > 0$, choose k so that

$$(\forall n > k) \quad |\eta - x_n| < \epsilon.$$

Then $\eta + \epsilon > x_n$ for all but finitely many indices n; hence $\eta + \epsilon \geq \limsup x_n$. And $\eta - \epsilon < x_n$ for all but finitely many indices n; so $\eta - \epsilon \leq \liminf x_n$. We have therefore

$$(\forall \epsilon > 0) \quad \eta - \epsilon \leq \liminf x_n \leq \limsup x_n \leq \eta + \epsilon.$$

We can conclude that $\liminf x_n = \limsup x_n = \eta$.

12–4.6. Improper convergence. It is convenient sometimes to write $x_n \to +\infty$ or $\lim x_n = +\infty$ to mean that $\liminf x_n = \limsup x_n = +\infty$. In view of 12–4.2, this is the same as $\liminf x_n = +\infty$. Similarly, one writes $x_n \to -\infty$ or $\lim x_n = -\infty$ to mean that $\limsup x_n = -\infty$.

This extended notion of convergence is sometimes known as improper convergence. Proposition 12–4.5 shows that it is a perfectly natural extension of the limit notion. Recall, however, the convention that one does not write $\lim x_n$, without context to the contrary, unless he is also asserting that the limit exists in the sense of Definition 12–1.3. This convention remains in force. One never writes $\lim x_n$ when the extended definition is meant without calling attention to the possibility of an infinite limit. At the same time one regularly allows superior and inferior limits to take values in **R***.

Sometimes the improper convergence $x_n \to +\infty$ is expressed as "x_n diverges to $+\infty$."

In this section we have considered only real sequences. This is natural, since the inferior and superior limits are defined in terms of order. But improper convergence can also be defined for complex sequences. If $\{z_n\}$ is a sequence of complex numbers, $z_n \to \infty$ or $\lim z_n = \infty$ means $|z_n| \to +\infty$. One can formalize this by introducing a set **C***, called the extended complex plane or Riemann sphere, consisting of **C** and a quasi-number ∞. This construction is important in the theory of functions of a complex variable, but we shall not discuss it. We should note, however, that if $\{x_n\}$ is a real sequence and therefore also a complex sequence, $x_n \to \infty$ is not the same thing as $x_n \to +\infty$. Take, for example, $x_n = (-1)^n n$: $|x_n| = n \to +\infty$, so $x_n \to \infty$; but neither $x_n \to +\infty$ nor $x_n \to -\infty$ is true. Thus the quasi-complex-number ∞ must not be confused with either of the quasi-real-numbers $+\infty$ or $-\infty$. Nevertheless, the plain symbol ∞ is often used in place of $+\infty$ when dealing with improper convergence of real sequences.

12–4.7. There are many inequalities relating the superior and inferior limits to the arithmetic operations, far too many to list. We will prove one as a sample of the technique involved. Suppose $\{x_n\}$ and $\{y_n\}$ are real sequences such that $\lim \inf x_n$ and $\lim \sup y_n$ are finite. Then

$$\lim \inf x_n + \lim \sup y_n \leq \lim \sup (x_n + y_n).$$

Let $\xi = \lim \inf x_n$ and $\eta = \lim \sup y_n$. Pick any $\epsilon > 0$. We know that $x_n > \xi - \epsilon$ for all but a finite number of indices, and that $y_n > \eta - \epsilon$ for infinitely many indices. Therefore, $x_n + y_n > \xi + \eta - 2\epsilon$ for infinitely many indices. Hence $\lim \sup (x_n + y_n) \geq \xi + \eta - 2\epsilon$. Since ϵ was chosen arbitrarily,

$$\lim \sup (x_n + y_n) \geq \xi + \eta,$$

as claimed.

Since inequalities of this type are usually quoted without proof or reference, students must acquire facility in handling them. The basic tools are the elementary arithmetic inequalities and the characterizations of the inferior and superior limits given in Proposition 12–4.4.

It is easy to prove that if $x_n \leq y_n$ for all n, then

$$\lim \sup x_n \leq \lim \sup y_n.$$

If it should happen that $\lim y_n$ exists, the conclusion becomes

$$\lim \sup x_n \leq \lim y_n$$

(which is stronger than $\lim \inf x_n \leq \lim y_n$). If $\lim x_n$ exists, then

$$\lim x_n \leq \lim \inf y_n.$$

EXERCISES

1. Find the superior and inferior limits of the sequences in Exercise 5, p. 167. Which of them converge improperly? Which of the sequences in Exercise 6 converge improperly as complex sequences?

2. Give examples to show that $\lim \sup x_n$ may fall in either L or U in the notation of the discussion following 12–4.4.

3. In the following assume that $\{x_n\}$ and $\{y_n\}$ are bounded sequences of real numbers. Prove:

(a) $\lim \sup (x_n + y_n) \leq \lim \sup x_n + \lim \sup y_n$
(b) $\lim \sup (x_n - y_n) \leq \lim \sup x_n - \lim \inf y_n$
(c) If $(\forall n) \, x_n > 0, \, y_n > 0$, then

$$\lim \inf x_n y_n \geq (\lim \inf x_n)(\lim \inf y_n).$$

(d) If $\lim x_n < 0$, then

$$\lim \sup x_n y_n = (\lim x_n)(\lim \inf y_n).$$

In (a), (b), and (c) show by example that strict inequality is possible.

4. Let $\{a_n\}$ be a sequence in $C - \{0\}$. Prove that if $\lim \sup |a_{n+1}|/|a_n| < 1$, then $\lim a_n = 0$.

5. Let φ be the order isomorphism from R^* to $\{r \in R \mid 0 \le r \le 1\}$ constructed in the exercise on p. 175. Show that $\{x_n\}$ converges to $a \in R^*$ if and only if $\{\varphi(x_n)\}$ converges to $\varphi(a)$.

12-5. CRITERIA FOR THE EXISTENCE OF LIMITS

Since taking limits plays such an important role in analysis, it is important to develop criteria for the existence of limits which do not require knowing in advance what the limits are. These criteria are consequences of the completeness of the real numbers.

12-5.1. Theorem. A bounded monotone sequence of real numbers is convergent.

Proof. Consider a bounded increasing sequence $\{x_n\}$. Let $\xi = \sup x_n$. We shall prove that $x_n \to \xi$.

Let ϵ be any positive number. Since $\xi - \epsilon$ is not an upper bound for the sequence, we can choose an integer m such that $x_m > \xi - \epsilon$. For any integer $n > m$, we have

$$\xi - \epsilon < x_m \le x_n \le \xi,$$

because x_n increases and ξ is an upper bound for the sequence. Thus $|\xi - x_n| < \epsilon$. This proves that $x_n \to \xi$.

Similarly, a bounded decreasing sequence converges to its infimum.

12-5.2. Corollary. A monotone sequence of real numbers is convergent in the extended real number system R^*.

12-5.3. Theorem. Let $\{z_n\}$ be a sequence of complex numbers. A necessary and sufficient condition for $\{z_n\}$ to converge is that

$$(\forall \epsilon > 0)(\exists M \in R)(\forall p, q > M) \quad |x_p - x_q| < \epsilon. \tag{1}$$

Proof. Suppose that $z_n \to \zeta$. Let ϵ be a given positive number. By the definition of convergence we can choose a number M such that $(\forall n > M) |\zeta - z_n| < \epsilon/2$. Then if both p and q exceed M,

$$|z_p - z_q| = |(\zeta - z_p) - (\zeta - z_q)| \le |\zeta - z_p| + |\zeta - z_q| < \epsilon.$$

Thus (1) is true for any convergent sequence.

To prove the converse, we begin with the special case of a real sequence $\{z_n\}$ which satisfies (1). Let ϵ be any positive number. Choose M so that

$$(\forall p, q > M) \quad |z_p - z_q| < \epsilon.$$

Fix an integer $m > M$. Then $|z_m - z_n| < \epsilon$ or $z_m - \epsilon < z_n < z_m + \epsilon$ for all but a finite number of indices n. By 12–4.4(h) and (d),

$$z_m - \epsilon \le \lim \inf z_n$$

and

$$\lim \sup z_n \le z_m + \epsilon.$$

Therefore, $\lim \sup z_n - \lim \inf z_n \le 2\epsilon$. Since ϵ was any positive number, $\lim \sup z_n \le \lim \inf z_n$. The opposite inequality is always valid; so $\lim \sup z_n = \lim \inf z_n$. By Proposition 12–4.5, $\{z_n\}$ converges.

Now consider a complex sequence $\{z_n\}$ satisfying (1). From the inequality

$$|\text{Re } z_p - \text{Re } z_q| = |\text{Re } (z_p - z_q)| \le |z_p - z_q|$$

it follows that the real sequence $\{\text{Re } z_n\}$ satisfies condition (1). Similarly, the real sequence $\{\text{Im } z_n\}$ satisfies condition (1). The work of the preceding paragraph shows that both $\{\text{Re } z_n\}$ and $\{\text{Im } z_n\}$ converge. Therefore $\{z_n\}$ converges.

Condition (1) is known as *Cauchy's criterion for convergence*. A sequence satisfying it is often called a *Cauchy sequence*.

Cauchy's criterion does *not* read:

$$(\forall \epsilon > 0)(\exists M \in \mathbf{R})(\forall n > M) \quad |z_{n+1} - z_n| < \epsilon.$$

It requires that *all* pairs of terms beyond a certain point to be close and not merely that consecutive terms of the sequence be close. The latter condition is not sufficient for convergence as the simple example $z_n = \sqrt{n}$ shows. (Here $z_{n+1} - z_n = \sqrt{n + 1} - \sqrt{n} = 1/(\sqrt{n + 1} + \sqrt{n}) \to 0$.)

Recall that the definition of convergence can be written in several equivalent forms. In the same way the Cauchy criterion can be put in several forms; for example,

$$(\forall \epsilon > 0)(\exists m \in \mathbf{N})(\forall p, q \ge m) \quad |z_p - z_q| \le \epsilon.$$

Another variant is

$$(\forall \epsilon > 0)(\exists m \in \mathbf{N})(\forall p > m) \quad |z_p - z_m| < \epsilon. \tag{2}$$

This is equivalent to (1), because if $(\forall p > m) |z_p - z_m| < \epsilon$, then

$$(\forall p, q > m) \quad |z_p - z_q| < 2\epsilon.$$

These alternative forms are sometimes quite useful.

EXERCISES

Let $\{a_n\}$ be a sequence of complex numbers. Let $\{S_n\}$ be defined recursively as follows: $S_1 = a_1, S_{n+1} = S_n + a_{n+1}$ for $n \in \mathbf{N}$. The sequence $\{S_n\}$ will be denoted by F-$\sum a_n$ and called an *infinite series*. The terms S_n are called *partial sums* of F-$\sum a_n$.

1. Determine, with proof, which of the following infinite series are convergent:

(a) F-$\sum \dfrac{(-1)^n}{n}$ (b) F-$\sum \dfrac{1}{n2^n}$ (c) F-$\sum \dfrac{1}{n}$ (d) F-$\sum \dfrac{1}{n^2}$

2. For what complex values of λ and ρ does F-$\sum \lambda \rho^n$ converge?

3. **The comparison test.** Suppose that $\{b_n\}$ is a real sequence and F-$\sum b_n$ converges. Prove that if

$$(\forall n) \quad |a_n| \leq b_n,$$

then F-$\sum a_n$ converges.

4. **The ratio test.** Suppose that $\{a_n\}$ is a sequence of nonzero complex numbers. Prove that if $\lim \sup |a_{n+1}|/|a_n| < 1$, then F-$\sum a_n$ converges, and if $\lim \inf |a_{n+1}|/|a_n| > 1$, then F-$\sum a_n$ diverges.

5. Suppose that $\{c_n\}$ is a monotone decreasing sequence of real numbers and that $\lim c_n = 0$. Let $\{B_n\}$ be a bounded sequence of complex numbers. Prove that F-$\sum (B_{n+1} - B_n)c_n$ converges. (*Note:* The case $B_n = (-1)^n/2$ is the standard result for alternating series.)

12–6. SUBSEQUENCES

We shall define the notion of subsequence and prove several theorems concerning the existence of convergent subsequences of a given real or complex sequence. Since convergence is a strong condition to impose on a sequence, it is perhaps surprising that every bounded sequence of complex numbers has a convergent subsequence.

12–6.1. Definition. Let $\{x_n\}$ be a sequence in an arbitrary set X. The sequence $\{y_n\}$ is called a *subsequence* of $\{x_n\}$ if and only if there exists a function f from $\mathbf{N}$ to $\mathbf{N}$ such that $\lim f(n) = +\infty$ and $(\forall n)\, y_n = x_{f(n)}$. We shall usually denote this subsequence by $\{x_{f(n)}\}$.

The condition imposed on f is designed to ensure that the "infinite part" of a subsequence depends only on the "infinite part" of the original sequence. Frequently the condition is strengthened by the demand that f be strictly increasing, which implies that $f(n) \rightarrow +\infty$.

Undoubtedly the most natural way to conceive of a subsequence is to imagine that from the original sequence $x_1, x_2, x_3, \ldots$ we drop terms, being sure to retain infinitely many, and then renumber those remaining in the obvious way. This leads to the requirement that f be strictly increasing. The definition given is more general, however. The sequence informally described by

$$x_1, x_2, x_2, x_3, x_3, x_3, x_4, \ldots,$$

where x_n is repeated n times, is a subsequence of $\{x_n\}$ as here defined. (The function f is given by $f(n) = k$ if $\frac{1}{2}k(k - 1) < n \leq \frac{1}{2}k(k + 1)$.) It is in fact true that we could get along perfectly well with the more restrictive definition, but the definition given leads to the appropriate generalization of subsequence in the general theory of convergence.

The most common notation for a subsequence is $\{x_{n_k}\}$ where $\{n_k\}$ is the sequence $\{f(k)\}$ of the definition above. Comparing the notations $\{x_n\}$ and $\{x_{n_k}\}$, we see that n serves as a dummy index in the first and as the symbol for a function in the second. This causes no confusion until a second subsequence of $\{x_n\}$ appears, which must be denoted by $\{x_{m_k}\}$. Confusion may arise if the different roles played by n are not adequately described. The functional notation also enjoys a substantial typographical advantage over the second-order subscript notation, which becomes cumbersome when subsequences of subsequences appear.

12–6.2. Proposition. If $\{y_n\}$ is a subsequence of $\{x_n\}$ and $\{z_n\}$ is a subsequence of $\{y_n\}$, then $\{z_n\}$ is a subsequence of $\{x_n\}$.

We omit the proof of this important, although nearly obvious, fact.

Our interest in subsequences in this chapter will focus on the convergence of subsequences of a given sequence. If a sequence of complex numbers converges, then passing to a subsequence will not alter the situation. This is our first result.

12–6.3. Proposition. If $\{z_n\}$ is a convergent sequence of complex numbers, then any subsequence of $\{z_n\}$ converges to the same limit.

Proof. Let $z_n \to \zeta$ and $\{z_{f(n)}\}$ be a subsequence of $\{z_n\}$. We must show that $z_{f(n)} \to \zeta$.

Let $\epsilon > 0$ be given. Choose a number M such that $(\forall n > M)\,|\zeta - z_n| < \epsilon$. Since $f(n) \to +\infty$, we can choose a number P such that $(\forall n > P)\,f(n) > M$. Then

$$(\forall n > P) \quad |\zeta - z_{f(n)}| < \epsilon.$$

This shows that $z_{f(n)} \to \zeta$.

It is clear that the proposition is also true for improperly convergent sequences.

12–6.4. Proposition. If $\{x_{f(n)}\}$ is a subsequence of the real sequence $\{x_n\}$, then

$$\liminf x_n \leq \liminf x_{f(n)} \leq \limsup x_{f(n)} \leq \limsup x_n.$$

We leave the proof to the reader. Note that 12–6.3 follows immediately from 12–6.4 and 12–4.5.

12–6.5. Proposition. Let $\{x_n\}$ be any sequence of real numbers. There is a subsequence of $\{x_n\}$ which converges to $\limsup x_n$ (if the latter is infinite, the subsequence will converge improperly, of course) and a subsequence which converges to $\liminf x_n$.

Proof. We shall prove only that there is a subsequence converging to lim sup x_n. Suppose first that $\xi = $ lim sup x_n is finite. We define a function from N to N recursively. Since $\xi - 1 < $ lim sup x_n, there are infinitely many indices n for which $x_n > \xi - 1$. Let $f(1)$ be the least of these. There are infinitely many indices n for which $x_n > \xi - \frac{1}{2}$. Some of these exceed $f(1)$. Let $f(2)$ be the least integer m such that $m > f(1)$ and $x_m > \xi - \frac{1}{2}$. In general, let $f(k + 1)$ be the least integer p such that $p > f(k)$ and $x_p > \xi - 1/p$. Since f is strictly increasing by construction, $\{x_{f(n)}\}$ is a subsequence of $\{x_n\}$.

Let k be any integer. Then $x_{f(n)} > \xi - 1/k$ for all $n \geq k$, and therefore, lim inf $x_{f(n)} \geq \xi - 1/k$. Since k was any integer, lim inf $x_{f(n)} \geq \xi$. But

$$\text{lim sup } x_{f(n)} \leq \text{lim sup } x_n = \xi$$

by 12–6.4. Hence lim inf $x_{f(n)} = $ lim sup $x_{f(n)} = \xi$. By Proposition 12–4.5,

$$\text{lim } x_{f(n)} = \xi.$$

If lim sup $x_n = +\infty$, then we can choose a strictly increasing unbounded subsequence as follows: Let $f(1)$ be the least integer m such that $x_m > 1$. Let $f(k + 1)$ be the least integer p such that $x_p > k + 1$, $x_p > x_{f(k)}$, and $p > f(k)$. (The last condition is automatically satisfied, to be sure, but there is no need to prove this.)

If lim sup $x_n = -\infty$, then lim $x_n = -\infty$. Obviously, every sequence is a subsequence of itself.

This proposition sheds more light on the nature of the inferior and superior limits. It is obvious from 12–4.4(c) that no subsequence of $\{x_n\}$ converges to a number larger than lim sup x_n. Hence lim sup x_n is the largest number to which a subsequence of $\{x_n\}$ converges. Similarly, lim inf x_n is the smallest number to which a subsequence of $\{x_n\}$ converges.

12–6.6. The Bolzano-Weierstrass theorem. Any bounded sequence of complex numbers has a convergent subsequence.

Proof. For a bounded real sequence $\{x_n\}$ this follows immediately from the last proposition since lim sup x_n is finite.

Let $\{z_n\}$ be a bounded sequence of complex numbers. Then $\{\text{Re } z_n\}$ is a bounded sequence of real numbers. Let $\{\text{Re } z_{f(n)}\}$ be a convergent subsequence. Now $\{\text{Im } z_{f(n)}\}$ is a bounded sequence of real numbers, so we can choose a subsequence $\{\text{Im } z_{f(g(n))}\}$ which converges. Since $\{\text{Re } z_{f(g(n))}\}$ is a subsequence of a convergent sequence, it converges. Thus $\{z_{f(g(n))}\}$ is a convergent subsequence of $\{z_n\}$, since both its real and imaginary parts converge.

12–6.7. Proposition. Let $\{z_n\}$ be a bounded sequence of complex numbers. If $\{z_n\}$ is not convergent, it contains two convergent subsequences with different limits.

Proof. This can be proved by refining the argument of the last theorem, but a different argument is instructive. By the theorem, we can select a subsequence which converges, say to ζ. If $\{z_n\}$ does not converge to ζ, then for some $\epsilon > 0$ there are infinitely many indices n such that $|\zeta - z_n| > \epsilon$. If we enumerate this infinite set of indices by the function f (that is, apply 11–3.2), we obtain a subsequence $\{z_{f(n)}\}$ such that

$$(\forall n) \quad |\zeta - z_{f(n)}| > \epsilon.$$

Now this sequence has a convergent subsequence, and this subsequence certainly does not converge to ζ; it is the required second convergent subsequence of $\{z_n\}$.

The contrapositive form of this proposition is important and merits a formal statement.

12–6.8. Corollary. Let $\{z_n\}$ be a bounded sequence of complex numbers. If every convergent subsequence of $\{z_n\}$ has the same limit, then $\{z_n\}$ converges.

EXERCISES

1. Give an example of a sequence $\{x_n\}$ and a subsequence $\{x_{f(n)}\}$ such that

$$\liminf x_n < \liminf x_{f(n)} < \limsup x_{f(n)} < \limsup x_n.$$

2. Prove that, if improper convergence is allowed, any sequence of complex numbers has a convergent subsequence.

3. Prove that if a sequence of complex numbers does not converge to ∞, it has a convergent subsequence.

4. Find all complex numbers which are limits of convergent subsequences of $\{i^n + 1/n^2\}$. Do the same for the sequence $\{i^n + (-1)^n + (-1)^{n(n-1)/2}\}$.

5. There is a bijection from N to the set Q of all rational numbers. This function may be regarded as a sequence $\{x_n\}$. Prove that for every real number r there is a subsequence of $\{x_n\}$ which converges to r.

CHAPTER 13

SUMS AND PRODUCTS

As we all know, the sum of a finite set of numbers is independent of the order in which the numbers are added. This is a consequence of the associativity and commutativity of addition. The same is true for products. We shall state this fact as a formal theorem. Then we shall investigate infinite sums and products. Finally we shall develop the decimal system of representation for real numbers.

13–1. FINITE SUMS AND PRODUCTS

We shall prove a general theorem on associative and commutative binary operations. The required facts about sums and products will then be a special case of this theorem, which, roughly, says that associated with any commutative and associative binary operation there is a function which combines any finite number of elements all at once without regard to their order.

13–1.1. Theorem. Let S be a set endowed with an associative and commutative binary operation $*$ (written medially) having an identity element e. Let $\mathcal{P}$ be the set of all ordered pairs $\langle f, A \rangle$ such that f is a function, range $f \subseteq S$, and A is a finite subset of domain f. There is a unique function $\bigstar$ from $\mathcal{P}$ to S such that

$$\bigstar(f, \emptyset) = e, \tag{1}$$

and

$$\bigstar(f, A \cup \{x\}) = \bigstar(f, A) * f(x) \tag{2}$$

whenever $\langle f, A \rangle \in \mathcal{P}$ and $x \in$ domain $f - A$. Furthermore, this function has the following properties:

(a) if A and B are disjoint, then $\bigstar(f, A \cup B) = \bigstar(f, A) * \bigstar(f, B)$;
(b) if $(\forall x \in A)\, f(x) = g(x)$, then $\bigstar(f, A) = \bigstar(g, A)$;
(c) if $(\forall x \in A)\, h(x) = f(x) * g(x)$, then $\bigstar(h, A) = \bigstar(f, A) * \bigstar(g, A)$;
(d) if φ is a bijection from B to A, then $\bigstar(f, A) = \bigstar(f \circ \varphi, B)$.

Proof. We begin with a fixed function f such that range $f \subseteq S$. Let $\mathcal{R}$ be the set of all finite subsets of domain f. For each $n \in \mathbf{N}^*$ let $\mathcal{Q}_n$ be the set of all subsets of domain f having cardinal at most n.

186

We shall define recursively a sequence $\{G_n\}$ of functions such that for each n,

$$\text{domain } G_n = \mathcal{Q}_n, \tag{3}$$

$$G_n(\emptyset) = e, \tag{4}$$

and

$$G_n(A \cup \{x\}) = G_n(A) * f(x) \tag{5}$$

whenever $A \in \mathcal{Q}_{n-1}$ and $x \in \text{domain } f - A$.

Define G_1 by $G_1(\emptyset) = e$, $G_1(\{x\}) = f(x)$ for all $x \in \text{domain } f$. Then G_1 has properties (3), (4), and (5).

Suppose that G_n has been defined and that it satisfies (3), (4), and (5). Let A be a subset of domain f having cardinal $n + 1$. If $x, y \in A$, $x \neq y$, then

$$\begin{aligned}
G_n(A - \{x\}) * f(x) &= (G_n(A - \{x, y\}) * f(y)) * f(x) \\
&= (G_n(A - \{x, y\}) * f(x)) * f(y) \\
&= G_n(A - \{y\}) * f(y),
\end{aligned}$$

where the second equality involves two applications of the associative law and one of the commutative law. Therefore, we may define G_{n+1} by $G_{n+1}(A) = G_n(A)$ if $A \in \mathcal{Q}_n$ and $G_{n+1}(A) = G_n(A - \{x\}) * f(x)$, where x is any member of A, if card $A = n + 1$. (We are really saying that if we adjoin to G_n all ordered pairs of the form $\langle A, G_n(A - \{x\}) * f(x)\rangle$, where $A \subseteq \text{domain } f$, card $A = n + 1$, and $x \in A$, the result is a function.) It is clear that G_{n+1} satisfies (3), (4), and (5). This completes the inductive construction.

Since G_{n+1} is always an extension of G_n, there is a function H which is an extension of them all. (As a set of ordered pairs, $H = \bigcup_n G_n$.) It is clear that H has the following properties:

$$\text{domain } H = \mathcal{R}, \tag{6}$$

$$H(\emptyset) = e, \tag{7}$$

and

$$H(A \cup \{x\}) = H(A) * f(x) \tag{8}$$

whenever $A \in \mathcal{R}$, and $x \in \text{domain } f - A$.

If H' is another function having these three properties, it is easy to show by induction that H' agrees with G_n on $\mathcal{Q}_n$; hence $H' = H$.

Thus we have proved that there is a unique function having the properties (6), (7), and (8) for a fixed f. There remains only the purely technical problem of assembling all these functions for different f's into a single function. To this end let $\star$ be the set of all $\langle\langle f, A\rangle, s\rangle$, where f is a function with range in S, A is a finite subset of domain f, and $(\exists H) H(A) = s$, where H satisfies (6), (7), and (8). Now $\star$ is obviously the required function.

Properties (a) through (d) are easily proved by induction on the cardinal of the sets involved. We shall give the details only for (d). It is obviously true when $B = \emptyset$. Assume it is true for sets of cardinal n and suppose that card $B = n + 1$. Pick $x \in B$. Now φ (restricted) is a bijection from $B - \{x\}$ to $A - \{\varphi(x)\}$, both of which have cardinal n. Therefore,

$$☆(f, A) = ☆(f, A - \{\varphi(x)\}) * f(\varphi(x))$$
$$= ☆(f \circ \varphi, B - \{x\}) * f(\varphi(x)) = ☆(f \circ \varphi, B).$$

(A careful look reveals that Property (b) is used in this calculation.) Thus (d) holds for sets of cardinal $n + 1$. By induction it holds for all finite sets.

This theorem can be called a general associative-commutative law. Most texts dealing with foundations prove a thorem of this type. The point of the theorem is not merely to justify the omission of parentheses and rearrangement of operands in small computations. Such procedures could always be justified by filling in the details. Rather the theorem is meant to justify definitions and calculations which involve infinitely many rearrangements at once. For example, in the next section we define a function as the supremum of infinitely many finite sums.

To justify all types of reasoning with finite sums and products would require a vast expansion of the list of properties (a) through (d). This is an area in which reasoning is usually left informal, since formalization is straightforward but tedious. The theorem is therefore to be interpreted more as a sample of the kinds of formal proofs that are usually omitted than as a fundamental result on which our subsequent work will be based.

A set endowed with an associative binary operation is called a *semigroup*. Hence the S of the theorem is a commutative semigroup with identity. According to Proposition 8–1.5, the identity element is unique. It is easy to modify the theorem slightly to cover commutative semigroups which are not assumed to have an identity.

The subsets of any fixed set form a commutative semigroup with identity, using either the $\cup$ or the $\cap$ operation. Applying the theorem, we are led to indexed unions and intersections for finite families of sets. It turns out that these operations can be defined directly, even for infinite families of sets. In general, a binary operation cannot be given a reasonable extension which combines more than a finite number of operands. Therefore, one cannot expect to define the extension until a theory of counting is available.

When $S = C$ and the operation in question is addition, it is customary to denote the function ☆ with an indexed notation using the symbol $\sum$. We write $\sum_{x \in A} f(x)$ in place of $☆(f, A)$. When A is a set of consecutive integers, say

$$A = \{i \in I \mid p \leq i \leq q\},$$

one usually writes $\sum_{i=p}^{q} f(i)$. There are numerous variations on these notations which retain the symbol $\sum$ but omit some of the symbols that describe the range of summation. These notations are also used for the extension of any other semi-group operation denoted by $+$. (The symbol $+$ is traditionally reserved for a commutative associative operation.)

The use of an indexed notation is very convenient. It enables us to write 13–1.1(d), for example, in the form

$$\sum_{x \in A} f(x) = \sum_{y \in B} f(\varphi(y)).$$

Properties (a) and (c) can be readily extended to

(i) $\sum_{i=1}^{n} \sum_{x \in A_i} f(x) = \sum_{x \in \cup A_i} f(x)$, and

(ii) $\sum_A (\lambda f(x) + \mu g(x)) = \lambda \sum_A f(x) + \mu \sum_A g(x)$,

if the sets A_i are disjoint, and λ and μ are any complex numbers. Obviously, (i) would be valid if restated for an abstract semigroup, but (ii) involves a second operation, multiplication, and the distributive law. We shall mention only three other formulas among the many involving finite sums:

(iii) if $(\forall x \in A)$ $f(x) \leq g(x)$, then $\sum_{x \in A} f(x) \leq \sum_{x \in A} g(x)$;

(iv) $|\sum_{x \in A} f(x)| \leq \sum_{x \in A} |f(x)|$;

(v) if $(\forall x \in A)$ $f(x) = \lambda$, then $\sum_{x \in A} f(x) = \lambda$ card A.

The last equation in (v) is usually written $\sum_{x \in A} \lambda = \lambda$ card A. The fact that the dummy index x is missing in the notation for the constant function can be misleading and often results in the computational error $\sum_{x \in A} \lambda = \lambda$.

When $S = \mathbf{C}$ and the operation is multiplication, one uses similar notations with the symbol $\prod$ instead of $\sum$. The symbol $\prod$ is often used in conjunction with an operation denoted by juxtaposition or $\cdot$ in other semigroups, even noncommutative semigroups. In the latter case the order of combination must be taken into account, of course, and the theorem similar to 13–1.1 is much weaker. See Exercise 3.

13–1.2. Definition. Let X be any set and let f be a function from X to $\mathbf{C}$. The *support* of f, often denoted by spt f, is $\{x \in X \mid f(x) \neq 0\}$. (When the set X is a topological space, the support of a function is defined in a slightly different way.)

13–1.3. Proposition. Let X be any set and let f and g be functions from X to $\mathbf{C}$ having finite support. Let λ and μ be complex numbers and define h by

$$h(x) = \lambda f(x) + \mu g(x).$$

Then h has finite support; in fact, spt $h \subseteq$ spt $f \cup$ spt g.

We shall omit the proof. The function h defined here is often denoted by $\lambda f + \mu g$.

13–1.4. Proposition. Let X be a fixed set. Let $\mathfrak{F}$ be the set of all functions from X to $\mathbf{C}$ having finite support. Then there is a unique function Ψ from $\mathfrak{F}$ to $\mathbf{C}$ with the following properties:

(a) if spt $f = \{x\}$, then $\Psi(f) = f(x)$;   (b) $\Psi(\lambda f + \mu g) = \lambda\Psi(f) + \mu\Psi(g)$.

Proof. Let $\Psi(f) = \sum_{x\in\text{spt}f} f(x)$. It is clear that (a) holds. Furthermore, $\Psi(f) = \sum_{x\in A} f(x)$ for any finite subset A of X such that $A \supseteq \text{spt} f$. Hence

$$\Psi(\lambda f + \mu g) = \sum_{\text{spt}f\cup\text{spt}g} (\lambda f(x) + \mu g(x))$$
$$= \lambda\sum_{\text{spt}f} f(x) + \mu\sum_{\text{spt}g} g(x) = \lambda\Psi(f) + \mu\Psi(g).$$

The uniqueness follows immediately by induction on the cardinal of spt f, since we can represent any function f with card (spt f) $= k + 1$ as the sum of two functions f_1 and f_2, where card (spt f_1) $= k$ and spt f_2 is a single point.

It is customary to write $\sum_{x\in X} f(x)$ for $\Psi(f)$. Here we are explicitly allowing X to be infinite. The sum in question is often described as *effectively finite*, because all but a finite number of its formal summands are zero. When a sum is formally infinite, but effectively finite, authors usually call attention to this fact.

EXERCISES

1. With the notation of 13-1.1 prove that

$$\text{☆}(f, A \cup B) * \text{☆}(f, A \cap B) = \text{☆}(f, A) * \text{☆}(f, B).$$

2. Restate Theorem 13-1.1 for cases where S has no identity element.

3. Let S be a semigroup with an operation denoted by juxtaposition. Let $\mathfrak{F}$ be the set of functions from $\mathbf{I}$ to S. Let $K = \{\langle p,q\rangle \in \mathbf{I} \times \mathbf{I} \mid p < q\}$. Prove that there is a unique function $\prod$ from $\mathfrak{F} \times K$ to S such that

(a) $\prod(f, p, p + 1) = f(p)$, and
(b) $\prod(f, p, q)\prod(f, q, r) = \prod(f, p, r)$.

(The usual notation is $\prod_{i=p}^{q-1} f(i)$ instead of $\prod(f, p, q)$.) Suppose that S is actually a group; that is, S has a two-sided identity element e and every element has an inverse: $(\forall s \in S)(\exists t) \, st = e$. Prove that the function $\prod$ above has a unique extension over $\mathfrak{F} \times \mathbf{I} \times \mathbf{I}$ such that (a) and (b) remain valid.

4. Suppose that for each fixed x in the finite set A, $\{f_n(x)\}$ is convergent. Prove that

$$\lim_n \sum_{x\in A} f_n(x) = \sum_{x\in A} \lim_n f_n(x).$$

5. Suppose that A is a finite set and that for each fixed x in A, $\{f_n(x)\}$ is a bounded sequence of real numbers. Prove that

$$\limsup_n \sum_{x\in A} f_n(x) \leq \sum_{x\in A} \limsup_n f_n(x).$$

6. Suppose that f is a function from X to $\mathbf{C}$ having finite support. Let $\{Y(i) \mid i \in I\}$ be a family of mutually disjoint subsets of X. Show that

$$\sum_{i\in I} \sum_{x\in Y(i)} f(x) = \sum_{x\in\cup Y(i)} f(x),$$

all sums being effectively finite.

13–2. INFINITE SERIES

We shall develop the usual theory of infinite series. Many of the theorems have already appeared as exercises in Chapter 12. At the end of this section we shall take up the theory of unordered infinite sums and obtain from it the facts about absolutely convergent double series.

13–2.1. Definition. Let $\{a_n\}$ be a sequence of complex numbers. The *infinite series* F-$\sum_{n=1}^{\infty} a_n$ is the sequence $\{A_n\}$ given by $A_n = \sum_{k=1}^{n} a_k$. The numbers a_n are called the *terms* of F-$\sum_{n=1}^{\infty} a_n$, and the numbers A_n are called its *partial sums*. If F-$\sum_{n=1}^{\infty} a_n$ is convergent (that is, if $\{A_n\}$ is convergent), we shall designate its limit by $\sum_{n=1}^{\infty} a_n$. This number is called the *sum* of the series.

These notations may be abbreviated in many ways; for example, F-$\sum a_n$ or $\sum_n a_n$. The informal notation

$$a_1 + a_2 + a_3 + \cdots$$

is often useful. It is sometimes convenient to enumerate the terms of an infinite series starting with 0 or some other integer. When this is done, it is customary to enumerate the partial sums in the same way. Thus the partial sums of F-$\sum_{n=k}^{\infty} a_n$ are $A_k = a_k$, $A_{k+1} = a_k + a_{k+1}, \ldots$

The terminology given here is standard, but the definition and notation are not. Most classical books define an infinite series to be a notation, not a mathematical object. Some modern books define it to be an ordered pair of sequences, the sequence of terms and the sequence of partial sums. The distinction between an infinite series and its sum is important, so it seems worthwhile to distinguish them in the notation, at least at first. The F is intended to remind us that F-$\sum a_n$ is a formal infinite sum. We shall not write $\sum a_n$ unless F-$\sum a_n$ is convergent. Later, we will drop the distinction and allow $\sum a_n$ to do double duty, sometimes denoting a series and sometimes a number.

One often sees the statement $\sum a_n = +\infty$, which means, of course, that $A_n \to +\infty$ in $\mathbf{R}^*$.

13–2.2. An example. We shall discuss the infinite series

$$1 - \tfrac{1}{2} + \tfrac{1}{3} - \tfrac{1}{4} + \tfrac{1}{5} - \cdots, \tag{1}$$

more formally expressed as F-$\sum a_n$ where $a_n = (-1)^{n-1}/n$. If we denote the partial sums of this series by $A_1, A_2, A_3, \ldots$, then we can easily prove that

$$A_1 > A_3 > A_5 > \cdots$$

and

$$A_2 < A_4 < A_6 < \cdots$$

and that all even partial sums are less than all odd ones. Therefore we have

$$(\forall n \geq 2m) \quad A_{2m} \leq A_n \leq A_{2m+1}.$$

This shows that

$$(\forall p, q \geq 2m) \quad |A_p - A_q| \leq A_{2m+1} - A_{2m} = \frac{1}{2m+1}.$$

Thus the partial sums satisfy Cauchy's criterion and the series is convergent.

Now let us rearrange the series by taking two positive terms, then one negative term, two positive terms, one negative term, etc. from the original series. This leads to

$$1 + \tfrac{1}{3} - \tfrac{1}{2} + \tfrac{1}{5} + \tfrac{1}{7} - \tfrac{1}{4} + \tfrac{1}{9} + \tfrac{1}{11} - \cdots \tag{2}$$

or F-$\sum b_n$, where

$$b_{3k+1} = \frac{1}{4k+1}, \qquad b_{3k+2} = \frac{1}{4k+3}, \qquad b_{3k+3} = -\frac{1}{2k+2}$$

for $k = 0, 1, 2, \ldots$. Note that every term of (1) appears exactly once in (2) and vice versa. Denoting the partial sums of this series by $B_1, B_2, \ldots$, we find that

$$(\forall n \geq 3k) \qquad B_{3k} \leq B_n < B_{3k-1} \quad \text{if} \quad k > 0.$$

Therefore

$$(\forall p, q \geq 3k) \quad |B_p - B_q| < B_{3k-1} - B_{3k} = \frac{1}{2k}.$$

Again Cauchy's criterion is satisfied and the series is convergent.

It is clear that $\lim A_n \leq A_5 < A_3 = \tfrac{5}{6}$, while $\lim B_n \geq B_6 > B_3 = \tfrac{5}{6}$. Therefore, $\lim A_n \neq \lim B_n$, or $\sum a_n \neq \sum b_n$. Thus, in spite of the commutativity of addition, the rearrangement of the summands has altered the value of the sum!

While terminology suggests that finding the sum of an infinite series is a generalized addition process, the example shows that this process does not share the usual properties of addition. It is necessary, therefore, to study the effects of formal manipulations on the convergence of series. Since the convergence of an infinite series is, by definition, the convergence of its sequence of partial sums, the theorems on limits of sequences can all be translated directly into theorems on the convergence of infinite series.

13–2.3. Proposition. The infinite series F-$\sum a_n$ converges if and only if both F-$\sum \operatorname{Re} a_n$ and F-$\sum \operatorname{Im} a_n$ converge. If so,

$$\operatorname{Re}(\textstyle\sum a_n) = \sum \operatorname{Re} a_n \qquad \text{and} \qquad \operatorname{Im}(\textstyle\sum a_n) = \sum \operatorname{Im} a_n.$$

This follows immediately from 12–2.4.

13–2.4. Proposition. If F-$\sum a_n$ and F-$\sum b_n$ are convergent infinite series and λ and μ are complex numbers, then F-$\sum(\lambda a_n + \mu b_n)$ is convergent and

$$\textstyle\sum(\lambda a_n + \mu b_n) = \lambda \sum a_n + \mu \sum b_n.$$

This is an immediate consequence of 12–2.2 and 12–2.1(a).

13–2.5. Proposition. Suppose F-$\sum b_n$ is obtained from F-$\sum a_n$ by omitting the first k terms and renumbering; that is, $(\forall n)\, b_n = a_{n+k}$. Then F-$\sum a_n$ converges if and only if F-$\sum b_n$ converges. If they converge, then

$$\sum_{n=1}^{\infty} a_n = \sum_{n=1}^{k} a_n + \sum_{n=1}^{\infty} b_n.$$

Proof. If $\{A_n\}$ and $\{B_n\}$ are the partial sums of the two series, then

$$A_{p+k} = \sum_{n=1}^{k} a_n + B_p$$

for any positive integer p. Hence the proposition follows from 12–1.8 and 12–2.1(a).

This proposition is often written:

$$\sum_{n=1}^{\infty} a_n = \sum_{n=1}^{k} a_n + \sum_{n=k+1}^{\infty} a_n.$$

The number $\sum_{n=k+1}^{\infty} a_n$ is called the *remainder* of the series *after k terms*. If we denote this remainder by R_k, it is clear that $R_k \to 0$. This fact is occasionally quite useful.

13–2.6. Corollary. Suppose F-$\sum a_n$ and F-$\sum b_n$ can be obtained from each other by modifying, inserting, or omitting a finite number of terms. Then F-$\sum a_n$ converges if and only if F-$\sum b_n$ converges.

Proof. The hypothesis implies that if we omit appropriate numbers of terms from the beginnings of the two series, the resulting series are identical. The result then follows by two applications of 13–2.5.

This corollary means that the convergence of an infinite series is independent of any finite number of terms; it is a property of the "infinite part" of the sequence of terms.

The next two propositions concern two ways of modifying a series which do not affect the order in which the terms are to be added. Their statements are somewhat technical. The first says that if we add consecutive terms of one series to make the terms of a new series, then we do not spoil the convergence. The second says that convergence is unaffected if we intercalate zero terms into a series.

13–2.7. Proposition. Suppose that f is a strictly increasing function from N^* to N^* with $f(0) = 0$ and that

$$b_n = \sum_{k=f(n-1)+1}^{f(n)} a_k \qquad \text{for all } n \geq 1.$$

If F-$\sum a_n$ converges, then so does F-$\sum b_n$, and $\sum a_n = \sum b_n$.

Proof. Under the hypothesis the partial sums of F-$\sum b_n$ form a subsequence of the partial sums of F-$\sum a_n$.

Note, however, that it can happen that F-$\sum b_n$ converges while F-$\sum a_n$ diverges. For example, from the divergent series

$$1 - 1 + 1 - 1 + 1 - 1 + \cdots$$

we obtain the convergent series $0 + 0 + 0 + \cdots$ by adding consecutive pairs of terms.

13–2.8. Proposition. Suppose f is a strictly increasing function from $\mathbf{N}$ to $\mathbf{N}$ and that $b_n = a_{f^{-1}(n)}$ if $n \in$ range f, and $b_n = 0$ otherwise. Then F-$\sum a_n$ converges if and only if F-$\sum b_n$ converges. If they converge, then $\sum a_n = \sum b_n$.

Proof. Under this hypothesis the partial sums of either series form a subsequence of the partial sums of the other (under the definition of subsequence given in 12–6.1).

This result is often applied tacitly when we go from a series like

$$1 + 0 - \tfrac{1}{3} + 0 + \tfrac{1}{5} + 0 - \tfrac{1}{7} + 0 + \cdots$$

to

$$1 - \tfrac{1}{3} + \tfrac{1}{5} - \tfrac{1}{7} + \cdots.$$

13–2.9. Proposition. A series of nonnegative real terms converges if and only if its partial sums are bounded. If so, its sum is the least upper bound of its partial sums.

Proof. Since the partial sums form an increasing sequence in $\mathbf{R}$, this is 12–5.1.

13–2.10. Proposition. Suppose that $(\forall n)\, 0 \le a_n \le b_n$. If F-$\sum b_n$ converges, then so does F-$\sum a_n$, and $\sum a_n \le \sum b_n$ with equality holding if and only if $(\forall n)\, a_n = b_n$. If F-$\sum a_n$ diverges, then so does F-$\sum b_n$.

Proof. If $\{A_n\}$ and $\{B_n\}$ are the sequences of partial sums of the two series, then it is clear that $(\forall n)\, A_n \le B_n$, whence sup $A_n \le$ sup B_n. The first two conclusions now follow from 13–2.9. If $a_k < b_k$ for a single integer k, then $A_k < B_k$ and

$$\sum_{n=k+1}^{\infty} a_n \le \sum_{n=k+1}^{\infty} b_n$$

by the part already proved. If we add and apply 13–2.5, we get

$$\sum_{n=1}^{\infty} a_n < \sum_{n=1}^{\infty} b_n.$$

This establishes the stated criterion for equality. The last statement is simply the contrapositive of the first.

This proposition is the basis of the all-important comparison method for investigating the convergence of series which we shall discuss in 13–2.15.

13–2.11. Proposition. The series F-$\sum a_n$ converges if and only if

$$(\forall \epsilon > 0)(\exists M \in \mathbf{R})(\forall p > M)(\forall q \geq p) \quad \left| \sum_{n=p}^{q} a_n \right| < \epsilon. \tag{3}$$

Proof. This is just Cauchy's criterion for the convergence of the sequence of partial sums.

13–2.12. Theorem. If F-$\sum |a_n|$ converges, then F-$\sum a_n$ converges and

$$\left| \sum a_n \right| \leq \sum |a_n|.$$

Equality holds if and only if there exists a nonzero complex number λ such that $(\forall n)\ \lambda a_n \geq 0$.

Proof. The inequality

$$\left| \sum_{n=p}^{q} a_n \right| \leq \sum_{n=p}^{q} |a_n|$$

implies immediately that if F-$\sum |a_n|$ satisfies (3), then so does F-$\sum a_n$. Proposition 13–2.11 then implies that if F-$\sum |a_n|$ converges, so does F-$\sum a_n$.

We could prove the inequality by a limit argument from the known inequality for finite sums, but the following reasoning involves a very interesting technical device. There is a complex number μ such that $|\mu| = 1$ and $|\sum a_n| = \mu \sum a_n$. Then we have

$$\left| \sum a_n \right| = \mathrm{Re}(\mu \sum a_n) = \sum \mathrm{Re}(\mu a_n) \leq \sum |\mu a_n| = \sum |a_n|.$$

Equality can hold here only if $(\forall n)\ \mathrm{Re}(\mu a_n) = |\mu a_n|$; that is, only if $(\forall n)\ \mu a_n \geq 0$.

Now suppose λ is a nonzero complex number such that $(\forall n)\ \lambda a_n \geq 0$. Then $|\sum \lambda a_n| = \sum \lambda a_n = \sum |\lambda a_n|$. Therefore

$$|\lambda| \left| \sum a_n \right| = |\lambda| \sum |a_n|$$

and, dividing out $|\lambda|$, we have

$$\left| \sum a_n \right| = \sum |a_n|.$$

13–2.13. Definition. The infinite series F-$\sum a_n$ is said to converge *absolutely* if and only if F-$\sum |a_n|$ converges.

This terminology is justified by the previous theorem which says that an absolutely convergent series is convergent. A series which converges, but not absolutely, is said to be *conditionally* convergent.

13–2.14. Theorem. Suppose that F-$\sum b_n$ is a convergent series of nonnegative real numbers. If for all sufficiently large n, $|a_n| \leq b_n$, then F-$\sum a_n$ is absolutely convergent.

Proof. This is an immediate consequence of 13–2.6 and 13–2.10.

When the conditions of this proposition are satisfied, the series F-$\sum b_n$ is said to *dominate* F-$\sum a_n$. This criterion for absolute convergence is known as the *comparison test*.

13–2.15. One of the commonest problems in analysis is to decide whether a given series converges or diverges. There is no general method for solving such problems. In fact, one of the most celebrated problems in all mathematics can be put in the following form:

Consider F-$\sum \mu(n)/n^\alpha$, where $\mu(n) = 0$ if n is divisible by the square of a prime and $\mu(n) = (-1)^p$ if n is the product of p distinct primes. (The first few values of μ, called the Möbius function, are $\mu(1) = 1$, $\mu(2) = -1$, $\mu(3) = -1$, $\mu(4) = 0$, $\mu(5) = -1$, $\mu(6) = 1, \ldots$) For what values of α does this series converge?

By the comparison test, the series converges absolutely for $\alpha > 1$. There is evidence to suggest that it converges for $\alpha > \frac{1}{2}$, but it has never been proved to converge for any $\alpha < 1$. The conjecture that it does converge for all $\alpha > \frac{1}{2}$ is equivalent to a problem raised by Riemann in 1859, now known as the Riemann hypothesis.

Nevertheless there are some routine tests to try, which in practice often settle the matter. First, one should check to see whether the terms approach zero. This is a necessary but not sufficient condition for convergence (necessity by 13–2.11; the standard example used to show that it is not sufficient is the *harmonic* series, F-$\sum(1/n)$, which diverges although its terms tend to zero). Hence, if we can prove that the terms do not approach zero, the problem is solved.

The next thing to try is the *ratio test*. Consider the behavior of the ratio $|a_{n+1}|/|a_n|$ as n increases. (If some of the a's are zero, we cannot form this ratio. Since convergence depends only on the "infinite part" of the series, any finite number of zero terms can simply be neglected. If there are infinitely many zero terms, the series should be "closed up" by suppressing the zero terms and renumbering those remaining. This doesn't affect the convergence according to 13–2.8.)

Suppose that $|a_{n+1}|/|a_n| \geq 1$ for all sufficiently large n. This happens, in particular, if $\lim \inf |a_{n+1}|/|a_n| > 1$. Then for some integer k, we have

$$|a_k| \leq |a_{k+1}| \leq |a_{k+2}| \leq \cdots$$

Thus the terms do not approach zero and the series diverges.

Suppose $\rho = \lim \sup |a_{n+1}|/|a_n| < 1$. Pick a number μ satisfying $\rho < \mu < 1$. There is an integer k such that $|a_{n+1}|/|a_n| < \mu$ for all $n \geq k$. It follows that $|a_n| \leq |a_k|\mu^{n-k}$ for all $n \geq k$. Hence the series is dominated by the convergent geometric series F-$\sum_n |a_k|\mu^{n-k}$ and converges absolutely.

If $\lim |a_{n+1}|/|a_n| = 1$, as often happens, the next step is to try to compare the series with F-$\sum(1/n)^\alpha$, which converges for $\alpha > 1$ and diverges for $\alpha \leq 1$. Naturally, one tries the critical case $\alpha = 1$ first. If $\lim \inf n|a_n| > \beta > 0$, then F-$\sum|a_n|$ dominates F-$\sum(\beta/n)$ and diverges. Of course, this does not mean that F-$\sum a_n$ diverges, only that it does not converge absolutely.

Raabe's test is a systematic method for comparing a series with F-$\sum \lambda/n^\alpha$. It is in many respects analogous to the ratio test. Let

$$\sigma_n = n\left(1 - \frac{|a_{n+1}|}{|a_n|}\right).$$

If $\sigma_n \le 1$ for all sufficiently large n (in particular, if lim sup $\sigma_n < 1$), then there exists an integer k such that $n|a_{n+1}| \ge (n-1)|a_n|$ for all $n \ge k$. This implies that lim inf $n|a_n| > 0$, so F-$\sum a_n$ does not converge absolutely.

Suppose, on the other hand, that lim inf $\sigma_n > \alpha > 1$. It is not hard to establish the inequality $(1+x)^\alpha(1-\alpha x) < 1$ for all positive x. (Look at the derivative with respect to x.) Setting $x = 1/n$, we obtain $(n+1)^\alpha(1-\alpha/n) < n^\alpha$ for all n. Now choose k so that $\sigma_n > \alpha$ for all $n \ge k$. Then we have $|a_{n+1}| < (1-\alpha/n)|a_n|$ and therefore

$$(n+1)^\alpha|a_{n+1}| < (n+1)^\alpha(1-\alpha/n)|a_n| < n^\alpha|a_n|$$

for all $n \ge k$. It follows that

$$\text{F-}\sum_n \frac{k^\alpha|a_k|}{n^\alpha}$$

dominates F-$\sum a_n$. Thus the latter converges absolutely.

Even if a series is actually dominated by a geometric series, its convergence may not be demonstrable with either the ratio test or Raabe's test. For example, consider the series

$$\frac{1}{2} + \frac{1}{9} + \frac{1}{8} + \frac{1}{3^4} + \frac{1}{2^5} + \cdots$$

($a_n = (\frac{1}{2})^n$ if n is odd, $a_n = (\frac{1}{3})^n$ if n is even). This is clearly dominated by F-$\sum(\frac{1}{2})^n$, but both tests fail. Such "artificial" series do not usually cause any difficulty, however.

If a series does not converge absolutely, or if one cannot prove that it does, then the outlook is grim. Favorable cases include alternating series and others that can be put in the form given in Exercise 5, p. 182. Blocking the terms together, as in 13–2.7, may sometimes clarify the situation, but one must remember that this process may convert a divergent series into a convergent one.

The establishment of absolute convergence by comparison with a known series of positive terms is not only important for deciding convergence. It is essential in computation. As we noted in Chapter 12, it is of no help in computation to know that a sequence converges unless we have some estimate of how rapidly it converges. Suppose that we want to compute $\sum_{n=1}^\infty (1/n^2 2^n)$. If we evaluate the sum of the first k terms, we can be sure that the remainder $\sum_{n=k+1}^\infty (1/n^2 2^n)$ is less than

$$\sum_{n=k+1}^\infty \frac{1}{(k+1)^2 2^n} = \frac{1}{(k+1)^2 2^k}.$$

If we wanted actually to find the answer within 10^{-6} we could take $k = 13$ and be certain that the actual sum of the series differs from the sum of the first 13 terms by less than 10^{-6}.

Series which converge absolutely have another very important property. They can be rearranged arbitrarily without affecting either convergence or the value of the sum. In other words, an infinite form of the commutative law is valid. Before proving this, let us show that a convergent, but not absolutely convergent, series always has the opposite property. Thus the curious result of 13-2.2 is typical of conditionally convergent series.

Suppose we are given a conditionally convergent series of real terms. Let α and β be any two real numbers subject only to the condition $\alpha \leq \beta$. We shall show that we can arrange the terms in such an order that α is the inferior limit of the sequence of partial sums and β is the superior limit.

Let F-$\sum a_n$ be the series whose terms are the negative terms of the given series in the same order. Let F-$\sum b_n$ be the series whose terms are the nonnegative terms of the given series. If both of these new series converge, then the original series would have been absolutely convergent. If F-$\sum a_n$ converges while F-$\sum b_n$ diverges, then the original series would have diverged to $+\infty$. It is also impossible for F-$\sum a_n$ to diverge and F-$\sum b_n$ to converge. Thus both new series are divergent. Furthermore, because the original series converges, $a_n \to 0$ and $b_n \to 0$. Let $\{A_n\}$ and $\{B_n\}$ be the partial-sum sequences of F-$\sum a_n$ and F-$\sum b_n$, respectively.

We now reshuffle the a's and the b's together to obtain the required rearrangement of the original series. First, take just enough a's to make the partial sum less than α. Then take just enough b's to bring the partial sum above β. Then take just enough a's to again depress the partial sum below α, and again take just enough b's to bring the partial sum above β. Continue in this alternating fashion. Each cycle requires at least one a and at least one b, so eventually all the terms are used. This gives us the required rearrangement.

To describe this process a little more precisely, let us define two increasing sequences of positive integers by induction. Let r_1 be the least integer such that $A_{r_1} < \alpha$. Let s_1 be the least integer such that $A_{r_1} + B_{s_1} > \beta$. If r_k and s_k have been defined, then let r_{k+1} be the least integer such that $A_{r_{k+1}} + B_{s_k} < \alpha$, and let s_{k+1} be the least integer such that $A_{r_{k+1}} + B_{s_{k+1}} > \beta$.

The final rearranged series, written informally, is to be

$$a_1 + a_2 + \cdots + a_{r_1} + b_1 + b_2 + \cdots + b_{s_1} + a_{r_1+1} + a_{r_1+2} + \cdots$$
$$+ a_{r_2} + b_{s_1+1} + b_{s_1+2} + \cdots + b_{s_2} + a_{r_2+1} + \cdots$$

Let us denote the partial sums of this new series by $\{C_n\}$. Now,

$$C_{r_k+s_k} = A_{r_k} + B_{s_k} > \beta;$$

therefore, $\limsup C_n \geq \beta$. If $\epsilon > 0$ is given, we can choose k so that $b_n < \epsilon$ for all $n \geq s_k$. Suppose that $n \geq r_k + s_k$. There is a unique integer m such that $m \geq k$ and

$$r_m + s_m \leq n < r_{m+1} + s_{m+1}.$$

It follows from the construction that

$$C_n \leq C_{r_m+s_m} \leq \beta + b_{s_m} < \beta + \epsilon.$$

Hence lim sup $C_n \leq \beta + \epsilon$. Since ϵ was chosen arbitrarily, this gives

$$\text{lim sup } C_n \leq \beta.$$

Finally, lim sup $C_n = \beta$, as required. The proof that lim inf $C_n = \alpha$ is similar. It is clear that every number between α and β is the limit of a subsequence of $\{C_n\}$.

A slight modification of the argument will show that we could allow α or β to be infinite. The reader will find it interesting to discover the possibilities which may occur when a convergent series of complex terms is rearranged.

As we remarked above, an absolutely convergent series remains convergent with the same sum no matter how it is rearranged. We shall now prove this in full generality by developing the notion of an unordered infinite sum.

13-2.16. Let X be a fixed set and consider the set $\mathfrak{F}$ of all functions from X to $\mathbf{C}$. We shall combine and compare members of $\mathfrak{F}$ in a *pointwise* manner. Thus if $f, g \in \mathfrak{F}$ and $\lambda, \mu \in \mathbf{C}$, then $\lambda f + \mu g$ denotes the function defined by

$$(\lambda f + \mu g)(x) = \lambda f(x) + \mu g(x) \qquad \text{for } x \in X.$$

(This definition makes $\mathfrak{F}$ into a *vector space*.) Similarly, we shall write fg or $f \cdot g$ for the function defined by

$$(fg)(x) = f(x)g(x) \qquad \text{for } x \in X,$$

and $|f|$ for the function defined by

$$|f|(x) = |f(x)|.$$

The set of real-valued members of $\mathfrak{F}$ becomes partially ordered if we define $f \leq g$ to mean $(\forall x) f(x) \leq g(x)$. The corresponding strong order relation will be denoted by $<$, as usual, but it is *not* calculated pointwise, because $f < g$ does *not* mean $(\forall x) f(x) < g(x)$. If $\{f_n\}$ is a sequence in $\mathfrak{F}$ and $f \in \mathfrak{F}$, we shall write $f_n \to f$ *pointwise* to mean that $(\forall x) f_n(x) \to f(x)$. (The word *pointwise* is usually included in the notation because there are many other commonly used senses of $f_n \to f$. In the other contexts, particularly in the addition of functions or the multiplication by numbers, it is usually omitted, because no other interpretation is at all common.)

Pointwise addition and multiplication of functions are both associative and commutative, and multiplication distributes over addition. We shall occasionally use 0 to denote the constant function with value 0, for example, in $f_n \to 0$ *point-wise*. The theory of Section 13–1 applies to $\mathfrak{F}$ regarded as a semigroup under addition, hence unordered finite sums are meaningful in $\mathfrak{F}$. It is trivial to verify that

$$\left(\sum_{n=1}^{k} f_n\right)(x) = \sum_{n=1}^{k} f_n(x).$$

We shall need the characteristic functions of subsets of X. If $Y \subseteq X$, then $\chi(Y)$, called the *characteristic function* of Y, is defined by

$$\chi(Y)(x) = 1 \quad \text{if } x \in Y,$$

and

$$\chi(Y)(x) = 0 \quad \text{if } x \notin Y.$$

If Y and Z are disjoint subsets of X, then $\chi(Y \cup Z) = \chi(Y) + \chi(Z)$. This fact can be extended immediately by induction to:

(i) If $\{Y_j \mid j \in J\}$ is a finite family of mutually disjoint subsets of X, then
$$\chi\left(\bigcup_j Y_j\right) = \sum_j \chi(Y_j).$$

Suppose that $f \in \mathfrak{F}$. We would like to assign a meaning, in some cases at least, to $\sum_{x \in X} f(x)$. We have already done this when X or $X \cap \mathrm{spt}\, f$ is finite. To do this we needed only the associative and commutative properties of the addition of complex numbers. To extend the definition, we will need some sort of limit process. If it is to resemble the sum of an infinite series, $\sum_{x \in X} f(x)$ must be a number which can be well approximated by the sums $\sum_{x \in A} f(x)$, where A is a *finite* subset of X.

Suppose there exists a complex number ζ with the following property:

$$(\forall \epsilon > 0)(\exists A)(\forall B \supseteq A) \quad |\zeta - \textstyle\sum_B f(x)| < \epsilon, \tag{4}$$

where A and B are understood to range over the *finite* subsets of X. This means that ζ can be well-approximated by the finite sum $\sum_A f(x)$ and increasing A will make very little difference. It seems reasonable, therefore, that we should write $\sum_{x \in X} f(x)$ for ζ.

Compare this condition with the definition of convergence of an infinite series. Then X is N and A and B must be not merely finite subsets of N but actually segments of the integers. In the absence of an order structure for X we could hardly transfer the old definition more precisely.

Before stating this definition formally we should check two things. First, if $X \cap \mathrm{spt}\, f$ is finite, then (4) is satisfied with $\zeta = \sum_{X \cap \mathrm{spt} f} f(x)$, so the new definition will agree with the old in this case. Second, we must be sure that there is at most one number ζ which fulfills condition (4). We shall prove this in a manner analogous to 12–1.5.

Assume that (4) is true and that

$$(\forall \epsilon > 0)(\exists A)(\forall B \supseteq A) \quad |\eta - \textstyle\sum_B f(x)| < \epsilon.$$

We shall prove $\zeta = \eta$. Suppose $\zeta \neq \eta$. Then we could choose A_1 and A_2 such that

$$(\forall B \supseteq A_1) \quad |\zeta - \textstyle\sum_B f(x)| < \tfrac{1}{3}|\zeta - \eta|$$

and

$$(\forall B \supseteq A_2) \quad |\eta - \textstyle\sum_B f(x)| < \tfrac{1}{3}|\zeta - \eta|.$$

Then

$$|\zeta - \eta| \leq |\zeta - \textstyle\sum_{A_1 \cup A_2} f(x)| + |\eta - \textstyle\sum_{A_1 \cup A_2} f(x)| < \tfrac{2}{3}|\zeta - \eta|,$$

which is a contradiction.

13–2.17. Definition. Let f be a function from X to $\mathbf{C}$. We shall say that f is *summable* if and only if there is a number ζ which satisfies (4), and this number ζ will be called the *sum of f over X*. It will be denoted by $\sum_{x \in X} f(x)$. This notation will be abbreviated in various ways, such as $\sum_X f$. Suppose that $Y \subseteq X$. We shall say that f is *summable over Y* if and only if f restricted to Y is summable.

The following proposition is important as a matter of housekeeping, but its proof is routine and we shall omit it.

13–2.18. Proposition. Let $f \in \mathfrak{F}$ and $Y \subseteq X$. Then f is summable over Y if and only if $f \cdot \chi(Y)$ is summable over X. If so $\sum_Y f = \sum_X f \cdot \chi(Y)$.

The partial sums of a function $\{\sum_A f \mid A \text{ finite}\}$ form a family of numbers of which the index class is not linearly ordered like the integers, but only partially ordered (by inclusion). Nevertheless we can define the notion of limits for such families. The proof of the uniqueness of limits depends on the fact that given any two members of the index class, there is an index larger than both. In the following proposition this fact is once again the key to the proof.

13–2.19. Proposition. If f and g are summable and $\lambda, \mu \in \mathbf{C}$, then $\lambda f + \mu g$ is summable and

$$\textstyle\sum_X (\lambda f(x) + \mu g(x)) = \lambda \sum_X f(x) + \mu \sum_X g(x).$$

Proof. Suppose $\zeta = \sum_X f(x)$ and $\eta = \sum_X g(x)$. Let ϵ be a positive number. Choose finite subsets A_1 and A_2 of X such that

$$(\forall B \supseteq A_1) \quad |\zeta - \textstyle\sum_B f(x)| < \epsilon/2(1 + |\lambda|)$$

and

$$(\forall B \supseteq A_2) \quad |\eta - \textstyle\sum_B g(x)| < \epsilon/2(1 + |\mu|).$$

Then

$$(\forall B \supseteq A_1 \cup A_2) \quad |\lambda\zeta + \mu\eta - \textstyle\sum_B (\lambda f(x) + \mu g(x))| < \epsilon.$$

13–2.20. Lemma. Suppose that $\{a_j \mid j \in J\}$ is a family of complex numbers such that the sums $\sum_{j \in K} a_j$ are bounded as K varies over the finite subsets of J. Then the sums $\sum_{j \in K} |a_j|$ are also bounded; in fact,

$$\sup_K \{\sum_K |a_j|\} \leq 4 \sup_K \{|\sum_K a_j|\}. \tag{5}$$

Proof. Let $S = \sup_K \{|\sum_K a_j|\}$. Let L be any finite subset of J. Let

$$L_1 = \{j \in L \mid \text{Re } a_j \geq 0\} \quad \text{and} \quad L_2 = \{j \in L \mid \text{Re } a_j < 0\}.$$

Then

$$\sum_{L_1} |\text{Re } a_j| = |\sum_{L_1} \text{Re } a_j| = |\text{Re } \sum_{L_1} a_j| \leq |\sum_{L_1} a_j| \leq S.$$

Similarly, $\sum_{L_2} |\text{Re } a_j| \leq S$. Hence $\sum_L |\text{Re } a_j| \leq 2S$. By the same kind of reasoning, $\sum_L |\text{Im } a_j| \leq 2S$. Finally,

$$\sum_L |a_j| \leq \sum_L (|\text{Re } a_j| + |\text{Im } a_j|) \leq 4S.$$

Since L is any finite subset of J, (5) follows.

It is interesting to determine the least number which can replace 4 in (5).

13–2.21. Theorem. Let $f \in \mathfrak{F}$. Then f is summable if and only if the sums $\sum_{x \in C} |f(x)|$ are bounded as C varies over the finite subsets of X. Moreover, if f is summable,

$$|\sum_{x \in X} f(x)| \leq \sup_C \sum_C |f(x)|.$$

Proof. Suppose that f is summable; say, $\mathfrak{s} = \sum_X f(x)$. Let ϵ be any positive number. Choose a finite set A such that

$$(\forall B \supseteq A) \quad |\mathfrak{s} - \sum_B f(x)| < \epsilon.$$

If K is any finite subset of $X - A$, then

$$|\sum_K f(x)| = |(\mathfrak{s} - \sum_A f(x)) - (\mathfrak{s} - \sum_{A \cup K} f(x))| < 2\epsilon.$$

Hence, by the lemma, if L is any finite subset of $X - A$,

$$\sum_L |f(x)| < 8\epsilon.$$

Now, if C is any finite subset of X, then

$$\sum_C |f(x)| \leq \sum_A |f(x)| + \sum_{C-A} |f(x)| < \sum_A |f(x)| + 8\epsilon,$$

so the sums $\sum_C |f(x)|$ are bounded. Furthermore,

$$|\mathfrak{s}| \leq |\mathfrak{s} - \sum_A f(x)| + \sum_A |f(x)| < \epsilon + \sup_C \sum_C |f(x)|.$$

Since ϵ was chosen arbitrarily, $|\mathfrak{s}| \leq \sup_C \sum_C |f(x)|$.

Conversely, suppose that $\sup_C \sum_C |f(x)| = M < \infty$. For each integer n, choose a set C_n such that $\sum_{C_n} |f(x)| > M - 1/n$. Then if D is any finite subset of $X - C_n$, $\sum_D |f(x)| < 1/n$.

Let $\lambda_n = \sum_{C_n} f(x)$. For any p and q, we have

$$|\lambda_p - \lambda_q| = |\sum_{C_p - C_q} f(x) - \sum_{C_q - C_p} f(x)|$$

$$\leq \sum_{C_p - C_q} |f(x)| + \sum_{C_q - C_p} |f(x)| < \frac{1}{q} + \frac{1}{p}.$$

Hence $\{\lambda_n\}$ is a Cauchy sequence. By 12–5.3, $\{\lambda_n\}$ converges, say to $\mathfrak{z}$. Then

$$|\mathfrak{z} - \lambda_q| = \lim_p |\lambda_p - \lambda_q| \leq \lim_p \sup \left(\frac{1}{q} + \frac{1}{p}\right) = \frac{1}{q}.$$

Therefore, if B is a finite set containing C_q, then

$$|\mathfrak{z} - \sum_B f(x)| \leq |\mathfrak{z} - \sum_{C_q} f(x)| + \sum_{B - C_q} |f(x)| < \frac{1}{q} + \frac{1}{q}.$$

Thus if a positive ϵ is given, and $q > 2/\epsilon$, then

$$(\forall B \supseteq C_q) \quad |\mathfrak{z} - \sum_B f(x)| < \epsilon.$$

Hence f is summable; in fact, $\sum_X f(x) = \mathfrak{z}$.

13–2.22. Corollary. If f is summable, then spt f is countable.

Proof. In the notation of the previous proof, it is clear that

$$(\forall x \in X - C_n) \quad |f(x)| < \frac{1}{n}.$$

Hence spt $f \subseteq \bigcup_n C_n$, a countable set.

Although the definition of an unordered sum appears to enable us to add uncountably many numbers together, this is an illusion, because all but countably many of them must be zero.

13–2.23. Corollary. Suppose f is summable and g is a function such that $|g| \leq |f|$. Then g is summable. In particular, $|f|$ is summable, and

$$\sum_X |f| = \sup \{\sum_C |f(x)| \,|\, C \text{ finite}\}.$$

Proof. All but the last of these statements are immediate consequences of the theorem. The last equation follows from the fact that, for any $\epsilon > 0$, a finite set A can be found which satisfies both

$$|\sum_X |f| - \sum_A |f|| < \epsilon \quad \text{and} \quad \sum_A |f| > \sup_C \sum_C |f| - \epsilon.$$

This corollary tells us that the convergence of unordered sums is always absolute (compare this with 13–2.14).

13–2.24. Corollary. If f and g are summable real-valued functions and $f < g$, then $\sum_X f < \sum_X g$.

Proof. Since $0 < g - f = |g - f|$,
$$\sum_X(g - f) > 0.$$

By 13–2.19, $\sum_X f < \sum_X g$. (We are using the fact that $\sum_X f$ and $\sum_X g$ are real. See Exercise 9.)

13–2.25. Corollary. If f is summable and $Y \subseteq X$, then f is summable over Y.

Proof. This follows immediately from 13–2.18 and 13–2.23.

13–2.26. Lemma. Suppose f is summable. For any positive number ϵ there is a finite subset A of spt f such that
$$|\sum_X f - \sum_Z f| < \epsilon \tag{6}$$
for any (not necessarily finite) subset Z of X which contains A.

Proof. Choose B so that $|\sum_X f - \sum_C f| < \epsilon/2$ for all finite supersets C of B. Let $A = B \cap \text{spt} f$.
Suppose that $Z \supseteq A$. Pick a finite subset D of Z such that
$$|\sum_Z f - \sum_E f| < \frac{\epsilon}{2}$$
for all finite sets E satisfying $D \subseteq E \subseteq Z$. Then we have
$$|\sum_X f - \sum_{B \cup D} f| < \frac{\epsilon}{2}, \qquad |\sum_Z f - \sum_{A \cup D} f| < \frac{\epsilon}{2},$$
and
$$\sum_{B \cup D} f = \sum_{A \cup D} f,$$
the last since $(B \cup D) \cap \text{spt} f = (A \cup D) \cap \text{spt} f$. Combining these we have (6).
Note how the summability of f over Z enters the proof. We could not even state this lemma without 13–2.25.

13–2.27. Theorem. Let $f \in \mathfrak{F}$. Suppose $\{Y_j \mid j \in J\}$ is a family of mutually disjoint subsets of X such that $\text{spt} f \subseteq \bigcup Y_j$. Then f is summable if and only if

(a) f is summable over Y_j for each $j \in J$, and
(b) the function $\tau : j \to \sum_{x \in Y_j} |f(x)|$ is summable over J. If so, then
$$\sum_{x \in X} f(x) = \sum_{j \in J} \sum_{x \in Y_j} f(x), \tag{7}$$
and
$$\sum_{x \in X} |f(x)| = \sum_{j \in J} \sum_{x \in Y_j} |f(x)|.$$

This is the general associative-commutative law for unordered infinite sums. Granting that f is summable over X, we can find $\sum_X f$ by breaking up the set X at pleasure, summing f over each part and totaling the results, exactly as we would do for finite sums. Moreover, we can decide whether f is summable by attempting to do the same thing for $|f|$ (or any larger function). Roughly speaking, f is summable over X if and only if $\sum_J \sum_{Y_j} |f|$ is sensible. Often one defines $\sum_Z |f(x)|$ as ∞ if f is not summable over Z; then, with the obvious convention on how to add ∞, f is summable if and only if $\sum_J \sum_{Y_j} |f| < \infty$.

Proof. Suppose that f is summable over X and let $\zeta = \sum_X f$. Since f is also summable over each of the sets Y_j, we can define σ by

$$\sigma(j) = \sum_{Y_j} f = \sum_X f \cdot \chi(Y_j).$$

We shall first prove (7), that is, $\sum_J \sigma(j) = \zeta$.

Let $\epsilon > 0$ be given. Pick a finite set $A \subseteq \operatorname{spt} f$ such that

$$|\zeta - \sum_Z f(x)| < \epsilon \tag{8}$$

for any (possibly infinite) superset Z of A. Let $K = \{j \in J \mid Y_j \cap A \neq \emptyset\}$. Since A is finite and the Y's are disjoint, K is finite. Since

$$A \subseteq \operatorname{spt} f \subseteq \bigcup Y_j, \quad A \subseteq \bigcup_{j \in K} Y_j.$$

Let L be any finite set such that $K \subseteq L \subseteq J$. Set $Z = \bigcup_{j \in L} Y_j$. Then $Z \supseteq A$; hence (8) is true. Also,

$$\sum_L \sigma(j) = \sum_L \sum_X f \cdot \chi(Y_j) = \sum_X f \cdot \chi(Z) = \sum_Z f,$$

where the second equality is by (i) on p. 200 and 13–2.19 extended to finite sums. Thus $|\zeta - \sum_L \sigma| < \epsilon$.

By definition then, σ is summable over J and (7) holds. The proof to this point is applicable to any summable function f, in particular, to $|f|$. Hence the function τ is summable over J and the last equation of the theorem is established.

Conversely, suppose f is summable over each of the sets Y_j and that τ is summable over J. Let C be any finite subset of X. Let

$$H = \{j \in J \mid C \cap Y_j \neq \emptyset\}.$$

Then H is finite and $C \cap \operatorname{spt} f \subseteq \bigcup_H Y_j$. Therefore,

$$\sum_C |f| = \sum_{C \cap \operatorname{spt} f} |f| = \sum_{j \in H} \sum_{C \cap Y_j \cap \operatorname{spt} f} |f|$$
$$\leq \sum_H \sum_{Y_j} |f| = \sum_H \tau \leq \sum_J \tau.$$

Thus the sums $\sum_C |f|$ are bounded and f is summable by 13–2.21.

13–2.28. Corollary. Let f be a function of countable support. Let φ be an injection of $\mathbf{N}$ into X such that range $\varphi \supseteq \operatorname{spt} f$. Then f is summable if and only if

F-$\sum_{n=1}^{\infty} f(\varphi(n))$ is absolutely convergent. If so,

$$\sum_X f(x) = \sum_{n=1}^{\infty} f(\varphi(n)).$$

Proof. In the theorem let $J = N$ and let Y_n be the one-point set $\{\varphi(n)\}$.

13–2.29. Corollary. Let $\{a_n\}$ be a sequence of complex numbers. Regarded as a function, a is summable over N if and only if F-$\sum_{n=1}^{\infty} a_n$ is absolutely convergent. If so,

$$\sum_N a_n = \sum_{n=1}^{\infty} a_n.$$

13–2.30. Corollary. Let φ be a bijection from N to itself. If F-$\sum_{n=1}^{\infty} a_n$ is absolutely convergent, then so is F-$\sum_{n=1}^{\infty} a_{\varphi(n)}$, and

$$\sum_{n=1}^{\infty} a_n = \sum_{n=1}^{\infty} a_{\varphi(n)}.$$

13–2.31. Theorem. Let $\{f_n\}$ be a sequence of functions which converges pointwise to g. Suppose there exists a summable function h such that $(\forall n)\, |f_n| \leq h$. Then g is summable and

$$\sum_X g = \lim_n \sum_X f_n.$$

In other words, the formula

$$\sum_X \lim_n f_n = \lim_n \sum_X f_n$$

is valid whenever there is a single summable function h which dominates every term of the sequence $\{f_n\}$. The theorem is therefore known as the dominated convergence theorem. It is a special case of the Lebesgue dominated convergence theorem in general integration theory.

This is our first theorem concerned with the important question of when it is permissible to interchange two limit operations. To emphasize this aspect of the theorem we might write the formula

$$\lim_A \lim_n \sum_A f_n = \lim_n \lim_A \sum_A f_n,$$

where A is to be a finite subset of X and $\lim_A$ refers to the limiting process (4) used in defining sums over X. In obtaining the left member of this last equation, we have used the fact that $\lim_n \sum_A f_n = \sum_A \lim_n f_n$ for *finite* sets A. This is an immediate consequence of 12–2.1(a). It is easy to give examples in which the interchange of limits and sums is invalid.

Proof. The hypothesis and 13–2.23 make it quite clear that each of the functions f_n is summable. Temporarily, let us make the additional assumption that $g = 0$.

The hypothesis then becomes $f_n \to 0$ pointwise, and the desired conclusion is $\lim_n \sum_X f_n = 0$. In view of the obvious inequality

$$\liminf_n |\sum_X f_n| \geq 0,$$

it will be sufficient to prove

$$\limsup_n |\sum_X f_n| \leq 0.$$

The idea of the proof is to use h to divide the sum into two sums. One sum is finite; for this the interchange is trivial. The other sum is (presumably) infinite, but because of the domination by h, it is negligible.

Let ϵ be a given positive number. Choose a finite set A such that

$$\sum_A h(x) > \sup_C \sum_C h(x) - \epsilon.$$

Then we have

$$\sum_D |f_n(x)| \leq \sum_D h(x) < \epsilon$$

for any n and any finite subset D of $X - A$. If B is any finite subset of X, then

$$\sum_B |f_n(x)| \leq \sum_A |f_n(x)| + \sum_{B-A} |f_n(x)| \leq \sum_A |f_n(x)| + \epsilon.$$

Hence

$$|\sum_X f_n(x)| \leq \sup_B \sum_B |f_n(x)| \leq \sum_A |f_n(x)| + \epsilon.$$

Therefore,

$$\limsup_n |\sum_X f_n| \leq \sum_A \limsup_n |f_n(x)| + \epsilon = \epsilon,$$

because A is finite and $f_n(x) \to 0$ for each fixed x. Since ϵ was arbitrary, this shows that $\limsup_n |\sum_X f_n| \leq 0$, and finishes the proof in the special case.

Now let us drop the hypothesis $g = 0$. We note that

$$|g(x)| = \lim_n |f_n(x)| \leq h(x)$$

for all x. Thus $|g| \leq h$, and g is summable. The functions $f_n - g$ tend to zero pointwise and are dominated by $2h$. Then the special case applies and we see that $\sum_X (f_n - g) \to 0$, whence $\sum_X f_n \to \sum_X g$.

13–2.32. Double series. *A double sequence is an array*

$$
\begin{array}{llll}
a_{0,0} & a_{0,1} & a_{0,2} & a_{0,3} \quad \cdots \\
a_{1,0} & a_{1,1} & a_{1,2} & a_{1,3} \quad \cdots \\
a_{2,0} & a_{2,1} & a_{2,2} & \cdots \\
\cdots
\end{array}
$$

infinite both downward and to the right. Technically, it is a function defined on

N* × N*. (Or on **N × N**, of course. For our present purposes it is more natural to number from zero.)

If we are given a double sequence $\{a_{m,n}\}$ of complex numbers, we may wish to add them up. There are several natural ways to go about this. For each row, we could try to add the entries of that row. Then, granting that these series converge, we could add up the results. In formal language, for each fixed m, consider F-$\sum_{n=0}^{\infty} a_{m,n}$. If the series converges for every m, consider F-$\sum_{m=0}^{\infty} \sum_{n=0}^{\infty} a_{m,n}$. If this series converges, its sum is called the *sum by rows* of the double series $\sum_{m,n} a_{m,n}$. Similarly, we can define the *sum by columns*.

Another way would be to sum the finite diagonals from lower left to upper right to get the sequence $a_{0,0}$, $a_{1,0} + a_{0,1}$, $a_{2,0} + a_{1,1} + a_{0,2}$, ..., and then sum these numbers. That is, consider

$$\text{F-}\sum_{k=0}^{\infty} \sum_{i=0}^{k} a_{k-i,i}.$$

It is natural to consider the rectangular partial sums

$$A_{p,q} = \sum_{m=0}^{p} \sum_{n=0}^{q} a_{m,n}, \tag{9}$$

and hope that $\lim_n A_{n,n}$ exists, or perhaps $\lim_n A_{2n,n}$. The so-called *double limit* of these sums might exist. This is a number ζ such that

$$(\forall \epsilon > 0)(\exists M \in \mathbf{R})(\forall p, q > M) \quad |\zeta - A_{p,q}| < \epsilon.$$

(There is at most one such number, of course.) When such a number does exist, it is classically defined to be the sum of the double series $\sum_{m,n} a_{m,n}$. Note that the sum by rows is $\lim_p \lim_q A_{p,q}$ and the sum by columns is $\lim_q \lim_p A_{p,q}$. Thus the problem of interchanging limits appears again in connection with summing a double series.

It is a remarkable fact that there is almost no consistency between these various methods of summing the double series $\sum_{m,n} a_{m,n}$. Consider the double sequence

$$
\begin{array}{ccccccc}
1 & 0 & 0 & 0 & 0 & \cdots \\
-1 & 1 & 0 & 0 & 0 & \cdots \\
0 & -1 & 1 & 0 & 0 & \cdots \\
0 & 0 & -1 & 1 & 0 & \cdots \\
& \cdots
\end{array}
$$

($a_{n,n} = 1$, $a_{n+1,n} = -1$, all others $= 0$). The sum by rows is 1, the sum by columns is 0, but F-$\sum_{k=0}^{\infty} \sum_{i=0}^{k} a_{k-i,i}$ does not converge, and the classical double sum does not exist!

On the other hand, if the double sequence $\{a_{m,n}\}$ is summable over **N* × N***, then all methods of summation lead to the same result. The only requirement is that every term should enter the sum exactly once.

13–2.33. Theorem. Let $\{a_{m,n}\}$ be a double sequence of complex numbers. If any one of the series

$$\sum_{m=0}^{\infty}\sum_{n=0}^{\infty}|a_{m,n}|, \qquad \sum_{n=0}^{\infty}\sum_{m=0}^{\infty}|a_{m,n}|, \qquad \sum_{k=0}^{\infty}\sum_{i=0}^{k}|a_{k-i,i}| \qquad (10)$$

converges, then all of the sums

$$\sum_{m=0}^{\infty}\sum_{n=0}^{\infty}a_{m,n}, \qquad \sum_{n=0}^{\infty}\sum_{m=0}^{\infty}a_{m,n}, \qquad \sum_{k=0}^{\infty}\sum_{i=0}^{k}a_{k-i,i}, \qquad (11)$$

and the classical double sum exist and are equal.

In (10) we abandoned the F-$\sum$ notation. Note that the criterion for the summability of $\{a_{m,n}\}$ is *not* the convergence of one of the series

$$\sum_{m=0}^{\infty}\left|\sum_{n=0}^{\infty}a_{m,n}\right|, \qquad \sum_{n=0}^{\infty}\left|\sum_{m=0}^{\infty}a_{m,n}\right|, \qquad \text{or} \qquad \sum_{k=0}^{\infty}\left|\sum_{i=0}^{k}a_{k-i,i}\right|. \qquad (12)$$

Proof. By 13–2.27, the convergence of any one of the three series (10) is sufficient to prove that the function a is summable over $\mathbf{N}^* \times \mathbf{N}^*$. Then each of the series (11) converges to $\zeta = \sum_{\mathbf{N}^*\times\mathbf{N}^*} a_{m,n}$. Moreover, given an $\epsilon > 0$, we can find a subset C of $\mathbf{N}^* \times \mathbf{N}^*$ such that $|\zeta - \sum_D a_{m,n}| < \epsilon$ for any superset D of C. There is an integer r such that $C \subseteq \{\langle m, n\rangle \mid m \leq r, n \leq r\}$. Then we have

$$(\forall p, q \geq r) \quad |\zeta - A_{p,q}| < \epsilon,$$

where $A_{p,q}$ is the rectangular partial sum defined in (9). Hence the classical double sum is also ζ.

13–2.34. The Cauchy product of two series. Suppose F-$\sum_{n=0}^{\infty} a_n$ and F-$\sum_{n=0}^{\infty} b_n$ are two convergent series; say, $\zeta = \sum_{n=0}^{\infty} a_n$ and $\eta = \sum_{n=0}^{\infty} b_n$. Then

$$\zeta\eta = \left(\sum_{m=0}^{\infty} a_m\right)\left(\sum_{n=0}^{\infty} b_n\right) = \sum_{n=0}^{\infty}\left(\sum_{m=0}^{\infty} a_m\right)b_n = \sum_{n=0}^{\infty}\sum_{m=0}^{\infty} a_m b_n$$

by two applications of 13–2.4. (Note the change of dummy indices to avoid confusing the two sums.) Similarly,

$$\zeta\eta = \sum_{m=0}^{\infty}\sum_{n=0}^{\infty} a_m b_n.$$

Thus $\zeta\eta$ can be obtained either as the row sum or as the column sum of the double series $\sum_{m,n} a_m b_n$. If the terms of this double series are collected along the diagonals, the resulting series

$$\text{F-}\sum_{k=0}^{\infty}\sum_{i=0}^{k} a_{k-i}b_i$$

is called the *Cauchy product* of F-$\sum a_n$ and F-$\sum b_n$. Evidently, the Cauchy product can be formed regardless of whether the original series converge or not. The significance of this method of collecting terms is apparent when the original series are *power series*; that is, $a_n = c_n z^n$ and $b_n = d_n z^n$. Then the Cauchy product is

$$\text{F-}\sum_{k=0}^{\infty} \left(\sum_{i=0}^{k} c_{k-i} d_i \right) z^k.$$

As we might expect, the Cauchy product of two convergent series need not be convergent (see Exercise 15). However, if the two original series are absolutely convergent, the situation is more favorable.

13–2.35. Proposition. The Cauchy product of two absolutely convergent series F-$\sum a_n$ and F-$\sum b_n$ is absolutely convergent and

$$\sum_{k=0}^{\infty} \sum_{i=0}^{k} a_{k-i} b_i = \left(\sum_{n=0}^{\infty} a_n \right) \left(\sum_{n=0}^{\infty} b_n \right). \tag{13}$$

Proof. The absolute convergence of F-$\sum a_n$ and F-$\sum b_n$ leads immediately to the fact that $\sum_{m=0}^{\infty} \sum_{n=0}^{\infty} |a_m b_n|$ is finite. Hence the double sequence $\{a_m b_n\}$ is summable. Therefore, its sum by diagonals (the left member of (13)) is absolutely convergent and has the same value as the sum by rows, which is the right member of (13).

A more sophisticated theorem asserts that the Cauchy product of an absolutely convergent series and a convergent series is convergent and that (13) holds.

EXERCISES

1. In the example of 13–2.2 show that $B_{3k} = A_{4k} + \frac{1}{2} A_{2k}$. What can you conclude about $\sum a_n$ and $\sum b_n$?

2. Prove the inequalities

$$\sum_{n=1}^{k} \frac{1}{n^2} + \frac{1}{k+1} < \sum_{n=1}^{\infty} \frac{1}{n^2} < \sum_{n=1}^{k} \frac{1}{n^2} + \frac{1}{k}$$

for any positive integer k.

3. Let A_k be the kth partial sum of the alternating harmonic series (1) and let S be its sum. Prove the inequalities

$$A_{2k-1} - \frac{1}{4k-1} < S < A_{2k} + \frac{1}{4k+1}$$

for every positive integer k.

*4. Let α be a real number. Prove that

$$\text{F-}\sum_{n=1}^{\infty} \frac{1}{n^{\alpha}}$$

is convergent if $\alpha > 1$, divergent if $\alpha \leq 1$.

*5. Consider the binomial expansion of $(1 + (-1))^m$, that is, the series

$$1 - m + \frac{m(m-1)}{1 \cdot 2} - \frac{m(m-1)(m-2)}{1 \cdot 2 \cdot 3} + \cdots$$

or F-$\sum a_n$, where

$$a_0 = 1 \quad \text{and} \quad a_{n+1} = -a_n \frac{m-n}{n+1} \quad \text{for all } n.$$

For which values of m does it converge?

6. Prove that if $(\forall n) \, a_n \leq b_n$ and F-$\sum b_n$ is convergent, then either F-$\sum a_n$ is convergent or it diverges to $-\infty$.

7. Disentangling two convergent series. Let $\{a_n\}$ be a sequence of complex numbers. Suppose φ is a strictly increasing function from $\mathbf{N}$ to $\mathbf{N}$ and that F-$\sum a_{\varphi(n)}$ converges. Let $c_n = a_n$ if $n \notin$ range φ, $c_n = 0$ if $n \in$ range φ. Prove that F-$\sum a_n$ converges if and only if F-$\sum c_n$ converges, and if so,

$$\sum a_n = \sum c_n + \sum a_{\varphi(n)}.$$

8. Prove that F-$\sum a_n$ converges absolutely if and only if both F-$\sum$ Re a_n and F-$\sum$ Im a_n converge absolutely.

9. Prove that a complex-valued function f is summable over X if and only if its real and imaginary parts are both summable, and if so,

$$\text{Re} \sum_X f(x) = \sum_X \text{Re} f(x), \qquad \text{Im} \sum_X f(x) = \sum_X \text{Im} f(x).$$

10. Suppose that f is summable over X and g is a bounded function on X. Prove that fg is summable over X.

11. Let f be a bijection from $\mathbf{N}$ to $\mathbf{N}$ such that $\{f(n) - n\}$ is bounded. Prove that if F-$\sum a_n$ is convergent, then F-$\sum a_{f(n)}$ is convergent, and $\sum a_n = \sum a_{f(n)}$.

12. Give an example of a sequence of summable functions $\{f_n\}$ which converges pointwise to a summable function, yet $\lim_n \sum_X f_n(x) \neq \sum_X \lim_n f_n(x)$.

13. Give an example of a double series for which all the double sums (12) are convergent and yet $\{a_{m,n}\}$ is not summable.

14. Suppose $\sum_{m,n} a_{m,n}$ is a double series for which the classical double sum exists and each row sum exists. Prove that the sum by rows converges and equals the classical sum (Pringsheim's theorem). Show, however, that it is possible for the double sum to exist without any row sums existing.

15. Show that the Cauchy product of the convergent series

$$\text{F-}\sum_{n=0}^{\infty} \frac{(-1)^n}{\sqrt{n+1}}$$

with itself is divergent.

16. By looking at the sum by rows and the sum by columns of a suitable double series, show that

$$\sum_{n=1}^{\infty} \frac{nz^n}{1 + (-z)^n} = \sum_{n=1}^{\infty} \frac{z^n}{(1 + (-z)^n)^2}$$

provided $|z| < 1$.

13–3. INFINITE PRODUCTS

It is natural to extend the notion of product to allow for an infinite number of factors. The obvious thing to do is consider the infinite sequence of partial products. (We could also consider unordered infinite products, but we leave this to the reader.) However, because of the special properties of zero with respect to multiplication, the most obvious definition of a convergent infinite product is not the valuable one. In this section we shall define convergence of infinite products and establish the most useful criterion for identifying a convergent infinite product.

Suppose $\{\rho_n\}$ is a sequence of complex numbers. We want to define the infinite product $\prod_{n=1}^{\infty} \rho_n$. Form the partial products $P_k = \prod_{n=1}^{k} \rho_n$. If we should define the infinite product as simply $\lim_k P_k$ whenever the latter exists, the theory would not be closely analogous to the theory of infinite series. If one of the ρ's should be zero, then $P_k = 0$ for all large k, and $\lim P_k = 0$. If we suppress this zero factor from the product, the resulting infinite product might not converge. Thus the convergence of an infinite product would depend on single terms and not, as in the case of infinite series, only on the "infinite part" of the sequence $\{\rho_n\}$. Suppose for a moment that none of the ρ's vanish, and $\lim P_k = \zeta$. There is a significant difference between the cases $\zeta = 0$ and $\zeta \neq 0$. In the latter case we can reason that

$$\lim \rho_n = \frac{\lim P_n}{\lim P_{n-1}} = 1.$$

In the former case this is not valid and, indeed, we might have $\rho_n = \frac{1}{2}$ for all n, or ρ_n might oscillate wildly as n increases. To put the matter in another way, when we are obtaining a limit by successive multiplications, we might hope to find a partial product which has a small *relative* error, and this is impossible if $\lim P_k = 0$ but no ρ is zero. To avoid both difficulties we shall admit a finite number of zero factors and demand that the product of the nonzero factors converge to a number other than zero.

13–3.1. Definition. Let $\{\rho_n\}$ be a sequence of complex numbers. For each pair of integers k, m with $0 \leq k < m$, let $P(k, m) = \prod_{n=k+1}^{m} \rho_n$. The *formal infinite product* F-$\prod_{n=1}^{\infty} \rho_n$ is the function P. It is said to be convergent if and only if for some k, $\lim_m P(k, m)$ exists and is not zero. Otherwise it is divergent. If it is convergent, we denote $\lim_m P(0, m)$ by $\prod_{n=1}^{\infty} \rho_n$. If for every k, $\lim_m P(k, m) = 0$, then F-$\prod_{n=1}^{\infty} \rho_n$ is said to *diverge to zero*.

Note the relation $P(k_1, k_2)P(k_2, m) = P(k_1, m)$. It follows that, if $\lim_m P(k, m)$ exists for one value of k, it exists for smaller values of k; in particular, $\lim_m P(0, m)$ exists. It is obvious that $\lim_m P(k, m) \neq 0$ if and only if $\rho_{k+1} \neq 0$ and

$$\lim_m P(k + 1, m) \neq 0.$$

Thus the choice of k in the definition is unimportant except that there must be no zero factor with index larger than k. It is an immediate consequence of the definition that $\prod_{n=1}^{\infty} \rho_n = 0$ if and only if at least one factor is zero.

The admission of a finite number of zero factors is very convenient when the factors are functions. Consider the infinite product

$$\left(1 - \frac{z^2}{1^2}\right)\left(1 - \frac{z^2}{2^2}\right)\left(1 - \frac{z^2}{3^2}\right) \cdots \tag{1}$$

$(\rho_n = 1 - z^2/n^2)$. We shall see presently that this product converges for all complex numbers z. (Unfortunately we will not be able to prove the remarkable fact that it converges to $(\sin \pi z)/\pi z$ for $z \neq 0$.) Denoting the product by $\varphi(z)$, we see that $\varphi(z) = 0$ if and only if one of the factors vanishes; that is, if and only if z is a nonzero integer.

We shall omit a number of elementary propositions concerning infinite products analogous to 13–2.5 ff. and take up the Cauchy criterion for convergence as it applies to infinite products.

13–3.2. Proposition. In order that $\text{F-}\prod_{n=1}^{\infty} \rho_n$ should converge it is necessary and sufficient that (in the notation of 13–3.1)

$$(\forall \epsilon > 0)(\exists m)(\forall q > m) \quad |P(m, q) - 1| < \epsilon. \tag{2}$$

In particular, it is necessary but not sufficient that $\rho_n \to 1$.

Proof. Suppose that $\text{F-}\prod \rho_n$ converges. Choose k so that

$$\lim_n P(k, n) = \zeta \neq 0.$$

Let ϵ be a given positive number. We shall assume, without loss of generality, that $\epsilon < 2$. Choose $m > k$ so that

$$(\forall q \geq m) \quad |\zeta - P(k, q)| < \tfrac{1}{4}|\zeta|\epsilon.$$

Then

$$|\zeta| \leq |\zeta - P(k, m)| + |P(k, m)| < \tfrac{1}{4}|\zeta|\epsilon + |P(k, m)| < \tfrac{1}{2}|\zeta| + |P(k, m)|,$$

whence $|P(k, m)| > \tfrac{1}{2}|\zeta|$. Now, if $q > m$, then

$$\tfrac{1}{2}|\zeta| \, |P(m, q) - 1| \leq |P(k, m)| \, |P(m, q) - 1| = |P(k, q) - P(k, m)|$$
$$\leq |\zeta - P(k, m)| + |\zeta - P(k, q)| < \tfrac{1}{2}|\zeta|\epsilon.$$

Thus $(\forall q > m) |P(m, q) - 1| < \epsilon$. This proves that (2) is necessary.

The simple necessary condition for convergence, $\rho_n \to 1$, is a consequence of (2), but it can be derived more simply, because $\rho_n = P(k, n)/P(k, n - 1)$ if $n > k + 1$, so

$$\lim \rho_n = \frac{\lim_n P(k, n)}{\lim_n P(k, n - 1)} = 1.$$

Conversely, suppose that (2) is true. Let k be an integer such that

$$(\forall m > k) \quad |P(k, m) - 1| < \tfrac{1}{2}. \tag{3}$$

Then $(\forall m > k)\,|P(k, m)| < 2$.

We shall show that $\{P(k, n) \mid n = k + 1, k + 2, \ldots\}$ is a Cauchy sequence. Let ϵ be a positive number. Choose m so that $m > k$ and

$$(\forall q > m) \quad |P(m, q) - 1| < \frac{\epsilon}{2}.$$

Then if $q > m$,

$$|P(k, q) - P(k, m)| = |P(k, m)|\,|P(m, q) - 1| < 2 \cdot \frac{\epsilon}{2} = \epsilon.$$

Therefore,

$$(\forall \epsilon > 0)(\exists m)(\forall q > m) \quad |P(k, q) - P(k, m)| < \epsilon.$$

This is an alternative form of the Cauchy criterion. Hence $\lim_m P(k, m)$ exists. By (3),

$$|\lim_m P(k, m) - 1| \le \tfrac{1}{2};$$

hence $\lim_m P(k, m) \ne 0$. This proves that $\text{F-}\prod_{n=1}^{\infty} \rho_n$ converges.

It is now evident that the convergence of $\text{F-}\prod \rho_n$ is related to the speed with which $\rho_n \to 1$. For this reason we shall in the future consider products in the form $\text{F-}\prod(1 + a_n)$. Now the problem becomes how $a_n \to 0$. We shall see that $\text{F-}\prod(1 + a_n)$ is significantly related to $\text{F-}\sum a_n$.

13-3.3. Lemma. Let $\{b_n \mid 1 \le n \le k\}$ be a finite sequence of complex numbers such that $\sum_{n=1}^{k} |b_n| \le \tfrac{1}{2}$. Then

$$\left| \prod_{n=1}^{k} (1 + b_n) - 1 \right| \le 2 \sum_{n=1}^{k} |b_n|.$$

Proof. This is obvious for $k = 1$. Suppose it is true for $k = m - 1$. Writing $\prod$ for $\prod_{n=1}^{m-1} (1 + b_n)$ and $\sum$ for $\sum_{n=1}^{m-1} |b_n|$, we obtain

$$|(1 + b_m)\prod - 1| = |(1 + b_m)(\prod - 1) + b_m| \le |1 + b_m|\,|\prod - 1| + |b_m|$$

$$\le (1 + |b_m|)(2\textstyle\sum) + |b_m| = 2\textstyle\sum + (2\textstyle\sum + 1)|b_m|$$

$$\le 2(\textstyle\sum + |b_m|).$$

Here the third step is by the inductive hypothesis and the last because we are given $\sum \le \tfrac{1}{2}$. Rewriting this inequality, we have

$$\left| \prod_{n=1}^{m} (1 + b_n) - 1 \right| \le 2 \sum_{n=1}^{m} |b_n|.$$

This completes the inductive proof.

13-3.4. Theorem. If F-$\sum_{n=1}^{\infty} |a_n|$ converges, then F-$\prod_{n=1}^{\infty} (1 + a_n)$ converges.

Proof. Let ϵ be a positive number. We assume, without loss of generality, that $\epsilon < 1$. Choose an integer m such that $\sum_{n=m+1}^{\infty} |a_n| < \epsilon/2$. Then if $q > m$, the lemma gives

$$\left| \prod_{n=m+1}^{q} (1 + a_n) - 1 \right| \leq 2 \sum_{n=m+1}^{q} |a_n| < \epsilon.$$

Therefore, F-$\prod(1 + a_n)$ satisfies (2) and converges.

This theorem implies that the product (1) is convergent for any value of z, because F-$\sum |z^2/n^2|$ is convergent for any value of z.

A formal infinite product which satisfies the condition of this theorem is said to be absolutely convergent. By a slight modification of the preceding proof, it can be shown that an absolutely convergent product remains convergent to the same value if its terms are rearranged. Conversely, a formal infinite product which converges, but not absolutely, can be rearranged to become divergent. This is easy to show when the a's are real (Exercise 5), but is somewhat harder with complex a's.

The situation becomes transparent with the aid of logarithms. The positive real numbers endowed with the operation of multiplication and their usual order structure is isomorphic to the set of all real numbers endowed with the operation of addition and its usual order structure. The function log effects this isomorphism since log is an order-preserving bijection from the positive reals to **R** and

$$\log ab = \log a + \log b.$$

Assuming that the a's are real and exceed -1, this transformation converts the formal infinite product F-$\prod_{n=1}^{\infty} (1 + a_n)$ into the infinite series F-$\sum_{n=1}^{\infty} \log (1 + a_n)$ and the convergence of the former is equivalent to the convergence of the latter.

If F-$\sum \log (1 + a_n)$ converges absolutely, then it remains convergent to the same sum when its terms are rearranged. Correspondingly, F-$\prod(1 + a_n)$ remains convergent with the same product when its factors are rearranged. But if

$$\text{F-}\sum \log (1 + a_n)$$

converges only conditionally, then rearrangement of its terms may spoil the convergence or alter the value of the sum. Similarly, rearrangement of the factors of F-$\prod(1 + a_n)$ may spoil the convergence or alter the value of the product.

Absolute convergence of F-$\sum \log (1 + a_n)$ is equivalent to the convergence of F-$\sum |a_n|$, because of the inequalities

$$\tfrac{1}{2}|a_n| \leq |\log (1 + a_n)| \leq 2|a_n|, \tag{4}$$

which hold for small values of $|a_n|$.

For complex numbers, the situation is slightly more complicated. The additive structure of the complex numbers is not quite isomorphic to the multiplicative structure of $\mathbf{C} - \{0\}$. Nevertheless, they are nearly isomorphic (we shall study the situation in Chapter 15), and there exists an extension of the logarithm function to complex numbers with, say, positive real parts which has the appropriate properties. Since F-$\prod(1 + a_n)$ cannot converge unless all the partial products $P(k, m)$ for large k and m are close to 1, the question of convergence can once again be transformed into the question of convergence of F-$\sum \log (1 + a_n)$. The inequalities (4) remain valid for small $|a_n|$, hence the question of absolute convergence of the series is again equivalent to the convergence of F-$\sum |a_n|$.

Unfortunately, the convergence of F-$\prod(1 + a_n)$ is not in general equivalent to the convergence of F-$\sum a_n$. (See Exercises 7 and 8.)

EXERCISES

1. Suppose that $(\forall n)\ \rho_n \neq 0$ and F-$\prod \rho_n$ converges. Prove that F-$\prod(1/\rho_n)$ converges and that

$$\prod_{n=1}^{\infty} \frac{1}{\rho_n} = \left(\prod_{n=1}^{\infty} \rho_n \right)^{-1}.$$

2. By a direct analysis of the partial products show that Wallis' infinite product

$$\frac{2 \cdot 2 \cdot 4 \cdot 4 \cdot 6 \cdot 6}{1 \cdot 3 \cdot 3 \cdot 5 \cdot 5 \cdot 7} \cdots$$

$(\rho_{2k-1} = 2k/(2k - 1),\ \rho_{2k} = 2k/(2k + 1))$ is convergent.

3. Show that

$$\varphi(z) = z \left(1 - \frac{z^2}{1^2} \right) \left(1 - \frac{z^2}{2^2} \right) \left(1 - \frac{z^2}{3^2} \right) \cdots$$

(the product (1) multiplied by z) defines a function φ such that $\varphi(z + 1) = -\varphi(z)$ and $\varphi(1 - z) = \varphi(z)$.

4. Suppose that $\{a_n\}$ is a decreasing sequence of positive numbers and that

$$\frac{a_{n+1}}{a_n} = 1 - \frac{1}{n} + \epsilon_n,$$

where $n\epsilon_n \to 0$ (in which case Raabe's test fails to decide the convergence of F-$\sum a_n$). Show that if F-$\sum |\epsilon_n|$ converges, then F-$\sum a_n$ diverges.

*Deduce the following frequently applied result: If a_{n+1}/a_n is a rational function of n, and

$$\lim_n n \left(1 - \frac{a_{n+1}}{a_n} \right) = 1,$$

then F-$\sum a_n$ diverges.

5. Suppose that $\{a_n\}$ is a sequence of nonnegative numbers and F-$\sum a_n$ diverges. Show that F-$\prod(1 + a_n)$ and F-$\prod(1 - a_n)$ both diverge, the latter to zero provided that $a_n < 1$ for sufficiently large n. Hence show that if $\{b_n\}$ is a real sequence such that F-$\sum |b_n|$ diverges, then there is a bijection φ from $\mathbf{N}$ to $\mathbf{N}$ such that F-$\prod(1 + b_{\varphi(n)})$ diverges.

6. Show that $F\text{-}\prod_{n=0}^{\infty}(1 + (-1)^n/(2n + 1))$ converges.

7. Show that $F\text{-}\prod_{n=0}^{\infty}(1 + (-1)^n/\sqrt{n + 1})$ diverges and thus disprove the conjecture that if $F\text{-}\sum a_n$ converges, then $F\text{-}\prod(1 + a_n)$ converges.

*8. Suppose that $\{a_n\}$ is a real sequence and $F\text{-}\sum a_n$ converges. Use the inequalities

$$t - t^2 \leq \log(1 + t) \leq t - \tfrac{1}{3}t^2 \qquad \text{for} \quad -\tfrac{1}{2} \leq t \leq \tfrac{1}{2}$$

to prove that $F\text{-}\prod(1 + a_n)$ converges if $F\text{-}\sum a_n^2$ converges and diverges to zero if $F\text{-}\sum a^2$ diverges. By modifying the infinite product of Exercise 7, show that there is a convergent infinite product $F\text{-}\prod(1 + b_n)$ such that $F\text{-}\sum b_n$ diverges.

13-4. NUMERATION AND CALCULATION

We shall develop the well-known decimal representation of real numbers as well as the representations in other bases. As a by-product we shall prove that card $\mathbf{R} = 2^{\aleph_0}$.

13-4.1. Lemma. For every real number x there is a unique integer y such that $y \leq x < y + 1$. If $x \geq 0$, then $y \geq 0$.

Proof. This follows easily from the Archimedean property of the real numbers and 8-5.6.

13-4.2. Definition. If m is an integer greater than 1, we shall denote by D_m the set $\{i \in \mathbf{I} \mid 0 \leq i < m\}$. (This is the set of "digits" when numbers are expressed in the base m.)

13-4.3. Proposition. Let m be an integer greater than 1. Let x be a real number, $0 \leq x < 1$. There is a unique sequence $\{a_n\}$ in D_m such that $x = \sum_{n=1}^{\infty} a_n m^{-n}$ and $a_n \neq m - 1$ for infinitely many indices n.

Proof. For each $p \in \mathbf{N}^*$, let y_p be the integer such that $y_p \leq m^p x < y_p + 1$. Then $0 \leq x - m^{-p}y_p < m^{-p}$, so $\lim_p m^{-p}y_p = x$.
For $p \in \mathbf{N}$, set $a_p = y_p - my_{p-1}$. Then $a_p \in \mathbf{I}$. We know that

$$my_{p-1} \leq m^p x < my_{p-1} + m.$$

Hence $my_{p-1} \leq y_p \leq my_{p-1} + m - 1$. Therefore, $a_p \in D_m$. Furthermore, it follows by induction on p that

$$y_p = \sum_{n=1}^{p} a_n m^{p-n}.$$

Hence

$$\sum_{n=1}^{p} a_n m^{-n} = m^{-p}y_p,$$

and therefore $\sum_{n=1}^{\infty} a_n m^{-n} = x$.

If $(\forall n > k)\, a_n = m - 1$, then

$$m^k x - y_k = \sum_{n=1}^{\infty} a_n m^{k-n} - \sum_{n=1}^{k} a_n m^{k-n} = \sum_{n=k+1}^{\infty} (m-1) m^{k-n} = 1,$$

contradicting the choice of y_k. Hence $a_n \neq m - 1$ for infinitely many n. This proves the existence of a sequence of the required type.

Suppose $\{b_n\}$ is a sequence in D_m such that $x = \sum_{n=1}^{\infty} b_n m^{-n}$ and $b_n \neq m - 1$ for infinitely many indices n. For any $p \in \mathbf{N}^*$, we have

$$\sum_{n=1}^{p} b_n m^{p-n} \leq m^p x < \sum_{n=1}^{p} b_n m^{p-n} + \sum_{n=p+1}^{\infty} (m-1) m^{p-n} = \sum_{n=1}^{p} b_n m^{p-n} + 1.$$

Since $\sum_{n=1}^{p} b_n m^{p-n}$ is an integer, it follows that $y_p = \sum_{n=1}^{p} b_n m^{p-n}$. Hence

$$a_p = y_p - m y_{p-1} = \sum_{n=1}^{p} b_n m^{p-n} - \sum_{n=1}^{p-1} b_n m^{p-n} = b_p.$$

Thus the sequence $\{a_n\}$ is unique.

13–4.4. Corollary. The cardinal of $\mathbf{R}$ is $2^{\aleph_0}$.

Proof. Let $J = \{x \in \mathbf{R} \mid 0 \leq x \leq 1\}$. It is a consequence of Exercise 4, p. 140 and the Schröder-Bernstein theorem that card $\mathbf{R} = $ card J.

Let A be the set of functions (sequences) from $\mathbf{N}$ to $D_2 = \{0, 1\}$. By definition card $A = 2^{\aleph_0}$. Setting $m = 2$ in 13–4.3, we find that $\{a_n\} \to \sum_{n=1}^{\infty} a_n 2^{-n}$ is a surjection from A to J. Setting $m = 3$, we see that $\{a_n\} \to \sum_{n=1}^{\infty} a_n 3^{-n}$ is an injection from A to J. It follows that card $A = $ card J.

13–4.5. Definition. Let m be an integer greater than 1. An *m-adic sequence* is a function f from $\mathbf{I}$ to D_m which takes the value 0 at all sufficiently large positive integers; that is,

$$(\exists k)(\forall i > k)\quad f(i) = 0.$$

An m-adic sequence will be called *standard* if and only if for infinitely many negative integers its value is not $m - 1$; that is,

$$(\forall k)(\exists i < k)\quad f(i) \neq m - 1.$$

13–4.6. Theorem. Let m be an integer greater than 1. The mapping $f \to \sum_{\mathbf{I}} f(i) m^i$ is a bijection from the set of all standard m-adic sequences to the nonnegative real numbers.

Proof. To begin with we must note that if f is an m-adic sequence, the function $i \to f(i) m^i$ is summable over $\mathbf{I}$. It is dominated by the function g defined by $g(i) = (m - 1) m^i$ for $i \leq k$ and $g(i) = 0$ for $i > k$, where k is an integer such that $(\forall i > k)\, f(i) = 0$. It is easy to prove that g is summable; in fact, $\sum_{\mathbf{I}} g = m^{k+1}$.

Let y be a nonnegative number. We must prove that y is the image of exactly one standard m-adic sequence. There is an integer p such that $p > y$. Since $m^p > p$, $y < m^p$.

By 13–4.3 there is a sequence $\{a_n\}$ in D_m such that $m^{-p}y = \sum_{n=1}^{\infty} a_n m^{-n}$, and $a_n \neq m - 1$ for infinitely many indices n. Let $f(i) = a_{p-i}$ for $i < p$, and $f(i) = 0$ for $i \geq p$. Then f is a standard m-adic sequence.

Let $\varphi(n) = p - n$ for $n \in \mathbf{N}$. Since φ is an injection from $\mathbf{N}$ to $\mathbf{I}$ and

$$\text{range } \varphi \supseteq \text{spt } f,$$

Corollary 13–2.28 tells us that

$$\sum_{\mathbf{I}} f(i)m^i = \sum_{n=1}^{\infty} f(\varphi(n))m^{\varphi(n)} = m^p \sum_{n=1}^{\infty} a_n m^{-n} = m^p(m^{-p}y) = y.$$

Thus y is the image of at least one standard m-adic sequence.

Suppose g is a standard m-adic sequence and $y = \sum_{\mathbf{I}} g(i)m^i$. If $q \geq p$, then $g(q)m^q \leq y < m^q$. Since $g(q)$ is a nonnegative integer, $g(q) = 0$. Thus spt $g \subseteq$ range φ, so

$$y = \sum_{n=1}^{\infty} g(\varphi(n))m^{\varphi(n)} = m^p \sum_{n=1}^{\infty} g(p - n)m^{-n}.$$

Since g is standard, $g(p - n) \neq m - 1$ for infinitely many integers n. By 13–4.3, $g(p - n) = a_n = f(p - n)$ for all $n \in \mathbf{N}$, or $g(i) = f(i)$ for $i < p$. We already know that $g(i) = f(i) = 0$ for $i \geq p$; hence $f = g$. Thus y is the image of exactly one standard m-adic sequence.

The unique standard m-adic sequence f for which $y = \sum f(i)m^i$ is called the *standard m-adic representation of y*. The sum is frequently written $y = \sum_{i=-\infty}^{k} f(i)m^i$, where the upper summation index reminds us that the support of f is bounded above. Sometimes it is convenient to number in the other direction: $a_i = f(-i)$. Then $y = \sum_{i=-k}^{\infty} a_i m^{-i}$.

Numbers which can be put in the form nm^{-p}, where n and p are integers are called *m-adic rational* numbers. The nonnegative m-adic rationals are precisely the numbers for which the standard m-adic representation has finite support. There are convenient practical algorithms to combine or compare m-adic rationals directly in terms of their standard representations. That is, we can easily compute the standard representations of $x + y$ and xy, or decide whether $x < y$, directly from the standard representations of x and y, provided that x and y are m-adic rationals.

13–4.7. Proposition. The mapping $f \rightarrow \sum_{\mathbf{I}} f(i)m^i$ is a bijection from the set of all nonstandard m-adic sequences to the set of positive m-adic rationals.

We shall omit the proof of this proposition, which follows from the fact that $\sum_{i=-\infty}^{k} (m - 1)m^i = m^{k+1}$.

Thus we see that all nonnegative numbers, except the positive m-adic rationals, have exactly one m-adic representation, which is standard, while each positive m-adic rational has two m-adic representations, the standard one, which has finite support, and the nonstandard one, which "terminates in $(m - 1)$'s."

We find that m-adic representations are of particular importance for two values of m. When m is two, the word m-adic is rendered *dyadic*. Dyadic representations are widely used in computing machines and occasionally in theoretical work. The other important value of m is ten.

We define the symbols 2, 3, 4, 5, 6, 7, 8, and 9 formally by $2 = 1 + 1$, $3 = 2 + 1$, $4 = 3 + 1$, $5 = 4 + 1$, $6 = 5 + 1$, $7 = 6 + 1$, $8 = 7 + 1$, and $9 = 8 + 1$. In what follows, the symbols 0, 1, 2, 3, 4, 5, 6, 7, 8, and 9 will be called the *digital representatives* of the corresponding integers. When $m = 9 + 1$, the word m-adic is rendered *decimal*. With the aid of the standard decimal representation f of a nonnegative decimal rational x we obtain the *standard English numeral for x* as follows. Let $k = \sup(\operatorname{spt} f \cup \{0\})$ and $l = \inf(\operatorname{spt} f \cup \{0\})$. We write the digital representatives of $f(k)$, $f(k - 1)$, $f(k - 2)$, ..., $f(l)$ consecutively in this order. Then if l is negative, we insert a period, called the *decimal point*, between $f(0)$ and $f(-1)$. The standard English numeral for a negative decimal rational y is the numeral for $|y|$ preceded by the symbol $-$. The standard English numeral for -2^{-5} is therefore -0.03125.

Certain departures from these rules are generally accepted. When $k = 0$ and $f(0) = 0$, this zero is often not written (e.g., .04 for 5^{-2}). Extra zeros are often written to the right of $f(l)$ (with a decimal point, of course, if $l = 0$). Rarely (except in computer outputs) are extra zeros written to the left of $f(k)$. When $k - l$ is large, the numeral is often broken up into short blocks. The usual scheme is blocks of three separated by commas to the left of the decimal point and blocks of five separated by spaces to the right of the decimal point.

Another well-known numeration scheme is the exponential form. Let x be a positive decimal rational having f as the standard decimal representation. Let $k = \sup \operatorname{spt} f$ and $l = \inf \operatorname{spt} f$. We write in order the digital representatives of $f(k)$, $f(k - 1)$, ..., $f(l)$. If $k \neq l$, we put a decimal point between $f(k)$ and $f(k - 1)$. If $k \neq 0$, this sequence is followed by $\times 10^k$, where k is written as a standard English numeral. In the exponential form 2^{-5} is

$$3.125 \times 10^{-2}.$$

It happens that the m-adic expansion f of a rational number is always eventually periodic $\big($that is, $(\exists p > 0)(\exists k)(\forall m < k)\, f(m - p) = f(m)\big)$. Sometimes this fact is used to define a standard numeral for each nonnegative rational $\big($e.g., $\frac{3}{14} = 0.2\dot{1}4285\dot{7}$, where the dots indicate that $(\forall m < -1)\, f(m - 6) = f(m)\big)$. An expression like 3.14159^+ is only an informal representation of a certain decimal sequence. It cannot be interpreted as a genuine numeral (that is, an unambiguous symbol for a definite number). The "equation" $\pi = 3.14159^+$ is really the double inequality $3.14159 < \pi < 3.14160$.

It is possible to compute directly with irrational numbers; analysis, in fact, is largely devoted to establishing arithmetic relations between numbers defined by various limiting processes. However, most of what we usually call computation is concerned with approximate calculations in which numbers are, if necessary, "rounded off" to a nearby decimal rational (dyadic rational in many computers). This rounding-off process, valuable as it is in keeping computations within reasonable bounds, nevertheless introduces serious complications into the whole computational process.

Under the traditional rules of rounding off, the statement

$$\xi = 0.92 \text{ to 2 decimals}$$

implies $0.915 \le \xi \le 0.925$. (According to one standard rule, it would be equivalent to $0.915 < \xi \le 0.925$.) These rules lead to a curious anomaly. If we know that $1.24995 < \eta < 1.25005$, then we know $\eta = 1.2500$ to 4 decimal places, but we do not know η to 1 decimal place!

To avoid this difficulty, it seems better to define "compute ξ to k decimals" as meaning "prove inequalities of the form $a \le \xi \le b$," where a and b are decimal rationals expressed as standard numerals, and $b - a \le 10^{-k}$.

EXAMPLE. Compute $\sum_{n=1}^{\infty} (10^{-n}/n^2)$ to 5 decimals. We have

$$0.100000 = \frac{1}{1^2 10^1},$$

$$0.002500 = \frac{1}{2^2 10^2},$$

$$0.000111 < \frac{1}{3^2 10^3} < 0.000112,$$

$$0.000006 < \frac{1}{4^2 10^4} < 0.000007,$$

and

$$0 \le \sum_{n=5}^{\infty} \frac{1}{n^2 10^n} < \frac{1}{25 \cdot 10^4} \sum_{n=5}^{\infty} \frac{1}{10^{n-4}} = \frac{1}{25 \cdot 10^4 \cdot 9} < 0.000001.$$

Therefore,

$$0.102617 < \sum_{n=1}^{\infty} \frac{1}{n^2 10^n} < 0.102620.$$

We could report this as the answer or the less valuable double inequality

$$0.10261 \le \sum_{n=1}^{\infty} \frac{1}{n^2 10^n} \le 0.10262.$$

This illustrates how numerical computation in the real number system is based on the arithmetic of inequalities. One could, of course, avoid rounding-off problems in this calculation by carrying rational numbers throughout, recording the numerators and denominators separately as standard numerals, but this would make the work longer. Although the field operations can be carried out explicitly using rational numbers, it is usually easier to compute using rounded decimals, because the simplicity of the algorithms more than compensates for the complications of rounding-off.

EXERCISES

1. State and prove a theorem giving a criterion for deciding whether $x < y$ directly from the m-adic expansions of x and y.

2. Compute

$$\sum_{n=0}^{\infty} \frac{1}{n!} \ (= e) \quad \text{and} \quad \sum_{n=0}^{\infty} \frac{(-1)^n}{n!} \ (= e^{-1})$$

to 10 decimals. Recall that $n!$ is defined by $0! = 1$, $(n + 1)! = (n + 1)(n!)$.

3. Compute

$$\sum_{n=1}^{\infty} \frac{1}{n(3 + 4i)^n}$$

to 3 decimals (i.e., compute both the real and the imaginary parts to 3 decimals).

4. Using the inequality of Exercise 3, p. 210, compute

$$\sum_{n=1}^{\infty} \frac{(-1)^{n-1}}{n} \ (= \log 2)$$

to 3 decimals.

CHAPTER 14

THE TOPOLOGY OF METRIC SPACES

The notions of convergence and continuity play a central role in that branch of mathematics known as analysis. The most primitive form of these concepts concerns the convergence of sequences of real numbers and the continuity of real functions of a real variable.

Considerations of the geometry of space and functions of several variables make it necessary to extend these ideas to $\mathbf{R}^n$, whereupon it becomes clear that the arguments require only a few facts concerning the distances between points. When these facts are abstracted as the postulates for a class of configurations we are led to the notion of a metric space. The theory of metric spaces covers almost all of the convergence and continuity arguments that occur in analysis.

Particularly in Sections 14–2 through 14–5 there are many entirely routine propositions whose proofs are left without comment as exercises for the reader.

14–1. METRIC SPACES

A metric space is a set of objects, usually called points, between which a measure of distance is defined. The postulates single out those properties of ordinary distance which are important in convergence and continuity arguments.

14–1.1. Definition. A *metric space* is a configuration $\langle S, \rho \rangle$, where ρ is a function from $S \times S$ to $\mathbf{R}$ such that for all $s, t, u \in S$,

(a) $\rho(s, s) = 0$;
(b) if $s \neq t$, then $\rho(s, t) > 0$;
(c) $\rho(s, t) = \rho(t, s)$; and
(d) $\rho(s, u) \leq \rho(s, t) + \rho(t, u)$.

The function ρ is called the *distance function* or *metric* for S.

These postulates are all satisfied by geometric distance if expressed in some fixed unit. Thus, (a) and (b), the distance between any two points is a nonnegative number and actually positive if the points are different. Distances are the same measured in either direction (c) and, if measured via some third point, they are not diminished (d). The last postulate is called the *triangle law* because it reflects the geometrical fact that length of one side of a triangle is less than the sum of the lengths of the other two.

Following the usual practice, we shall often refer to a metric space by the name of its underlying set without mention of the metric.

The reference to the real number system in the definition implies that a metric space is actually a configuration involving two basic sets, S and $\mathbf{R}$, and the structure of $\mathbf{R}$ in addition to the function ρ. Since the postulates for $\mathbf{R}$ are categorical, we focus our attention only on the more significant components of the configuration, S and ρ.

It is occasionally convenient to allow the metric to take the value ∞ ($= +\infty$). Since we do not have addition defined for ∞, in the triangle law we make the obvious interpretation that $\infty + a = a + \infty = \infty$ for all $a \in \mathbf{R} \cup \{\infty\}$. Although very few changes would be necessary in the theorems to follow, no author would use a metric taking the value ∞ without explicitly calling attention to this fact.

The real and complex numbers become metric spaces if we define $\rho(x, y) = |x - y|$. In verifying this, only the triangle law presents the slightest difficulty and it is an immediate consequence of formula 10–3.7(h). Unless there is explicit context to the contrary, this metric is invariably meant whenever metric concepts concerning numbers are under discussion.

14–1.2. Definition. Two metric spaces $\langle S, \rho \rangle$ and $\langle T, \sigma \rangle$ are said to be *isometric* if and only if there exists a bijection f from S to T such that

$$(\forall s_1, s_2 \in S) \quad \sigma\big(f(s_1), f(s_2)\big) = \rho(s_1, s_2). \tag{1}$$

Such a function f is called an *isometry*.

This is, of course, nothing else than the familiar concept of isomorphism for the class of metric spaces; however, the word *isometry* is almost invariably used in this context.

As usual, isometry is an equivalence relation in the class of all metric spaces, and all our subsequent theorems and definitions will be intrinsic to the class of metric spaces. It turns out, however, that there is another, more inclusive equivalence relation, homeomorphism, which is more important than isometry. This means that we have not yet abstracted the truly essential aspects of convergence and continuity in the postulates for a metric space. The essence is really captured by another class of configurations known as topological spaces, which we shall define later. Associated with each metric space there is a topological space. Most of the properties of metric spaces which we shall study are topological; that is, they are intrinsic with respect to the associated topological spaces.

Trivial cases aside, a set S will always have many distinct metrics which lead to the same topological space, and for many purposes it does not matter which of these metrics is used. This explains why some definitions (for example, 14–1.5) are given in different ways by different authors.

14–1.3. Definition. A metric space $\langle T, \sigma \rangle$ is called a *subspace* of $\langle S, \rho \rangle$ if and only if $T \subseteq S$ and σ is the restriction of ρ to $T \times T$.

It is obvious that every subset of S is the underlying set of a *unique* subspace of $\langle S, \rho \rangle$. Consequently, it is customary to regard any subset of S as a subspace without explicit mention of the metric. If T is a subset of S, then we may refer to T as either a subset or a subspace of S. The choice is one of emphasis. If we refer to the subspace T, we are focusing our attention primarily on T itself. If we refer to the subset T, we are considering T in relation to S.

The relation of being a subspace is transitive; hence, in regarding the subset T as a subspace, we need not consider whether the metric in T is derived directly from S or from the metric of some subspace U, where $T \subseteq U \subseteq S$.

When $\mathbf{R}$ and $\mathbf{C}$ are regarded as metric spaces as in 14–1.1, $\mathbf{R}$ is a subspace of $\mathbf{C}$.

An interesting way of viewing the postulates for a metric space is to note that they assert that every three-point subspace is isometric to a subset of $\mathbf{C}$.

14–1.4. Proposition. Let $\langle S, \rho \rangle$ and $\langle T, \sigma \rangle$ be metric spaces. Define a function θ on $(S \times T) \times (S \times T)$ by

$$\theta(\langle s_1, t_1 \rangle, \langle s_2, t_2 \rangle) = \sup \{\rho(s_1, s_2), \sigma(t_1, t_2)\}. \tag{2}$$

Then θ is a metric for $S \times T$.

14–1.5. Definition. If $\langle S, \rho \rangle$ and $\langle T, \sigma \rangle$ are metric spaces, then $\langle S \times T, \theta \rangle$, where θ is defined by (2), is called the *direct product* of the spaces S and T.

Warning. There are other plausible ways to define a metric in $S \times T$; for example,

$$\theta'(\langle s_1, t_1 \rangle, \langle s_2, t_2 \rangle) = \rho(s_1, s_2) + \sigma(t_1, t_2) \tag{3}$$

or

$$\theta''(\langle s_1, t_1 \rangle, \langle s_2, t_2 \rangle) = \big(\rho(s_1, s_2)^2 + \sigma(t_1, t_2)^2\big)^{1/2}. \tag{4}$$

Letting $S = T = \mathbf{R}$ in (4), we recognize the usual way to assign a metric in $\mathbf{R}^2$. Except in trivial cases the metrics θ, θ', and θ'' are all different, but it turns out that the three different metrics all lead to the same topological space. Therefore, for most considerations it makes no difference which definition is used.

If $\langle U, \tau \rangle$ is a third metric space, then the usual bijection from $(S \times T) \times U$ to $S \times (T \times U)$ is an isometry of the metric structures assigned to these sets. Consequently, we may treat direct products of metric spaces as if they were associative. The metric in $S_1 \times S_2 \times \cdots \times S_n$ is then given by

$$\rho(x, y) = \sup \{\rho_i(x_i, y_i) \mid i = 1, 2, \ldots, n\}, \tag{5}$$

where $x = \langle x_1, x_2, \ldots, x_n \rangle$, $y = \langle y_1, y_2, \ldots, y_n \rangle$, and ρ_i is the metric in S_i. Similar considerations are valid if we prefer to define the metric of a direct product by (3) or (4).

14–1.6. Examples. Perhaps the most important examples of metric spaces which arise in analysis are the spaces $\mathbf{R}^n$ and $\mathbf{C}^n$. It is customary to define the metric for these spaces by the Pythagorean formula (4) instead of (2), but it really makes very little difference. With the Pythagorean metric, $\mathbf{R}^2$ is isometric to $\mathbf{C}$ under the usual identification ($\langle a, b \rangle \to a + bi$) and $\mathbf{C}^n$ is isometric to $\mathbf{R}^{2n}$.

Of course, any subset of $\mathbf{R}^n$ can be regarded as a subspace, but some are of such importance that they have acquired a standard name. We shall mention only the spheres. The *n-sphere* S^n is the subspace of $\mathbf{R}^{n+1}$ consisting of those points $\langle x_1, x_2, \ldots, x_{n+1} \rangle$ such that

$$x_1^2 + x_2^2 + \cdots + x_{n+1}^2 = 1.$$

More abstract than $\mathbf{R}^n$ but of almost equal importance are the function spaces. Let S be any set and consider the set $B(S)$ of all bounded real-valued functions with domain S. $B(S)$ becomes a metric space if we define

$$\rho(f_1, f_2) = \sup\{|f_1(s) - f_2(s)| \mid s \in S\}. \tag{6}$$

We might equally well take the set of all bounded complex-valued functions. The same formula defines a metric in the set of *all* real- or complex-valued functions if we admit the possibility of infinite distances.

If S is the unit interval [0, 1], then the set of continuous real-valued functions on S is a subset of $B(S)$ and is often considered with the metric given by (6). Other metrics analogous to (3) and (4) can be defined by means of integration:

$$\rho'(f_1, f_2) = \int_0^1 |f_1(t) - f_2(t)| \, dt, \tag{7}$$

$$\rho''(f_1, f_2) = \left(\int_0^1 |f_1(t) - f_2(t)|^2 \, dt \right)^{1/2}. \tag{8}$$

Unlike the previous situation, these different metrics define spaces which are topologically quite different, so it does matter which metric you use.

The assignment of a metric to a set of functions gives this set an intuitively geometric character. The success of the theory of metric spaces in analysis can be attributed in no small measure to the remarkable insight into the nature of functions which has come from exploiting the geometric point of view.

14–1.7. Definition. Let $\langle S, \rho \rangle$ be a metric space, p be a point of S, and ϵ a positive number. The *ball of radius ϵ about p* is the set

$$\{x \in S \mid \rho(p, x) < \epsilon\}.$$

We shall denote this set by $B(p, \epsilon)$ without mentioning either S or ρ whenever no confusion is likely. This is frequently done even when several different metric spaces are under discussion; only the argument p serves to tell us which space is meant.

Many authors write *sphere* in place of *ball*, but this usage is disappearing, since *sphere* is firmly established in algebraic topology as referring to the "skin" of a ball in euclidean space as defined above.

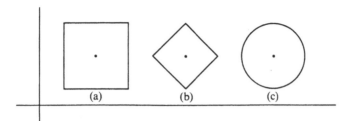

(a)   (b)   (c)

FIG. 14–1. Balls in $\mathbf{R}^2$: (a) $\theta(x, y) = \max\{|x_1 - y_1|, |x_2 - y_2|\}$,
(b) $\theta'(x, y) = |x_1 - y_1| + |x_2 - y_2|$,
(c) $\theta''(x, y) = \sqrt{(x_1 - y_1)^2 + (x_2 - y_2)^2}$.

Most of the properties of metric spaces which we shall study will be definable in terms of the balls in the space; in fact, in terms of the balls of small radius as suggested by the choice of ϵ to represent the radius.

It is worth stopping to consider the geometric appearance of the balls in $\mathbf{R}$, $\mathbf{R}^2$, and $\mathbf{R}^3$. Identifying these with euclidean space via the usual coordinates, we see that in $\mathbf{R}$, $B(p, \epsilon) = (p - \epsilon, p + \epsilon)$, an open interval about p, while in $\mathbf{R}^2$ (see Fig. 14–1) it is an edgeless square with sides of length 2ϵ parallel to the coordinate axes. If, however, the metric in $\mathbf{R}^2$ is defined by (3), then $B(p, \epsilon)$ is a square of side $\epsilon\sqrt{2}$ with diagonals parallel to the axes. If the metric is defined by (4), it is a circular disk. In three dimensions (see Fig. 14–2), the corresponding possibilities are a cube of edge 2ϵ, an octahedron of edge $\epsilon\sqrt{2}$, or a spherical ball of radius ϵ, in each case, centered at p. The essential property of a ball is that it contains a "solid" portion of the space near p.

When studying metric spaces it is convenient to define certain distances associated with sets.

14–1.8. Definition. Let $\langle S, \rho \rangle$ be a metric space and let A be a nonvoid subset of S. The *distance* of a point s *from the set A,* usually denoted by $\rho(s, A)$, is given by

$$\rho(s, A) = \inf\{\rho(s, a) \mid a \in A\}. \tag{9}$$

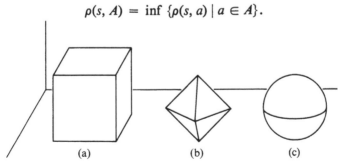

(a)   (b)   (c)

FIG. 14–2. Balls in $\mathbf{R}^3$: (a) $\theta(x, y) = \max\{|x_1 - y_1|, |x_2 - y_2|, |x_3 - y_3|\}$,
(b) $\theta'(x, y) = |x_1 - y_1| + |x_2 - y_2| + |x_3 - y_3|$,
(c) $\theta''(x, y) = \sqrt{(x_1 - y_1)^2 + (x_2 - y_2)^2 + (x_3 - y_3)^2}$.

14–1.9. Definition. Let $\langle S, \rho \rangle$ be a metric space and let A be a nonvoid subset of S. The *diameter* of A is given by

$$\text{diam } A = \sup \{\rho(s_1, s_2) \mid s_1, s_2 \in A\}. \tag{10}$$

The diameter of A may be ∞, of course. The diameter of the empty set is usually taken to be zero.

14–1.10. Definition. A subset A of a metric space is said to be *bounded* if and only if it is empty or its diameter is finite. A function with range in a metric space is said to be bounded if and only if its range is bounded.

Note that a subset of $\mathbf{R}$ or $\mathbf{C}$ is bounded in this new sense if and only if it is bounded in the sense of 6–4.1 or 10–3.8.

EXERCISES

1. Show that in any metric space diam $B(x, r) \leq 2r$. Give an example to show that inequality is possible.

2. If A and B are bounded subsets of a metric space, show that $A \cup B$ is also bounded. Moreover, if $A \cap B \neq \emptyset$, show that diam $(A \cup B) \leq$ diam $A +$ diam B.

3. Suppose $q \in B(p, \epsilon)$. Show that for some δ, $B(q, \delta) \subseteq B(p, \epsilon)$.

4. The definition of $S_1 \times S_2 \times \cdots \times S_n$ given on p. 225 is informal. How can it be stated formally?

5. Show that the function of 14–1.8 satisfies the inequalities

$$\rho(x, A) \leq \rho(x, y) + \rho(y, A) \quad \text{and} \quad |\rho(x, A) - \rho(y, A)| \leq \rho(x, y).$$

6. Let ρ be the usual metric for $\mathbf{R}$, and let θ, θ', and θ'' be the metrics for $\mathbf{R}^n$ derived from (2), (3), and (4), respectively. Show that for all $x, y \in \mathbf{R}^n$,

$$\theta(x, y) \leq \theta''(x, y) \leq \theta'(x, y) \quad \text{and} \quad \theta'(x, y) \leq \sqrt{n}\, \theta''(x, y) \leq n\theta(x, y).$$

7. Let S be any set. A function ρ defined on $S \times S$ with nonnegative real values satisfying postulates (a), (c), and (d) for a metric is called a *pseudometric* for S. Show that $\rho(s, t) = 0$ defines an equivalence relation on S, and then define a metric on the quotient set. (This is an important construction.)

8. Suppose $\langle S, \rho \rangle$ is a metric space. Define

$$\sigma(s, t) = \min \{\rho(s, t), 1\}, \qquad \tau(s, t) = \frac{\rho(s, t)}{1 + \rho(s, t)}.$$

Show that both σ and τ are metrics for S. Either of these metrics makes S bounded. (If ρ is allowed to take the value ∞, set $\infty/(1 + \infty) = 1$.)

9. Suppose $\{\rho_i \mid i \in I\}$ is a family of metrics for a set S. Show that

$$\sigma(x, y) = \sup_i \rho_i(x, y)$$

defines a new metric for S (possibly taking the value ∞, of course).

Suppose, in addition, that I is a linearly ordered set and that for each x, $y \in S$, $i \rightarrow \rho_i(x, y)$ is weakly order-preserving. Show that $\tau(x, y) = \inf_i \rho_i(x, y)$ is a pseudo-metric for S (see Exercise 7 for definition).

10. Suppose that every three-point subset of a metric space S is isometric to a subset of $\mathbf{R}$. Does it follow that S is isometric to a subset of $\mathbf{R}$?

14–2. CONVERGENCE

The definition (12–1.3) of a convergent sequence of numbers may be generalized immediately to sequences in a metric space.

14–2.1. Definition. Let $\{x_n\}$ be an (infinite) sequence of points in a metric space $\langle S, \rho \rangle$, and let p be a point of S. The sequence $\{x_n\}$ is said to *converge to p*, in symbols, $x_n \rightarrow p$, if and only if

$$(\forall \epsilon > 0)(\exists k \in \mathbf{N})(\forall n \geq k) \quad \rho(x_n, p) < \epsilon. \tag{1}$$

The sequence $\{x_n\}$ is said to converge if and only if there exists a p in S to which it converges. If no such p exists, then $\{x_n\}$ is said to *diverge*.

The definition can be expressed directly in terms of the convergence of sequences of real numbers as follows: $x_n \rightarrow p$ if and only if $\rho(x_n, p) \rightarrow 0$.

The convergence of a sequence depends only on its "infinite tail"; that is, if we modify (or even omit) a finite number of terms in a convergent sequence, the sequence will still converge to the same limit, and if we modify a finite number of terms in a divergent sequence, it remains divergent.

Evidently, a constant sequence ($x_n = p$, for all n) converges to the common value of its terms, and the same applies to sequences which are eventually constant (that is, $(\exists k)(\forall n \geq k) \, x_n = p$).

14–2.2. Proposition. Let $\{x_n\}$ be a sequence in a metric space $\langle S, \rho \rangle$. There is at most one point p of S such that $\{x_n\}$ converges to p.

Proof. Suppose that $x_n \rightarrow p$ and $x_n \rightarrow q$, where $p \neq q$. Then $\rho(p, q) > 0$ and we can apply the definition of convergence with ϵ replaced by $\frac{1}{2}\rho(p, q)$. There are integers j and k such that

$$(\forall n \geq j) \, \rho(x_n, p) < \tfrac{1}{2}\rho(p, q), \quad \text{and} \quad (\forall n \geq k) \, \rho(x_n, q) < \tfrac{1}{2}\rho(p, q).$$

Let $m = \max \{j, k\}$. Then $\rho(p, x_m) = \rho(x_m, p) < \frac{1}{2}\rho(p, q)$ and $\rho(x_m, q) < \frac{1}{2}\rho(p, q)$. Therefore,

$$\rho(p, q) \leq \rho(p, x_m) + \rho(x_m, q) < \tfrac{1}{2}\rho(p, q) + \tfrac{1}{2}\rho(p, q) = \rho(p, q)$$

by the triangle law, which is a contradiction. We conclude that it is impossible for $\{x_n\}$ to converge to two different points.

14–2.3. Notation. If $\{x_n\}$ is a convergent sequence in a metric space, then the unique point p to which it converges is called the limit of the sequence and is denoted by $\lim_{n \to \infty} x_n$. The latter is often abbreviated $\lim_n x_n$ or $\lim x_n$. We shall extend the convention introduced in 12–1.6 that the use of the notation $\lim_{n \to \infty} x_n$ without context to the contrary includes the assertion that $\{x_n\}$ converges.

14–2.4. Proposition. Suppose $x_n \to p$ in a metric space. If $\{x_{g(n)}\}$ is any subsequence of $\{x_n\}$, then $x_{g(n)} \to p$.

14–2.5. Proposition. Suppose that $\{x_n\}$ is a sequence in a metric space S and let p be a point of S. If for every subsequence $\{x_{g(n)}\}$ there is a subsubsequence $\{x_{g(h(n))}\}$ which converges to p, then $x_n \to p$.

14–2.6. Proposition. Let $\langle S, \rho \rangle$ and $\langle T, \sigma \rangle$ be metric spaces. Let $\{x_n\}$ be a sequence in S, and $\{y_n\}$ be a sequence in T. The sequence $\{\langle x_n, y_n \rangle\}$ converges in $S \times T$ if and only if $\{x_n\}$ converges in S and $\{y_n\}$ converges in T. Moreover,

$$\lim \langle x_n, y_n \rangle = \langle \lim x_n, \lim y_n \rangle \tag{2}$$

when the limits exist.

Proof. Suppose that $x_n \to p$ in S and $y_n \to q$ in T. We shall prove that $\langle x_n, y_n \rangle \to \langle p, q \rangle$ in $S \times T$.

Let $\epsilon > 0$ be given. Applying the definition of convergence, we can find integers j and k such that

$$(\forall n \geq j) \quad \rho(x_n, p) < \epsilon \quad \text{and} \quad (\forall n \geq k) \quad \sigma(y_n, q) < \epsilon.$$

Let $m = \max \{j, k\}$. Recalling the definition of the metric θ for $S \times T$, we have

$$\theta(\langle x_n, y_n \rangle, \langle p, q \rangle) = \max \{\rho(x_n, p), \sigma(y_n, q)\} < \epsilon,$$

provided that $n \geq m$. Thus we have proved

$$(\forall \epsilon > 0)(\exists m \in \mathbf{N})(\forall n \geq m) \quad \theta(\langle x_n, y_n \rangle, \langle p, q \rangle) < \epsilon;$$

that is, $\langle x_n, y_n \rangle \to \langle p, q \rangle$.

This proves one-half of the double implication and formula (2). The proof of the reverse implication is omitted.

We have given the proof by using the metric 14–1(2) for $S \times T$. The result is equally valid if 14–1(3) or (4) is used.

14–2.7. Proposition. Let $\langle S, \rho \rangle$ be a metric space, T be a subspace of S, and $\{x_n\}$ a sequence in T. Then $\{x_n\}$ converges in T if and only if $\{x_n\}$ converges in S, and its limit lies in T. If so, the limit calculated in T and the limit calculated in S are the same.

14–2.8. Definition. Let E be a subset of a metric space S. A point p of S is said to be *adherent* to E if and only if there exists a sequence $\{x_n\}$ in E such that $x_n \to p$.

Since Definition 14–2.8 does not exclude sequences which are constant, we have the following.

14–2.9. Proposition. Every point of E adheres to E.

14–2.10. Theorem. Let E be a subset of a metric space $\langle S, \rho \rangle$ and let p be a point of S. A necessary and sufficient condition for p to be adherent to E is that every ball about p meets E; in symbols,

$$(\forall \epsilon > 0) \quad B(p, \epsilon) \cap E \neq \emptyset.$$

Proof. Suppose that p adheres to E. We choose a sequence $\{x_n\}$ in E which converges to p. If $\epsilon > 0$ is given, then for all sufficiently large indices n,

$$x_n \in B(p, \epsilon) \cap E.$$

Thus $B(p, \epsilon) \cap E \neq \emptyset$. This shows that the condition is necessary.

Now, suppose that the condition is satisfied. For each positive integer n, choose $x_n \in B(p, 1/n) \cap E$. Obviously, $\{x_n\}$ is a sequence in E, and $x_n \to p$, because $\rho(p, x_n) < 1/n$. Thus p adheres to E.

Every point of a set E adheres to E trivially. It is often desirable to distinguish those points which adhere to E nontrivially. A point p may adhere to E both for the trivial reason that $p \in E$ and because some nontrivial sequence in E converges to p. In this case we want to count p as nontrivially adherent, so we make the following definition.

14–2.11. Definition. Let E be a subset of a metric space S. A point p of S is said to be a *cluster point* of E if and only if p adheres to $E - \{p\}$.

EXERCISES

1. Show that a point p adheres to the set E if and only if the distance from p to E (see 14–1.8) is zero.

2. Let $\{x_n\}$ be a sequence in a metric space S and let p be a point of S. One says that p *adheres to the sequence* $\{x_n\}$ if and only if every ball about p contains infinitely many of the terms x_n. Show that p adheres to $\{x_n\}$ if and only if there is a subsequence $\{x_{g(n)}\}$ such that $x_{g(n)} \to p$.

3. Let $\langle S, \rho \rangle$ and $\langle T, \sigma \rangle$ be metric spaces, $E \subseteq S$, and f a function from E to T. If $p \in S$, define $\lim_{s \to p} f(s)$. (We do not require that $p \in E$, but some restriction should be imposed on p.)

4. A *double sequence* is a family with index set $\mathbf{N} \times \mathbf{N}$. A double sequence $\{x_{m,n}\}$ in a metric space $\langle S, \rho \rangle$ is said to converge to ξ, and we write $x_{m,n} \to \xi$ as $m, n \to \infty$, if

and only if

$$(\forall \epsilon > 0)(\exists k \in N)(\forall m, n \geq k) \quad \rho(x_{m,n}, \xi) < \epsilon.$$

Such a point ξ is unique if it exists (prove this) and is therefore designated $\lim_{m,n\to\infty} x_{m,n}$. This double limit should not be confused with the iterated limit, $\lim_{m\to\infty} \lim_{n\to\infty} x_{m,n}$, which means $\lim_{m\to\infty} y_m$, where $y_m = \lim_{n\to\infty} x_{m,n}$ (assuming that these limits exist, of course). The two iterated limits, $\lim_{m\to\infty} \lim_{n\to\infty} x_{m,n}$ and $\lim_{n\to\infty} \lim_{m\to\infty} x_{m,n}$, and the double limit are almost independent of one another. Determine the exact interrelationships of these three limits; that is, find what patterns of existence and nonexistence are possible and what equalities must hold between the various limits when they exist.

5. Suppose that $\{x_{m,n}\}$ is a convergent double sequence in a metric space with limit p. Prove that $x_{k,k} \to p$ and $x_{k,k^2} \to p$.

14–3. CLOSURE, CLOSED SETS, AND OPEN SETS

Let E be a subset of a metric space S. We shall frequently be interested in the set of all points adherent to E. This set is called the closure of E and denoted by $\overline{E}$. Sets E for which $\overline{E} = E$ are called closed sets. The complements of closed sets are called open.

The most important properties of metric spaces are those that can be described in terms of the open and closed sets. In this section we shall develop the main properties of these classes of sets and of the closure operation.

14–3.1. Definition. A subset E of a metric space is said to be *closed* if and only if E contains every point adherent to E.

It is easy to verify that a closed interval in $\mathbf{R}$ (that is, a set of the form $\{x \mid a \leq x \leq b\}$) is indeed closed. So, too, are the infinite closed intervals $\{x \mid a \leq x\}$, $\{x \mid x \leq a\}$, and $\mathbf{R}$.

Note that closed sets are exactly those for which the extra condition in Proposition 14–2.7 is always satisfied.

The adjective *closed* is often applied to a subspace. Thus we may refer to the closed subspace T of S; we mean the closed subset T of S regarded as a subspace. Other adjectives defined for subsets of a metric space are also applied to subspaces in the same way.

14–3.2. Theorem. Let S be a metric space. Both S and $\emptyset$ are closed. If $F_1, F_2, \ldots, F_k$ are closed subsets of S, then so is $F_1 \cup F_2 \cup \cdots \cup F_k$. If $\{F_\alpha \mid \alpha \in A\}$ is any nonvoid family of closed subsets of S, then $\bigcap_\alpha F_\alpha$ is closed.

Proof. If $\{x_n\}$ is a convergent sequence in S, then $\lim x_n$ is in S by the definition of convergence, so S is closed. No point adheres to $\emptyset$, so $\emptyset$ is closed.

Let $F_1, F_2, \ldots, F_k$ be a finite sequence of closed subsets of S. Suppose p adheres to

$$E = F_1 \cup F_2 \cup \cdots \cup F_k.$$

Let $\{x_n\}$ be a sequence in E which converges to p. At least one of the sets

$$Q_i = \{n \in \mathbf{N} \mid x_n \in F_i\}, \qquad i = 1, 2, \ldots, k,$$

is infinite, since $\mathbf{N} = \bigcup_i Q_i$. Say Q_j is infinite. There is a subsequence $\{x_{g(n)}\}$ of $\{x_n\}$ such that $g(n) \in Q_j$, that is, $x_{g(n)} \in F_j$ for all n. Now, $\{x_{g(n)}\}$, being a subsequence of a convergent sequence, converges to p; thus p adheres to F_j. Since F_j is closed, $p \in F_j$, and therefore $p \in E$. This shows that E is closed.

Now consider a nonvoid family of closed sets $\{F_\alpha\}$. Suppose that p adheres to $\bigcap_\alpha F_\alpha$. There is a sequence $\{x_n\}$ lying in $\bigcap_\alpha F_\alpha$ which converges to p. Since this sequence is in each F_α, p adheres to each of these sets. Since they are closed, $p \in F_\alpha$ for all α. Thus $p \in \bigcap_\alpha F_\alpha$ and this proves that $\bigcap_\alpha F_\alpha$ is closed.

If we interpret the intersection of the empty family of closed sets as S, then the conclusion is still valid.

14–3.3. Corollary. Every finite subset of a metric space is closed.

Proof. In view of the theorem we need only prove that a set with one element is closed, and this is obvious.

14–3.4. Definition. Let E be a subset of a metric space. The set of all points adherent to E is called the *closure* of E and is denoted* by $\overline{E}$.

Closure is thus a function from $\mathfrak{P}(S)$ to itself.

14–3.5. Theorem. The closure of any set is closed.

Proof. Let E be any set in a metric space $\langle S, \rho \rangle$. We wish to show that $\overline{E}$ is closed. Suppose that p is adherent to $\overline{E}$. Let ϵ be any positive number. By Proposition 14–2.10, $B(p, \epsilon) \cap \overline{E}$ contains a point q. Now, $\rho(p, q) < \epsilon$, so $\delta = \epsilon - \rho(p, q)$ is a positive number. Since q is adherent to E, there is a point r in $B(q, \delta) \cap E$. We know that $\rho(p, r) \leq \rho(p, q) + \rho(q, r) < \epsilon$, so $r \in B(p, \epsilon) \cap E$, and the latter set is not void. Applying Proposition 14–2.10 in the other direction, we find that p is adherent to E; that is, $p \in \overline{E}$. Thus we have shown that $\overline{E}$ is closed.

14–3.6. Proposition. Let S be a metric space. The closure function from $\mathfrak{P}(S)$ to itself has the following properties:

(a) $\overline{E} \supseteq E$,
                    (b) E is closed $\Leftrightarrow \overline{E} = E$,

(c) $\overline{\overline{E}} = \overline{E}$,
                    (d) $E \subseteq F \Rightarrow \overline{E} \subseteq \overline{F}$,

(e) $\overline{E}$ is the smallest closed set containing E,

(f) $\overline{E \cup F} = \overline{E} \cup \overline{F}$,
         (g) $\overline{E_1 \cup E_2 \cup \cdots \cup E_k} = \overline{E}_1 \cup \overline{E}_2 \cup \cdots \cup \overline{E}_k$,

(h) $\overline{\bigcap_\alpha E_\alpha} \subseteq \bigcap_\alpha \overline{E}_\alpha$,
      (i) $\overline{\bigcup_\alpha E_\alpha} \supseteq \bigcup_\alpha \overline{E}_\alpha$.

---

* Because of the expense involved in setting overbars in type, notations like cl E are often employed.

Proof. Formula (a) follows immediately from the definition of closure and Proposition 14–2.9. The double implication

$$E \text{ is closed} \Leftrightarrow \overline{E} \subseteq E$$

is a paraphrase of Definition 14–3.1. With (a) this gives (b). Now (c) follows from Theorem 14–3.5.

Suppose $E \subseteq F$. If p adheres to E, there is a sequence in E which converges to p. This sequence is also in F, so p adheres to F. Thus $\overline{E} \subseteq \overline{F}$, and (d) is established.

Now, if F is any closed set containing E, then $\overline{E} \subseteq \overline{F} = F$. Since $\overline{E}$ is closed by Theorem 14–3.5, $\overline{E}$ is the smallest closed set containing E.

To obtain (f) we note that $\overline{E} \subseteq \overline{E \cup F}$ and $\overline{F} \subseteq \overline{E \cup F}$ by (d); therefore

$$\overline{E} \cup \overline{F} \subseteq \overline{E \cup F}.$$

But $E \cup F \subseteq \overline{E} \cup \overline{F}$ by (a) and $\overline{E} \cup \overline{F}$ is closed by Theorems 14–3.2 and 14–3.5; hence $\overline{E \cup F} \subseteq \overline{E} \cup \overline{F}$ by (e). Thus (f) is proved and (g) follows by induction on k.

Suppose $\{E_\alpha\}$ is any family of sets. Then $E_\alpha \subseteq \overline{E_\alpha}$ for all α, so

$$\bigcap_\alpha E_\alpha \subseteq \bigcap_\alpha \overline{E_\alpha}.$$

The latter is closed; hence $\overline{\bigcap_\alpha E_\alpha} \subseteq \bigcap_\alpha \overline{E_\alpha}$. Also, for any β, $E_\beta \subseteq \bigcup_\alpha E_\alpha$; therefore, $\overline{E_\beta} \subseteq \overline{\bigcup_\alpha E_\alpha}$, and thus $\bigcup_\alpha \overline{E_\alpha} \subseteq \overline{\bigcup_\alpha E_\alpha}$. This establishes (h) and (i), and completes the proof.

In both (h) and (i) inequality is possible. For intersections it can happen even for two sets. Let $S = \mathbf{R}$, $E_1 = \mathbf{Q}$ (the set of rational numbers), and $E_2 = \mathbf{R} - \mathbf{Q}$. Then

$$\overline{E_1 \cap E_2} = \overline{\emptyset} = \emptyset \quad \text{and} \quad \overline{E_1} \cap \overline{E_2} = \mathbf{R} \cap \mathbf{R} = \mathbf{R}.$$

For unions, equality occurs when there are only a finite number of sets, (g), but often fails when there are infinitely many. Let E_n be the one-point set $\{1/n\}$; then $\overline{E_n} = E_n$ and $0 \in \overline{\bigcup_n E_n} - \bigcup_n \overline{E_n}$.

Suppose E is a subset of a subspace T of a metric space S. If we say that E is closed, it makes a difference whether we mean closed as a subset of T or closed as a subset of S. For example, the set $\mathbf{Q}$ of rational numbers is closed as a subset of $\mathbf{Q}$, but not as a subset of $\mathbf{R}$. The next two results explain the situation.

14–3.7. Proposition. Let E be a subset of a subspace T of a metric space S. Then E is closed as a subset of T if and only if there is a closed subset F of S such that $E = T \cap F$. Moreover,

$$\overline{E}^T = T \cap \overline{E}^S, \tag{1}$$

where $\overline{E}^T$ represents the closure of E calculated in the space T and $\overline{E}^S$ is the closure calculated in the space S.

Proof. We shall first establish the formula $\bar{E}^T = T \cap \bar{E}^S$. Suppose that $p \in \bar{E}^T$. Then we can select a sequence $\{x_n\}$ in E such that $x_n \to p$ in T. By 14–2.7, $x_n \to p$ in S as well, so $p \in \bar{E}^S$. Thus $\bar{E}^T \subseteq T \cap \bar{E}^S$. Conversely, suppose that $q \in T \cap \bar{E}^S$. Then we can select a sequence $\{y_n\}$ in E such that $y_n \to q$ in S. Since $E \subseteq T$, $\{y_n\}$ is a sequence in T, and by Proposition 14–2.7 again $y_n \to q$ in T, whence $q \in \bar{E}^T$. Thus $T \cap \bar{E}^S = \bar{E}^T$.

Suppose E is closed as a subset of T. Then $E = \bar{E}^T = T \cap \bar{E}^S$, and a closed subset F of S exists such that $E = T \cap F$; namely, $F = \bar{E}^S$. On the other hand, suppose that $E = T \cap F$, where F is closed in S. Then $\bar{E}^S \subseteq \bar{F}^S = F$ by 14–3.6(d) and

$$\bar{E}^T = T \cap \bar{E}^S \subseteq T \cap F = E.$$

Therefore, E is closed as a subset of T.

14–3.8. Corollary. A subset E of a closed subspace T of a metric space S is closed in T if and only if it is closed in S.

14–3.9. Definition. Let S be a metric space. A subset G of S is said to be *open* if and only if

$$(\forall p \in G)(\exists \epsilon > 0) \quad B(p, \epsilon) \subseteq G.$$

Open intervals in $\mathbf{R}$ (i.e., sets of the form $\{x \mid a < x < b\}$, $\{x \mid a < x\}$, $\{x \mid x < a\}$ or $\mathbf{R}$) are indeed open. The reader is cautioned against the fallacy that every set is either open or closed. Some sets, like $\mathbf{R}$ and $\emptyset$, are both open and closed; others, like $\mathbf{Q}$ are neither open nor closed.

14–3.10. Proposition. Every ball in a metric space is open.

Proof. Let $B(q, \delta)$ be a ball in the metric space $\langle S, \rho \rangle$. If $p \in B(q, \delta)$, then $\rho(p, q) < \delta$ and so $\epsilon = \delta - \rho(p, q)$ is a positive number. If $r \in B(p, \epsilon)$, then

$$\rho(q, r) \leq \rho(q, p) + \rho(p, r) < \rho(p, q) + \epsilon = \delta;$$

hence $r \in B(q, \delta)$. Thus $B(p, \epsilon) \subseteq B(q, \delta)$. This proves that $B(q, \delta)$ is open.

14–3.11. Theorem. Let S be a metric space. Both $\emptyset$ and S are open. If

$$G_1, G_2, \ldots, G_n$$

are open subsets of S, then so is $G_1 \cap G_2 \cap \cdots \cap G_n$. If $\{G_\alpha \mid \alpha \in A\}$ is any family of open subsets of S, then $\bigcup_\alpha G_\alpha$ is also open.

Proof. The definition of open applies to $\emptyset$ as a logical triviality and to S as a consequence of the definition of a ball. We turn now to the second statement. Let p be a point in $G_1 \cap G_2 \cap \cdots \cap G_n$. For each index i, G_i is open and $p \in G_i$, so we can choose a positive ϵ_i such that $B(p, \epsilon_i) \subseteq G_i$. Let $\epsilon = \inf \{\epsilon_1, \epsilon_2, \ldots, \epsilon_n\}$.

Then

$$B(p, \epsilon) \subseteq G_1 \cap G_2 \cap \cdots \cap G_n,$$

whence the latter is open.

Let $\{G_\alpha\}$ be a family of open sets and suppose that $p \in \bigcup_\alpha G_\alpha$. There is an index β such that $p \in G_\beta$. Since the latter is open, there is a positive ϵ such that

$$B(p, \epsilon) \subseteq G_\beta.$$

But then $B(p, \epsilon) \subseteq \bigcup_\alpha G_\alpha$. This proves the last statement of the Theorem.

14-3.12. Theorem. A subset of a metric space is closed if and only if its complement is open.

Proof. Let E be a closed subset of a metric space S. Let p be any point in $S - E$. By the definition of closed, p does not adhere to E. Hence, by Proposition 14-2.10 there is a positive ϵ such that $B(p, \epsilon) \cap E = \emptyset$, or $B(p, \epsilon) \subseteq S - E$. This shows that $S - E$ is open.

Now suppose that $S - E$ is open. Let p adhere to E. We must prove that $p \in E$. If $p \notin E$, then $p \in S - E$, and since the latter is open, we can choose a positive ϵ such that $B(p, \epsilon) \subseteq S - E$ or $B(p, \epsilon) \cap E = \emptyset$. But this shows that p does not adhere to E, which is a contradiction.

The correspondence between open and closed sets established by complementation enables us to translate any theorems about open sets into theorems about closed sets or vice versa. Thus the conclusions of Theorem 14-3.11 correspond to those of Theorem 14-3.2 with the aid of the set-theoretic identities

$$S - (E_1 \cup E_2 \cup \cdots \cup E_n) = (S - E_1) \cap (S - E_2) \cap \cdots \cap (S - E_n)$$

and

$$S - \bigcap_\alpha E_\alpha = \bigcup_\alpha (S - E_\alpha).$$

If E is a subset of a subspace T of a metric space S, then E may be open in T but not open in S. Hence when a set is described as open it is important to know in what space it is open. If $E = \mathbf{R}$, then E is open as a subset of $\mathbf{R}$, but not as a subset of $\mathbf{C}$. The situation is similar to that for closed sets, and the following proposition is dual to 14-3.7.

14-3.13. Proposition. Let E be a subset of a subspace T of a metric space S. Then E is open as a subset of T if and only if there is an open subset G of S such that $E = T \cap G$.

The notion dual to that of adherent point is that of interior point.

14-3.14. Definition. Let E be a subset of a metric space S. A point p of S is said to be *interior* to E if and only if there exists a positive ϵ such that $B(p, \epsilon) \subseteq E$.

With this definition we can restate Definition 14–3.9 in a form dual to the definition of *closed*.

14–3.15. Proposition. A subset E of a metric space is open if and only if every point of E is interior to E.

Corresponding to the closure of a set is the interior of a set.

14–3.16. Definition. Let E be a subset of a metric space S. The *interior* of E, denoted by Int E, is the set of all interior points of E.

Thus, Int is a function from $\mathfrak{P}(S)$ to itself. Its properties are summarized in the following proposition.

14–3.17. Proposition. The interior of any set is open. The interior function from $\mathfrak{P}(S)$ to itself has the following properties:

(a) $S - \overline{E} = \text{Int } (S - E)$,
(b) $S - \text{Int } E = \overline{S - E}$,
(c) $\text{Int } E \subseteq E$,
(d) E is open $\Leftrightarrow \text{Int } E = E$,
(e) $\text{Int Int } E = \text{Int } E$,
(f) $E \subseteq F \Rightarrow \text{Int } E \subseteq \text{Int } F$,
(g) $\text{Int } E$ is the largest open set contained in E,
(h) $\text{Int } (E \cap F) = \text{Int } E \cap \text{Int } F$,
(i) $\text{Int } (E_1 \cap E_2 \cap \cdots \cap E_n) = \text{Int } E_1 \cap \text{Int } E_2 \cap \cdots \cap \text{Int } E_n$,
(j) $\text{Int } (\bigcup_\alpha E_\alpha) \supseteq \bigcup_\alpha \text{Int } E_\alpha$,
(k) $\text{Int } (\bigcap_\alpha E_\alpha) \subseteq \bigcap_\alpha \text{Int } E_\alpha$.

Note that properties (c) through (k) are dual to 14–3.6(a) through (i).

The duality of interior and closure is obscure when subspaces are involved. In particular, the casually written dual to formula (1), p. 234, is false: it need not be true that $\text{Int}_T E = T \cap \text{Int}_S E$. However, this relation is valid whenever T is open as a subset of S.

Our next proposition is self-dual.

14–3.18. Proposition. If G is an open set and F is a closed set, then $G - F$ is open and $F - G$ is closed.

The following proposition is often handy in simplifying complicated expressions. The essential point of the proof is to show that $G \cap \overline{X} \subseteq \overline{G \cap X}$, which depends on the fact that G is open.

14–3.19. Proposition. Let G be an open set in a metric space S. If X is any subset of S, then $G \cap \overline{G \cap X} = G \cap \overline{X}$.

The relation "p is interior to E" is often expressed the other way around.

14–3.20. Definition. Let p be a point in a metric space S. A subset E of S is called a *neighborhood* of p if and only if p is interior to E.

The most commonly used neighborhoods of a point p are the balls $B(p, \epsilon)$, but it is clear that any set larger than a neighborhood of p is itself a neighborhood of p. Neighborhoods can frequently play the same role as balls. For example, we can rephrase some of the definitions and propositions of this section in such terms.

14–3.21. Proposition. A sequence $\{x_n\}$ in a metric space converges to p if and only if every neighborhood of p contains all but a finite number of the x_n (that is, for every neighborhood U of p, $(\exists k \in N)(\forall n \geq k)\, x_n \in U$).

14–3.22. Proposition. Let E be a set in a metric space. A point p adheres to E if and only if every neighborhood of p meets E (that is, for every neighborhood U of p, $U \cap E \neq \emptyset$).

14–3.23. Proposition. A set is open if and only if it is a neighborhood of each of its points.

An obvious but important property of neighborhoods is the following.

14–3.24. Proposition. The intersection of two (or any finite number of) neighborhoods of p is again a neighborhood of p.

14–3.25. Definition. The *boundary* of a set E in a metric space S is $\overline{E} \cap \overline{S - E}$.

The boundary of a set is evidently closed. If E is an open set, then the boundary of E is just what we would ordinarily regard as the boundary. For example, the boundary in $\mathbf{R}^2$ of the open unit disk $\{\langle x, y \rangle \mid x^2 + y^2 < 1\}$ is the unit circle $\{\langle x, y \rangle \mid x^2 + y^2 = 1\}$. When E is a complicated set, the boundary need not be a "thin" set at all. The boundary of $\mathbf{Q}$ in $\mathbf{R}$ is all of $\mathbf{R}$.

14–3.26. Definitions. A point p of a metric space S is said to be *isolated* if and only if $\{p\}$ is an open set. The space S is said to be *discrete* if and only if all of its points are isolated.

If p is an isolated point, then for some $\epsilon > 0$, $B(p, \epsilon) \subseteq \{p\}$; hence for every point $q \neq p$, $\rho(p, q) \geq \epsilon$. Conversely, if inf $\{\rho(p, q) \mid q \neq p\} > 0$, then p is isolated. Note that $\{p\}$ is always closed, but this is no barrier to its being open. If $\mathbf{N}$ is taken as a space (with the metric of $\mathbf{R}$) it is discrete. So too is the space $J = \{1/n \mid n \in \mathbf{N}\}$; it is irrelevant that 0 adheres to J, since 0 has been dismissed when we take J as the space. In a discrete space any set is open, since it is a union of one-point sets, all of which are open. Furthermore, it is closed, since its complement is open. A space is discrete if and only if the only convergent sequences in it are those which are eventually constant. To put it in another way, discrete spaces are those in which the limit concept is trivial.

14–3.27. Definition. A subset E of a metric space S is said to be *dense* if and only if $\bar{E} = S$.

14–3.28. Proposition. Let E be a subset of a metric space S. E is dense if and only if, for every nonempty open set G of S, $E \cap G \neq \emptyset$.

14–3.29. Definition. A subset E of a metric space is said to be *nowhere dense* if and only if Int $\bar{E} = \emptyset$.

Such a set was formerly said to be *nondense*. Since there is danger of confusing *nondense* with the much weaker *not dense*, the term *nowhere dense* is preferable.

EXERCISES

1. If S is a metric space, show that the diagonal of $S \times S$ is closed.

2. Let X and Y be nonvoid sets in the metric spaces S and T, respectively. Show that $X \times Y$ is a closed subset of $S \times T$ if and only if X and Y are both closed.

3. Let E be any subset of a metric space. Show that diam $\bar{E} = $ diam E.

4. If p is a point of a metric space $\langle S, \rho \rangle$, and ϵ is a positive number, then the set

$$C(p, \epsilon) = \{x \mid \rho(p, x) \leq \epsilon\}$$

is often called the *closed ball of radius ϵ about p.* Show that it is indeed a closed set, but that it need not be true that $C(p, \epsilon) = \overline{B(p, \epsilon)}$.

5. Let X and Y be nonvoid subsets of the metric spaces S and T, respectively. Show that $X \times Y$ is open in $S \times T$ if and only if X and Y are both open sets.

6. Let E be a set in a metric space. The set of all cluster points of E is called the *derived set of E* and is denoted by E'. Prove: (a) $\bar{E} = E \cup E'$; (b) $p \in E'$ if and only if every neighborhood of p contains at least two points of E; (c) $p \in E'$ if and only if there is a sequence $\{x_n\}$ in E such that $x_n \to p$ and $(\forall m \neq n)\, x_m \neq x_n$; (d) E' is closed.

7. Under what circumstances does a metric space contain a minimal dense subset? Show that a minimal dense subset is always a minimum dense subset.

14–4. CONTINUOUS FUNCTIONS

In the study of ordered sets we are more interested in the order-preserving functions from one ordered set to another than in the functions which do not take account of the order structures on the domain and range. In the study of metric spaces, we are also particularly interested in those functions from one metric space to another which are suitably related to the metric structures on the domain and range. The most interesting class turns out to be the continuous functions, which are defined by a simple generalization of the familiar definition of a continuous function from **R** to **R**.

14–4.1. Definition. Let $\langle S, \rho \rangle$ and $\langle T, \sigma \rangle$ be metric spaces and let f be a function from S to T. The function f is said to be *continuous at the point p* of S if and only if

$$(\forall \epsilon > 0)(\exists \delta > 0)(\forall q \in S) \quad \rho(p, q) < \delta \Rightarrow \sigma(f(p), f(q)) < \epsilon.$$

The function f is said to be *continuous* if and only if it is continuous at each point of S.

This is the familiar definition of continuity when applied to real functions of a real variable. It means that a good approximation to $f(p)$ is obtained when we consider $f(q)$ for q sufficiently near p.

Evidently any constant function is continuous, and if S is discrete, then every function from S to T is continuous.

A continuous function from one space to another is often referred to as a *map*, although this is a different use of the term from that described in 3–4.11.

The following proposition simply recasts the definition in terms of balls.

14–4.2. Proposition. Let f be a function from one metric space S to another T and let $p \in S$. In order for f to be continuous at p it is necessary and sufficient that

$$(\forall \epsilon > 0)(\exists \delta > 0) \quad f(B(p, \delta)) \subseteq B(f(p), \epsilon).$$

Note that $B(p, \delta)$ is a ball in S while $B(f(p), \epsilon)$ is a ball in T.

When a function is defined only on a subset E of S, then it is said to be continuous if and only if it is continuous as a function from the *subspace* E to T. Hence it makes no sense to talk of continuity at a point of S not in E. This is the customary modern usage; it disagrees with an older usage for real functions defined on $\mathbf{R}$. The function $f \colon x \to 1/x$ is defined only on the set $\mathbf{R} - \{0\}$ and it is continuous at every point of this set. Classical terminology would permit the assertion that f is discontinuous at 0. This would be expressed under modern conventions as follows: "f has no continuous extension to 0."

14–4.3. Theorem. Let f be a function from one metric space $\langle S, \rho \rangle$ to another $\langle T, \sigma \rangle$. Suppose that $p \in S$. Then f is continuous at p if and only if, for every sequence $\{x_n\}$ such that $x_n \to p$, $f(x_n) \to f(p)$. And f is continuous if and only if, for every convergent sequence $\{x_n\}$ in S,

$$\lim f(x_n) = f(\lim x_n). \tag{1}$$

Proof. Suppose f is continuous at p and $x_n \to p$. We shall prove that $f(x_n) \to f(p)$.

Let $\epsilon > 0$ be given. By the definition of continuity, there is a $\delta > 0$ such that

$$(\forall q \in S) \quad \rho(p, q) < \delta \Rightarrow \sigma(f(p), f(q)) < \epsilon.$$

Since $x_n \to p$, we can choose an integer k so that

$$(\forall n \geq k) \quad \rho(p, x_n) < \delta.$$

We have then

$$(\forall n \geq k) \quad \sigma(f(p), f(x_n)) < \epsilon.$$

Therefore $f(x_n) \to f(p)$.

Conversely, suppose that f is not continuous at p. We will construct a sequence $\{x_n\}$ such that $x_n \to p$ but $f(x_n) \nrightarrow f(p)$.

Because f is not continuous at p, there is a positive ϵ such that

$$(\forall \delta > 0)(\exists q \in S) \quad \rho(p, q) < \delta \quad \text{and} \quad \sigma(f(p), f(q)) \geq \epsilon.$$

For each n, take $\delta = 1/n$ and choose x_n so that

$$\rho(p, x_n) < 1/n \quad \text{and} \quad \sigma(f(p), f(x_n)) \geq \epsilon.$$

Then $\{x_n\}$ is the desired sequence.

The second part of the theorem is an immediate consequence of the first.

The criterion (1) for continuity can be paraphrased as follows: f is continuous if and only if it commutes with lim. This must be interpreted with some care, however, because for a continuous f, $\lim f(x_n)$ may exist when $\lim x_n$ does not.

14-4.4. Proposition. Let f and g be two continuous functions from one metric space S to another T. Suppose that E is a dense set in S and that f and g agree on E (that is, $(\forall x \in E) f(x) = g(x)$). Then $f = g$.

Proof. Let p be any point of S. Since E is dense, there is a sequence $\{x_n\}$ in E such that $x_n \to p$. Since f and g are both continuous,

$$f(p) = \lim f(x_n) = \lim g(x_n) = g(p).$$

Thus f and g agree at every point of S.

14-4.5. Proposition. Let f be a function from one metric space $\langle S, \rho \rangle$ to another $\langle T, \sigma \rangle$. Then f is continuous if and only if for every open set G in T, $f^{-1}(G)$ is open in S.

Proof. Suppose first that f is continuous and that G is an open set in T. If $p \in f^{-1}(G)$, then $f(p) \in G$. Since G is open, there is a positive ϵ such that $B(f(p), \epsilon) \subseteq G$. Since f is continuous, by 14-4.2 there is a positive number δ such that $f(B(p, \delta)) \subseteq B(f(p), \epsilon)$. Then $B(p, \delta) \subseteq f^{-1}[f(B(p, \delta))] \subseteq f^{-1}(G)$. Hence $f^{-1}(G)$ is open.

Conversely, suppose that $f^{-1}(G)$ is open for every open set G. Let $p \in S$ and $\epsilon > 0$ be given. Since $B(f(p), \epsilon)$ is an open set, $f^{-1}[B(f(p), \epsilon))]$ is open. Since

$p \in f^{-1}[B(f(p), \epsilon))]$, there is a positive number δ such that

$$B(p, \delta) \subseteq f^{-1}[B(f(p), \epsilon))].$$

Then $f(B(p, \delta)) \subseteq B(f(p), \epsilon)$. This shows that f is continuous by 14–4.2.

14–4.6. Corollary. Let f be a function from one metric space S to another T. Then f is continuous if and only if for every closed set X in T, $f^{-1}(X)$ is a closed set in S.

14–4.7. Proposition. Let f be a function from one metric space S to another T. Then f is continuous if and only if for every set E in S, $f(\overline{E}) \subseteq \overline{f(E)}$.

Proof. Suppose f is continuous and $E \subseteq S$. Then clearly $E \subseteq f^{-1}(\overline{f(E)})$, and the latter is closed by 14–4.6. Therefore, $\overline{E} \subseteq f^{-1}(\overline{f(E)})$, whence $f(\overline{E}) \subseteq \overline{f(E)}$.

Conversely, suppose that $(\forall E \subseteq S) f(\overline{E}) \subseteq \overline{f(E)}$. Let X be any closed set in T. Taking $E = f^{-1}(X)$, we have

$$f(\overline{f^{-1}(X)}) \subseteq \overline{f(f^{-1}(X))} \subseteq \overline{X} = X.$$

Therefore, $\overline{f^{-1}(X)} \subseteq f^{-1}(X)$, hence $f^{-1}(X)$ is closed. Now 14–4.6 shows that f is continuous.

14–4.8. Proposition. Let f be a function from one metric space to another. Then f is continuous at p if and only if for every neighborhood U of $f(p)$, $f^{-1}(U)$ is a neighborhood of p.

14–4.9. Theorem. Let S, T, and U be metric spaces. Let f be a function from S to T and let g be a function from T to U. If f is continuous at p and g is continuous at $f(p)$, then $g \circ f$ is continuous at p. If f and g are both continuous, then so is $g \circ f$.

Proof. Suppose that f is continuous at p and g is continuous at $f(p)$. We shall apply 14–4.3. Let $x_n \to p$. Since f is continuous at p, $f(x_n) \to f(p)$. Since g is continuous at $f(p)$, $g(f(x_n)) \to g(f(p))$; that is, $(g \circ f)(x_n) \to (g \circ f)(p)$. Therefore, $g \circ f$ is continuous at p.

The second statement of the theorem is an immediate consequence of the first, but there is another proof which is interesting.

Let H be any open set in U. Since g is continuous, $g^{-1}(H)$ is an open set in T. Since f is continuous, $f^{-1}(g^{-1}(H))$ is open in S. But $(g \circ f)^{-1}(H) = f^{-1}(g^{-1}(H))$; hence $g \circ f$ is continuous by Proposition 14–4.5.

14–4.10. Proposition. Let S be a metric space and let T be a subspace of S. Then the identity function from T to S is continuous.

14–4.11. Proposition. Let S be a metric space and let T be a subspace of S. If f is a continuous function from S to another metric space, then f restricted to T is also continuous.

While this proposition is easily proved directly, it is worth noting that f restricted to T is $f \circ i$, where i is the identity from T to S. Hence the result follows from the last two propositions.

Beware, however, of the plausible converse. If f restricted to T is continuous, it need not be true that f is continuous even at points of T. Let $S = \mathbf{R}$ and $T = \mathbf{Q}$. Define $f(x) = 0$ for $x \in \mathbf{Q}$, $f(x) = 1$ for $x \in \mathbf{R} - \mathbf{Q}$. Then f is discontinuous at every point of $\mathbf{R}$, yet its restriction to $\mathbf{Q}$ is continuous (being constant).

14–4.12. Proposition. Let S, T, and U be metric spaces, where T is a subspace of U. Let f be a function from S to T. Then f is continuous as a function from S to U if and only if it is continuous as a function from S to T.

14–4.13. Proposition. Let $\langle S, \rho \rangle$, $\langle T, \sigma \rangle$, and $\langle U, \tau \rangle$ be metric spaces. Let f be a function from S to T and g be a function from S to U. Then the function

$$h: x \rightarrow \langle f(x), g(x) \rangle$$

from S to $T \times U$ is continuous at p if and only if f is continuous at p and g is continuous at p. And h is continuous if and only if both f and g are continuous.

This proposition asserts that the continuity of a function into a direct product is calculated coordinate by coordinate. The similar statement concerning functions from a direct product is false. Suppose that f is a function from $S \times T$ to U. It may happen that f is discontinuous, although the maps $x \rightarrow f(x, y)$ for every y in T and $y \rightarrow f(x, y)$ for every x in S are all continuous. A well-known example is the map of $\mathbf{R} \times \mathbf{R}$ into $\mathbf{R}$ defined by

$$f(x, y) = \frac{xy}{x^2 + y^2} \quad \text{for} \quad \langle x, y \rangle \neq \langle 0, 0 \rangle,$$

$$f(0, 0) = 0,$$

which is discontinuous at $\langle 0, 0 \rangle$, although all of the maps $x \rightarrow f(x, y)$ and $y \rightarrow f(x, y)$ are continuous.

14–4.14. Definitions. Let S and T be metric spaces. A function f from S to T is said to be *open* if and only if for every open set G in S, $f(G)$ is an open set in T. A function f from S to T is said to be *closed* if and only if for every closed set E in S, $f(E)$ is a closed set in T.

14–4.15. Proposition. The composition of two open functions is open and the composition of two closed functions is closed.

14–4.16. Proposition. The arithmetic operations are continuous. More precisely, addition, subtraction, and multiplication are continuous functions from $\mathbf{C} \times \mathbf{C}$ to $\mathbf{C}$, and division is a continuous function from $\mathbf{C} \times (\mathbf{C} - \{0\})$ to $\mathbf{C}$.

Proof. This follows immediately from 12–2.1 with the aid of 14–4.3 and 14–2.6.

14–4.17. Definitions. Let D be a subset of $\mathbf{C}^n$. A function f from D to $\mathbf{C}$ is called a *monomial* function if and only if

$$f(x_1, x_2, \ldots, x_n) = \lambda x_1^{e_1} x_2^{e_2} \cdots x_n^{e_n}$$

for all $x = \langle x_1, x_2, \ldots, x_n \rangle \in D$, where $\lambda \in \mathbf{C}$ and $e_1, e_2, \ldots, e_n \in \mathbf{N}^*$. A function g from D to $\mathbf{C}$ is called a *polynomial* function if and only if

$$g(x_1, x_2, \ldots, x_n) = \sum_{i=1}^{k} f_i(x_1, x_2, \ldots, x_n)$$

for all $x \in D$, where $f_1, f_2, \ldots, f_k$ are monomial functions. A function h from D to $\mathbf{C}$ is called a *rational* function if and only if

$$h(x_1, x_2, \ldots, x_n) = \frac{g_1(x_1, x_2, \ldots, x_n)}{g_2(x_1, x_2, \ldots, x_n)}$$

for all $x \in D$, where g_1 and g_2 are polynomial functions.

14–4.18. Corollary. Every polynomial function in n variables is continuous from $\mathbf{C}^n$ to $\mathbf{C}$.

14–4.19. Corollary. Every rational function in n variables is continuous.

Proof. Let h be defined on D by

$$h(x_1, x_2, \ldots, x_n) = \frac{g_1(x_1, x_2, \ldots, x_n)}{g_2(x_1, x_2, \ldots, x_n)}, \tag{2}$$

where g_1 and g_2 are polynomial functions. Let $S = \{y \in \mathbf{C}^n \mid g_2(y) = 0\}$. Then $D \subseteq \mathbf{C}^n - S$ (otherwise (2) is not sensible). According to 14–4.18 and 14–4.13, $x \to \langle g_1(x), g_2(x) \rangle$ is a continuous map of D into $\mathbf{C} \times (\mathbf{C} - \{0\})$. The composition of this function with division is h.

14–4.20. Corollary. Let $f_1, f_2, \ldots, f_n$ be continuous functions from a metric space to the real or complex numbers. Let P be a polynomial function in n variables. Then $x \to P(f_1(x), f_2(x), \ldots, f_n(x))$ is a continuous function. If Q is a rational function of n variables, then $x \to Q(f_1(x), f_2(x), \ldots, f_n(x))$ is a continuous function on its domain of definition.

This result is often appealed to tacitly. For example, if f and g are continuous real-valued functions on a metric space, then the set $S = \{x \mid f(x) < g(x)\}$ is an open set. This is because S is the inverse image of the open interval $(0, \infty)$ under the continuous function $x \to g(x) - f(x)$. Similarly, $\{x \mid f(x)^2 = g(x) + f(x)\}$ is closed.

14–4.21. Proposition. Let $\langle S, \rho \rangle$ be a metric space. Then ρ is a continuous function from $S \times S$ to $\mathbf{R}$.

Proof. From the triangle inequality we find that

$$\rho(x, y) \geq \rho(x, z) - \rho(y, z) \quad \text{and} \quad \rho(y, x) \geq \rho(y, z) - \rho(x, z),$$

whence $|\rho(x, z) - \rho(y, z)| \leq \rho(x, y)$, for any three points x, y, and z in S.

Let θ denote the metric for $S \times S$. Let $\langle p, q \rangle$ be any point of $S \times S$. We will prove that ρ is continuous at $\langle p, q \rangle$. Let a positive ϵ be given. If

$$\theta(\langle x, y \rangle, \langle p, q \rangle) < \frac{\epsilon}{2},$$

then

$$|\rho(x, y) - \rho(p, q)| \leq |\rho(x, y) - \rho(p, y)| + |\rho(p, y) - \rho(p, q)|$$

$$\leq \rho(x, p) + \rho(y, q) < \frac{\epsilon}{2} + \frac{\epsilon}{2} = \epsilon.$$

This is another fact that is often used without explicit reference. It occurs perhaps most commonly in the context of convergent sequences; say, $x_n \to p$ and $y_n \to q$, then $\rho(x_n, y_n) \to \rho(p, q)$.

EXERCISES

1. If A is a nonempty set in a metric space $\langle S, \rho \rangle$, show that the function $x \to \rho(x, A)$ (see 14–1.8) is continuous from S to $\mathbf{R}$.

2. Prove Proposition 14–4.16 directly from the definition of continuity at a point without appeal to the results of Chapter 12.

3. Show that the projection functions from a direct product into the factors are both open and continuous. Show that they need not be closed.

4. Let $\langle S, \rho \rangle$ be a metric space and let p be a point of S. For each point s of S, define a function f_s from S to $\mathbf{R}$ by

$$f_s(t) = \rho(s, t) - \rho(p, t).$$

Show that f_s is a bounded function. Regarding the set B of all bounded functions from S to $\mathbf{R}$ as a metric space, as in 14–1.6, show that $s \to f_s$ is a continuous function from S to B. Show that it is in fact an isometry (from S to its range regarded as a subspace of B).

5. Let S and T be metric spaces. For each $i = 1, 2, \ldots, n$ let f_i be a continuous function from a closed subset E_i of S to T. Given that the functions f_i have a common extension, prove that their least common extension is continuous.

14–5. UNIFORM CONTINUITY AND UNIFORM CONVERGENCE

The word *uniform* is used in many contexts involving the limit concept to imply that a certain choice can be made independent of some other choice on which it would normally depend.

To say that a function f from one metric space S to another is continuous means that

$$(\forall p \in S)(\forall \epsilon > 0)(\exists \delta > 0) \quad f(B(p, \delta)) \subseteq B(f(p), \epsilon). \tag{1}$$

In general we must expect that the choice of δ will depend on both p and ϵ. It may happen, however, that it need not depend on p; in other words, the stronger

$$(\forall \epsilon > 0)(\exists \delta > 0)(\forall p \in S) \quad f(B(p, \delta)) \subseteq B(f(p), \epsilon) \tag{2}$$

may be true.

14–5.1. Definition. Let f be a function from one metric space $\langle S, \rho \rangle$ to another $\langle T, \sigma \rangle$. Then f is said to be *uniformly continuous* if and only if (2) holds.

Condition (2) is obtained from (1) by rearranging the quantifiers. Permuting consecutive quantifiers of the same type never affects the truth of a proposition, but moving a universal quantifier to the right past an existential quantifier will often convert a true proposition into a false one. (See p. 18.) On the other hand, (2) implies (1); hence a uniformly continuous function is continuous.

When condition (2) is rewritten directly in terms of the metric, we get

$$(\forall \epsilon > 0)(\exists \delta > 0)(\forall p, q \in S) \quad \rho(p, q) < \delta \Rightarrow \sigma(f(p), f(q)) < \epsilon,$$

and when S and T are $\mathbf{R}$ or $\mathbf{C}$, this becomes

$$(\forall \epsilon > 0)(\exists \delta > 0)(\forall x, y) \quad |x - y| < \delta \Rightarrow |f(x) - f(y)| < \epsilon.$$

The function $x \to x + 1$ from $\mathbf{R}$ to $\mathbf{R}$ is uniformly continuous and, less trivially, so is $x \to x/(1 + x^2)$, since we can prove that

$$|x - y| < \epsilon \Rightarrow \left| \frac{x}{1 + x^2} - \frac{y}{1 + y^2} \right| < \epsilon.$$

But even a simple function like $x \to x^2$ is not uniformly continuous. In fact,

$$(\exists \delta > 0)(\forall x, y) \quad |x - y| < \delta \Rightarrow |x^2 - y^2| < 1$$

is false. For, given δ, we can let

$$x = \frac{1}{\delta} + \frac{1}{2}\delta, \qquad y = \frac{1}{\delta},$$

and obtain $|x - y| < \delta$ and $|x^2 - y^2| > 1$.

However, if we consider $x \to x^2$ restricted to some bounded interval, say $[-A, +A]$, then it is uniformly continuous, because

$$|x - y| < \frac{\epsilon}{2A + 1} \Rightarrow |x^2 - y^2| < \epsilon,$$

provided $x, y \in [-A, +A]$. We shall later prove (14–8.10 and 14–8.17) that every continuous function on a bounded closed interval in $\mathbf{R}$ is uniformly continuous.

This example brings up a common situation. A function f defined on a metric space S is continuous on all of S and the continuity is uniform on a certain subset

E of S. The latter clause means that

$$(\forall \epsilon > 0)(\exists \delta > 0)(\forall p \in E) \quad f(B(p, \delta)) \subseteq B(f(p), \epsilon);$$

written in terms of the metrics, it becomes

$$(\forall \epsilon > 0)(\exists \delta > 0)(\forall p \in E)(\forall q \in S) \quad \rho(p, q) < \delta \Rightarrow \sigma(f(p), f(q)) < \epsilon. \quad (3)$$

This is stronger than the assertion "f restricted to E is uniformly continuous," which is

$$(\forall \epsilon > 0)(\exists \delta > 0)(\forall p \in E)(\forall q \in E) \quad \rho(p, q) < \delta \Rightarrow \sigma(f(p), f(q)) < \epsilon. \quad (4)$$

When one reads that a function defined on S is uniformly continuous on E, there is some ambiguity as to whether (3) or (4) is meant, but the weaker condition (4) is most common.

Another context for the word *uniform* is that of uniform convergence. Suppose $\{f_n\}$ is a sequence of functions from a set S to a metric space $\langle T, \sigma \rangle$. What should we mean by the convergence of this sequence? There are two natural possibilities. Let g be another function from S to T. We might define $f_n \to g$ to mean that

$$(\forall p \in S) \quad f_n(p) \to g(p).$$

Written out in full, this becomes

$$(\forall p \in S)(\forall \epsilon > 0)(\exists k \in \mathbb{N})(\forall n \geq k) \quad \sigma(f_n(p), g(p)) < \epsilon. \quad (5)$$

As we remarked in Chapter 13, this notion is known as *pointwise convergence*. Another possibility is to make the set of all functions from S to T into a metric space by defining

$$\tau(h_1, h_2) = \sup_p \sigma(h_1(p), h_2(p)) \quad (6)$$

(τ may be infinite, of course) and then to interpret $f_n \to g$ as convergence in this space. When this is written out in full, it becomes

$$(\forall \epsilon > 0)(\exists k \in \mathbb{N})(\forall n \geq k) \quad \sup_p \sigma(f_n(p), g(p)) < \epsilon, \quad (7)$$

which is almost the same as, and logically equivalent to,

$$(\forall \epsilon > 0)(\exists k \in \mathbb{N})(\forall n \geq k)(\forall p \in S) \quad \sigma(f_n(p), g(p)) < \epsilon. \quad (8)$$

(Here (7) $\Rightarrow$ (8) directly, but the reverse implication requires some manipulation of the ϵ's, because $(\forall p \in S) \sigma(f_n(p), g(p)) < \epsilon$ gives only $\sup_p \sigma(f_n(p), g(p)) \leq \epsilon$.)

We see that (8) is obtained from (5) by moving the universal quantifier $(\forall p \in S)$ to the right past an existential quantifier. In (5) the choice of k may be dependent on both p and ϵ; in (8) it must depend on ϵ alone. Convergence in this sense is called *uniform convergence*. We shall state these definitions formally for reference.

14–5.2. Definitions. Let $\{f_n\}$ be a sequence of functions from a set S to a metric space $\langle T, \sigma \rangle$ and let g be a function from S to T. The sequence $\{f_n\}$ is said to *converge to g pointwise* if and only if (5) holds. It is said to *converge to g uniformly* if and only if (8) holds.

The notion of uniform convergence of a sequence of functions is, generally speaking, more useful than that of pointwise convergence. This is due in part to the fact that uniform convergence is just convergence in a suitable metric space, and can therefore be subsumed under the general theory of convergence in metric spaces, whereas pointwise convergence is not in general equivalent to convergence in any metric on the space of functions.

It frequently happens that the set S is itself a metric space, in which case it is natural to inquire about the behavior of sequences of continuous functions. The substantial superiority of uniform convergence over pointwise convergence is shown in the following theorem.

14–5.3. Theorem. Let $\{f_n\}$ be a sequence of continuous functions from a metric space $\langle S, \rho \rangle$ to a metric space $\langle T, \sigma \rangle$. Suppose that $\{f_n\}$ converges uniformly to a function g from S to T. Then g is continuous.

This can be conveniently summarized as follows: The uniform limit of continuous functions is continuous.

Proof. Let q be any point of S and let $\epsilon > 0$ be given. Choose k so large that

$$(\forall n \geq k)(\forall p \in S) \quad \sigma(f_n(p), g(p)) < \tfrac{1}{3}\epsilon. \tag{9}$$

Since f_k is continuous at q, we can choose $\delta > 0$ so that

$$(\forall p \in S) \quad \rho(p, q) < \delta \Rightarrow \sigma(f_k(p), f_k(q)) < \tfrac{1}{3}\epsilon. \tag{10}$$

Now, if $\rho(q, r) < \delta$, we have

$$\sigma(g(q), g(r)) \leq \sigma(g(q), f_k(q)) + \sigma(f_k(q), f_k(r)) + \sigma(f_k(r), g(r)).$$

Each term on the right-hand side is less than $\tfrac{1}{3}\epsilon$, the first and third by (9) and the second by (10). Thus,

$$\rho(q, r) < \delta \Rightarrow \sigma(g(q), g(r)) < \epsilon.$$

This proves that g is continuous.

As suggested above, the situation is not so favorable for pointwise convergence. If we define $f_n(x) = (1 + x^2)^{-n}$ for $x \in \mathbf{R}$, then $\{f_n\}$ is a sequence of continuous functions which converges pointwise to a discontinuous function.

14–5.4. Corollary. The set of all continuous functions from S to T is a closed subset of the set of all functions from S to T when the latter is assigned the metric (6).

EXERCISES

1. In Section 14–1 three different metrics, θ, θ', and θ'', were defined on the direct product $S \times T$ of two metric spaces. Show that the identity map i is uniformly continuous from $\langle S \times T, \theta \rangle$ to $\langle S \times T, \theta' \rangle$. Do the same for each of the other five ways of assigning different metrics to the domain and range of i.

2. Let $\{f_n\}$ be a sequence of uniformly continuous functions from a metric space $\langle S, \rho \rangle$ to a metric space $\langle T, \sigma \rangle$. Suppose that $\{f_n\}$ converges uniformly to a function g from S to T. Prove that g is uniformly continuous.

3. Suppose that $\{f_i \mid i \in I\}$ is a family of continuous functions from one metric space to another. In the definition of the continuity of f_i at p, the choice of δ presumably depends on i, p, and ϵ. If it can be made independent of i, the family is called *equicontinuous*. Write the definition of an equicontinuous family of functions with quantifiers. Prove the following theorems:

(a) If an equicontinuous sequence of functions converges pointwise, then the limit is continuous.

(b) If a sequence of continuous functions converges uniformly, then the sequence is equicontinuous.

4. Suppose that $\{f_n\}$ is a sequence of continuous functions from a metric space S to $\mathbf{R}$ such that $\{f_n(x)\}$ is a Cauchy sequence uniformly for $x \in S$. Explain what this means and prove that there exists a continuous function f from S to $\mathbf{R}$ such that $f_n \to f$ uniformly.

14–6. HOMEOMORPHISM

There is an equivalence relation for metric spaces both broader and more important than isometry. A consideration of this relation, homeomorphism, shows the way to a new class of mathematical configurations, the topological spaces. Although we shall make no serious attempt to investigate this new class as such, it is important to acquire a feeling for the topological, as opposed to the metric, properties of spaces.

14–6.1. Definition. Two metric spaces $\langle S, \rho \rangle$ and $\langle T, \sigma \rangle$ are said to be *homeomorphic* if and only if there exists a bijection φ from S to T such that

$$(\forall E \subseteq S) \quad E \text{ is open in } S \Leftrightarrow \varphi(E) \text{ is open in } T.$$

Any such function φ is called a *homeomorphism*. An injective function f from S to T is also called a homeomorphism if it is a homeomorphism in the previous sense from S to $f(S)$ when $f(S)$ is regarded as a subspace of T.

With the aid of Proposition 14–4.5 the definition can be restated as follows.

14–6.2. Proposition. Let f be a bijection from one metric space to another. A necessary and sufficient condition for f to be a homeomorphism is that both f and its inverse be continuous.

A function satisfying this condition is often referred to as *bicontinuous*.

The relation "*S* is homeomorphic to *T*" between metric spaces is an equivalence relation, which is also called homeomorphism. The usage here precisely parallels that of *isomorphism*. In fact, homeomorphism is isomorphism in a particular context.

Let S_1 and S_2 be two metric spaces, and let $\mathfrak{I}_1$ and $\mathfrak{I}_2$ be the sets of all open subsets of S_1 and S_2, respectively. We can regard $\langle S_1, \mathfrak{I}_1 \rangle$ as a configuration, where $\mathfrak{I}_1 \in \mathfrak{P}^2(S_1)$. Similarly, $\langle S_2, \mathfrak{I}_2 \rangle$ is a configuration of the same type. The metric spaces S_1 and S_2 are homeomorphic if and only if the configurations $\langle S_1, \mathfrak{I}_1 \rangle$ and $\langle S_2, \mathfrak{I}_2 \rangle$ are isomorphic in the usual sense for configurations.

This suggests that we should try to find restrictive postulates for configurations of the type $\langle S, \mathfrak{I} \rangle$, where $\mathfrak{I} \in \mathfrak{P}^2(S)$, which pick out precisely those configurations which represent a metric space and the set of all its open subsets. No simple postulate system is known which accomplishes this, though we can make a start. Theorem 14–3.11 tells us that if $\langle S, \mathfrak{I} \rangle$ does in fact come from a metric space, then

(a) $S \in \mathfrak{I}, \emptyset \in \mathfrak{I}$;
(b) if $S_1 \in \mathfrak{I}$ and $S_2 \in \mathfrak{I}$, then $S_1 \cap S_2 \in \mathfrak{I}$;
(c) if $S_\alpha \in \mathfrak{I}$ for all α, then $\bigcup_\alpha S_\alpha \in \mathfrak{I}$.

Since these properties are certainly intrinsic for the class of configurations under discussion, they can be imposed as restrictive postulates. The resulting configurations are called *topological spaces*; in other words, a topological space is a pair $\langle S, \mathfrak{I} \rangle$ such that $\mathfrak{I} \in \mathfrak{P}^2(S)$ and (a), (b), and (c) hold. A set $\mathfrak{I}$ satisfying these conditions is called a *topology for S*. The word *topology* is also used to describe the study of topological spaces.

Associated with every metric space *S* is a topological space having *S* as its basic set, the topology being the set of all open subsets of *S*. This gives us a function from the set of all metrics for a set *S* to the set of all topologies for *S*. Unfortunately, this function is neither injective nor surjective (assuming that *S* has more than one point).

There are many different metrics for a set *S* which produce the same topology. Metrics which produce the same topology for a set are said to be *topologically equivalent*. The metrics ρ and σ for *S* are topologically equivalent if and only if the identity function is a homeomorphism from $\langle S, \rho \rangle$ to $\langle S, \sigma \rangle$. If ρ is a metric for *S* and λ is a positive number, then $\lambda\rho$ (that is, $\langle s, t \rangle \to \lambda\rho(s, t)$) is another metric for *S* which is topologically equivalent to ρ. Moreover, both

$$\sigma(s, t) = \min\{1, \rho(s, t)\} \quad \text{and} \quad \tau(s, t) = \frac{\rho(s, t)}{1 + \rho(s, t)}$$

define metrics for *S* which are equivalent to ρ. (See Exercise 8, p. 228.) The three metrics we considered $(14–1(2), (3), \text{ and } (4))$ for the direct product of two metric spaces are topologically equivalent (see Exercise 6, p. 228).

Every set with more than one point has topologies which do not arise from a metric. For example, $\{\emptyset, S\}$ is always a topology for S and it does not arise from a metric, since a one-point set in a metric space is always closed and its complement open. A topological space which does arise from a metric space is called *metrizable*. As we might expect, metrizable spaces have some desirable properties not possessed by all topological spaces. Topological spaces are usually restricted by the imposition of postulates in addition to (a), (b), and (c). Often these additional postulates are designed to ensure some of the more desirable properties of metric spaces.

The word *homeomorphism* is used in place of *isomorphism* for general topological spaces. Two metric spaces are homeomorphic in the sense of 14–6.1 if and only if the associated topological spaces are homeomorphic.

Much, but not all, of the theory we have developed for metric spaces is applicable with no essential change to topological spaces. Many of the changes would replace theorems by definitions and vice versa. Thus, if $\langle S, \mathfrak{J} \rangle$ is a topological space, we *define* a subset E of S to be open if and only if $E \in \mathfrak{J}$. A closed set in a topological space is *defined* as the complement of an open set. A function f from one topological space $\langle S, \mathfrak{J}_1 \rangle$ to another $\langle S_2, \mathfrak{J}_2 \rangle$ is defined to be continuous if and only if

$$(\forall G \in \mathfrak{J}_2) \quad f^{-1}(G) \in \mathfrak{J}_1.$$

A set U is called a neighborhood of the point p if and only if there is an open set G such that $p \in G \subseteq U$. Proposition 14–3.21 tells us how the definition of convergence for sequences may be extended to a general topological space. All of these new definitions are consistent with the old definitions for metric spaces. However, an important breakdown occurs. Let E be a set in a topological space, and let F be the set of all points which are limits of sequences in E. It need not be true that F is a closed set. That this is always true in metrizable spaces is one of the essential properties which make metrizable spaces simpler than general topological spaces.

A property of a metric space is said to be *topological* if and only if it is intrinsic as a property of the associated topological space. We can recognize topological properties because they are the properties which can be defined by reference to the open sets. The convergence of sequences is topological, because $x_n \to p$ if and only if every open set containing p contains all but a finite number of the x_n. It is not important that we *did not* define convergence in this way; it is sufficient that we *could have*. The continuity of functions from one metric space to another is topological, because Proposition 14–4.5 gives us a criterion for continuity which refers only to open sets.

The importance of the homeomorphism concept lies in the fact that, of two homeomorphic spaces, one may be easier to deal with than another. A closed triangular region in the plane is homeomorphic to a closed circular disk; hence any topological fact about one of these spaces is also true of the other. The plane itself is homeomorphic both to the open unit disk and to the sphere S^2 less one point. These three spaces present quite different views of the structure of the "infinite" part of the plane.

No effective procedure is known for deciding whether or not two spaces are homeomorphic. It is often possible to tell by inspection in the case of subsets of the plane or three-dimensional space, but a proof of homeomorphism must ultimately rest on the construction of a bicontinuous function. Proof of non-homeomorphism requires us to find a topological property of one space that is not shared by the other. The subject known as algebraic topology is concerned with defining and computing invariants of topological spaces which may be useful in this connection.

Topology is full of surprises. Sometimes the surprise is in the difficulty of proving something that is intuitively obvious. Sometimes the intuitively obvious is false. Let S be a subset of the plane which is homeomorphic to a circle (that is, a 1-sphere). From our long experience with drawing closed curves on paper, we are sure that $\mathbf{R}^2 - S$ falls into two parts, the "inside" and the "outside." Moreover, the "inside" is homeomorphic to an open circular disk. This is true, but the proof is hard. Intuitively, it is also obvious that the same is true concerning a spherical surface in $\mathbf{R}^3$. If T is a subset of $\mathbf{R}^3$ homeomorphic to the sphere, then $\mathbf{R}^3 - T$ falls into two parts, the "inside" and the "outside." This is true. But the "inside" need not be homeomorphic to an open spherical ball.

Among the nontrivial theorems on homeomorphism we find the following, which is quite in accord with our intuition. A nonvoid open subset of $\mathbf{R}^m$ is not homeomorphic to any subset of $\mathbf{R}^n$ unless $m \leq n$. From this it follows that a nonvoid open subset of $\mathbf{R}^m$ is not homeomorphic to an open subset of $\mathbf{R}^n$ unless $m = n$.

EXERCISES

1. Prove that any two open intervals in $\mathbf{R}$ are homeomorphic.

2. Prove that any two bounded closed intervals in $\mathbf{R}$ are homeomorphic.

3. Prove that a triangle (union of three line segments) is homeomorphic to a circle.

4. Prove that an open circular disk is homeomorphic to an open triangular region in $\mathbf{R}^2$.

5. Prove that the rational numbers are not homeomorphic to the set

$$\{0, 1, \tfrac{1}{2}, \tfrac{1}{3}, \ldots, 1/n, \ldots\}.$$

6. Suppose that S_1 and S_2 and T_1 and T_2 are two pairs of homeomorphic metric spaces. Prove that $S_1 \times T_1$ is homeomorphic to $S_2 \times T_2$.

7. Let ρ_1 and ρ_2 be two metrics for a set S. Show that ρ_1 and ρ_2 are topologically equivalent if and only if the set of sequences in S which converge in $\langle S, \rho_1 \rangle$ is the same as the set of sequences which converge in $\langle S, \rho_2 \rangle$.

8. Let S and T be metric spaces and let f be a function from S to T. Prove that f is continuous if and only if the function $s \rightarrow \langle s, f(s) \rangle$ is a homeomorphism from S to f (where f is regarded as a subspace of $S \times T$).

9. Let ρ_1 and ρ_2 be two topologically equivalent metrics for a set S. Define

$$\sigma(s, t) = \sup \{\rho_1(s, t), \rho_2(s, t)\}.$$

Show that σ is a metric for S which is topologically equivalent to ρ_1 and ρ_2.

10. Let $\langle S, \rho \rangle$ be a metric space and let σ be another metric for S satisfying

$$(\forall s, t) \quad \sigma(s, t) \geq \rho(s, t).$$

Show that σ is equivalent to ρ if and only if it is a continuous function on $S \times S$ (with the metric induced by ρ). Hence show that if f is any continuous function from S to $\mathbf{R}$, then

$$\sigma(s, t) = \rho(s, t) + |f(s) - f(t)|$$

defines a metric for S which is equivalent to ρ.

11. Is the set of all rational numbers homeomorphic to the set of all nonnegative rational numbers?

14-7. COMPLETE SPACES

Cauchy's criterion for convergence can readily be stated for general metric spaces. While it remains necessary, it is not sufficient for convergence. For example, take a sequence of rational numbers which converges to an irrational number, say $\sqrt{2}$ in $\mathbf{R}$. This sequence satisfies Cauchy's criterion, and continues to do so if we regard it instead as a sequence in the space $\mathbf{Q}$ of rational numbers, but it no longer converges.

A metric space is called complete if every sequence in it which satisfies Cauchy's criterion actually converges. Complete spaces have, so to speak, all their cracks filled in. We shall prove that in any metric space the cracks can be filled up by adjoining new points to give a complete metric space in much the same way that $\mathbf{R}$ arises from $\mathbf{Q}$. The section concludes with three theorems of great importance in analysis.

14-7.1. Definition. Let $\langle S, \rho \rangle$ be a metric space. A sequence $\{x_n\}$ in S is said to be a *Cauchy sequence* if and only if it satisfies the following condition, known as Cauchy's criterion:

$$(\forall \epsilon > 0)(\exists k \in \mathbf{N})(\forall m, n \geq k) \quad \rho(x_m, x_n) < \epsilon.$$

14-7.2. Proposition. A convergent sequence in a metric space is a Cauchy sequence.

14-7.3. Definition. A metric space S is said to be *complete* if and only if every Cauchy sequence in S converges in S.

Completeness is a property of a space, but we shall also apply it to subsets of a space, reversing the convention which allows us to refer to a *closed subspace*. A complete subset T of a metric space S is a subset which is complete when regarded as a subspace.

While such properties as convergence and continuity are topological, this is not true of completeness, because the definition of a Cauchy sequence involves the metric too intimately to be given in terms of open sets alone. Completeness is, of course, intrinsic to metric spaces.

To show that completeness is not topological, we note that $(-1, +1)$ and $\mathbf{R}$ are homeomorphic spaces and that $\mathbf{R}$ is complete while $(-1, +1)$ is not. To examine the situation more closely, consider the function $x \to x/\sqrt{1 - x^2}$ which is a homeomorphism from $(-1, +1)$ to $\mathbf{R}$. It carries the nonconvergent Cauchy sequence $\{1 - 1/n\}$ in $(-1, +1)$ into $\{(n - 1)/\sqrt{2n - 1}\}$, which is not a Cauchy sequence in $\mathbf{R}$. The difficulty lies in the fact that although $x \to x/\sqrt{1 - x^2}$ is continuous, it is not uniformly continuous (see Exercise 5).

Theorem 12–5.3 can now be restated, "$\mathbf{R}$ and $\mathbf{C}$ are complete." Note that we are using the word *complete* in a sense different from that of 6–5.2. There the notion was a property of ordered sets. It is clear from the work in Chapter 12 that completeness as an ordered set and completeness as a metric space express the same fundamental property of $\mathbf{R}$. The term *complete* was originally used to describe this property of $\mathbf{R}$ and was subsequently generalized to apply to ordered sets with one meaning and to metric spaces with another.

14–7.4. Proposition. The direct product of two complete metric spaces is complete.

Proof. Let $\langle S, \rho \rangle$ and $\langle T, \sigma \rangle$ be complete metric spaces, and let θ be the metric for $S \times T$ defined in 14–1.5. Suppose that $\{\langle x_n, y_n \rangle\}$ is a Cauchy sequence in $S \times T$. From the fact that

$$\rho(x_m, x_n) \leq \theta(\langle x_m, y_m \rangle, \langle x_n, y_n \rangle)$$

it follows immediately that $\{x_n\}$ is a Cauchy sequence in S. Since S is complete, $\{x_n\}$ converges. Similarly, $\{y_n\}$ converges in T. By 14–2.6, $\{\langle x_n, y_n \rangle\}$ converges in $S \times T$. Therefore, $S \times T$ is complete.

The proposition remains true no matter which of the three metrics 14–1(2), (3), or (4) we assign to $S \times T$.

14–7.5. Corollary. $\mathbf{R}^n$ and $\mathbf{C}^n$ are complete.

As the example of the rational numbers shows, in general we cannot expect a subspace of a complete space to be complete. The situation is explained in the next two propositions.

14–7.6. Proposition. A closed subspace of a complete metric space is complete.

14–7.7. Proposition. Let T be a complete subspace of a metric space S. Then T is closed in S.

The latter proposition is more than a converse of the former, for it asserts that T is closed without regard to whether S is complete. It means that complete spaces possess an absolute closure property, for whenever the complete space T, or an isometric replica of it, appears as a subset of a metric space it will be closed.

Proof. Suppose that p adheres to T. Choose a sequence $\{x_n\}$ in T such that $x_n \to p$. Now, $\{x_n\}$ is a Cauchy sequence and, since T is complete, it converges to a point q of T. This shows that $p = q$ and therefore that $p \in T$. Thus T is closed.

14–7.8. Proposition. Let $\langle T, \sigma \rangle$ be a complete metric space and let S be any set. Let B be the set of all functions from S to T, and let τ be the (possibly infinite) metric for B defined by

$$\tau(f, g) = \sup \{\sigma(f(x), g(x)) \mid x \in S\}. \tag{1}$$

Then $\langle B, \tau \rangle$ is a complete metric space.

Proof. Suppose that $\{f_n\}$ is a Cauchy sequence in B. For each fixed $x \in S$, the sequence $\{f_n(x)\}$ is a Cauchy sequence in T, since $\sigma(f_m(x), f_n(x)) \leq \tau(f_m, f_n)$. Since T is complete, we can define g from S to T by

$$g(x) = \lim f_n(x).$$

We must prove that $f_n \to g$ in B. Let a positive number ϵ be given. Choose k so large that

$$(\forall m, n \geq k) \quad \tau(f_m, f_n) < \epsilon.$$

For any x and any $m, n \geq k$, we have $\sigma(f_m(x), f_n(x)) < \epsilon$, so

$$\sigma(g(x), f_n(x)) = \lim_{m \to \infty} \sigma(f_m(x), f_n(x)) \leq \epsilon.$$

Therefore, $\tau(g, f_n) \leq \epsilon$ for any $n \geq k$. Thus $f_n \to g$ in B. This proves that $\langle B, \tau \rangle$ is a complete metric space.

14–7.9. Corollary. Let $\langle S, \rho \rangle$ be a metric space and let $\langle T, \sigma \rangle$ be a complete metric space. Let C be the space of all continuous functions from S to T, with the (possibly infinite) metric τ given by (1). Then $\langle C, \tau \rangle$ is complete.

Proof. According to 14–5.3, C is a closed subspace of $\langle B, \tau \rangle$. A closed subspace of a complete space is complete.

14–7.10. Proposition. Let $\langle S, \rho \rangle$ be a metric space. S is complete if and only if it has the following property: For every decreasing sequence $\{F_n\}$ of nonvoid closed subsets of S for which diam $F_n \to 0$, the intersection $\bigcap_n F_n$ is not void.

Note that $\bigcap F_n$ can contain at most one point.

Proof. Assume first that S is complete and let $\{F_n\}$ be a decreasing sequence of nonvoid closed subsets of S such that diam $F_n \to 0$. Choose a point x_n in F_n. If $m \geq k$ and $n \geq k$, then $x_m \in F_k$ and $x_n \in F_k$; therefore, $\rho(x_m, x_n) \leq$ diam F_k. Since diam F_k tends to zero as k increases, $\{x_n\}$ is a Cauchy sequence. Let p be its limit. For any integer k, p is also the limit of the sequence $\{x_{n+k}\}$ which is in F_k; hence p adheres to F_k. Since F_k is closed, $p \in F_k$. This shows that $p \in \bigcap F_n$; that is, the intersection is not void.

Now suppose S has the stated property. Let $\{x_n\}$ be any Cauchy sequence in S. Consider the sets $E_k = \{x_n \mid n \geq k\}$. Evidently $\{\bar{E}_k\}$ is a decreasing sequence of closed sets. For any $\epsilon > 0$, we can choose k such that $(\forall m, n \geq k)\,\rho(x_m, x_n) < \epsilon$; whence diam $\bar{E}_k =$ diam $E_k \leq \epsilon$. Thus, diam $\bar{E}_k \to 0$. Let p be a point in $\bigcap \bar{E}_n$. Since $\rho(x_n, p) \leq$ diam $\bar{E}_n$, we see that $\rho(x_n, p) \to 0$; hence $\{x_n\}$ is convergent. This proves that S is complete.

In analysis one is often faced with the problem of interchanging the order of limiting processes. Typically, we have a double sequence $\{x_{m,n}\}$ (that is, a family with index set $\mathbf{N} \times \mathbf{N}$) in a metric space and we would like to know that

$$\lim_{m \to \infty} \lim_{n \to \infty} x_{m,n} = \lim_{n \to \infty} \lim_{m \to \infty} x_{m,n}.$$

Simple examples show that this is not always true even assuming the existence of these iterated limits (Exercise 4, p. 231). One can also consider the so-called *double limit* which is a point p such that

$$(\forall \epsilon > 0)(\exists k \in \mathbf{N})(\forall m, n \geq k) \quad \rho(x_{m,n}, p) < \epsilon.$$

This limit is unique if it exists and is denoted by $\lim_{m,n \to \infty} x_{m,n}$. The Moore-Smith theorem provides the most usable condition for the equality of these three limits.

14-7.11. The Moore-Smith theorem. Let $\{x_{m,n}\}$ be a double sequence in a complete metric space $\langle T, \sigma \rangle$. Assume that

(a) $\lim_{n \to \infty} x_{m,n}$ exists uniformly in m, and
(b) $\lim_{m \to \infty} x_{m,n}$ exists for each n.

Then

$$\lim_{m \to \infty} \lim_{n \to \infty} x_{m,n}, \quad \lim_{n \to \infty} \lim_{m \to \infty} x_{m,n}, \quad \text{and} \quad \lim_{m,n \to \infty} x_{m,n}$$

all exist and are equal.

Proof. Let $y_m = \lim_n x_{m,n}$ and $z_n = \lim_m x_{m,n}$. In (a) the phrase *uniformly in m* means

$$(\forall \epsilon > 0)(\exists k \in \mathbf{N})(\forall n \geq k)(\forall m \in \mathbf{N}) \quad \sigma(y_m, x_{m,n}) < \epsilon.$$

This is really a special case of Definition 14-5.2, where we let $S = \mathbf{N}$, let f_n be the function $m \to x_{m,n}$, and let $g = y$.

We shall show first that $\{y_m\}$ is a Cauchy sequence. Let $\epsilon > 0$ be given. By the definition of uniform convergence, choose k so that

$$(\forall n \geq k)(\forall m \in \mathbf{N}) \quad \sigma(y_m, x_{m,n}) < \frac{\epsilon}{4}. \tag{2}$$

Now choose $r \in \mathbf{N}$ so that

$$(\forall p \geq r) \quad \sigma(z_k, x_{p,k}) < \frac{\epsilon}{4}.$$

(We are using here the convergence of the single sequence $\{x_{m,k} \mid m \in \mathbf{N}\}$.) Now, if $p, q \geq r$, we find that

$$\sigma(y_p, y_q) \leq \sigma(y_p, x_{p,k}) + \sigma(x_{p,k}, z_k) + \sigma(z_k, x_{q,k}) + \sigma(x_{q,k}, y_q).$$

Here the right-hand terms are each less than $\epsilon/4$. Hence $\{y_m\}$ is a Cauchy sequence.

Since T is complete, let $y_m \to w$. We shall prove that $\lim_{m,n\to\infty} x_{m,n} = w$. Let $\epsilon > 0$ be given. Choose k so that (2) holds and s so that

$$(\forall m \geq s) \quad \sigma(y_m, w) < \frac{\epsilon}{2}.$$

Then, if $m, n \geq \sup \{k, s\}$, we have

$$\sigma(x_{m,n}, w) \leq \sigma(x_{m,n}, y_m) + \sigma(y_m, w) < \frac{\epsilon}{4} + \frac{\epsilon}{2} < \epsilon.$$

This proves that $\lim_{m,n\to\infty} x_{m,n} = w$.

Finally, we must show that $z_n \to w$. For a given ϵ we have proved that $\sigma(x_{m,n}, w) < \epsilon$ for all sufficiently large m and n. Letting $m \to \infty$, this gives us $\sigma(z_n, w) \leq \epsilon$ for n sufficiently large, which shows that $z_n \to w$.

14–7.12. Corollary. Let $\{x_{m,n}\}$ be a double sequence in a metric space $\langle T, \sigma \rangle$. Assume 14–7.11(a) and (b). If one of the three limits

$$\lim_{m\to\infty} \lim_{n\to\infty} x_{m,n}, \quad \lim_{n\to\infty} \lim_{m\to\infty} x_{m,n}, \quad \text{and} \quad \lim_{m,n\to\infty} x_{m,n}$$

exists, then they all exist and are equal.

Proof. The completeness of T enters the proof of the theorem only to establish the existence of the point w. If we assume that $\lim_m \lim_n x_{m,n} = \lim_m y_m$ exists, then the second half of the previous proof proves the corollary directly. If we assume that the other iterated limit or the double limit exists, the proof requires some modification which we leave to the reader.

It frequently happens that we have a continuous function defined on a dense subset E of a space S and we would like to extend it continuously over all of S. For example, we can define a^r for $a > 0$ and rational exponents r. Can we extend this continuously to define a^x for $a > 0$ and all real exponents x? The following theorem is useful in this connection, although it does not directly handle the problem we have just posed.

14-7.13. Theorem. Let E be a dense subset of a metric space $\langle S, \rho \rangle$ and let $\langle T, \sigma \rangle$ be a complete metric space. Let f be a uniformly continuous function from E to T. There exists a unique continuous function g from S to T which extends f, and this function is uniformly continuous.

Proof. Let $p \in S$. Since E is dense in S, we can choose a sequence $\{x_n\}$ in E such that $x_n \to p$. We shall define $g(p) = \lim_n f(x_n)$, but first we must verify some facts.

Let us show that $\{f(x_n)\}$ is a Cauchy sequence in T. Let $\epsilon > 0$ be given. Since f is uniformly continuous, we can choose δ so that

$$(\forall x, y \in E) \quad \rho(x, y) < \delta \Rightarrow \sigma(f(x), f(y)) < \epsilon. \tag{3}$$

Since $\{x_n\}$ is a Cauchy sequence in S, we can choose k so that

$$(\forall m, n \geq k) \quad \rho(x_m, x_n) < \delta.$$

This gives

$$(\forall m, n \geq k) \quad \sigma(f(x_m), f(x_n)) < \epsilon.$$

Therefore $\{f(x_n)\}$ is a Cauchy sequence.

Since T is complete, $\lim_n f(x_n)$ exists. Before we can set $g(p) = \lim_n f(x_n)$ we must consider that we might have chosen a different sequence $\{y_n\}$ in E converging to p, in which case $\lim_n f(y_n)$ might conceivably be different. This does not happen, however, for if $x_n \to p$ and $y_n \to p$, then the sequence $\{z_n\}$, where $z_{2n-1} = x_n$, $z_{2n} = y_n$ (that is, the sequence $x_1, y_1, x_2, y_2, x_3, \ldots$) converges to p, whence $\{f(z_n)\}$ is convergent in T and $\{f(x_n)\}$ and $\{f(y_n)\}$ are both subsequences of the same convergent sequence, and therefore $\lim f(x_n) = \lim f(y_n)$. Thus we can define $g(p) = \lim f(x_n)$, where $\{x_n\}$ is any sequence in E such that $x_n \to p$. If $p \in E$, we can choose $\{x_n\}$ to be a constant sequence, whence it follows that $g(p) = f(p)$. Thus g is an extension of f.

Now we shall show that g is uniformly continuous. Let $\epsilon > 0$ be given. Choose δ so that (3) holds. If $p, q \in S$ and $\rho(p, q) < \delta$, we can choose sequences $\{x_n\}$ and $\{y_n\}$ in E so that $x_n \to p$ and $y_n \to q$. Since the metric is continuous, $\rho(x_n, y_n) < \delta$ for all large n; hence $\sigma(f(x_n), f(y_n)) < \epsilon$ for all large n. Letting $n \to \infty$, we find that $\sigma(g(p), g(q)) \leq \epsilon$. Thus we have proved that

$$(\forall \epsilon > 0)(\exists \delta > 0)(\forall p, q \in S) \quad \rho(p, q) < \delta \Rightarrow \sigma(g(p), g(q)) \leq \epsilon,$$

which establishes the uniform continuity of g.

That g is unique, even in the class of continuous functions, follows from 14-4.4.

As we pointed out at the beginning of this section, the rational numbers are an incomplete space. This, of course, follows from Proposition 14-7.7 as well, because the rationals are a subspace but not a closed subspace of **R**. This suggests a way in which we can choose any number of examples of incomplete spaces; any nonclosed subspace of any space will serve. Going in the other direction, we may wonder whether every incomplete metric space appears as a nonclosed subspace of some larger complete space. This is in fact true as we shall now prove. The idea of the proof is clear. Given the metric space T which is not complete, for each nonconvergent Cauchy sequence we adjoin a point to serve as its limit. However, there will be different Cauchy sequences which must have the same limit point, so we must proceed with some care to take account of these equivalent Cauchy sequences. In what follows we do not actually embed T in a complete space S, but instead construct an entirely new space S which contains a subspace T' isometric to T. This is good enough for most purposes, but if we really require a complete space of which T is a subspace, we can define an appropriate metric for $(S - T') \cup T$.

14–7.14. Theorem. Let $\langle T, \rho \rangle$ be a metric space. There exists a triple $\langle S, \tau, \varphi \rangle$ such that $\langle S, \tau \rangle$ is a complete metric space and φ is an isometry from T to a dense subset T' of S. The triple $\langle S, \tau, \varphi \rangle$ is unique in the following sense. If $\langle U, \pi, \psi \rangle$ is another such triple, then there exists an isometry θ from S to U such that $\psi = \theta \circ \varphi$.

Proof. We shall give here a proof which tackles the problem directly. A shorter proof can be given (see Exercise 6).

Let C be the set of all Cauchy sequences of T. It is convenient to denote the sequence $\{x_n\}$ simply by x.

Define a function σ from $C \times C$ to **R** by

$$\sigma(x, y) = \lim_{j \to \infty} \rho(x_j, y_j). \qquad (4)$$

The limit on the right always exists because

$$|\rho(x_m, y_m) - \rho(x_n, y_n)| \leq \rho(x_m, x_n) + \rho(y_m, y_n).$$

Since x and y are Cauchy sequences, it follows that $\{\rho(x_n, y_n)\}$ is a Cauchy sequence in the complete space **R**. The function σ has the following properties:

$$\sigma(x, x) = 0, \qquad \sigma(x, y) = \sigma(y, x),$$

and

$$\sigma(x, z) \leq \sigma(x, y) + \sigma(y, z)$$

for any x, y, and z in C. In other words, σ is a pseudometric for C.

The set $\{\langle x, y \rangle \mid \sigma(x, y) = 0\}$ is an equivalence relation in the set C. Let S be the corresponding quotient space, and denote the quotient map by an overbar. If $\sigma(x, x') = 0$ and $\sigma(y, y') = 0$, then

$$\sigma(x, y) \leq \sigma(x, x') + \sigma(x', y') + \sigma(y, y') = \sigma(x', y');$$

similarly, $\sigma(x', y') \leq \sigma(x, y)$ and so $\sigma(x, y) = \sigma(x', y')$. Hence we can define τ on $S \times S$ by $\tau(\bar{x}, \bar{y}) = \sigma(x, y)$. Then τ is a metric for S.

We shall prove that $\langle S, \tau \rangle$ is complete. Suppose $\{s_n\}$ is a Cauchy sequence in S. For each n choose $x(n)$ in C such that $\overline{x(n)} = s_n$. Now, $x(n)$ is a Cauchy sequence in T, so we can choose an integer $k(n)$ so that

$$(\forall j \geq k(n)) \quad \rho\big(x(n)_j, x(n)_{k(n)}\big) < \frac{1}{n}.$$

Define $w_n = x(n)_{k(n)}$. We claim that w is a Cauchy sequence in T and that $s_n \to \bar{w}$ in S.

Let ϵ be a positive number. Choose an integer q such that $q > 4/\epsilon$ and

$$(\forall m, n \geq q) \quad \tau(s_m, s_n) < \frac{\epsilon}{4}.$$

Now

$$\rho(w_m, w_n) \leq \rho\big(x(m)_{k(m)}, x(m)_j\big) + \rho\big(x(m)_j, x(n)_j\big) + \rho\big(x(n)_j, x(n)_{k(n)}\big)$$

$$< \frac{1}{m} + \rho\big(x(m)_j, x(n)_j\big) + \frac{1}{n}$$

for any j greater than $k(m)$ and $k(n)$. If $m \geq q$ and $n \geq q$, then

$$\lim_j \rho\big(x(m)_j, x(n)_j\big) = \sigma\big(x(m), x(n)\big) = \tau(s_m, s_n) < \frac{\epsilon}{4},$$

and so, for sufficiently large j, $\rho\big(x(m)_j, x(n)_n\big) < \epsilon/4$. Combining the last two inequalities, we find $\rho(w_m, w_n) < 3\epsilon/4 < \epsilon$, provided that $m \geq q$ and $n \geq q$. This proves that w is a Cauchy sequence in T. Moreover,

$$\rho\big(w_m, x(n)_m\big) \leq \rho(w_m, w_n) + \rho\big(x(n)_{k(n)}, x(n)_m\big) \leq \frac{3\epsilon}{4} + \frac{1}{m} < \epsilon$$

if $m \geq q, n \geq q$, and $m \geq k(n)$. Letting $m \to \infty$, we now find that

$$\tau(\bar{w}, s_n) = \sigma\big(w, x(n)\big) = \lim_m \rho\big(w_m, x(n)_m\big) \leq \epsilon$$

for $n \geq q$. This proves that $s_n \to \bar{w}$ in S.

For each p in T, let $v(p)$ denote the constant sequence defined by $v(p)_n = p$. It is obviously a Cauchy sequence. Set $\varphi(p) = \overline{v(p)}$. This defines an isometry φ from T to a subspace T' of S. To prove that T' is dense in S, suppose that $s \in S$. Choose $x \in C$ so that $\bar{x} = s$. Then

$$\tau\big(\varphi(x_n), s\big) = \sigma\big(v(x_n), x\big) = \lim_j \rho\big(v(x_n)_j, x_j\big) = \lim_j \rho(x_n, x_j).$$

Since x is a Cauchy sequence in T, it follows that $\tau\big(\varphi(x_n), s\big) \to 0$ as $n \to \infty$, or $\varphi(x_n) \to s$ in S. This proves that T' is dense, and completes the proof of the existence of the triple $\langle S, \tau, \varphi \rangle$.

Suppose that $\langle U, \pi \rangle$ is a complete metric space and that ψ is an isometry from T to a dense subset T'' of U. For each $x \in C$, $\{\psi(x_n)\}$ is a Cauchy sequence in U, so we may define $\xi(x) = \lim_n \psi(x_n)$. If $x, y \in C$, then

$$\pi\big(\xi(x), \xi(y)\big) = \lim_n \pi\big(\psi(x_n), \psi(y_n)\big) = \lim_n \rho(x_n, y_n) = \sigma(x, y)$$

(the second equality because ψ is an isometry). In particular, $\sigma(x, y) = 0$ implies that $\xi(x) = \xi(y)$. By Theorem 5–2.1 there is a function θ from S to U such that $\xi(x) = \theta(\bar{x})$ for all $x \in C$. It satisfies the condition

$$\pi\big(\theta(\bar{x}), \theta(\bar{y})\big) = \pi\big(\xi(x), \xi(y)\big) = \sigma(x, y) = \tau(\bar{x}, \bar{y});$$

that is, θ is an isometry. This implies that θ is injective. Since T'' is dense in U, every point of U has the form $\lim \psi(x_n)$ for some Cauchy sequence x in T; therefore θ is surjective. If $p \in T$,

$$\psi(p) = \lim_n \psi\big(v(p)_n\big) = \xi\big(v(p)\big) = \theta\big(\overline{v(p)}\big) = \theta\big(\varphi(p)\big).$$

Therefore $\psi = \theta \circ \varphi$, which completes the proof of the theorem.

We remark that the isometry θ is unique, because it is determined on the dense set T' by the condition $\psi = \theta \circ \varphi$.

A familiar method for calculating square roots goes as follows. If a and x_0 are positive numbers and we define a sequence recursively by

$$x_{n+1} = \frac{1}{2}\left(x_n + \frac{a}{x_n}\right),$$

then $x_n \to \sqrt{a}$. This iterative method is often useful both for existence proofs and actual calculation. Its ultimate basis is contained in the following theorem and its extensions.

14–7.15. The contraction fixed-point theorem. Let $\langle S, \rho \rangle$ be a complete metric space and let λ be a positive number less than 1. Suppose that f is a function from S to itself which satisfies the condition

$$(\forall x, y \in S) \quad \rho\big(f(x), f(y)\big) \le \lambda \rho(x, y). \tag{5}$$

Then there is a unique point p of S which is left fixed by f (that is, $f(p) = p$). For any x in S, the sequence $x, f(x), f(f(x)), \ldots$ converges to p, and

$$\rho(x, p) \le \frac{\rho(x, f(x))}{1 - \lambda}.$$

Proof. If p and q are fixed points of f, then (5) gives $\rho(p, q) \le \lambda \rho(p, q)$, whence $\rho(p, q) = 0$. Thus f has at most one fixed point.

Consider any point x of S and define the sequence $\{x_n\}$ recursively by

$$x_0 = x, \quad x_{n+1} = f(x_n).$$

Then, for any n,

$$\rho(x_{n+1}, x_n) \le \lambda\rho(x_n, x_{n-1}).$$

By induction on n,

$$\rho(x_{n+1}, x_n) \le \lambda^n\rho(x_1, x_0).$$

Therefore, for $m > n$,

$$\rho(x_m, x_n) \le \sum_{k=n}^{m-1} \rho(x_{k+1}, x_k) \le \sum_{k=n}^{m-1} \lambda^k\rho(x_1, x_0) = A(\lambda^n - \lambda^m),$$

where we have written A for $\rho(x_1, x_0)/(1 - \lambda)$. This shows that $\{x_n\}$ is a Cauchy sequence. Since S is complete, let $x_n \to p$. If we let $m \to \infty$, the last inequality becomes $\rho(p, x_n) \le A\lambda^n$. Now

$$\rho(f(p), x_{n+1}) = \rho(f(p), f(x_n)) \le \lambda\rho(p, x_n) \le A\lambda^{n+1};$$

this shows that $x_n \to f(p)$ and therefore $f(p) = p$. Thus we have found at least one point left fixed by f. Since x was arbitrary, the other two statements follow.

A function which satisfies the conditions of this theorem is often called a *contraction*.

Suppose that f is a function from some metric space to itself and that p is a fixed point of f. If there is a ball B about p such that $f(B) \subseteq B$, and for all x in B the sequence $x, f(x), f(f(x)), \ldots$ converges to p, then p is called an *attractive* fixed point. If, on the other hand, for some sufficiently small ball B about p and every x in $B - \{p\}$ the sequence $x, f(x), f(f(x)), \ldots$ contains a point not in B, then p is called a *repulsive* fixed point. Evidently, the iterative method will be useful in calculating an attractive fixed point provided we can get a good enough start, while it is not likely to be of help in calculating a repulsive fixed point.

Our next theorem shows that the fixed point of a contraction changes continuously when the contraction itself is changed continuously.

14–7.16. Theorem. Suppose that $\langle S, \rho \rangle$ and $\langle T, \sigma \rangle$ are metric spaces and that S is complete. Let g be a continuous function from $S \times T$ to S such that for some positive number λ less than one

$$(\forall t \in T)(\forall x, y \in S) \quad \rho(g(x, t), g(y, t)) \le \lambda\rho(x, y).$$

If $\varphi(t)$ is defined to be the unique point p in S for which $g(p, t) = p$, then φ is a continuous function from T to S.

Proof. Let t_0 be any point of T and let ϵ be a given positive number. Since g is continuous at $\langle \varphi(t_0), t_0 \rangle$, there is a positive δ so small that

$$\sigma(t, t_0) < \delta \Rightarrow \rho[g(\varphi(t_0), t), g(\varphi(t_0), t_0)] = \rho[g(\varphi(t_0), t), \varphi(t_0)] < \epsilon(1 - \lambda).$$

Assume that $\sigma(t_1, t_0) < \delta$. Find $\varphi(t_1)$ by iteration starting from $\varphi(t_0)$. By the last statement of the previous theorem we have

$$\rho\big(\varphi(t_0), \varphi(t_1)\big) \leq \frac{\rho[\varphi(t_0), g(\varphi(t_0), t_1)]}{1 - \lambda} < \epsilon.$$

This shows that φ is continuous at t_0. Since t_0 was arbitrary, φ is continuous.

14-7.17. An Application. The power of these theorems is well illustrated by the following argument. Suppose ψ is a function from $\mathbf{C}$ to $\mathbf{C}$ for which

$$|\psi(x) - \psi(y)| \leq \lambda|x - y|,$$

where $\lambda < 1$. Then the function $\theta: x \to x + \psi(x)$ is a homeomorphism from $\mathbf{C}$ to $\mathbf{C}$.

The inequality immediately implies that ψ is continuous and therefore so is θ. Moreover, $\theta(x) = \theta(y)$ gives $x - y = \psi(y) - \psi(x)$, so $|\psi(x) - \psi(y)| = |x - y|$; hence $|x - y| = 0$, and $x = y$. Thus θ is injective.

Consider the function g from $\mathbf{C} \times \mathbf{C}$ to $\mathbf{C}$ given by $g(x, y) = y - \psi(x)$. For any fixed y, Theorem 14-7.15 asserts that $x \to g(x, y)$ has a unique fixed point $\varphi(y)$. This satisfies $\varphi(y) = y - \psi(\varphi(y))$ or $\theta(\varphi(y)) = y$, which shows that θ is surjective with right inverse φ. Since we have already seen that θ is injective, φ is its inverse. Theorem 14-7.16 tells us that φ is continuous.

14-7.18. The Baire category theorem. A nonvoid open set in a complete metric space is not the union of a countable family of nowhere dense subsets.

Proof. Let G be a nonempty open set in a complete metric space S. Let $\{E_n\}$ be an infinite sequence of nowhere dense subsets of S. (If the countable family is actually finite, we can either adjoin more sets, say $E_n = \emptyset$ for all large n, or allow the inductive construction given below to terminate.) We shall construct a point p of G not in $\bigcup_n E_n$.

There exist sequences $\{\epsilon_n\}$ of positive numbers and $\{x_n\}$ in S such that

$$\epsilon_n \to 0, \qquad \overline{B(x_1, \epsilon_1)} \subseteq G - \overline{E}_1,$$

and

$$(\forall n) \quad \overline{B(x_{n+1}, \epsilon_{n+1})} \subseteq B(x_n, \epsilon_n) - \overline{E}_{n+1}. \tag{6}$$

We can find such sequences by inductive choice. Since Int $\overline{E}_1 = \emptyset$, $\overline{E}_1 \not\supseteq G$ and the open set $G - \overline{E}_1$ is not void. Hence we can choose $\epsilon_1 < 1$ and x_1 so that

$$\overline{B(x_1, \epsilon_1)} \subseteq G - \overline{E}_1.$$

Then, since Int $\overline{E}_2 = \emptyset$, $\overline{E}_2 \not\supseteq B(x_1, \epsilon_1)$, and the open set $B(x_1, \epsilon_1) - \overline{E}_2$ is not void. Hence we can choose $\epsilon_2 < \frac{1}{2}$ and x_2 so that

$$\overline{B(x_2, \epsilon_2)} \subseteq B(x_1, \epsilon_1) - \overline{E}_2.$$

If $\epsilon_1, \epsilon_2, \ldots, \epsilon_n$ and $x_1, x_2, \ldots, x_n$ have been chosen, then $B(x_n, \epsilon_n) - \overline{E}_{n+1}$ is a nonempty open set and we can choose $\epsilon_{n+1} < 1/(n + 1)$ and x_{n+1} so that (6) is again satisfied.

Now, $\{\overline{B(x_n, \epsilon_n)}\}$ is a decreasing sequence of closed sets with diameters tending to zero. Hence there is a point p in $\bigcap_n \overline{B(x_n, \epsilon_n)}$. This point p is not in any of the sets E_n, because E_n and $\overline{B(x_n, \epsilon_n)}$ are disjoint. On the other hand, $p \in G$ since $\overline{B(x_1, \epsilon_1)} \subseteq G$. Therefore, $G \neq \bigcup_n E_n$.

The term *category* appearing in the name of the theorem arises in the advanced theory of the nature of the subsets of a space. All subsets are divided into two categories. The first category consists of those sets which can be represented as the union of a countable family of nowhere dense sets, and the second category consists of those that cannot be so represented. It is evident that any subset of a first category set is itself a first category set. Hence the classification would be trivial if the whole space were a first category set. Of course this can happen—for example, the rational numbers can be represented as the union of a countable family of one-point sets each of which is nowhere dense as a subset of the rationals —but it does not happen in complete spaces.

While it is usually true that a one-point set is nowhere dense, it need not be, for the set may be open and the point isolated. This is obviously the only such case, but it must not be overlooked.

14-7.19. Corollary. In a complete metric space the intersection of a countable family of dense open sets is dense.

Proof. Let $\{G_n\}$ be a countable family of dense open sets in the complete metric space S. We shall prove that $\bigcap_n G_n$ is dense.

Let H be any open set and let $E_n = S - G_n$. Then E_n is closed, and by 14-3.17(a),

$$\text{Int } \overline{E}_n = \text{Int } E_n = S - \overline{G}_n = \emptyset,$$

so $\{E_n \cap H\}$ is a countable family of nowhere dense sets. Since $\bigcup_n (E_n \cap H) \neq H$, we have $H \cap \bigcap_n G_n \neq \emptyset$; thus $\bigcap_n G_n$ is dense by 14-3.28.

To illustrate the use of the theorem in analysis, we shall prove the following proposition.

14-7.20. Proposition. Let $\{f_n\}$ be a sequence of complex-valued continuous functions on a complete nonvoid metric space. Suppose that $\{f_n(x)\}$ is a bounded sequence for each $x \in S$. Then there is a nonvoid open set H in S on which the sequence $\{f_n\}$ is bounded (that is, $(\exists k \in \mathbf{R})(\forall x \in H)(\forall n) \, |f_n(x)| \leq k$).

Proof. Since the function f_n is continuous, the set $\{x \mid |f_n(x)| \leq m\}$ is closed for each pair of integers n and m. Hence

$$E_m = \{x \mid (\forall n)|f_n(x)| \leq m\} = \bigcap_n \{x \mid |f_n(x)| \leq m\}$$

is closed for each m. Because $\{f_n(x)\}$ is a bounded sequence for each x, $S = \bigcup_m E_m$. The Baire category theorem tells us that not all of the sets E_m are nowhere dense; say, Int $\bar{E}_k \neq \emptyset$. Since E_k is closed, this is the same as Int $E_k \neq \emptyset$, and we see that the sequence $\{f_n\}$ is bounded by k on the nonvoid open set $H = \text{Int } E_k$.

EXERCISES

1. Let $\{x_n\}$ be a Cauchy sequence in a metric space. Suppose that the subsequence $\{x_{g(n)}\}$ converges. Prove that $\{x_n\}$ converges.

2. Suppose that $\{x_n\}$ is a sequence in a metric space $\langle S, \rho \rangle$ such that F-$\sum_n \rho(x_n, x_{n+1})$ converges. Prove that $\{x_n\}$ is a Cauchy sequence.

3. Suppose that $\{x_n\}$ is a Cauchy sequence in a metric space $\langle S, \rho \rangle$. Prove that it has a subsequence $\{x_{g(n)}\}$ such that F-$\sum_n \rho(x_{g(n)}, x_{g(n+1)})$ converges.

4. Is the converse of Proposition 14–7.4 true?

5. Suppose that S and T are metric spaces and that f is a bijection from S to T such that both f and f^{-1} are uniformly continuous. Prove that if S is complete, T is complete, and vice versa.

Under these hypotheses, the metric spaces S and T are said to be *uniformly equivalent.* The relation of uniform equivalence stands between isometry and homeomorphism; that is, isometric $\Rightarrow$ uniformly equivalent $\Rightarrow$ homeomorphic.

6. Give another proof of Theorem 14–7.14 along the following lines. Use the results of Exercise 4, page 245, to show the existence of a triple $\langle S, \tau, \varphi \rangle$ and Theorem 14–7.13 to show its essential uniqueness.

7. A metric space $\langle S, \rho \rangle$ is said to be *convex* if and only if

$$(\forall x, y \in S)(\exists z \in S) \quad \rho(x, z) = \rho(z, y) = \tfrac{1}{2}\rho(x, y).$$

Let p and q be two points in a complete convex metric space. Say, $\rho(p, q) = a$. Prove that there is an isometry f from the interval $[0, a]$ to a subset of S such that $f(0) = p$ and $f(a) = q$.

8. Show that if $a > 0$ and $S = [\sqrt{a/2}, \infty)$, then the map $x \rightarrow \tfrac{1}{2}(x + a/x)$ is a contraction.

*9. Suppose f is a function from $\mathbf{R}$ to $\mathbf{R}$ with a continuous derivative. Suppose that $x_0 = f(x_0)$. Show that x_0 is an attractive fixed point of f if $|f'(x_0)| < 1$, and that in this case the condition of Theorem 14–7.15 is satisfied if S is taken to be a sufficiently small interval about x_0. Show that x_0 is a repulsive fixed point if $|f'(x_0)| > 1$.

*10. The real root of $x^3 + x - 1$ can be regarded as the fixed point of the map $x \rightarrow 1 - x^3$. Unfortunately, this is a repulsive fixed point by the criterion of the previous exercise. But it is also a fixed point of the map $x \rightarrow \frac{1}{5}(3x + 2 - 2x^3)$. Show that this is an attractive fixed point, and calculate the root starting from the guess $x = 0.7$.

11. Let the contraction condition of Theorem 14–7.15 be replaced by the somewhat weaker condition: If $x \neq y$, then $\rho(f(x), f(y)) < \rho(x, y)$. Show that f has at most one fixed point, and that if there is any point adherent (see Exercise 2, p. 231) to the sequence $x, f(x), f(f(x)), \ldots$, then that point is fixed and is actually the limit of the sequence.

12. Suppose S is a complete metric space and $S = \bigcup_n E_n$, where $\{E_n\}$ is a sequence of closed sets. Prove that $\bigcup_n \operatorname{Int} E_n$ is a dense open set.

14–8. COMPACT SPACES

The bounded closed intervals in **R** have some particularly desirable topological properties. For example, every continuous real-valued function defined on a bounded closed interval is bounded and achieves its least upper bound, and every sequence in a bounded closed interval has a convergent subsequence. There are other metric spaces with these properties; they are known as compact spaces. The topological properties of compact spaces resemble in many ways the properties of finite sets, even though they may be (and usually are) uncountable as sets. In a sense, compact spaces are the smallest kind of spaces.

There are several equivalent ways to define compactness for metric spaces, and one of the main theorems on this subject is the proof of their equivalence (14–8.5). We shall take the definition in a form which can be applied without change in the context of general topological spaces.

14–8.1. Definitions. A family $\{X_\alpha \mid \alpha \in A\}$ of sets is called a *covering* of the set Y if and only if $Y \subseteq \bigcup_A X_\alpha$. Suppose that the sets X_α are all subsets of a topological space. The covering is called *open* if and only if each X_α is open. A *subcovering* of the covering $\{X_\alpha \mid \alpha \in A\}$ is a covering $\{X_\alpha \mid \alpha \in B\}$, where $B \subseteq A$.

14–8.2. Definition. Let S be a metric space. S is said to be *compact* if and only if every open covering of S contains a finite subcovering. (Recall that a finite family is one having a finite index set.)

Since the definition refers only to the open sets of S, it is clear that compactness is a topological property of S; in other words, if one of two homeomorphic spaces is compact, so is the other.

Compactness is a property of a space. Consequently, when a subset T of a space S is said to be compact, it means that T regarded as a subspace is compact.

Suppose for a moment that S is a discrete space. Then each one-point set is open and the family of all one-point sets is an open covering of S. Clearly, this covering has no proper subcovering. Hence if S is compact and discrete, it is finite. Conversely, it is clear that every finite metric space is compact. Thus compactness and finiteness coincide for discrete spaces. In other words, when topological considerations are vacuous, compactness coincides with finiteness.

The Heine-Borel theorem, which we shall derive below as a consequence of our general results, asserts that if S is a bounded closed interval in $\mathbf{R}$ and $\{I_\alpha \mid \alpha \in A\}$ is a covering of S by open intervals (of $\mathbf{R}$), then there is a finite set $B \subseteq A$ such that $S \subseteq \bigcup_B I_\alpha$; that is, there is a finite subcovering. Consequently the property we have used to define compactness is known as the *Heine-Borel property*.

Before we prove the equivalence of the Heine-Borel property with some other characterizations of compactness we need two new concepts.

14–8.3. Definition. Let S be a metric space and let ϵ be a positive number. A subset E of S is said to be ϵ-*dense* if and only if

$$(\forall x \in S)(\exists p \in E) \quad \rho(x, p) < \epsilon.$$

14–8.4. Definition. A metric space S is said to be *totally bounded* if and only if for every positive ϵ it contains a finite ϵ-dense set.

An obviously equivalent condition is: For every positive ϵ, S can be covered by a finite number of balls of radius ϵ.

Let I be the open interval $(0, 1)$ in $\mathbf{R}$. Clearly, the set

$$E_n = \left\{ \frac{1}{n}, \frac{2}{n}, \ldots, \frac{n-1}{n} \right\}$$

is $(1/n)$-dense in I for each n. Therefore I is totally bounded. On the other hand, any totally bounded space is bounded: If E is a finite 1-dense set in S, then

$$\operatorname{diam} S \leq 2 + \operatorname{diam} E.$$

Hence $\mathbf{R}$ is not totally bounded. Since I and $\mathbf{R}$ are homeomorphic, we see that total boundedness is *not* a topological property of a space. (See, however, Exercise 4.)

14–8.5. Theorem. Let S be a metric space. S is compact if and only if it has *either* of the following properties:

(a) S is complete and totally bounded;
(b) every sequence in S has a convergent subsequence.

Proof. We shall prove the circular chain of implications

$$S \text{ compact} \Rightarrow (b) \Rightarrow (a) \Rightarrow S \text{ compact}.$$

(A) Suppose S is compact. Let $\{x_n\}$ be any sequence in S. We shall construct a convergent subsequence of $\{x_n\}$.

Let $\mathcal{G}$ be the set of all open sets G such that $\{n \mid x_n \in G\}$ is finite. We may regard $\mathcal{G}$ as a family. If $\mathcal{G}$ were a covering, then we could find a finite subcovering, say $\{G_1, G_2, \ldots, G_k\} \subseteq \mathcal{G}$. From $S = \bigcup_{i=1}^{k} G_i$ follows $\mathbf{N} = \bigcup_{i=1}^{k} \{n \mid x_n \in G_i\}$. Since the latter set is finite, this is absurd. Therefore, there is a point y not in $\bigcup_{\mathcal{G}} G$.

Then no ball centered at y is in $\mathcal{G}$, so $\{n \mid x_n \in B(y, 1/p)\}$ is infinite for every integer p.

Define a function g from $\mathbf{N}$ to $\mathbf{N}$ inductively as follows. Let $g(1)$ be the least integer for which $x_{g(1)} \in B(y, 1)$. Let $g(n + 1)$ be the least integer greater than $g(n)$ for which $x_{g(n+1)} \in B(y, 1/(n + 1))$. Since g is strictly increasing, $\{x_{g(n)}\}$ is a subsequence of $\{x_n\}$. Obviously $x_{g(n)} \to y$. This proves the first implication.

(B) Suppose that every sequence in S has a convergent subsequence. If $\{x_n\}$ is a Cauchy sequence in S, let y be the limit of some convergent subsequence of $\{x_n\}$. It is a routine matter to show that $x_n \to y$. Thus S is complete.

Suppose that S is not totally bounded. Choose an $\epsilon > 0$ for which it is impossible to cover S by a finite number of balls of radius ϵ. By induction we can find a sequence $\{x_n\}$ in S such that

$$(\forall n) \quad x_{n+1} \notin \bigcup_{i=1}^{n} B(x_i, \epsilon).$$

Now, $(\forall j, k) j \neq k \Rightarrow \rho(x_j, x_k) \geq \epsilon$. Hence no subsequence of this sequence is a Cauchy sequence, which contradicts the possibility of finding a convergent subsequence in $\{x_n\}$. This proves that (b) $\Rightarrow$ (a).

(C) Suppose that S is complete and totally bounded. We shall prove that S has the Heine-Borel property in its contrapositive form.

Let $\{G_\alpha \mid \alpha \in A\}$ be a family of open sets containing no finite covering of S. We shall find a point not in $\bigcup_A G_\alpha$.

We construct inductively a sequence $\{B_n\}$ such that for all n,

(i) B_n is a ball of radius $1/n$, and
(ii) $B_1 \cap B_2 \cap \cdots \cap B_n$ cannot be covered by any finite subfamily of $\{G_\alpha\}$.

Since S is totally bounded, let it be covered by a finite number of balls of radius 1. If each of these could be covered by a finite subfamily of $\{G_\alpha\}$, then S could be so covered. Hence at least one of these balls, say B_1, cannot be covered by a finite subfamily of $\{G_\alpha\}$.

Suppose we have found $B_1, B_2, \ldots, B_k$ such that (ii) is true for $n = k$. Cover S by a finite set of balls, say $\{C_1, C_2, \ldots, C_p\}$, of radius $1/(k + 1)$. If each of the sets $B_1 \cap B_2 \cap \cdots \cap B_k \cap C_i$, $i = 1, 2, \ldots, p$, could be covered by a finite subfamily of $\{G_\alpha\}$, then $B_1 \cap B_2 \cap \cdots \cap B_k$ could also be so covered. Hence we can choose $B_{k+1} \in \{C_1, C_2, \ldots, C_p\}$ so that (i) and (ii) hold for $n = k + 1$.

Now define $E_k = \overline{B}_1 \cap \overline{B}_2 \cap \cdots \cap \overline{B}_k$. Then $\{E_k\}$ is a decreasing sequence of closed sets, and diam $E_k \to 0$ (since diam $E_k \leq$ diam $\overline{B}_k =$ diam $B_k \leq 2/k$). By 14-7.10 there is a point x in $\bigcap_k E_k$. We shall prove that $x \notin \bigcup_A G_\alpha$.

Suppose, on the contrary, that $x \in \bigcup_A G_\alpha$; say $x \in G_\beta$, where $\beta \in A$. There is an $\epsilon > 0$ such that $B(x, \epsilon) \subseteq G_\beta$. If we choose k so that $2/k < \epsilon$, then we have $E_k \subseteq B(x, \epsilon)$, since $x \in E_k$ and diam $E_k \leq 2/k$. Thus E_k and, a fortiori, $B_1 \cap B_2 \cap \cdots \cap B_k$ can be covered by a single member of $\{G_\alpha\}$, namely, G_β. This contradicts (ii).

This concludes the proof of the theorem.

Because of its resemblance to the Bolzano-Weierstrass theorem (12–6.6), property 14–8.5(b) is known as the *Bolzano-Weierstrass property*. It is often taken as the defining property for compactness since it is frequently the handiest property for dealing with compact metric spaces. However, it is not equivalent to the Heine-Borel property in general topological spaces, and it turns out that the latter is the more valuable in the general case.

14–8.6. Proposition. Let T be a subset of a metric space S. Then T is compact if and only if every covering of T by open sets of S has a finite subcovering.

Proof. Suppose that T is compact. This means that T regarded as a subspace of S is compact. Let $\{G_\alpha \mid \alpha \in A\}$ be a covering of T by open sets of S. We must find a finite set $B \subseteq A$ such that $\{G_\alpha \mid \alpha \in B\}$ is a covering of T.

Now $\{G_\alpha \cap T \mid \alpha \in A\}$ is a covering of T by open sets of T (14–3.13). Since T is compact, there is a finite subcovering $\{G_\alpha \cap T \mid \alpha \in B\}$, and hence $\{G_\alpha \mid \alpha \in B\}$ is the required subcovering.

Conversely, suppose that every covering of T by open sets of S has a finite subcovering. Let $\{H_\alpha \mid \alpha \in A\}$ be a covering of T by open sets of T. By 14–3.13, each set H_α has the form $G_\alpha \cap T$, where G_α is open in S. This gives us a covering $\{G_\alpha \mid \alpha \in A\}$ of T by open sets of S. Our hypothesis tells us that there is a finite set $B \subseteq A$ such that $T \subseteq \bigcup_B G_\alpha$. Then

$$T = T \cap \bigcup_B G_\alpha = \bigcup_B (T \cap G_\alpha) = \bigcup_{\alpha \in B} H_\alpha.$$

Thus T is compact.

14–8.7. Proposition. A closed subspace of a compact metric space is compact. Conversely, a compact subspace of any metric space is closed and bounded.

Proof. Let E be a closed subspace of a compact metric space S. We shall show that E has the Bolzano-Weierstrass property (14–8.5(b)). If $\{x_n\}$ is a sequence in E, we can extract a subsequence $\{x_{g(n)}\}$ which converges in S. Since E is closed, it converges in E also (14–2.7 and 14–3.1).

A compact subspace of a metric space is complete and totally bounded (14–8.5). Every complete subspace of a metric space is closed (14–7.7) and every totally bounded space is bounded.

14–8.8. Proposition. Suppose S and T are metric spaces such that every bounded, closed subset of either S or T is compact. Then $S \times T$ has the same property.

Proof. Suppose E is a bounded, closed subset of $S \times T$. Say diam $E = d$. Let $\{\langle x_n, y_n \rangle\}$ be a sequence in E. Then $\{x_n\}$ is a sequence in S which falls in a set X of diameter at most d. Then $\overline{X}$ is compact by hypothesis, so we can find a subsequence $\{x_{f(n)}\}$ which converges in $\overline{X}$ and therefore in S. Similarly, the sequence $\{y_{f(n)}\}$ falls in a subset Y of T having diameter at most d, so a suitable subsequence $\{y_{f(g(n))}\}$ converges in T. Then $\{\langle x_{f(g(n))}, y_{f(g(n))} \rangle\}$ converges in

$S \times T$. Since E is closed, it converges in E. Thus E has the Bolzano-Weierstrass property and is compact. Note that the argument remains valid whether we use 14–1(2), (3), or (4) as the metric in $S \times T$.

14–8.9. Corollary. The direct product of any finite number of compact metric spaces is compact.

Proof. It is sufficient to prove that the direct product of two compact metric spaces is compact. Let S and T be compact metric spaces. By 14–8.8, every bounded, closed subset of $S \times T$ is compact. Since S and T are bounded, $S \times T$ is bounded. It is surely closed, so it is compact.

14–8.10. Proposition. A subset of $\mathbf{R}$ is compact if and only if it is closed and bounded. A nonvoid compact subset of $\mathbf{R}$ contains its supremum and its infimum.

Proof. Note that in $\mathbf{R}$ the term *bounded*, as defined in 6–4.1 using order, coincides with the term *bounded* as defined in 14–1.10 using the metric in $\mathbf{R}$. Let E be a closed, bounded subset of $\mathbf{R}$. Let $\{x_n\}$ be a sequence in E. By the Bolzano-Weierstrass theorem (12–6.6), there is a subsequence $\{x_{g(n)}\}$ which converges in $\mathbf{R}$. Since E is closed, it converges in E. Thus E is compact. The opposite implication of the first statement of the proposition has already been established in 14–8.7.
Suppose E is a nonvoid, bounded, closed subset of $\mathbf{R}$. Let $a = \sup E$. For any integer $n > 0$, $a - 1/n$ is not an upper bound for E, so we can choose x_n so that $x_n \in E$ and $x_n > a - 1/n$. Since we also have $x_n \le a$, this gives $|x_n - a| < 1/n$. Therefore, $x_n \to a$. Since E is closed, $a \in E$. Similarly, $\inf E \in E$.

14–8.11. Corollary. For any integer $n > 0$, a subset of $\mathbf{R}^n$ or $\mathbf{C}^n$ is compact if and only if it is closed and bounded.

Proof. For $\mathbf{R}^n$ this follows immediately from 14–8.10 and 14–8.8. For $\mathbf{C}$ we must first note that *bounded* as defined for subsets of $\mathbf{C}$ in 10–3.8 agrees with the term *bounded* as used in metric spaces. We may then apply 12–6.6 as in the proof of 14–8.10. Alternatively, we may note that $\mathbf{C}$ is isometric to $\mathbf{R}^2$ with the Pythagorean metric $(14$–$1(4))$. Finally, the conclusion for $\mathbf{C}^n$ follows from 14–8.8.

14–8.12. The Heine-Borel theorem. If $\{I_\alpha \mid \alpha \in A\}$ is a covering of a bounded, closed interval J in $\mathbf{R}$ by open intervals, then there is a finite subset B of A such that $J \subseteq \bigcup_B I_\alpha$.

Proof. Immediate from 14–8.10 and 14–8.6.

We turn now to the study of continuous functions defined on compact metric spaces.

14–8.13. Theorem. Let f be a continuous function from a compact metric space S to a metric space T. Then the range of f is a compact subset of T.

Proof. Let $\{G_\alpha \mid \alpha \in A\}$ be a covering of $f(S)$ by open subsets of T. Then $\{f^{-1}(G_\alpha) \mid \alpha \in A\}$ is an open covering of S by 14–4.5. Choose a finite subcovering $\{f^{-1}(G_\alpha) \mid \alpha \in B\}$ of S. Then $\{G_\alpha \mid \alpha \in B\}$ is a finite covering of $f(S)$. Therefore $f(S)$ is compact by 14–8.6.

14–8.14. Theorem. A continuous function f from a compact metric space S to $\mathbf{R}$ is bounded. Furthermore, if S is not void, there exist points s and t in S such that $f(s) = \sup \{f(x) \mid x \in S\}$ and $f(t) = \inf \{f(x) \mid x \in S\}$.

Proof. Since $f(S)$ is a compact subset of $\mathbf{R}$, this follows immediately from 14–8.10.

14–8.15. Theorem. Let S be a compact metric space and let f be a continuous bijection from S to a metric space T. Then f^{-1} is continuous and S is homeomorphic to T.

Proof. To avoid confusion we set $g = f^{-1}$. Let F be any closed set in S. If we can prove that $g^{-1}(F)$ is closed, it will follow that g is continuous by 14–4.6. Now, $g^{-1}(F) = f(F)$ and the latter is compact by 14–8.13 and therefore closed by 14–8.7. Hence f is a homeomorphism by 14–6.2.

14–8.16. Theorem. Let f be a continuous function from one metric space $\langle S, \rho \rangle$ to another $\langle T, \sigma \rangle$. Let E be a compact subset of S. Then the continuity of f is uniform on E.

Proof. We are going to prove

$$(\forall \epsilon > 0)(\exists \delta > 0)(\forall x \in E)(\forall y \in S) \quad \rho(x, y) < \delta \Rightarrow \sigma(f(x), f(y)) < \epsilon. \quad (1)$$

Note that this is stronger than the assertion, "f restricted to E is uniformly continuous." (See Section 14–5.)

Let $\epsilon > 0$ be given. By the definition of continuity we can choose, for each point $s \in S$, a positive number $\eta(s)$ such that

$$f[B(s, 2\eta(s))] \subseteq B(f(s), \epsilon/2). \quad (2)$$

Now, the family $\{B(s, \eta(s)) \mid s \in S\}$ is an open covering of S. Since E is compact, there is a finite subfamily, say $\{B(s, \eta(s)) \mid s \in F\}$, which covers E (14–8.6). Since F is finite,

$$\delta = \inf \{\eta(s) \mid s \in F\} = \eta(s_0)$$

for some $s_0 \in F$, and so $\delta > 0$.

Let $x \in E$ and $y \in S$ be given, and suppose that $\rho(x, y) < \delta$. Choose $t \in F$ so that $x \in B(t, \eta(t))$. Since $\delta \leq \eta(t)$, $y \in B(t, 2\eta(t))$. Hence by (2), both $f(x)$ and $f(y)$ are in $B(f(t), \epsilon/2)$, and therefore $\sigma(f(x), f(y)) < \epsilon$. This proves (1).

14-8.17. Corollary. Any continuous function from a compact metric space to a metric space is uniformly continuous.

14-8.18. Definition. A metric space S is said to be *locally compact* if and only if each point of S has a neighborhood which is compact.

If N is a compact neighborhood of p, then N is closed. If V is any neighborhood of p such that $V \subseteq N$, then $\overline{V} \subseteq N$, so $\overline{V}$ is compact by 14-8.7. Hence a space is locally compact if and only if for each point p all sufficiently small neighborhoods of p have compact closure.

This is our first use of the term *locally*. In general, when a space S is said to have a certain property locally, it means that every point has arbitrarily small neighborhoods which have this property.

Local compactness is evidently a topological property. $\mathbf{R}^n$ and $\mathbf{C}^n$ are locally compact spaces, as is any space in which all bounded closed sets are compact. On the other hand, spaces of continuous functions are locally compact only in trivial cases (see Exercise 13).

EXERCISES

1. Show that any subspace of a totally bounded space is totally bounded.

2. Show that the direct product of two totally bounded spaces is totally bounded.

3. Show that a metric space S is totally bounded if and only if every infinite sequence in S has a Cauchy subsequence.

4. Suppose that S is a totally bounded metric space, T is a metric space, and f is a bijection from S to T such that both f and f^{-1} are uniformly continuous. Prove that T is totally bounded. How much weaker can the hypothesis be? Compare this with Exercise 5, p. 265.

5. A family $\{X_\alpha \mid \alpha \in A\}$ is said to have the *finite-intersection property* if and only if $A \neq \emptyset$ and, for every finite nonvoid subset B of A, $\bigcap_B X_\alpha \neq \emptyset$. Prove that a space S is compact if and only if it has the following property: If $\{F_\alpha \mid \alpha \in A\}$ is any family of closed subsets of S having the finite intersection property, then $\bigcap_A F_\alpha \neq \emptyset$.

6. Suppose $\{X_\alpha \mid \alpha \in A\}$ is a family of nonvoid closed sets in a metric space. Suppose the family is linearly ordered by inclusion and that at least one of the X_α is compact. Prove that $\bigcap_A X \neq \emptyset$.

7. Prove that a space is compact if and only if it contains no closed, infinite, discrete subspace.

8. Suppose that $E_1, E_2, \ldots, E_k$ are compact subsets of a metric space S. Prove that $E_1 \cup E_2 \cup \cdots \cup E_k$ is compact.

9. Let S be a totally bounded metric space. Show that its completion (i.e., the space constructed in 14-7.14) is compact. For this reason totally bounded spaces are sometimes called *precompact* spaces.

10. Suppose f is a function from a metric space S to a compact metric space T. Prove that f is continuous if and only if f, regarded as a subset of $S \times T$, is closed. Show that the hypothesis that T is compact cannot be dropped.

11. Let F and G be respectively a closed subset and an open subset of a locally compact metric space. Show that $F \cap G$ is locally compact. Conversely, let E be a locally compact subset of a metric space S (not necessarily locally compact). Prove that $E = F \cap G$, where F is closed in S and G is open in S.

12. Prove that the direct product of two locally compact spaces is locally compact.

13. Let S be the space of continuous functions from $[0, 1]$ to $\mathbf{R}$ with the metric

$$\rho(f, g) = \sup \{|f(x) - g(x)| \mid x \in [0, 1]\}$$

(which is finite by 14–8.14). Show that S is not locally compact.

14. Prove that a compact metric space is not isometric with any of its proper subsets.

15. Let $\{G_\alpha\}$ be an open covering of a compact metric space S. Show that there is a positive number ϵ such that

$$(\forall x \in S)(\exists \alpha) \quad B(x, \epsilon) \subseteq G_\alpha.$$

Any such number is called a *Lebesgue number of the covering*.

16. Prove the following theorem. Let S and T be metric spaces and let f be a function from S to T. In order that f be continuous it is necessary and sufficient that, for every compact subset X of S, the restriction of f to X be continuous.

17. Prove the following theorem. Let S and T be complete metric spaces and suppose that f is a continuous function from a dense subset E of S to T. A necessary and sufficient condition that f should have a continuous extension to S (i.e., that there exist a continuous function g from S to T such that g restricted to E is f) is that, for every totally bounded subset X of E, f restricted to X be uniformly continuous.

14–9. SEPARABLE SPACES

If compact metric spaces are the counterparts of finite sets, then separable spaces are the counterparts of countable sets. Most spaces of importance in analysis are separable. We shall prove one major theorem to this effect.

14–9.1. Definition. A metric space is said to be *separable* if and only if it contains a countable, dense subset.

Separability is clearly a topological property. We shall see below that $\mathbf{R}^n$, $\mathbf{C}^n$, and all of their subspaces are separable.

14–9.2. Proposition. A metric space is separable if and only if for every positive ϵ it contains a countable, ϵ-dense subset.

Proof. Let S be a metric space. If S contains a countable, dense set X, then for every positive ϵ, X itself is a countable, ϵ-dense subset.

Conversely, suppose that for every positive ϵ, S contains a countable, ϵ-dense subset. For each positive integer n, let X_n be a countable subset of S which is $(1/n)$-dense. Then $X = \bigcup_n X_n$ is a countable, dense subset of S.

14–9.3. Corollary. Every totally bounded, in particular every compact, metric space is separable.

14–9.4. Proposition. Any subspace of a separable metric space is separable.

Proof. Let T be a subspace of a separable metric space $\langle S, \rho \rangle$. Let X be a countable, dense subset of S. Now, $X \times \mathbf{N}$ is countable. For each $\langle x, n \rangle \in X \times \mathbf{N}$ such that $T \cap B(x, 1/n)$ is not empty, pick a point of $T \cap B(x, 1/n)$. These points form a countable subset Y of T. We shall prove that Y is dense in T.

Let $t \in T$. Let $\epsilon > 0$ be given. Choose a positive integer m such that $m > 2/\epsilon$ and a point $x \in X$ such that $\rho(t, x) < 1/m$. Since $T \cap B(x, 1/m)$ is not empty, it contains a point $y \in Y$. Then $\rho(t, y) < 2/m < \epsilon$. Since ϵ was arbitrary, this shows that Y is dense in T.

14–9.5. Proposition. If a metric space is the union of a countable family of separable subspaces, it is separable.

14–9.6. Corollary. $\mathbf{R}^n$ and $\mathbf{C}^n$ are separable.

Proof. $\mathbf{R}^n = \bigcup_{k=1}^{\infty} \overline{B(0, k)}$, where 0 stands for $\langle 0, 0, \ldots, 0 \rangle$. Each of the sets $\overline{B(0, k)}$ is compact (14–8.8 and 14–8.10) and therefore separable (14–9.3). The same proof is valid for $\mathbf{C}^n$.

14–9.7. Proposition. The direct product of two separable spaces is separable.

Proof. If X is a countable, dense subset of S and Y is a countable, dense subset of T, then $X \times Y$ is a countable, dense subset of $S \times T$.

14–9.8. Definition. A family $\{G_\alpha \mid \alpha \in A\}$ of open sets in a metric space S is called a *base* for the open sets of S if and only if for every open set H of S there is a subset C of A such that $H = \bigcup_{\alpha \in C} G_\alpha$.

14–9.9. Theorem. A metric space S is separable if and only if there is a countable base for the open sets of S.

Proof. Suppose $\{G_\alpha \mid \alpha \in A\}$ is a countable base for the open sets of S. Pick one point from each nonvoid G. These points form a countable set X. To show that X is dense, let p be any point of S and let $\epsilon > 0$ be given. We must prove that $X \cap B(p, \epsilon) \neq \emptyset$. $B(p, \epsilon)$ is open, so let $C \subseteq A$ be chosen so that

$$B(p, \epsilon) = \bigcup_C G_\alpha.$$

Choose $\beta \in C$ so that $p \in G_\beta$. Then $X \cap B(p, \epsilon) \supseteq X \cap G_\beta$, and the latter is not void by the choice of X.

Conversely, suppose that X is a countable dense subset of S. We shall prove that $\{B(x, 1/n) \mid x \in X, n \in \mathbf{N}\}$ is a countable base for the open sets of S. It is a family of open sets and its index set, $X \times \mathbf{N}$, is countable. Let H be any open set of S. Let $C = \{\langle x, n \rangle \mid B(x, 1/n) \subseteq H\}$ and let $K = \bigcup_C B(x, 1/n)$. Obviously, $K \subseteq H$. Conversely, suppose that $p \in H$. Since H is open, there is a positive ϵ such that $B(p, \epsilon) \subseteq H$. Pick $m \in \mathbf{N}$ such that $2/m < \epsilon$. Since X is dense, we can choose $y \in X \cap B(p, 1/m)$. Then $B(y, 1/m) \subseteq B(p, 2/m) \subseteq B(p, \epsilon) \subseteq H$, so $\langle y, m \rangle \in C$. Therefore, $B(y, 1/m) \subseteq K$. Since $p \in B(y, 1/m)$, we have $p \in K$. Thus $H = K$. This proves that $\{B(x, 1/n) \mid x \in X, n \in \mathbf{N}\}$ is a base for the open sets of S.

14–9.10. Corollary. Let S be a separable metric space. The topology $\mathfrak{I}$ (set of all open subsets) of S has cardinal at most $\mathbf{c}$, the cardinal of $\mathbf{R}$.

Proof. Since S is separable, there is a countable base, say $\{G_n \mid n \in \mathbf{N}\}$, for its open sets (we may assume that the index set is $\mathbf{N}$). The function $C \rightarrow \bigcup_{n \in C} G_n$ is a surjection from $\mathfrak{P}(\mathbf{N})$ to $\mathfrak{I}$. Hence card $\mathfrak{I} \leq$ card $\mathfrak{P}(\mathbf{N}) =$ card $\mathbf{R} = \mathbf{c}$ by 11–5.8 and 13–4.4.

14–9.11. Corollary. Let S be a separable metric space. Then the cardinal of S is at most $\mathbf{c}$.

Proof. The function $x \rightarrow S - \{x\}$ is an injection from S to $\mathfrak{I}$, the topology of S. Therefore card $S \leq$ card $\mathfrak{I} \leq \mathbf{c}$.

14–9.12. The Lindelöf covering theorem. Let $\{G_\alpha \mid \alpha \in A\}$ be a family of open sets in a separable metric space S. There is a countable subset B of A such that

$$\bigcup_{\alpha \in B} G_\alpha = \bigcup_{\alpha \in A} G_\alpha.$$

Proof. Let $\{H_n \mid n \in \mathbf{N}\}$ be a base for the open sets of S. Let

$$J = \{n \mid (\exists \alpha \in A) H_n \subseteq G_\alpha\}.$$

For each $n \in J$, choose $g(n) \in A$ such that $H_n \subseteq G_{g(n)}$. Let $B = $ range g. Then B is a countable subset of A. Obviously, $\bigcup_B G_\alpha \subseteq \bigcup_A G_\alpha$. We need only prove the opposite inclusion.

Suppose $p \in \bigcup_A G_\alpha$. Choose an index β such that $p \in G_\beta$. Since $\{H_n\}$ is a base for the open sets, there is an integer k such that $p \in H_k \subseteq G_\beta$. Then

$$p \in G_{g(k)} \subseteq \bigcup_B G_\alpha.$$

This shows that $\bigcup_A G_\alpha \subseteq \bigcup_B G_\alpha$.

The following corollary points up the relation between separable and compact spaces.

14–9.13. Corollary. Any open covering of a separable metric space contains a countable subcovering.

For the final theorem of this section we state a set-theoretic lemma. The proofs of several preceding propositions have involved simpler applications of the same idea, and the details of those arguments involve the axiom of choice in much the same way.

14–9.14. Lemma. Let $\{D_\alpha \mid \alpha \in A\}$ be a countable family of subsets of a set F. There is a countable subset E of F such that $E \cap \bigcap_{\alpha \in B} D_\alpha \neq \emptyset$ for every nonvoid finite subset B of A such that $\bigcap_B D_\alpha \neq \emptyset$.

Proof. Since A is countable, the set $\mathfrak{Q}_0$ of all finite subsets B of A is countable, and so is the set $\mathfrak{Q}$ of all nonvoid finite subsets B of A for which $\bigcap_B D_\alpha \neq \emptyset$. Pick a function f from $\mathfrak{Q}$ to F such that $f(B) \in \bigcap_B D_\alpha$ for all $B \in \mathfrak{Q}$. Let $E = \text{range} f$.

14–9.15. Theorem. Let F be the space of continuous functions from a compact metric space $\langle S, \rho \rangle$ to a separable metric space $\langle T, \sigma \rangle$ endowed with the metric π defined by

$$\pi(f, g) = \sup \{\sigma(f(x), g(x)) \mid x \in S\}.$$

Then F is separable.

Proof. Note that π is finite by 14–8.14. Let $\epsilon > 0$ be given. We shall construct a countable set E which is ϵ-dense in F. By 14–9.2 this will prove that F is separable.

Let X and Y be countable sets, dense in S and T respectively. For each $n \in \mathbf{N}$, $x \in X$, and $y \in Y$, define

$$D_n(x, y) = \{f \in F \mid f(B(x, 1/n)) \subseteq B(y, \epsilon/4)\}. \tag{1}$$

For each n, the family $\{D_n(x, y) \mid \langle x, y \rangle \in X \times Y\}$ is countable, so let E_n be a countable set in F such that

$$E_n \cap \bigcap_{\langle x, y \rangle \in Z} D_n(x, y) \neq \emptyset$$

for every nonvoid finite subset Z of $X \times Y$ with $\bigcap_Z D_n(x, y) \neq \emptyset$. We shall prove that the countable set $E = \bigcup_n E_n$ is ϵ-dense in F.

Let $g \in F$. Since g is uniformly continuous, there is a positive integer n such that

$$(\forall s \in S)\ g(B(s, 1/n)) \subseteq B(g(s), \epsilon/8).$$

Since $\{B(x, 1/n) \mid x \in X\}$ is an open covering of the compact space S, there is a finite set $\{x_1, x_2, \ldots, x_k\} \subseteq X$ such that $S = \bigcup_{i=1}^k B(x_i, 1/n)$. For each $i = 1, 2, \ldots, k$, choose $y_i \in Y$ so that $\sigma(g(x_i), y_i) < \epsilon/8$. Then we have

$$g(B(x_i, 1/n)) \subseteq B(y_i, \epsilon/4), \tag{2}$$

whence by (1), $g \in D_n(x_i, y_i)$ for each i. Now $g \in \bigcap_{i=1}^{k} D_n(x_i, y_i)$. Since the latter is not void, we can choose $h \in E_n \cap \bigcap_{i=1}^{k} D_n(x_i, y_i)$. Since $h \in E$, the proof will be finished if we show that $\pi(g, h) < \epsilon$.

Let s be any point of S. There is an index i such that $s \in B(x_i, 1/n)$. Since $h \in D_n(x_i, y_i)$, $h(B(x_i, 1/n)) \subseteq B(y_i, \epsilon/4)$. Using this and (2), we find that $g(s)$ and $h(s)$ are both in $B(y_i, \epsilon/4)$, whence $\sigma(g(s), h(s)) < \epsilon/2$. Therefore,

$$\pi(g, h) = \sup \{\sigma(g(s), h(s))\} \leq \epsilon/2 < \epsilon.$$

The hypothesis that S is compact cannot be weakened. For example, the space $\mathcal{C}$ of bounded continuous functions from $\mathbf{R}$ to $\mathbf{R}$ is not separable. It is clear that there is a bounded continuous function from $\mathbf{R}$ to $\mathbf{R}$ having any prescribed (bounded) values at points of $\mathbf{N}$. For each subset J of $\mathbf{N}$, let f_J be a bounded continuous function with $f_J(x) = 1$ for $x \in J$, and $f_J(x) = 0$ for $x \in \mathbf{N} - J$. These functions satisfy the condition

$$\pi(f_J, f_K) \geq 1 \quad \text{if} \quad J \neq K.$$

Thus the balls $B(f_J, \frac{1}{2})$ for $J \in \mathfrak{P}(\mathbf{N})$ are disjoint. Surely no countable subset of $\mathcal{C}$ could meet each of these uncountably many disjoint open sets. Thus $\mathcal{C}$ is not separable.

EXERCISES

1. Let S be a metric space having a countable base for its open sets. Without using the concept of separability, show that the same is true for any subspace T of S.

2. Suppose f is a continuous surjection from a separable metric space S to a metric space T. Prove that T is separable.

3. Let S and T be metric spaces. Suppose that $\{G_\alpha \mid \alpha \in A\}$ and $\{H_\beta \mid \beta \in B\}$ are bases for the open sets of S and T respectively. Prove that $\{G_\alpha \times H_\beta \mid \langle \alpha, \beta \rangle \in A \times B\}$ is a base for the open sets of $S \times T$.

4. Show that in a separable metric space a family of mutually disjoint, nonvoid open sets is countable.

5. Suppose S is a metric space with the property that every open covering of S has a countable subcovering. Prove that S is separable.

6. Define *locally separable* metric space. Give an example of a nondiscrete, locally separable metric space which is not separable.

For each point x of a locally separable, but not separable, metric space $\langle S, \rho \rangle$, let

$$\varphi(x) = \sup \{r \in \mathbf{R} \mid B(x, r) \text{ is separable}\}.$$

Prove that $B(x, \varphi(x))$ is separable and that φ is a continuous function.

7. Let S be a set and let $\{f_n\}$ be a sequence of functions from S to $\mathbf{R}$. Define

$$\rho(s, t) = \sum_{n=1}^{\infty} \frac{1}{2^n} \frac{|f_n(s) - f_n(t)|}{1 + |f_n(s) - f_n(t)|}.$$

Prove that ρ is a pseudometric for S. Let $\langle T, \sigma \rangle$ be the associated metric space. (See Exercise 7, p. 228, for definitions.) Show that $\langle T, \sigma \rangle$ is separable. In particular, if ρ is a metric, $\langle S, \rho \rangle$ is separable.

14-10. CONNECTEDNESS

Intuitively we feel that a geometric line is like a piece of string. If we pull on one end, the whole thing moves because it is connected together. If we think of the line as being the metric space **R**, then **R** should be connected in some sense. On the other hand, $\mathbf{R} - \{0\}$ should be disconnected, since it falls into two parts, the positive numbers and the negative numbers. We shall define a topological property of a space which reflects our intuitive ideas of connectedness to a remarkable extent.

14-10.1. Definition. A metric space S is said to be *connected* if and only if S cannot be represented as the union of two disjoint, nonvoid, open subsets. Otherwise, S is said to be *disconnected*.

Since the definition is phrased entirely in terms of open sets, it is clear that connectedness is a topological property of a space.

The following proposition is immediate from the definition and the fact that the open sets in a space are the complements of the closed sets.

14-10.2. Proposition. A metric space S is connected if and only if S and $\emptyset$ are the only subsets of S which are both open and closed.

Another criterion for connectedness in terms of continuous functions will turn out to be particularly handy for our subsequent work.

14-10.3. Proposition. A metric space S is connected if and only if every continuous function from S to a discrete space is constant.

Proof. Suppose S is connected. Let f be a continuous function from S to a discrete space T. If S is empty, f is constant. Suppose S is not empty. Pick $p \in f(S)$. Since $\{p\}$ is both open and closed in $T, f^{-1}(p)$ is both open and closed in S. But the only subsets of S having this property are $\emptyset$ and S. Hence, $f^{-1}(p) = S$; that is, f has the constant value p.

Conversely, suppose S is disconnected. Say $S = G \cup H$, where G and H are disjoint, nonvoid open sets. Define $f(x) = 0$ for $x \in G$ and $f(x) = 1$ for $x \in H$. Since f is constant on a neighborhood (either G or H) of every point of S, f is continuous. Thus we have constructed a nonconstant continuous function from S to the discrete space $\{0, 1\}$.

Since connectedness is a property of a space, if we say that a subset E of a space S is connected, we mean that E, regarded as a subspace of S, is connected.

14-10.4. Proposition. Suppose E is a connected subset of a metric space S. If F is any set such that $E \subseteq F \subseteq \overline{E}$, then F is connected.

Proof. Let φ be a continuous function from F to a discrete space. Then φ restricted to E is a constant function, since E is connected. Let ψ be the constant function on F which extends φ restricted to E. Then φ and ψ are continuous functions on F which agree on a dense subset E of F. Therefore, $\varphi = \psi$; that is, φ is constant. This proves that F is connected.

14-10.5. Theorem. Let f be a continuous function from one metric space S to another T. If E is any connected subset of S, then $f(E)$ is a connected subset of T.

Proof. Suppose φ is a continuous function from $f(E)$ to a discrete space. Then $\varphi \circ f$ is a continuous function from E to a discrete space. Since E is connected, $\varphi \circ f$ is constant. Therefore φ is constant. Hence $f(E)$ is connected.

14-10.6. Proposition. Let S and T be nonvoid metric spaces. Then $S \times T$ is connected if and only if S and T are both connected.

Proof. Suppose $S \times T$ is connected. Since T is not void, the projection $\langle s, t \rangle \to s$ from $S \times T$ to S is surjective. It is also continuous. Hence S is connected by 14-10.5. Similarly, T is connected.

Now suppose that both S and T are connected. Let φ be any continuous function from $S \times T$ to a discrete space. For each fixed $t \in T$, $s \to \varphi(s, t)$ is a continuous function from S to a discrete space; therefore $\varphi(s, t)$ is independent of s. Similarly, it is independent of t. Thus φ is constant. Therefore $S \times T$ is connected.

14-10.7. Definition. An *interval* in $\mathbf{R}$ is a subset of $\mathbf{R}$ having one of the following forms:

$$(a, b) = \{x \mid a < x < b\},$$
$$(a, +\infty) = \{x \mid a < x\},$$
$$(a, b] = \{x \mid a < x \leq b\},$$

$$(-\infty, b) = \{x \mid x < b\},$$
$$(-\infty, +\infty) = \mathbf{R},$$
$$(-\infty, b] = \{x \mid x \leq b\},$$

$$[a, b) = \{x \mid a \leq x < b\},$$
$$[a, +\infty) = \{x \mid a \leq x\},$$
$$[a, b] = \{x \mid a \leq x \leq b\},$$

where $a, b \in \mathbf{R}$ and $a < b$.

Sometimes the condition $a < b$ is not required, in which case $[a, a] = \{a\}$ and $(a, a) = \emptyset$. Intervals of the form $(-\infty, +\infty)$, $(a, +\infty)$, $(-\infty, b)$, or (a, b) are known as *open* intervals. They are indeed open subsets of $\mathbf{R}$. Similarly, intervals of the form $(-\infty, +\infty)$, $[a, +\infty)$, $(-\infty, b]$, or $[a, b]$ are known as *closed* intervals; they are, of course, closed subsets of **R**. The remaining two types, $[a, b)$ and $(a, b]$ are called *half-open*.

14–10.8. Proposition. A subset of **R** containing more than one point is connected if and only if it is an interval.

Proof. Suppose E is a connected subset of **R** containing at least two points. Let $a = \inf E$ and $b = \sup E$. It may be that $a = -\infty$ or $b = +\infty$, but in any case $a < b$. We shall prove that $E \supseteq (a, b)$. If not, there is a number $x \in (a, b) - E$. Then

$$E = \big(E \cap (-\infty, x)\big) \cup \big(E \cap (x, +\infty)\big).$$

Both $E \cap (-\infty, x)$ and $E \cap (x, +\infty)$ are open subsets of E. If $E \cap (-\infty, x)$ were void, we would have $E \subseteq [x, +\infty)$, whence $a \geq x$, contrary to the fact that $x \in (a, b)$. Similarly, $E \cap (x, +\infty)$ is not void. But this contradicts the hypothesis that E is connected. Hence $E \supseteq (a, b)$. If $y \in E$, then $a \leq y \leq b$ by the definition of a and b. Thus $(a, b) \subseteq E \subseteq [a, b]$, whence $E = (a, b)$, $E = [a, b)$, $E = (a, b]$, or $E = [a, b]$ (if a or b is infinite, the argument must be changed very slightly). Thus every connected subset of **R** having more than one point is an interval.

Conversely, let I be an interval in **R**. Suppose $I = G \cup H$, where G and H are disjoint nonvoid sets open relative to I. Say $a \in G$, $b \in H$. We shall assume that $a < b$ (otherwise, interchange the names of G and H). Since I is an interval $[a, b] \subseteq I$, we shall find a point x between a and b at which G "abuts" H and prove that x can fall in neither G nor H. This contradicts $I = G \cup H$.

Let $x = \sup \big(G \cap (-\infty, b)\big)$. Since $a \in G \cap (-\infty, b)$ and b is an upper bound for this set, $a \leq x \leq b$, so $x \in I$.

Suppose $x \in G$. Then $x \neq b$, so $x < b$. The ball of radius ϵ about x in I is the set $I \cap (x - \epsilon, x + \epsilon)$. Since G is open in I, $I \cap (x - \epsilon, x + \epsilon) \subseteq G$ for some positive ϵ. We can pick y so that $x < y < x + \epsilon$ and $y < b$. Then $y \in [a, b] \subseteq I$; hence $y \in I \cap (x - \epsilon, x + \epsilon) \subseteq G$, and finally $y \in G \cap (-\infty, b)$. By the definition of x, $x \geq y$, which is a contradiction.

Suppose $x \in H$. Then for some positive δ, $I \cap (x - \delta, x + \delta) \subseteq H$. Now, $x - \delta$ is not an upper bound for $G \cap (-\infty, b)$. So pick $z \in G \cap (-\infty, b)$ with $z > x - \delta$. Then $z \leq x$ by the definition of x and $z \in G \subseteq I$; hence

$$z \in I \cap (x - \delta, x + \delta) \subseteq H,$$

which is a contradiction.

Thus we have proved that x is in I but in neither G nor H, contrary to the fact that $I = G \cup H$. This proves that I is connected.

The completeness of **R** (in the order sense) is essential to the second part of the proof. An interval in **Q**, for example, is not connected.

14-10.9. The intermediate-value theorem. Let f be a continuous function from a metric space S to $\mathbf{R}$. Let E be a connected subset of S and let $a, b \in E$. If $f(a) < f(b)$ and r is any number such that $f(a) < r < f(b)$, then there is a point $x \in E$ such that $f(x) = r$.

Proof. $f(E)$ is a connected subset of $\mathbf{R}$, and therefore an interval. Since $f(a)$ and $f(b)$ are both in $f(E)$, any intermediate number r is also in $f(E)$.

14-10.10. Corollary. Let f be a continuous function from the interval $[a, b]$ to $\mathbf{R}$. Then f takes every value between $f(a)$ and $f(b)$.

We can use our characterization of connected sets in $\mathbf{R}$ to prove the nonexistence of certain homeomorphisms.

14-10.11. Proposition. The 1-sphere is not homeomorphic to any subset of $\mathbf{R}$.

Proof. The 1-sphere is what we usually call the unit circle in $\mathbf{R}^2$ (14-1.6). Denote it by S. Then S is connected, and for every point p of S, $S - \{p\}$ is connected (being homeomorphic to an interval in $\mathbf{R}$). If S is homeomorphic to a subset E of $\mathbf{R}$, E must be a connected set with more than one point; that is, E must be an interval. But every interval contains a point x such that $E - \{x\}$ is disconnected. Therefore S is not homeomorphic to an interval.

14-10.12. Corollary. No nonvoid open subset of $\mathbf{R}^n$, $n \geq 2$, is homeomorphic to any subset of $\mathbf{R}$.

Proof. Clearly, every nonvoid open subset of $\mathbf{R}^n$, $n \geq 2$, contains a set homeomorphic to a 1-sphere, for example a circle lying in a two-dimensional plane which meets the open set.

14-10.13. Proposition. Let $\{E_\alpha \mid \alpha \in A\}$ be a family of connected subsets of a metric space having a common point. Then $\bigcup_A E_\alpha$ is connected.

Proof. Let φ be a continuous function from $\bigcup_A E_\alpha$ to a discrete space. Choose a point $p \in \bigcap_A E_\alpha$. If $x \in \bigcup_A E_\alpha$, then there is an index $\beta \in A$ such that $x \in E_\beta$. Since E_β is connected, φ is constant on E_β; therefore $\varphi(x) = \varphi(p)$. Thus φ has the constant value $\varphi(p)$ on $\bigcup_A E_\alpha$. Hence $\bigcup_A E_\alpha$ is connected.

14-10.14. Proposition. Let S be a metric space. Let $p \in S$. Among the connected subsets of S which contain p there is a largest. If X_p denotes the largest connected subset of S containing p, then X_p is closed. Moreover, the set $\{X_p \mid p \in S\}$ is a partition of S.

Proof. The union of all the connected subsets of S which contain p is a connected set X_p by the preceding proposition. X_p is clearly the largest connected subset of S containing p. Note that X_p is not void, since $\{p\}$ is a connected subset of S.

Since $\overline{X}_p$ is also connected by 14–10.4, $\overline{X}_p \subseteq X_p$. Hence X_p is closed. Obviously, $\{X_p \mid p \in S\}$ is a covering of S. To prove that it is a partition of S, we need only show that for any p and q, either $X_p = X_q$ or $X_p \cap X_q = \emptyset$. If $X_p \cap X_q \neq \emptyset$, then $X_p \cup X_q$ is a connected set (by 14–10.13) which contains p; so $X_p \cup X_q \subseteq X_p$. This gives $X_q \subseteq X_p$. Similarly, $X_p \subseteq X_q$. Thus $X_p = X_q$.

We know that any partition of S is the set of equivalence classes of some equivalence relation in S. In this case, the relation is "s and t are in a connected subset of S."

14–10.15. Definition. If S is a metric space, the maximal connected subsets of S are called *components* of S.

If $T \subseteq S$, the components of T refer to the components of T regarded as a subspace of S. The proposition tells us that the components of any set T are closed relative to T. A nonempty set is connected if and only if it has exactly one component. The following proposition, which is an immediate consequence of the preceding one, provides the most common way of using the notion of components.

14–10.16. Proposition. If U is a connected subset of a set X, then U lies wholly in a single component of X.

14–10.17. Definition. A set is *totally disconnected* if and only if each of its components consists of a single point.

Any discrete space is totally disconnected. A less trivial example is the set $\mathbf{Q}$ of rational numbers. Since a connected subset of $\mathbf{R}$ which contains more than one point is an interval, and $\mathbf{Q}$ contains no interval, $\mathbf{Q}$ contains no connected set bigger than a single point. Thus $\mathbf{Q}$ is totally disconnected. Although the definition of connectedness is in terms of splitting a space into open subsets, when it is finally split into its components, none of them need be open.

14–10.18. Definition. A metric space S is said to be *locally connected* if and only if every point x of S has arbitrarily small connected neighborhoods; that is, for every $x \in S$ and every $\epsilon > 0$, there is a connected neighborhood U of x such that $U \subseteq B(x, \epsilon)$.

Although we allowed a reference to the metric to appear in the definition, we did not have to. We might have said, "Every neighborhood of x contains a connected neighborhood of x." Therefore the property of local connectedness is topological.

The space $\mathbf{R}^n$ is locally connected. If $x = \langle x_1, x_2, \ldots, x_n \rangle$, the ball $B(x, \epsilon)$ is the direct product $X_{i=1}^{n} (x_i - \epsilon, x_i + \epsilon)$ of intervals in $\mathbf{R}$. This set is connected by 14–10.8 and 14–10.6. Since the property is topological and $\mathbf{C}^n$ is homeomorphic to $\mathbf{R}^{2n}$, $\mathbf{C}^n$ is also locally connected.

14–10.19. Proposition. A metric space S is locally connected if and only if every component of every open set in S is itself open in S.

Proof. Suppose every component of every open set in S is open in S. Let $x \in S$ and $\epsilon > 0$ be given. $B(x, \epsilon)$ is an open set of S; so the component U of $B(x, \epsilon)$ containing x is open in S. U is a connected neighborhood of x with $U \subseteq B(x, \epsilon)$. Thus S is locally connected.

Conversely, suppose that S is locally connected. Let G be any open set in S. Let X be any component of G. We must prove that X is a neighborhood of each of its points. Say $x \in X$. There is a positive ϵ such that $B(x, \epsilon) \subseteq G$, and there is a connected neighborhood U of x such that $U \subseteq B(x, \epsilon)$. Then U is a connected subset of G; hence U is part of the largest connected subset of G containing x; that is, $U \subseteq X$. Thus X is a neighborhood of x.

14–10.20. Corollary. An open subset of a locally connected space is locally connected.

Proof. Suppose that S is a locally connected space and G is an open subset of S. Let H be an open subset of G. Then H is also open in S. Hence the components of H are open in S and therefore in G.

14–10.21. Proposition. Every open subset of $\mathbf{R}$ is the union of a countable family of mutually disjoint open intervals.

Proof. If G is an open subset of $\mathbf{R}$, its components are connected open subsets of $\mathbf{R}$; that is, they are open intervals. Since the components are mutually disjoint and each contains a rational number, there are only a countable number of components.

14–10.22. An example. Let the set of all rational numbers be enumerated; that is, let $\{x_n\}$ be a sequence whose range is $\mathbf{Q}$. Let ϵ be a positive number. Set $I_n = (x_n - \epsilon/2^n, x_n + \epsilon/2^n)$ and $G = \bigcup_{n=1}^{\infty} I_n$. Being a union of open intervals, G is open in $\mathbf{R}$. Since G contains all rational numbers, it is dense in $\mathbf{R}$. But G is far from being all of $\mathbf{R}$; it contains no closed interval of length as large as 2ϵ. For suppose $[a, b] \subseteq G$. Since $[a, b]$ is a compact set covered by $\{I_n\}$, a finite subcovering can be found. It is easy to prove that a finite number of open intervals of total length λ cannot cover any closed interval of length λ or more. Since the length of I_n is $2\epsilon/2^n$, the sum of the lengths of the intervals in the finite subcovering of $[a, b]$ is less than $\sum_{n=1}^{\infty} (2\epsilon/2^n) = 2\epsilon$. Hence $b - a < 2\epsilon$. A refinement of this argument shows that the sum of the lengths of the intervals in any disjoint collection of intervals in G is at most 2ϵ. Thus only a very small portion of the line is in G, even though the latter is dense and open. The complement of G in $\mathbf{R}$ contains no interval, since G is dense. Therefore it is totally disconnected. Given an explicit enumeration of $\mathbf{Q}$ and a definite ϵ, it is an interesting problem to find even one member of $\mathbf{R} - G$.

14–10.23. Definition. A metric space S is said to be *pathwise connected* or *arcwise connected* if and only if for any two points p and q of S there is a continuous function f from $[0, 1]$ to S such that $f(0) = p$ and $f(1) = q$. Such a function f is called a *path* from p to q.

Intuitively, a path represents a continuous motion of a point from p to q. The term *arc* is sometimes used as a synonym for *path*, but it is more often reserved for a set homeomorphic to $[0, 1]$.

Let us note that we can distinguish the endpoints of $[0, 1]$ from the interior points topologically. A point $x \in [0, 1]$ is an endpoint if and only if $[0, 1] - \{x\}$ is connected. Therefore, if A is a set homeomorphic to $[0, 1]$, there are just two points p and q of A such that $A - \{p\}$ and $A - \{q\}$ are connected. These points are naturally referred to as the endpoints of A. With this in mind, it is clear that a space S should be called arcwise connected only if for every two distinct points p and q in S there is an arc A in S having p and q as endpoints. The reason for using the terms *arcwise connected* and *pathwise connected* interchangeably is that the definitions for them are equivalent although this is difficult to prove.

14–10.24. Definition. A metric space S is said to be *locally pathwise connected* if and only if every point of S has arbitrarily small, pathwise connected neighborhoods.

Evidently **R** is locally pathwise connected. When $\mathbf{R}^n$ is interpreted as usual as geometric space, the balls are convex and therefore any point can be connected to each point in a small neighborhood of itself by a straight-line segment. Therefore $\mathbf{R}^n$ is locally pathwise connected and so is $\mathbf{C}^n$.

14–10.25. Proposition. A connected, locally pathwise connected metric space is pathwise connected.

Proof. Let S be a metric space. If we set

$$E = \{\langle p, q \rangle \mid p \text{ and } q \text{ can be connected by a path in } S\},$$

then E is readily seen to be an equivalence relation in S. If every point of S has a pathwise connected neighborhood, then the equivalence class of each point is a neighborhood of that point; in other words, the equivalence classes are open. If S is connected, it cannot be represented as the union of two or more mutually disjoint, nonvoid open sets (since any one of them, together with the union of all the rest, would be a division of S into two mutually disjoint, nonvoid open sets). Hence there can be only one equivalence class if S is connected and locally pathwise connected. This means that S is pathwise connected.

The equivalence classes that were defined in the preceding proof are known as the *path components* of S. By an argument similar to the proof of 14–10.19, one can show that a space is locally pathwise connected if and only if the path components of its open sets are open. It is also clear that the open sets of a locally pathwise connected space are themselves locally pathwise connected. The similarities between the properties of connectedness and pathwise connectedness might lead one to suspect that they are equivalent. This is not true.

14–10.26. The sine curve (Fig. 14–3). Let $X = \{\langle t, \sin \pi/t \rangle \mid 0 < t \leq 2\}$ in $\mathbf{R}^2$. Let φ be the first projection from $\mathbf{R}^2$ to $\mathbf{R}$; that is, $\varphi(x, y) = x$. Since φ carries X homeomorphically onto the interval $(0, 2]$, X is connected. Therefore $\overline{X}$ is also connected. But $\overline{X}$ is neither locally connected nor pathwise connected.

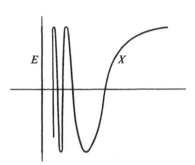

FIG. 14–3.   The sine curve.

Let $E = \{\langle 0, y \rangle \mid -1 \leq y \leq 1\}$. Then $\overline{X} = X \cup E$. Let U be a connected set of diameter less than 2 in $\overline{X}$ which meets E. We shall prove that $U \subseteq E$.

Let

$$V = X \cap (\mathbf{R} \times \{1\}) = \{2, \tfrac{2}{5}, \tfrac{2}{9}, \tfrac{2}{13}, \ldots\} \times \{1\},$$

and

$$W = X \cap (\mathbf{R} \times \{-1\}) = \{\tfrac{2}{3}, \tfrac{2}{7}, \tfrac{2}{11}, \ldots\} \times \{-1\}.$$

V is the set of maximum points of the curve X, and W is the set of minimum points. Since every point of V is at least at distance 2 from every point of W (exactly 2 if the metric is 14–1(2)), U does not meet both V and W. Hence either $U \subseteq \overline{X} - V$ or $U \subseteq \overline{X} - W$; say the former. Now $\varphi(U)$ is a connected subset of $\varphi(\overline{X}) = [0, 2]$ which contains 0 but does not contain any point of

$$\varphi(V) = \{2, \tfrac{2}{5}, \tfrac{2}{9}, \ldots\}.$$

If $\varphi(U)$ contained more than one point, it would be an interval, either $[0, a)$ or $[0, a]$ for some $a > 0$. Such an interval would meet $\varphi(V)$. Therefore $\varphi(U) = \{0\}$. Similarly, $U \subseteq \overline{X} - W$ implies $\varphi(U) = \{0\}$. Therefore $U \subseteq \varphi^{-1}(0) = E$.

Since $E = \overline{X} - X$ does not contain any set open relative to $\overline{X}$, no point of E has a connected neighborhood of diameter less than 2. Therefore $\overline{X}$ is not locally connected. Furthermore, no path in $\overline{X}$ connects any point in E to any point of X. Suppose, on the contrary, that f is a continuous function from $[0, 1]$ to $\overline{X}$ such that $f(0) \in E$ and $f(1) \in X$. The set $f^{-1}(E)$ is a closed subset of $[0, 1]$ containing 0. Let t be its largest element. Since $f(1) \in X$, $t < 1$. Since f is continuous, there is a positive number δ such that $t + \delta < 1$ and diam $f([t, t + \delta]) < 2$. Now, $f([t, t + \delta])$ is a connected set in $\overline{X}$ which meets E. Therefore,

$$f([t, t + \delta]) \subseteq E.$$

This gives $t + \delta \in f^{-1}(E)$, contrary to the choice of t.

The space $\overline{X}$, or any space homeomorphic to it, is known among topologists as a *sine curve*. It is the simplest example of a space which is connected but not locally connected or pathwise connected.

The proof shows that the fact that $\overline{X}$ is not pathwise connected is closely related to the fact that it is not locally connected. This is no accident. It is a theorem (which the reader will find challenging) that a complete metric space which is both connected and locally connected is arcwise connected. It is possible for a space to be pathwise connected without being locally pathwise connected or even locally connected. If we attach to $\overline{X}$ an arc connecting $\langle 0, 1 \rangle$ to $\langle 2, 1 \rangle$, the resulting space is pathwise connected but not locally connected.

EXERCISES

1. Suppose G is an open and closed subset of a space S. Let X be a connected subset of S. Show that either $X \subseteq G$ or $X \cap G = \emptyset$.

2. If a space has only a finite number of components, show that they are all both open and closed.

3. If the metric of a metric space actually takes the value ∞, prove that the space is not connected.

4. Prove that the direct product of two nonvoid locally connected metric spaces S and T is locally connected if and only if both S and T are locally connected.

5. Prove that the direct product of two nonvoid pathwise connected metric spaces S and T is pathwise connected if and only if both S and T are pathwise connected.

6. Is a bounded closed interval in $\mathbf{R}$ homeomorphic to an open interval or to a half-open interval? Is a half-open interval homeomorphic to an open interval?

7. Show that the n-sphere (14–1.6) is pathwise connected for $n \geq 1$.

8. Show that the space $B(S)$ of bounded real-valued functions on a set S with the metric 14–1(6) is connected and locally pathwise connected. If S is a metric space, show that the same is true for $C(S)$, the space of continuous real-valued functions on S regarded as a subspace of $B(S)$.

9. Let G be a connected open set in $\mathbf{R}^n$, $n \geq 2$, and let p be a point of G. Show that $G - \{p\}$ is connected. If X is a countable subset of G, show that $G - X$ is connected.

10. Show that the intersection of a decreasing family of compact connected sets is connected. But show that in $\mathbf{R}^2$ the intersection of a decreasing family of bounded connected sets need not be connected.

11. Suppose S is a compact metric space containing a dense subset X homeomorphic to a half-open interval (the sine curve is such a space). Prove that $S - X$ is connected.

12. Consider in $\mathbf{R}^2$ the set E of points having exactly one rational coordinate. Is it connected? Is its complement connected?

CHAPTER 15

INTRODUCTION TO ANALYTIC FUNCTIONS

In Chapters 12, 13, and 14 we developed the most important techniques of analysis except for integration. Now we shall see how they are applied in the theory of analytic functions.

Analytic functions are those which are represented by their Taylor's series. Hence they include the familiar functions of elementary calculus. Taylor's series provide a natural way to extend the elementary functions from **R** to **R** to functions from **C** to **C**. In the extended forms their properties are far more transparent. A great deal of information about standard problems of calculus has been obtained by studying the complex extensions of various functions. For example, it has been shown that the indefinite integral

$$\int \frac{1}{x} e^x \, dx$$

is not a combination of the usual elementary functions, no matter how complicated. While we cannot prove such a sophisticated result as this here, we can develop some of the most important general theorems and study in some detail the exponential, sine, and cosine functions.

15–1. DIFFERENTIATION

We shall develop the idea of differentiation in a context more general than usual. Almost everything works out just as it does in elementary calculus.

15–1.1. Definition. Let φ be a function from a subset X of a metric space $\langle S, \rho \rangle$ to a metric space $\langle T, \sigma \rangle$. Suppose $p \in \overline{X} - X$. We shall say that $\lim_{x \to p} \varphi(x)$ exists if and only if there is a point $t \in T$ such that

$$(\forall \epsilon > 0)(\exists \delta > 0)(\forall x \in X) \quad \rho(x, p) < \delta \Rightarrow \sigma(\varphi(x), t) < \epsilon.$$

It is easy to prove that such a point t is unique if it exists. Hence, if it does, we denote it by $\lim_{x \to p} \varphi(x)$.

15–1.2. Definition. Let f be a function from a subset E of **C** to **C**. We shall say that f is *differentiable at the point* $x \in E$ if and only if x is not an isolated point of E and

$$\lim_{z \to x} \frac{f(z) - f(x)}{z - x}$$

exists. If so, we call this limit the *derivative* of f at x and denote it $f'(x)$. In this way we define a new function f', called the *derivative* of f, having as domain the set of all points at which f is differentiable. We say that f is *differentiable* if and only if it is differentiable at all points of its domain.

The process of differentiation is particularly important in two cases. The first is the familiar one in which E is an interval (or possibly the union of several intervals) in **R**. (Note that the definition includes one-sided differentiation at an endpoint of the interval E if it should have one.) In this case, the assumption of differentiability carries almost no additional consequences; for example, a differentiable function can have a discontinuous derivative.

The second important case occurs when E is an open subset of **C**. Surprisingly, differentiability in this case implies not only that the derivative is a continuous function, but also that it is differentiable, so that successive derivatives of all orders exist on the entire domain. Unfortunately, we shall not be able to prove these facts. Since differentiability has such strong consequences, it is not surprising that many simple functions from **C** to **C** are not differentiable. For example, $z \to \bar{z}$ is not, since we find that

$$\frac{\bar{z} - \bar{x}}{z - x}$$

is 1 when $z - x$ is real and -1 when $z - x$ is purely imaginary.

The following criterion for differentiability enables us to reduce several of the standard theorems on differentiation to theorems on continuity.

15–1.3. Theorem. Let f be a function from a subset E of **C** to **C**. Suppose that $x \in E$ and that x is not an isolated point of E. In order that f be differentiable at x it is necessary and sufficient that there exist a function $\bar{f}$ from E to **C** which is continuous at x and which satisfies

$$(\forall z \in E) \quad f(z) = f(x) + (z - x)\bar{f}(z). \tag{1}$$

If so, $f'(x) = \bar{f}(x)$.

Proof. There is always a unique function $\bar{f}$ satisfying (1) at all points of $E - \{x\}$. It is given by

$$\bar{f}(z) = \frac{f(z) - f(x)}{z - x} \quad \text{for} \quad z \in E - \{x\}.$$

The question is whether this function can be extended continuously to the point x.

Suppose first that f is differentiable at x. Define $\overline{f}(x) = f'(x)$. We must prove that $\overline{f}$ is continuous at x; i.e.,

$$(\forall \epsilon > 0)(\exists \delta > 0)(\forall z \in E) \quad |z - x| < \delta \Rightarrow |\overline{f}(z) - \overline{f}(x)| < \epsilon. \tag{2}$$

Since f is differentiable at x,

$$(\forall \epsilon > 0)(\exists \delta > 0)(\forall z \in E - \{x\}) \quad |z - x| < \delta \Rightarrow |\overline{f}(z) - f'(x)| < \epsilon. \tag{3}$$

In view of our definition of $\overline{f}(x)$, this surely implies (2).

Conversely, if $\overline{f}$ is defined and continuous at x, then (2) holds, whence (3) holds if we take $f'(x) = \overline{f}(x)$.

15–1.4. Theorem. If f is a function from a subset E of $\mathbf{C}$ to $\mathbf{C}$ and f is differentiable at x, then f is continuous at x. If f is differentiable, it is continuous.

Proof. This follows from (1) with the aid of 14–4.9, 14–4.13, and 14–4.16.

Recall that if f and g are functions from the same domain E to $\mathbf{C}$ and $\lambda, \mu \in \mathbf{C}$, then $\lambda f + \mu g$ stands for the function defined by

$$(\lambda f + \mu g)(x) = \lambda f(x) + \mu g(x) \qquad \text{for} \quad x \in E.$$

Similarly, fg and f/g are defined by

$$(fg)(x) = f(x)g(x) \qquad \text{for} \quad x \in E$$

and

$$(f/g)(x) = f(x)/g(x) \qquad \text{for} \quad x \in E - g^{-1}(0).$$

15–1.5. Theorem. Let $E \subseteq \mathbf{C}$. Suppose f and g are two functions from E to $\mathbf{C}$, both differentiable at x. Let λ and μ be complex numbers. Then $\lambda f + \mu g$ and fg are differentiable at x,

$$(\lambda f + \mu g)'(x) = \lambda f'(x) + \mu g'(x)$$

and

$$(fg)'(x) = f(x)g'(x) + g(x)f'(x).$$

Assuming also that $g(x) \neq 0$, then f/g is differentiable at x and

$$(f/g)'(x) = \frac{g(x)f'(x) - f(x)g'(x)}{g(x)^2}.$$

Proof. According to 15–1.3,

$$f(z) = f(x) + (z - x)\overline{f}(z)$$

and

$$g(z) = g(x) + (z - x)\overline{g}(z),$$

where $\bar{f}$ and $\bar{g}$ are continuous at x. Therefore,

$$(\lambda f + \mu g)(z) = (\lambda f + \mu g)(x) + (z - x)[\lambda \bar{f}(z) + \mu \bar{g}(z)],$$

$$(fg)(z) = (fg)(x) + (z - x)[f(x)\bar{g}(z) + g(x)\bar{f}(z) + (z - x)\bar{f}(z)\bar{g}(z)],$$

$$(f/g)(z) = (f/g)(x) + (z - x)\left[\frac{g(x)\bar{f}(z) - f(x)\bar{g}(z)}{g(x)g(z)}\right],$$

provided that in the last equation $g(x)g(z) \neq 0$. In all three formulas the expression in brackets defines a function (of z) continuous at x. Hence the theorem follows immediately from Theorem 15–1.3. We should note in the case of the quotient that, although the domain of f/g may be smaller than E, the hypothesis $g(x) \neq 0$, together with the continuity of g at x, assures that x is not isolated in the domain of f/g.

Both a constant function and the identity function are differentiable at each nonisolated point of their domain. Moreover, since the derivative of a constant function is 0 and that of the identity function is 1 wherever they are defined, the familiar formulas for differentiating polynomial and rational functions are all valid. For example, if $f(z) = z^n$ for $z \in E$, then $f'(z) = nz^{n-1}$ for all nonisolated points z in E. This is proved, of course, by induction on n. We have thus the following theorem.

15–1.6. Theorem. Any rational function from a subset of $\mathbf{C}$ to $\mathbf{C}$ is differentiable at any nonisolated point of its domain.

15–1.7. Theorem. Let f be a function from a subset D of $\mathbf{C}$ to $\mathbf{C}$. Let g be a function from a subset E of $\mathbf{C}$ to $\mathbf{C}$. Suppose x is a nonisolated point of domain $f \circ g$. If g is differentiable at x and f is differentiable at $g(x)$, then $f \circ g$ is differentiable at x and

$$(f \circ g)'(x) = f'(g(x))g'(x).$$

Proof. Abbreviate $g(x)$ by y and apply Theorem 15–1.3:

$$(\forall z \in D) \quad f(z) = f(y) + (z - y)\bar{f}(z),$$
$$(\forall z \in E) \quad g(z) = g(x) + (z - x)\bar{g}(z) = y + (z - x)\bar{g}(z).$$

Hence

$$(\forall t \in \text{domain } f \circ g) \quad f(g(t)) = f(y) + (t - x)\bar{g}(t)\bar{f}(g(t)).$$

This shows that $f \circ g$ is differentiable at x and that

$$(f \circ g)'(x) = \bar{g}(x)\bar{f}(g(x)) = g'(x)f'(g(x))$$

by 14–4.9 and 15–1.3 again.

The following result can easily be proved directly, but it is an immediate corollary of the last theorem if we take g to be the identity function of E.

15–1.8. Corollary. Suppose f is a function from a subset D of $\mathbf{C}$ to $\mathbf{C}$. Let f^* be the restriction of f to a subset E of D. If x is a nonisolated point of E and f is differentiable at x, then f^* is differentiable at x and $f^{*\prime}(x) = f'(x)$.

We can, of course, consider the derivative f'' of f', and its derivative f''', etc. Higher derivatives are usually denoted with an exponent in parentheses: $f^{(4)}$ instead of f'''', $f^{(5)} = f^{(4)\prime}$, and so on. We shall formalize this notation for future use.

15–1.9. Definition. If f is a function from a subset E of $\mathbf{C}$ to $\mathbf{C}$, then $f^{(0)} = f$ and $(\forall n \in \mathbf{N}^*)\, f^{(n+1)} = f^{(n)\prime}$.

Note that all these derivatives exist under our definition, but their domains may be successively smaller; in fact, they may have the empty set as domain.

15–1.10. Lemma. Suppose g is a differentiable function from $[0, 1]$ to $\mathbf{C}$. There exists a number $t \in [0, 1]$ such that $|g'(t)| \geq |g(1) - g(0)|$.

Proof. The reader will recognize this as an easy consequence of the mean-value theorem. The following proof may not be so familiar.

Let $A = |g(1) - g(0)|$. Put $a_0 = 0$ and $b_0 = 1$. Since $A \leq |g(\frac{1}{2}) - g(0)| + |g(1) - g(\frac{1}{2})|$, either $|g(\frac{1}{2}) - g(0)| \geq A/2$ or $|g(1) - g(\frac{1}{2})| \geq A/2$. If the former, put $a_1 = 0$, $b_1 = \frac{1}{2}$; otherwise, put $a_1 = \frac{1}{2}$, $b_1 = 1$. In either case

$$|g(b_1) - g(a_1)| \geq A/2 \quad \text{and} \quad b_1 - a_1 = \tfrac{1}{2}.$$

Continue defining the sequences $\{a_n\}$ and $\{b_n\}$ as follows. Suppose

$$|g(b_n) - g(a_n)| \geq A/2^n \quad \text{and} \quad b_n - a_n = 1/2^n.$$

If $|g(\frac{1}{2}(a_n + b_n)) - g(a_n)| \geq A/2^{n+1}$, put $a_{n+1} = a_n$ and $b_{n+1} = \frac{1}{2}(a_n + b_n)$. If $|g(\frac{1}{2}(a_n + b_n)) - g(a_n)| < A/2^{n+1}$, put $a_{n+1} = \frac{1}{2}(a_n + b_n)$ and $b_{n+1} = b_n$. Then in either case $|g(b_{n+1}) - g(a_{n+1})| \geq A/2^{n+1}$ and $b_{n+1} - a_{n+1} = 1/2^{n+1}$.

The sequence $\{a_n\}$ is weakly increasing and bounded, and the sequence $\{b_n\}$ is decreasing and bounded. Furthermore, $\lim a_n = \lim b_n$. Denote this limit by t. We have $a_n \leq t \leq b_n$ for all n.

By 15–1.3 there is a function $\bar{g}$ such that

$$g(b_n) - g(t) = (b_n - t)\bar{g}(b_n), \qquad g(a_n) - g(t) = (a_n - t)\bar{g}(a_n),$$

and

$$\lim \bar{g}(b_n) = \lim \bar{g}(a_n) = g'(t).$$

Then

$$\frac{g(b_n) - g(a_n)}{b_n - a_n} = \frac{b_n - t}{b_n - a_n}\bar{g}(b_n) + \frac{t - a_n}{b_n - a_n}\bar{g}(a_n),$$

whence

$$\frac{b_n - t}{b_n - a_n}|\bar{g}(b_n)| + \frac{t - a_n}{b_n - a_n}|\bar{g}(a_n)| \geq \frac{|g(b_n) - g(a_n)|}{b_n - a_n} \geq \frac{A/2^n}{1/2^n} = A. \quad (4)$$

If $|g'(t)| < A$, we should have, for large n, both $|\bar{g}(b_n)| < A$ and $|\bar{g}(a_n)| < A$, which contradicts (4). Therefore $|g'(t)| \geq A$.

15–1.11. Theorem. Let f be a differentiable function from a subset D of C to C. If the line segment L from z_0 to z_1 (i.e., $L = \{z_0 + t(z_1 - z_0) \mid 0 \leq t \leq 1\}$) lies in D, there is a point z of L such that $|f(z_1) - f(z_0)| \leq |z_1 - z_0| \, |f'(z)|$.

Proof. Define g by $g(s) = f(z_0 + s(z_1 - z_0))$ for $s \in [0, 1]$. The preceding lemma shows that

$$|f(z_1) - f(z_0)| = |g(1) - g(0)| \leq |g'(t)| = |z_1 - z_0| \, |f'(z_0 + t(z_1 - z_0))|$$

for some $t \in [0, 1]$. Take $z = z_0 + t(z_1 - z_0)$.

Note that for functions with complex values one cannot conclude that there is a number z such that

$$f(z_1) - f(z_0) = (z_1 - z_0)f'(z).$$

15–1.12. Theorem. Suppose f is a differentiable function from a connected open set D of C to C. If $f'(z) = 0$ for all $z \in D$, then f is constant.

Proof. Let z_0 be a point of D. Theorem 15–1.11 shows that $f(z_0) = f(z_1)$ for any z_1 such that the segment from z_0 to z_1 lies in D. Thus f is constant in a neighborhood of z_0. This shows that, for each $\lambda \in C$, the set $f^{-1}(\lambda)$ is open. Since D is connected, it cannot be decomposed nontrivially into the union of mutually disjoint open sets. Hence one of the sets $f^{-1}(\lambda)$ is all of D and the rest are void. This means that f is constant.

EXERCISES

1. Let f be an injective function from a subset E of C to C. Suppose f is differentiable at x and that $f'(x) \neq 0$. Let g be the inverse of f. Assuming that g is continuous at $f(x)$, prove that g is differentiable at $f(x)$ and $g'(f(x)) = 1/f'(x)$. Show, however, that the inverse of a differentiable function need not be continuous.

2. Consider functions from an interval $[a, b]$ in R to R. Prove the following standard theorems of calculus:

   (a) If f is differentiable at $x \in (a, b)$ and f has a local maximum at x (i.e., for all y in a neighborhood of x, $f(x) \geq f(y)$), then $f'(x) = 0$.
   (b) **Rolle's theorem.** Suppose f is continuous on $[a, b]$ and differentiable on (a, b). If $f(a) = f(b)$, then for some $x \in (a, b)$, $f'(x) = 0$.
   (c) **The Mean-value theorem.** If f is continuous on $[a, b]$ and differentiable on (a, b), then, for some $x \in (a, b)$, $f(b) - f(a) = (b - a)f'(x)$.

3. Suppose f is a differentiable function from a connected open subset D of C to C. Suppose there is a continuous function φ from D to C such that $f'(z) = \varphi(z)f(z)$ for all $z \in D$.

   (a) Prove that either f vanishes on all of D or f does not vanish at any point of D.

(b) Suppose f is not zero and g is another solution of the differential equation with domain D (i.e., $g'(z) = \varphi(z)g(z)$ for all $z \in D$). Prove that there is a number λ such that $g = \lambda f$.

15-2. POWER SERIES

In this section we shall develop the basic facts about the convergence of power series and begin our study of the functions defined by them. Incidentally, we shall prove the binomial theorem for positive integral exponents.

15-2.1. Definition. A *power series* is a family of infinite series

$$\left\{ \text{F-}\sum_{n=0}^{\infty} c_n(z - a)^n \mid z \in \mathbf{C} \right\},$$

where a and the sequence $\{c_n\}$ are in $\mathbf{C}$. The number a is called the *center* of the power series and the members of the sequence $\{c_n\}$ are called its *coefficients*.

We shall denote power series by such notations as $\sum_{n=0}^{\infty} c_n(z - a)^n$, $\sum c_n(z - a)^n$, $\sum_{n=1}^{\infty} b_n z^{2n}$, etc. It is ambiguous whether these notations mean the power series itself, the formal infinite series obtained by considering a fixed value for z, or the actual sum of such a series. This ambiguity must be resolved from the context. It is not at all unusual to interpret the same expression in different senses in consecutive sentences. When $\sum c_n z^n$ denotes a power series, the z which appears is a dummy variable since $\sum c_n t^n$ would refer to the same family of infinite series, that is, to the same power series. Recall that $(z - a)^0 = 1$ even if $z = a$, according to Definition 10–4.1.

Concerning an infinite family of infinite series the natural first question is, "Which of these series are convergent?" The answer is provided by the following theorems.

15-2.2. Theorem. Let $\sum_{n=0}^{\infty} c_n(z - a)^n$ be a power series. There exists a unique extended real number ρ, $0 \le \rho \le +\infty$, such that for all z

$$|z - a| < \rho \Rightarrow \sum_{n=0}^{\infty} c_n(z - a)^n \text{ converges absolutely}$$

and

$$|z - a| > \rho \Rightarrow \sum_{n=0}^{\infty} c_n(z - a)^n \text{ diverges.}$$

Proof. Obviously there can be at most one number ρ with the properties stated. Let ρ be the least upper bound of those nonnegative numbers λ for which the sequence $|c_0|$, $|c_1|\lambda$, $|c_2|\lambda^2$, ... is bounded; i.e., $\rho = \sup S$, where

$$S = \{\lambda \ge 0 \mid (\exists M)(\forall n) \, |c_n|\lambda^n \le M\}.$$

Here $\rho \ge 0$ because $0 \in S$, but ρ may be $+\infty$ because S may be unbounded.

If $|z - a| > \rho$, then $|z - a| \notin S$; so the terms of $\sum c_n(z - a)^n$ are unbounded and the series diverges.

If $|z - a| < \rho$, then $|z - a|$ is not an upper bound for S; hence we can choose $\lambda \in S$ so that $|z - a| < \lambda$. By the definition of S there is a number M such that

$$(\forall n) \quad |c_n|\lambda^n \leq M. \tag{1}$$

Then $|c_n(z - a)^n| \leq M(|z - a|/\lambda)^n$ for any n; so the series $\sum_{n=0}^{\infty} c_n(z - a)^n$ is dominated by the geometric series $\sum_{n=0}^{\infty} M(|z - a|/\lambda)^n$. The latter converges because $|z - a|/\lambda < 1$. Hence $\sum_{n=0}^{\infty} c_n(z - a)^n$ converges absolutely.

15-2.3. Definitions. The extended real number ρ associated with a power series in Theorem 15-2.2 is called the *radius of convergence* of the power series. If $0 < \rho < \infty$, the set $\{z \mid |z - a| = \rho\}$ is called the *circle of convergence*. If $\rho > 0$, the set $\{z \mid |z - a| < \rho\}$ is called the *disk of convergence*.

The last terms are not completely standard. The term *region of convergence* is often applied to what we have called the disk of convergence and one can find instances where the term *circle of convergence* refers to the same thing.

Note that the theorem gives no information about what happens on the circle of convergence. Indeed, nothing can be said about the convergence of a power series on the circle of convergence without additional information concerning the coefficients. It is readily checked using the ratio test that the radius of convergence is 1 for each of the following power series.

$$\sum_{n=0}^{\infty} z^n, \quad \sum_{n=1}^{\infty} \frac{1}{n} z^n, \quad \sum_{n=1}^{\infty} \frac{1}{n^2} z^n.$$

The first diverges for all z on the circle of convergence, the last converges for all z on the circle of convergence, while the second diverges for $z = 1$ and converges for all other z with $|z| = 1$. (To prove this apply Exercise 5, p. 182.)

15-2.4. Corollary. If the radius of convergence ρ of the power series

$$\sum_{n=0}^{\infty} c_n(z - a)^n$$

is positive and $\sigma > 1/\rho$ (if $\rho = +\infty$, this means $\sigma > 0$), there is a number M such that $(\forall n) |c_n| \leq M\sigma^n$.

Proof. Since $1/\sigma < \rho$, the series $\sum c_n (1/\sigma^n)$ converges and its terms are bounded, say by M. Then $|c_n| \leq M\sigma^n$ for all n.

15-2.5. Corollary. The radius of convergence of $\sum_{n=1}^{\infty} nc_n(z - a)^{n-1}$ is the same as that of $\sum_{n=0}^{\infty} c_n(z - a)^n$.

Proof. Let ρ be the radius of convergence of $\sum_{n=0}^{\infty} c_n(z - a)^n$. It was shown in the proof of the theorem that, if $|z - a| > \rho$, the terms of $\sum c_n(z - a)^n$ are unbounded. If so, the terms of $\sum nc_n(z - a)^{n-1}$ are also unbounded and the latter diverges. Suppose $|z - a| < \rho$. As shown in the proof of the theorem, $\sum c_n(z - a)^n$ is dominated by a convergent geometric series $\sum M(|z - a|/\lambda)^n$. Then $\sum nc_n(z - a)^{n-1}$ is dominated by the series $\sum(M/\lambda)n(|z - a|/\lambda)^{n-1}$, which converges by the ratio test. Thus we have proved that $\sum nc_n(z - a)^{n-1}$ converges for $|z - a| < \rho$ and diverges for $|z - a| > \rho$. This shows that the radius of convergence of $\sum nc_n(z - a)^{n-1}$ is ρ.

15–2.6. Theorem. The radius of convergence of $\sum_{n=0}^{\infty} c_n(z - a)^n$ is $\liminf |c_n|^{-1/n}$ (where $0^{-1/n}$ is interpreted as $+\infty$).

Proof. Let ρ be the radius of convergence of the power series.

Suppose λ is a nonnegative number less than $\liminf |c_n|^{-1/n}$. Except for a finite number of integers n, $\lambda < |c_n|^{-1/n}$ or $|c_n|\lambda^n < 1$ (note that this inequality certainly holds if $c_n = 0$), whence the sequence $\{|c_n|\lambda^n\}$ is bounded. Hence, as we showed in the proof of Theorem 15–2.2, $\lambda \leq \rho$. This proves that

$$\liminf |c_n|^{-1/n} \leq \rho.$$

Now suppose $\mu > \liminf |c_n|^{-1/n}$. Then, for infinitely many integers n, $|c_n\mu^n| > 1$. Since its terms do not approach 0, the series $\sum c_n\mu^n$ diverges. Hence $\mu \geq \rho$. Therefore, $\liminf |c_n|^{-1/n} \geq \rho$.

15–2.7. The factorial function and the binomial coefficients. The *factorial* function from the nonnegative integers to $\mathbf{R}$ is defined recursively by $0! = 1$, $(n + 1)! = (n + 1)(n!)$ for $n \in \mathbf{N}^*$. Evidently $n!$ is an integer for all n.

For integers k and n such that $0 \leq k \leq n$, the *binomial coefficients* are defined by

$$\binom{n}{k} = \frac{n!}{k!(n - k)!}.$$

If k and n are integers and $1 \leq k \leq n$, then

$$\frac{1}{k} + \frac{1}{n - k + 1} = \frac{n + 1}{k(n - k + 1)}.$$

Multiplying by $n!/(k - 1)!(n - k)!$, we obtain the identity

$$\binom{n}{k} + \binom{n}{k - 1} = \binom{n + 1}{k} \tag{2}$$

for integers k and n such that $1 \leq k \leq n$. (The binomial coefficient $\binom{n}{k}$ is often defined to be 0 for positive integers n and integers k not satisfying $0 \leq k \leq n$. With this definition (2) remains valid for $n \geq 0$ and all k.)

Since $\binom{n}{0} = \binom{n}{n} = 1$, it follows immediately by induction that the binomial coefficients are all positive integers.

15–2.8. The binomial theorem. For any complex numbers a and b and any positive integer n,

$$(a + b)^n = \sum_{k=0}^{n} \binom{n}{k} a^{n-k}b^k. \tag{3}$$

Proof. This is evidently true for $n = 0$. Assume that it is true for $n = p$. Then

$$a(a + b)^p = \sum_{k=0}^{p} \binom{p}{k} a^{p-k+1}b^k = a^{p+1} + \sum_{k=1}^{p} \binom{p}{k} a^{p+1-k}b^k$$

and

$$b(a + b)^p = \sum_{j=0}^{p} \binom{p}{j} a^{p-j}b^{j+1} = \sum_{k=1}^{p} \binom{p}{k-1} a^{p+1-k}b^k + b^{p+1}.$$

Adding, we obtain

$$(a + b)^{p+1} = a^{p+1} + \sum_{k=1}^{p} \left[\binom{p}{k} + \binom{p}{k-1}\right] a^{p+1-k}b^k + b^{p+1}$$

$$= \sum_{k=0}^{p+1} \binom{p+1}{k} a^{p+1-k}b^k,$$

which is (3) for $n = p + 1$. This establishes the theorem by induction.

15–2.9. Lemma. Let x be a complex number, let n be a positive integer, and let δ be a positive real number. Then, for all complex numbers h with $0 < |h| < \delta$,

$$\left|\frac{(x + h)^n - x^n}{h} - nx^{n-1}\right| \leq \frac{|h|}{\delta^2} (|x| + \delta)^n. \tag{4}$$

Proof. If $0 < |h| < \delta$, we have

$$\delta^2|(x + h)^n - x^n - nx^{n-1}h| = \delta^2 \left|\sum_{k=2}^{n} \binom{n}{k} x^{n-k}h^k\right|$$

$$\leq \delta^2|h|^2 \sum_{k=2}^{n} \binom{n}{k} |x|^{n-k}|h|^{k-2}$$

$$\leq |h|^2 \sum_{k=2}^{n} \binom{n}{k} |x|^{n-k}\delta^k \leq |h|^2(|x| + \delta)^n.$$

Dividing by $|h|\delta^2$, we obtain (4).

15–2.10. Theorem. Let $\sum c_n(z - a)^n$ be a power series with a positive radius of convergence. The function f defined on its disk of convergence D by

$$f(z) = \sum_{n=0}^{\infty} c_n(z - a)^n$$

is continuous and differentiable. Moreover, its derivative is given by

$$f'(z) = \sum_{n=1}^{\infty} nc_n(z - a)^{n-1}$$

for all $z \in D$.

Proof. There is no loss of generality if we consider only the case $a = 0$; then

$$f(z) = \sum c_n z^n.$$

Let ρ be the radius of convergence of the power series. Let z be given with $|z| < \rho$. We shall prove directly that $f'(z) = \sum nc_n z^{n-1}$. According to 15–2.5 the latter series is convergent.

Choose a positive number δ less than $\rho - |z|$. If $0 < |h| < \delta$,

$$\left| \frac{f(z + h) - f(z)}{h} - \sum_{n=1}^{\infty} nc_n z^{n-1} \right|$$

$$= \left| \frac{1}{h} \left(\sum_{n=0}^{\infty} c_n(z + h)^n - \sum_{n=0}^{\infty} c_n z^n \right) - \sum_{n=1}^{\infty} nc_n z^{n-1} \right|$$

$$= \left| \sum_{n=1}^{\infty} c_n \left(\frac{(z + h)^n - z^n}{h} - nz^{n-1} \right) \right|$$

$$\leq \sum_{n=1}^{\infty} |c_n| \left| \frac{(z + h)^n - z^n}{h} - nz^{n-1} \right|$$

$$\leq \frac{|h|}{\delta^2} \sum_{n=1}^{\infty} |c_n|(|z| + \delta)^n.$$

The last series converges since $|z| + \delta < \rho$, and its sum does not involve h. Hence the inequality proves that

$$\lim_{h \to 0} \frac{f(z + h) - f(z)}{h} = \sum_{n=1}^{\infty} nc_n z^{n-1}.$$

The theorem shows that the derivative of f can be found by differentiating the series for f term by term. Note that we consider the function defined by the power series only in the open disk of convergence.

EXERCISES

1. Show that the radius of convergence of $\sum |c_n|(z - a)^n$ is the same as that of $\sum c_n(z - a)^n$.

2. Find the radius of convergence of the power series $\sum c_n z^n$ when

(a) $c_0 = 1, c_{n+1} = \dfrac{2n^2 + 1}{n + 2} c_n$ 　　(b) $c_0 = 1, c_{n+1} = \dfrac{(-1)^n n + 2}{2n - 1} c_n$

(c) $c_n = \dbinom{2n}{n}$ 　　　　　　　(d) $c_{3k} = 0, c_{3k+1} = 2^{2k+1}, c_{3k+2} = 2^{2k+2}$

3. If the radius of convergence of $\sum a_n z^n$ is ρ, what can you say about the radius of convergence of $\sum c_n z^n$, where $c_n = \sum_{k=0}^{n} a_k$?

4. Using the extended definition of the binomial coefficients, prove that

$$\sum_{k=0}^{r} \binom{n}{k}\binom{m}{r - k} = \binom{n + m}{r}.$$

5. Prove that the function f of Theorem 15-2.10 has derivatives of all orders on all of D and that its kth derivative is given by

$$f^{(k)}(z) = \sum_{n=0}^{\infty} \frac{(n + k)!}{n!} c_{n+k}(z - a)^n.$$

15-3. ANALYTIC FUNCTIONS

Functions which can be defined by convergent power series are called analytic functions. In a sense they are polynomial functions of infinite degree and they share many of the regularity properties of polynomials.

15-3.1. Definition. Let f be a function from a subset E of $\mathbf{C}$ to $\mathbf{C}$. Let $\sum c_n(z - a)^n$ be a power series. We say that the power series *represents f on the set E'* if and only if

$$f(z) = \sum_{n=0}^{\infty} c_n(z - a)^n \tag{1}$$

for all $z \in E'$.

In order for (1) to make sense it is necessary that the series converge for all $z \in E'$. If E' is an open subset of $\mathbf{C}$, this implies that $E' \subseteq D$, the disk of convergence for the power series.

15-3.2. Definition. A function f from a subset E of $\mathbf{C}$ to $\mathbf{C}$ is said to be *analytic* if and only if E is open and for every point $a \in E$ there is a power series centered at a which represents f in some neighborhood of a.

It is not required that the power series should represent f at every point at which f is defined and the series converges.

The terms *holomorphic* and *regular* are widely used as synonyms for *analytic*.

It follows immediately from Theorem 15–2.10 that an analytic function is continuous and differentiable; in fact, its derivative is an analytic function. It is one of the most remarkable theorems of mathematics that, conversely, any differentiable function with domain open in $\mathbf{C}$ is representable by a power series near each point of its domain and is therefore an analytic function in the sense just defined. The proof of this theorem seems to depend essentially on the theory of complex line integrals and therefore we cannot give it here. Since the complex line integral is the most useful tool for acquiring information about analytic functions, most books on the subject make the study of integration the first order of business. With this background it is natural to define an analytic function as a differentiable function from an open set in $\mathbf{C}$ to $\mathbf{C}$. Sometimes it is demanded that the function have a continuous derivative although this is, in fact, redundant. It should be understood that the definition given here agrees with the more common one, although the equivalence is highly nontrivial. By jumping ahead to this definition we can develop some important facts about analytic functions using the results of Chapters 13 and 14.

Clearly, the requirement that a function be representable by a power series near each point of its domain is a strong one. It is a matter of elementary algebra that any polynomial function can be expanded in powers of $z - a$ for any a. Hence a polynomial function from $\mathbf{C}$ to $\mathbf{C}$ is analytic. If $a \neq 0$, we know that the geometric series

$$\sum_{n=0}^{\infty} \frac{(-1)^n}{a^{n+1}} (z - a)^n \tag{2}$$

converges to $1/z$ for $|z - a| < |a|$. Hence the reciprocal function is analytic. Differentiating (2) and using 15–1.5 and 15–2.10, we see that $-1/z^2$ can be represented by a power series centered at a for every $a \neq 0$. Similarly the other negative power functions are analytic. In the remainder of this section we shall develop some general criteria for analyticity.

We begin, however, by proving that the power series which represent an analytic function are unique.

15–3.3. Theorem. Let f be an analytic function and let $a \in$ domain f. There is only one power series with center a which represents f near a. It is

$$\sum_{n=0}^{\infty} \frac{f^{(n)}(a)}{n!} (z - a)^n. \tag{3}$$

Proof. Suppose f is represented on a neighborhood N of a by $\sum_{n=0}^{\infty} c_n(z - a)^n$. By Theorem 15–2.10 applied recursively (Exercise 5, p. 298), the kth derivative of f exists on N and is given by

$$f^{(k)}(z) = \sum_{n=0}^{\infty} \frac{(n + k)!}{n!} c_{n+k}(z - a)^n.$$

Setting $z = a$, we see that $k!c_k = f^{(k)}(a)$. Thus the power series with center a which represents f near a is (3) as claimed.

We recognize (3) as the familiar Taylor's series.

Our next two results are strictly routine and we shall omit their proofs.

15–3.4. Proposition. Let f be a function from an open subset D of $\mathbf{C}$ to $\mathbf{C}$. If f is analytic and G is an open subset of D, then f restricted to G is analytic. Conversely, suppose that each point p of D has a neighborhood N such that f restricted to N is analytic. Then f is analytic.

15–3.5. Corollary. Suppose $\{f_\alpha \mid \alpha \in A\}$ is a family of analytic functions which have a common extension. Their least common extension is analytic.

Arithmetic combinations of analytic functions are analytic. We state this formally for linear combinations and products.

15–3.6. Theorem. Suppose f and g are analytic functions with a common domain, and let $\lambda, \mu \in \mathbf{C}$. Then $\lambda f + \mu g$ and fg are analytic functions.

Proof. We give the proof only for the product. Suppose a is a point of the common domain of f and g. Let

$$f(z) = \sum_{n=0}^{\infty} b_n(z - a)^n \tag{4}$$

and

$$g(z) = \sum_{n=0}^{\infty} c_n(z - a)^n \tag{5}$$

be the power series developments of f and g at a. There is a positive number δ so small that (4) and (5) are both valid for $|z - a| < \delta$ with both series converging absolutely. Now, if $|z_0 - a| < \delta$, then

$$(fg)(z_0) = f(z_0)g(z_0) = \sum_{n=0}^{\infty} \left(\sum_{k=0}^{n} b_k c_{n-k} \right) (z_0 - a)^n$$

by Theorem 13–2.35. Thus fg is representable on a neighborhood of a by a power series centered at a. Therefore fg is analytic.

This argument explains why in Chapter 13 we considered the Cauchy product of two infinite series in preference to any other method of collecting terms in the formal product.

The composition of two analytic functions is also analytic. This follows from Theorem 15–1.7, of course, if we use the fact that differentiability is equivalent to analyticity. A proof using power series directly is not difficult, but we leave it as an exercise since we shall not need the result. As the composition of two analytic functions, therefore, the reciprocal of an analytic function is analytic. Finally, the result for products shows that the quotient of two analytic functions is analytic. When we take quotients, the domain must be restricted to avoid zeros of the denominator. Since the denominator is continuous, the set deleted is closed and the restricted domain is open.

To check the analyticity of a function f under Definition 15–3.2 we must establish the existence, for each a in domain f, of a power series representing f with center a. The following theorem reduces the burden of these existence proofs.

15–3.7. Theorem. Let $\sum c_n(z - a)^n$ be a power series with positive radius ρ of convergence. The function f defined on its disk D of convergence by

$$f(z) = \sum_{n=0}^{\infty} c_n(z - a)^n$$

is analytic. Moreover, if $b \in D$, the power series development of f centered at b has radius of convergence at least $\rho - |b - a|$ and represents f on the disk $E = \{z \mid |z - b| < \rho - |b - a|\}$.

Proof. We simply replace $(z - a)$ in the series by $((z - b) + (b - a))$, multiply out, and collect terms:

$$f(z) = \sum_{k=0}^{\infty} c_k(z - a)^k = \sum_{k=0}^{\infty} \sum_{m=0}^{k} c_k \binom{k}{m} (b - a)^{k-m} (z - b)^m$$

$$= \sum_{m=0}^{\infty} \sum_{k=m}^{\infty} c_k \binom{k}{m} (b - a)^{k-m} (z - b)^m.$$

The last step requires justification.

Fix b so that $|b - a| < \rho$ and z so that $|z - b| < \rho - |b - a|$. Consider the double series

$$\sum_{m=0}^{\infty} \sum_{n=0}^{\infty} c_{m+n} \binom{m + n}{m} (b - a)^n (z - b)^m \tag{6}$$

and the corresponding series of positive terms

$$\sum_{m=0}^{\infty} \sum_{n=0}^{\infty} |c_{m+n}| \binom{m + n}{m} |b - a|^n |z - b|^m. \tag{7}$$

According to 13–2.33, if the sum by diagonals of (7) exists, then all methods of summing (6) are convergent and produce the same sum.

The sum by diagonals of (7) is

$$\sum_{k=0}^{\infty} \sum_{p=0}^{k} |c_k| \binom{k}{p} |b - a|^{k-p} |z - b|^p = \sum_{k=0}^{\infty} |c_k|(|z - b| + |b - a|)^k,$$

which is convergent because $|z - b| + |b - a| < \rho$ (see Exercise 1, p. 298). The sum by diagonals of (6) is

$$\sum_{k=0}^{\infty} \sum_{p=0}^{k} c_k \binom{k}{p} (b-a)^{k-p}(z-b)^p = \sum_{k=0}^{\infty} c_k(z-a)^k = f(z).$$

Therefore,

$$f(z) = \sum_{m=0}^{\infty} \left[\sum_{n=0}^{\infty} c_{m+n} \binom{m+n}{m} (b-a)^n \right] (z-b)^m,$$

where the series in question are now known to converge. Thus f is represented on the disk E by a power series centered at b. The radius of convergence of this power series must be at least the radius of E, that is, at least $\rho - |b-a|$.

In particular, if $\rho = +\infty$, then $D = \mathbf{C}$ and the proof shows that $E = \mathbf{C}$. Hence all of the power series representing f do so on all of $\mathbf{C}$. Functions with this property are called *entire* functions.

Let f be a function from an open set E in $\mathbf{C}$ to $\mathbf{C}$. To decide whether f is analytic, we need not find a power series with each center in E. It is enough to show that E can be covered with open sets on each of which f can be represented by a power series. For example, if $f(z) = 1/z$, then to show that f is analytic on $\mathbf{C} - \{0\}$, it would be sufficient to check that it is represented by (2) in the cases $a = \pm n$, $\pm ni$, $n = 1, 2, 3, \ldots$, since the union of the corresponding disks is $\mathbf{C} - \{0\}$. By the Lindelöf covering theorem (14–9.12) only a countable family of open sets will ever be required.

The theorem is actually a special case of a more general theorem. If f is the limit of a sequence $\{g_n\}$ of analytic functions defined on an open set E which converges uniformly on each compact subset of E, then f is analytic. In our case the functions g_n are the partial sums of the power series; the proof of Theorem 15–2.2 shows that the convergence is uniform on each compact subset of the disk of convergence. The limit of a sequence of analytic functions is analytic under still weaker hypotheses, but proofs of these more general theorems must be left to the theory of complex integration. Nevertheless, these limit theorems illustrate the extraordinary regularity properties of analytic functions. There is a famous theorem of Weierstrass which shows that just the opposite is true for functions from $\mathbf{R}$ to $\mathbf{R}$: Every continuous function (differentiable or not) from a bounded closed interval in $\mathbf{R}$ to $\mathbf{R}$ is the uniform limit of a sequence of polynomial functions.

Next we shall show that an analytic function with connected domain is completely determined by its values near any single point.

15–3.8. Theorem. Suppose f is an analytic function with connected domain D. The set $f^{-1}(0)$ is either all of D or a discrete set closed relative to D.

Proof. Since f is a continuous function, $f^{-1}(0)$ is a closed subset of D. We shall prove that every point of $f^{-1}(0)$ is either an isolated point or an interior point.

Suppose $a \in f^{-1}(0)$. Since f is analytic, there is a power series $\sum c_n(z - a)^n$ which represents f on a neighborhood N of a. If every coefficient $c_n = 0$, then $f(z) = 0$ for all $z \in N$ and a is an interior point of $f^{-1}(0)$. Suppose some coefficient $c_n \neq 0$ and let k be the least integer such that $c_k \neq 0$. Here $k > 0$ since $c_0 = f(a) = 0$. We can write the series representation of f in the form

$$f(z) = (z - a)^k \sum_{n=0}^{\infty} c_{k+n}(z - a)^n \qquad (8)$$

for $z \in N$. Since the radius of convergence of this power series is positive, there are (15-2.4) positive numbers M and σ such that $|c_{k+n}| \leq M\sigma^n$ for all n. Put $\tau = 2 + M/|c_k|$ and let E be the disk of radius $1/\sigma\tau$ about a. For $z \in E$ we have

$$\left| \sum_{n=1}^{\infty} c_{k+n}(z - a)^n \right| \leq \sum_{n=1}^{\infty} M\sigma^n |z - a|^n \leq \sum_{n=1}^{\infty} M \left(\frac{1}{\tau} \right)^n = \frac{M}{\tau - 1} < |c_k|.$$

Hence,

$$\left| \sum_{n=0}^{\infty} c_{k+n}(z - a)^n \right| = \left| c_k + \sum_{n=1}^{\infty} c_{k+n}(z - a)^n \right| > 0.$$

It follows from (8) that $f(z) \neq 0$ for $z \in N \cap (E - \{a\})$. Thus

$$\{a\} = f^{-1}(0) \cap N \cap E$$

is a neighborhood of a in $f^{-1}(0)$; that is, a is an isolated point of $f^{-1}(0)$.

Having established that $f^{-1}(0)$ consists entirely of interior points and isolated points, we use the connectedness of D to prove that either $f^{-1}(0) = D$ or there are no interior points of $f^{-1}(0)$. Let $G = \text{Int}\, f^{-1}(0)$. Let $\overline{G}$ be the closure of G calculated in the space D. Since $f^{-1}(0)$ is closed in D, $\overline{G} \subseteq f^{-1}(0)$. Because G is an open set in $\mathbf{C}$, no point of $\overline{G}$ is isolated in $f^{-1}(0)$. Hence our preliminary result shows that $\overline{G} \subseteq \text{Int}\, f^{-1}(0) = G$. Thus G is both open and closed in D. Since D is connected, either $G = D$ or $G = \emptyset$. If $G = D$, $f^{-1}(0) = D$. If $G = \emptyset$, every point of $f^{-1}(0)$ is isolated and $f^{-1}(0)$ is discrete.

Although $f^{-1}(0)$ is a closed subset of D, it need not be a closed subset of $\mathbf{C}$; $\overline{f^{-1}(0)}$ might contain all of the boundary of D even though $f^{-1}(0)$ is discrete.

15-3.9. Corollary. Let f be a nonconstant analytic function with connected domain D. Then f' vanishes only at isolated points of D.

Proof. We know that f' is an analytic function with connected domain D. By Theorem 15-1.12 f' does not vanish on all of D. Hence it vanishes only at isolated points of D.

15-3.10. Theorem. If f and g are analytic functions having the same connected domain D and if $\{z \mid f(z) = g(z)\}$ has a cluster point in D, then $f = g$.

Proof. Consider the analytic function $h = f - g$. Then $\{z \mid f(z) = g(z)\} = h^{-1}(0)$. Say it has a cluster point a in D. Since $h^{-1}(0)$ is closed in D, $a \in h^{-1}(0)$, and a is not isolated in $h^{-1}(0)$. By the preceding theorem $h^{-1}(0) = D$, whence $f = g$.

This theorem tells us that an analytic function f with connected domain D is completely determined by its values on any infinite sequence of distinct points converging to a point of D. This is another surprising fact about analytic functions. Similarly, f is determined by its value and the values of its successive derivatives at any point a of D, for these values determine the power series development of f at a and therefore the values of f in a neighborhood of a. It should be clear that the domain must be connected to obtain any such conclusion, for we can splice together any two analytic functions with disjoint domains and obtain an analytic function.

These facts naturally raise two questions. Given a connected open set D and a sequence $\{x_n\}$ of distinct points converging to a point p of D, for what sequences $\{\lambda_n\}$ of numbers will there exist an analytic function f on D such that $f(x_n) = \lambda_n$ for all n? Obviously we cannot specify the λ's arbitrarily, because, if f exists, any infinite number of them determine all the rest. Again, for what sequences $\{\mu_n\}$ of numbers will there exist an analytic function g on D such that $g^{(n)}(p) = \mu_n$ for all n? These are both difficult problems which have not yet been fully answered. We cannot discuss them further here.

Consider the power series $\sum z^n$ which converges for $|z| < 1$ to $(1 - z)^{-1}$. According to Theorem 15-3.7 the power series for this function with center at $\frac{1}{2}i$ will have radius of convergence at least $\frac{1}{2}$. But, in fact, this development is

$$\sum_{n=0}^{\infty} \left(\frac{1}{1 - \frac{1}{2}i} \right)^{n+1} (z - \tfrac{1}{2}i)^n,$$

which converges for $|z - \frac{1}{2}i| < |1 - \frac{1}{2}i| = \frac{1}{2}\sqrt{5}$; that is, the radius of convergence is a good bit larger than the minimum guaranteed by the theorem. In this case we know that both series represent $(1 - z)^{-1}$ on their disks of convergence and hence both series represent the same function on the intersection of their disks of convergence. Theorem 15-3.7 tells us only that both series will represent the same function on the disk $|z - \frac{1}{2}i| < \frac{1}{2}$. Theorem 15-3.10 shows that it always works out this way.

Let $\sum c_n(z - a)^n$ be a power series with positive radius of convergence ρ_1. It defines an analytic function f on the disk $D_1 = \{z \mid |z - a| < \rho_1\}$. Pick $b \in D_1$. According to Theorem 15-3.7, f is represented by a power series $\sum d_n(z - b)^n$ on the disk $E = \{z \mid |z - b| < \rho_1 - |b - a|\}$ (see Fig. 15-1). Suppose this new power series has radius of convergence ρ_2. Then it defines an analytic function g on the disk $D_2 = \{z \mid |z - b| < \rho_2\}$. We shall see that f and g must agree on $D_1 \cap D_2$ although Theorem 15-3.7 tells us only that they must agree on E, and E may be strictly smaller than $D_1 \cap D_2$. Both f and g are analytic on the connected set $D_1 \cap D_2$ and they agree on the nondiscrete set E. Hence they agree on all of $D_1 \cap D_2$. The estimates originally used to prove that f and g agree on E break down for $z \in (D_1 \cap D_2) - E$, but the series $\sum c_n(z - a)^n$ and $\sum d_n(z - b)^n$ have the same sum nevertheless.

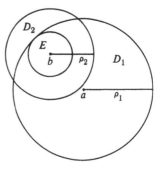

FIGURE 15-1

Since the functions f and g agree on $D_1 \cap D_2$, they have a common extension with domain $D_1 \cup D_2$ and this extension is analytic. When $D_1 \cup D_2$ is actually larger than D_1, this illustrates a phenomenon known as analytic continuation. Suppose we start with the power series $\sum z^n$ and denote by f_1 the function it defines on its disk of convergence D_1. Let us forget for a moment that f_1 is given by $(1 - z)^{-1}$. By expanding f_1 at the point $\frac{1}{2}i$ (see Fig. 15-2), we obtain a power series which defines a new function f_2 on its disk of convergence D_2. We know that f_1 and f_2 agree on $D_1 \cap D_2$. Now expand f_2 at the point i to obtain a new power series and a new function f_3 on the disk D_3; f_2 and f_3 will agree on the intersection $D_2 \cap D_3$. Since $2i \in D_3$, we can expand f_3 at $2i$ to obtain a new power series and a new function f_4 on the disk D_4. We can continue this process indefinitely, expanding f_4 at $4i$, f_5 at $8i$, etc. These functions have a common extension with domain $D_1 \cup \{z \mid \operatorname{Im} z > 0\}$. We can repeat this construction with centers in the lower half-plane, along the negative real axis, etc. Eventually through

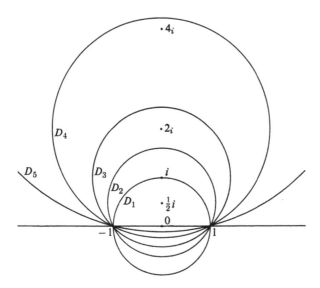

FIGURE 15-2

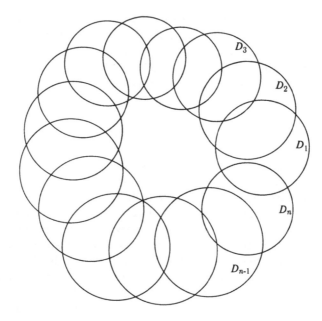

FIGURE 15–3

this continuation process we should discover that the original function f_1 has an extension which is analytic on $\mathbf{C} - \{1\}$.

Starting with any power series we can attempt to carry out a similar procedure. There is, of course, no guarantee that any of the power series obtained by expanding at new centers will actually extend the domain of the original function. Moreover, unless the circumstances are particularly favorable, we will not be able to carry out the algebra and actually find the various expansions.

Sometimes the continuation process leads to a more complicated situation than the one just described. Suppose we start with some other function f_1 defined on a disk D_1. By successively reexpanding we find functions $f_2, f_3, \ldots, f_n$ defined on the disks $D_2, D_3, \ldots, D_n$, respectively. The construction ensures that f_1 and f_2 will agree on $D_1 \cap D_2$, that f_2 and f_3 will agree on $D_2 \cap D_3$, etc., but it can happen, if the disks are disposed as illustrated in Fig. 15–3, that f_1 and f_n do not agree on $D_1 \cap D_n$. We shall meet this phenomenon in Section 15–6.

Let us return to the explicit function $f(z) = (1 - z)^{-1}$. This function is unbounded as $z \to 1$. Hence it is clear that there is no analytic function which extends f properly; in fact there isn't even a continuous one. The power series expansion of f at a cannot converge on a disk of radius larger than $|a - 1|$, for, if it did, it would define an analytic function g on a disk E including 1 and this function g would agree with f on all of the connected set $E \cap \text{domain} f = E - \{1\}$. The power series expansion of f at a is

$$\sum_{n=0}^{\infty} \left(\frac{1}{1-a}\right)^{n+1} (z - a)^n,$$

and its radius of convergence is indeed $|a - 1|$. This situation is typical. Suppose the radius of convergence of the power series $\sum c_n(z - a)^n$ is finite and positive. If f is the function defined by this power series on its disk of convergence, there is at least one point x of the circle of convergence at which f is singular, that is, x is not in the domain of any analytic extension of f. This implies that, if g is a function analytic on all of $\mathbf{C}$, then each of its power series developments represents it on all of $\mathbf{C}$. The study of analytic continuation is one of the many fascinating chapters in the theory of analytic functions.

EXERCISES

1. By finding its power series explicitly at every point, show that

$$f(z) = \frac{1}{z(z - 1)(z - 2)}$$

defines an analytic function on $\mathbf{C} - \{0, 1, 2\}$.

2. What is the radius of convergence of

$$\sum_{n=0}^{\infty} \frac{1}{n!} z^n?$$

What is the expansion of the corresponding function about the point a?

3. Show by direct calculation with power series that the composition of two analytic functions is analytic.

4. Let $\{f_n\}$ be a sequence of analytic functions, say

$$f_n(z) = \sum_{m=0}^{\infty} c_{n,m} z^m$$

for $|z| < R$. Suppose that $\lim_{n\to\infty} c_{n,m} = e_m$ for each m and $(\forall m, n) |c_{n,m}| < d_m$, where $\sum d_m z^m$ converges for $|z| < R$. Prove that $\{f_n\}$ converges pointwise to an analytic function g for $|z| < R$.

5. Use the result of Exercise 4 to prove that the infinite product

$$\prod_{n=1}^{\infty} \left(1 + \frac{z^2}{n^2}\right)$$

converges on $\mathbf{C}$ to an analytic function.

6. Show that

$$\prod_{n=1}^{\infty} \left(1 - \frac{1}{n^2 z^2}\right)$$

defines an analytic function f with domain $\mathbf{C} - \{0\}$. Note that $f^{-1}(0)$ is not closed in $\mathbf{C}$.

7. Suppose f is an analytic function with domain D. Prove that $f^{-1}(0)$ is the union of certain components of D and a discrete set closed in D.

8. Is there an analytic function defined on a neighborhood of 0 with derivatives given by
(a) $f^{(n)}(0) = (n!)^{3/2}$? (b) $f^{(n)}(0) = n^3 2^n$?
(c) $f^{(n)}(0) = n^{\sqrt{n}}$ for $n > 0$?

9. Is there an analytic function defined on a neighborhood of 0 such that for large n

(a) $f\left(\dfrac{1}{n}\right) = \dfrac{(-1)^n}{n}$?
(b) $f\left(\dfrac{1}{n}\right) = \dfrac{1}{n^{3/2}}$?

10. Suppose that f is defined for $|z| < R$ by $f(z) = \sum_{n=0}^{\infty} c_n z^n$. Prove that

$$f(z) = f(0) + 2 \sum_{n=0}^{\infty} \frac{1}{(2n+1)!} \left(\frac{z}{2}\right)^{2n+1} f^{(2n+1)}\left(\frac{z}{2}\right) \qquad \text{for } |z| < R.$$

11. Suppose that f is an analytic function with domain D and $f(a) = 0$. Show that there is an analytic function g with domain D such that

$$f(z) = (z - a)g(z) \quad \text{for } z \in D.$$

12. For each analytic function f with domain D define a function k_f from D to $\mathbf{N}^* \cup \{\infty\}$ as follows: If the expansion of f at a is $\sum c_n(z - a)^n$, then $k_f(a)$ is the least integer k such that $|c_k| \neq 0$. If there is no such integer, then $k_f(a) = \infty$. Prove that k_f is either 0 or ∞ except on a discrete set.

Let f and g be two analytic functions with the same domain D. Prove: In order that there be an analytic function h such that $f = gh$ it is necessary and sufficient that $k_f(a) \geq k_g(a)$ for all $a \in D$.

15–4. THE EXPONENTIAL AND CIRCULAR FUNCTIONS

In this section we shall define the exponential and circular functions and derive some of their most important properties. Although these functions, or at least their restrictions to real arguments, are familiar from elementary calculus, it should be noted that we define them for complex numbers directly in terms of power series without recourse to any geometrical notions.

15–4.1. Definition. The *exponential* function is defined by

$$\exp z = \sum_{n=0}^{\infty} \frac{1}{n!} z^n. \tag{1}$$

Since the power series in question is convergent for all z (by the ratio test, for example) the function exp is an entire analytic function.

The following properties of the exponential function are immediate consequences of the definition.

$$\exp 0 = 1, \tag{2}$$
$$\exp \bar{z} = \overline{\exp z}, \tag{3}$$
$$\exp' = \exp. \tag{4}$$

15–4.2. Theorem. For any two complex numbers a and b,

$$\exp (a + b) = \exp a \exp b. \tag{5}$$

Proof. Expanding exp at the point a, using 15–3.3, we find

$$\exp z = \sum_{n=0}^{\infty} \frac{\exp^{(n)} a}{n!} (z - a)^n = \sum_{n=0}^{\infty} \frac{\exp a}{n!} (z - a)^n$$

$$= (\exp a) \sum_{n=0}^{\infty} \frac{1}{n!} (z - a)^n = (\exp a) \exp (z - a).$$

Replacing z by $a + b$, we obtain (5).

15–4.3. Corollary. $0 \notin$ range exp.

Proof. By the theorem, $\exp z \exp (-z) = \exp 0 = 1$. Hence $\exp z \neq 0$.

15–4.4. Corollary. For any complex number a and integer n,

$$\exp na = (\exp a)^n.$$

15–4.5. Theorem. The exponential function, restricted to **R**, is an increasing bijection to the set of positive real numbers.

Proof. It is clear from the defining series (1) that $\exp x$ is real if x is real; moreover,

$$\text{if } x > 0, \text{ then } \exp x > 1 + x. \tag{6}$$

Since **R** is connected, while exp is continuous, $\exp \mathbf{R}$ is a connected subset of **R**, that is, an interval. Since this interval contains 1 but not 0, $\exp \mathbf{R} \subseteq (0, \infty)$. Thus $\exp x > 0$ if x is real.

Now, if $x < y$, $\exp (y - x) > 1$ by (6) and

$$\exp x < (\exp x) \exp (y - x) = \exp y.$$

Thus exp is strictly increasing for real arguments. Hence it is injective.

The inequality (6) shows that $\exp x \to +\infty$ as $x \to +\infty$. Since $\exp (-x) = 1/\exp x$, we see that $\exp x \to 0$ as $x \to -\infty$. This shows that the interval $\exp \mathbf{R} \supseteq (0, \infty)$. Therefore, $\exp \mathbf{R} = (0, \infty)$. This concludes the proof.

Next we shall study the exponential function for purely imaginary values. If we substitute ix, where x is real, for z in (1), the terms are alternately real and purely imaginary:

$$\exp ix = \sum_{n=0}^{\infty} \frac{(-1)^n}{(2n)!} x^{2n} + i \sum_{n=0}^{\infty} \frac{(-1)^n}{(2n + 1)!} x^{2n+1}. \tag{7}$$

Each of these power series defines an entire analytic function, and we shall consider these functions separately.

15–4.6. Definitions. The *sine* and *cosine* functions are defined for all $z \in C$ by

$$\sin z = \sum_{n=0}^{\infty} \frac{(-1)^n}{(2n+1)!} z^{2n+1} \tag{8}$$

and

$$\cos z = \sum_{n=0}^{\infty} \frac{(-1)^n}{(2n)!} z^{2n}. \tag{9}$$

Term-by-term differentiation shows that $\sin' = \cos$, and $\cos' = -\sin$. Later we shall need

$$\lim_{n \to \infty} n \sin \frac{z}{n} = z, \tag{10}$$

which follows easily from (8). The following identities are immediate.

$$\sin (-z) = -\sin z,$$
$$\cos (-z) = \cos z,$$
$$\exp iz = \cos z + i \sin z, \tag{11}$$
$$\sin z = \frac{1}{2i} (\exp iz - \exp (-iz)), \tag{12}$$
$$\cos z = \frac{1}{2} (\exp iz + \exp (-iz)). \tag{13}$$

Note that (11) does not assert that $\cos z = \text{Re} (\exp iz)$ for all z. As (7) shows, this is true if z is real, since in that case both $\cos z$ and $\sin z$ are real.

Combining (5) with (11), we find that

$$\exp i(a + b) = (\cos a + i \sin a)(\cos b + i \sin b)$$
$$= \cos a \cos b - \sin a \sin b + i(\sin a \cos b + \cos a \sin b).$$

Similarly,

$$\exp (- i(a + b)) = \cos a \cos b - \sin a \sin b - i(\sin a \cos b + \cos a \sin b).$$

Now (12) and (13) give the familiar addition formulas

$$\sin (a + b) = \sin a \cos b + \cos a \sin b \tag{14}$$

and

$$\cos (a + b) = \cos a \cos b - \sin a \sin b. \tag{15}$$

Another derivation of (15) is instructive. Restricting ourselves temporarily to real a and b, we have

$$\cos (a + b) = \text{Re} (\exp i(a + b))$$
$$= \text{Re} ((\cos a + i \sin a)(\cos b + i \sin b))$$
$$= \cos a \cos b - \sin a \sin b. \tag{16}$$

Let a be a fixed real number and consider the two entire analytic functions $\cos(a + z)$ and $\cos a \cos z - \sin a \sin z$. Equation (16) asserts that these functions agree for all real z. By 15–3.10 they agree for all z. Thus (15) is established for a real and b arbitrary. Now let b be a fixed complex number and consider the two entire functions $\cos(z + b)$ and $\cos z \cos b - \sin z \sin b$. These agree for real z, hence for all z, and (15) is proved.

This method of argument using Theorem 15–3.10 is sometimes known as the principle of the permanence of analytic identities. Roughly speaking, any identity between analytic functions which holds for real arguments, or arguments restricted to any set with a cluster point interior to the domain, holds everywhere.

If we replace b by $-a$ in (15) we obtain

$$\cos^2 a + \sin^2 a = 1. \tag{17}$$

Since both sin and cos take real values for real arguments, this implies the inequalities

$$-1 \leq \sin a \leq 1 \quad \text{and} \quad -1 \leq \cos a \leq 1 \tag{18}$$

for all real a. For other values of a, $\sin a$ and $\cos a$ are (usually) not real and these inequalities are meaningless. With absolute-value signs, they are false; both sin and cos are unbounded functions. In particular, it follows from (12) and (13) that they are unbounded on the imaginary axis.

In the next argument we shall need the following elementary fact.

15–4.7. Lemma. If α and β are complex numbers and $\alpha^2 = \beta^2$, then either $\alpha = \beta$ or $\alpha = -\beta$.

Proof. $(\alpha - \beta)(\alpha + \beta) = \alpha^2 - \beta^2 = 0$. Hence either

$$\alpha - \beta = 0 \quad \text{or} \quad \alpha + \beta = 0.$$

If $0 < x < \sqrt{6}$, the terms in the series for $\sin x$ are alternating in sign and strictly decreasing in absolute value. Therefore, $0 < \sin x < x$ for $0 < x < \sqrt{6}$. In view of (18) we have

$$\sin x < x \quad \text{for all positive } x. \tag{19}$$

On the other hand,

$$\sin \sqrt{10} < \sqrt{10}\left(1 - \frac{10}{3!} + \frac{100}{5!} - \frac{1000}{7!} + \frac{10000}{9!}\right) = -\frac{1520}{9!}\sqrt{10} < 0.$$

Since sin is a continuous function, there is at least one number x between $\sqrt{6}$ and $\sqrt{10}$ for which $\sin x = 0$. The set of such numbers is closed; so there is a least such number, which we shall hereafter denote by π. We have then

$$\sin x > 0 \quad \text{for} \quad 0 < x < \pi \tag{20}$$

and $\sin \pi = 0$.

The alternating series argument applied to the cosine shows that $\cos x < 1$ for $0 < x < \sqrt{12}$. Hence, in particular, $\cos \pi < 1$. Since $\cos^2 \pi = 1$ by (17), we find $\cos \pi = -1$. Therefore, $\exp \pi i = \cos \pi + i \sin \pi = -1$. Also

$$(\exp \tfrac{1}{2}\pi i)^2 = \exp \pi i = -1 = i^2.$$

Therefore $\exp \tfrac{1}{2}\pi i$ is either i or $-i$. Since $\sin \tfrac{1}{2}\pi > 0$, we see that $\exp \tfrac{1}{2}\pi i = i$, $\sin \tfrac{1}{2}\pi = 1$, $\cos \tfrac{1}{2}\pi = 0$.

Now that the values of $\sin \pi$ and $\cos \pi$ are known, (14) and (15) give

$$\sin (a + \pi) = -\sin a \quad \text{and} \quad \cos (a + \pi) = -\cos a. \tag{21}$$

Then (20) shows that

$$\sin x < 0 \quad \text{for} \quad \pi < x < 2\pi. \tag{22}$$

Since it is easily computed from the series that $\sin 3 > 0$, while we already know $3 < 2\pi$, we conclude that $3 < \pi$. This gives us the inequalities $3 < \pi < \sqrt{10}$. Although it is well known that π lies between 3.14159 and 3.14160, and its value has been computed to 100,000 decimal places, we shall make no effort to prove any more precise inequalities for π.

The relations (21) lead immediately to

$$\sin (a + 2\pi) = \sin a \quad \text{and} \quad \cos (a + 2\pi) = \cos a$$

for all $a \in \mathbf{C}$. Therefore, sin and cos are said to be *periodic* and 2π is said to be a *period* of these functions. It follows that $\exp (z + 2\pi i) = \exp z$; that is, $2\pi i$ is a period of exp.

It is obvious that any integral multiple of a period of any function is again a period, and this raises the question of whether $2\pi i$ is itself a multiple of some lesser period of exp. It is not. Not only is every period of exp a multiple of $2\pi i$ but if $\exp (z + \beta) = \exp z$ for even one value of z, then β is a multiple of $2\pi i$.

15–4.8. Theorem. If a is any complex number other than 0, then there are infinitely many complex numbers z such that $\exp z = a$. Among these there is exactly one, z_0, satisfying $-\pi < \operatorname{Im} z_0 \le \pi$. The others are the numbers of the form $z_0 + 2\pi n i$ where n is an integer.

Proof. Suppose first that $\exp z = 1$. Then $\exp \bar{z} = \overline{\exp z} = 1$; so

$$\exp (2\operatorname{Re} z) = \exp (z + \bar{z}) = 1.$$

By 15–4.5, $2 \operatorname{Re} z = 0$. Therefore, z is purely imaginary and we can put $z = iy$, where y is real. There is a unique integer n such that $2\pi n \le y < 2\pi(n + 1)$. Say $y = 2\pi n + u$, where $0 \le u < 2\pi$. Then

$$1 = \exp iy = \exp iu \exp 2\pi n i = \exp iu,$$

whence $\sin u = 0$ and $\cos u = 1$. Comparing with (20), (22), and the fact that $\cos \pi = -1$, we see that $u = 0$. Hence $z = 2\pi n i$. Since $\exp 2\pi n i = 1$ for every integer n, this proves the theorem for the special case $a = 1$.

Now suppose $a \neq 0$. By 15–4.5 there is a real number x such that $\exp x = |a|$. Since $\cos$ is continuous, $\cos 0 = 1$, and $\cos \pi = -1$, there is a real number v such that $0 \leq v \leq \pi$ and $\cos v = \operatorname{Re} a/|a|$. By (20) $\sin v \geq 0$ and $\sin^2 v = 1 - \cos^2 v = (\operatorname{Im} a)^2/|a|^2$. If $\operatorname{Im} a \geq 0$, then $\sin v = \operatorname{Im} a/|a|$; take $z_0 = x + iv$. If $\operatorname{Im} a < 0$, then $\sin v = -\operatorname{Im} a/|a|$ and $v < \pi$; in this case take $z_0 = x - iv$. In either case $\exp z_0 = a$. This proves the existence of the required solution z_0.

If z_1 is another solution, then $\exp (z_1 - z_0) = 1$, hence $z_1 - z_0 = 2\pi n i$ for some integer n. Thus every solution has the form stated. Conversely, every number $z_0 + 2\pi n i$ is a solution. Evidently z_1 does not satisfy $-\pi < \operatorname{Im} z_1 \leq \pi$ unless $n = 0$; so the uniqueness of z_0 follows.

From this result we can deduce some more facts about the circular functions.

15–4.9. Theorem. For every integer n, the function $\cos$ decreases strictly from 1 to -1 in the interval $[2n\pi, (2n + 1)\pi]$ and increases strictly from -1 to 1 in the interval $[(2n - 1)\pi, 2n\pi]$. The function $\sin$ increases strictly from -1 to 1 in the interval $[(2n - \frac{1}{2})\pi, (2n + \frac{1}{2})\pi]$ and decreases strictly from 1 to -1 in the interval $[(2n + \frac{1}{2})\pi, (2n + \frac{3}{2})\pi]$.

Proof. By virtue of the periodicity of $\sin$ and $\cos$, the formulas (21), and $\sin a = \cos (a - \pi/2)$ (from (15)), it is enough to show that $\cos$ decreases strictly from 1 to -1 in the interval $[0, \pi]$. We already know that $\cos 0 = 1$ and $\cos \pi = -1$.

Suppose $0 \leq a < a' \leq \pi$ and $\cos a = \cos a'$. Then using (17) and (20), we see that $\sin a = \sin a'$, whence $\exp ia = \exp ia'$; so $\exp i(a' - a) = 1$. Since $0 < a' - a \leq \pi$, this contradicts Theorem 15–4.8. We conclude that $\cos$ is injective in the interval $[0, \pi]$. Now suppose $0 \leq a < b \leq \pi$ and $\cos a < \cos b$. Since $-1 \leq \cos a$, continuity shows that there is an $a' \in (b, \pi]$ such that $\cos a' = \cos a$. Since $a' > a$, this contradicts our previous result. By elimination, we have $\cos a > \cos b$. Thus $\cos$ is strictly decreasing in $[0, \pi]$.

Finally, we shall prove the existence of integral roots of complex numbers.

15–4.10. Theorem. Let a be a nonzero complex number and let n be a positive integer. There exist exactly n complex numbers z satisfying $z^n = a$. If z_0 is any one of them, the others are $z_0 \exp (2\pi k i/n)$ for $k = 1, 2, \ldots, n - 1$.

Proof. Since $a \neq 0$, we can write $a = \exp \beta$ for some complex number β. Then $\exp (\beta/n)$ obviously satisfies $z^n = a$.

Say that z_0 and z_1 are two numbers with $z_0^n = z_1^n = a$. Neither is zero; hence we can find numbers γ and δ such that $z_0 = \exp \gamma$ and $z_1 = \exp \delta$. Then $\exp n\gamma = \exp n\delta$. By Theorem 15–4.8, there is an integer m such that $n\delta = n\gamma + 2\pi m i$.

We can write $m = qn + k$ where $q \in I$ and $0 \le k \le n - 1$. Then

$$\delta = \gamma + \frac{2\pi ki}{n} + 2\pi qi$$

and

$$z_1 = \exp \delta = \exp\left(\gamma + \frac{2\pi ki}{n}\right) = z_0 \exp \frac{2\pi ki}{n}.$$

This shows that every solution of $z^n = a$ has the stated form. On the other hand, all such numbers are indeed solutions.

EXERCISES

1. Show that $\sin \pi/6 = \frac{1}{2}$.

2. Show that every period of sin or cos is an integral multiple of 2π.

3. Under what conditions does $\cos w = \cos z$?

4. Define tan by $\tan z = \sin z/\cos z$. Find the domain and periods of tan. Show that tan determines a strictly increasing bijection from $(-\pi/2, \pi/2)$ to $\mathbf{R}$. Prove the inequality $\tan x > 2 \tan \frac{1}{2}x$ for $0 < x < \pi/2$ and use this to prove $\tan x > x$ in the same interval.

5. Define the other trigonometric functions and find their domains and periods.

6. Show that

$$f(z) = \frac{1}{z} + 2z \sum_{n=1}^{\infty} \frac{1}{z^2 - n^2}$$

defines a function analytic on $\mathbf{C} - \mathbf{I}$ which has period 1. (Actually this function is $\pi \cot \pi z$.)

7. Suppose f is a continuous function from $\mathbf{C}$ to a metric space. Let P be the set of periods of f. Show that P is a closed subset of $\mathbf{C}$ and that it has one of the following six forms:

(a) $P = \{0\}$.
(b) $P = \{n\alpha \mid n \in \mathbf{I}\}$, where $\alpha \ne 0$.
(c) $P = \{\lambda\alpha \mid \lambda \in \mathbf{R}\}$, where $\alpha \ne 0$.
(d) $P = \{m\alpha + n\beta \mid m, n \in \mathbf{I}\}$, where $\alpha \ne 0, \beta \ne 0$, and $\alpha/\beta \notin \mathbf{R}$.
(e) $P = \{\lambda\alpha + n\beta \mid \lambda \in \mathbf{R}, n \in \mathbf{I}\}$, where $\alpha \ne 0, \beta \ne 0$, and $\alpha/\beta \notin \mathbf{R}$.
(f) $P = \mathbf{C}$.

Show that the cases (c) and (e) cannot occur if f is analytic. Case (d) is also impossible if f is analytic, but possible if we allow the domain of f to omit a discrete set of points. Show that

$$f(z) = \sum_{m,n} \frac{1}{(z - m\alpha - n\beta)^3},$$

where the sum is over $\mathbf{I} \times \mathbf{I}$, defines an analytic function with domain $\mathbf{C} - \{m\alpha + n\beta\}$, provided $\alpha \ne 0, \beta \ne 0$, and $\alpha/\beta \notin \mathbf{R}$. Show that this function has both α and β as periods. Such a function is called *doubly periodic*.

8. Consider the analytic function defined for $|z| < 1$ by

$$f(z) = \sum_{n=0}^{\infty} z^{2^n}.$$

Prove that f is unbounded on the ray $\{\lambda \exp(2\pi ki/2^n) \mid 0 \leq \lambda < 1\}$ for any $n, k \in \mathbf{N}$. Deduce that f cannot be extended to be analytic (or even continuous) on any connected set containing the unit disk properly. Therefore, the unit circle is said to be the *natural boundary* of f.

15-5. THE MODULUS PRINCIPLE

One of the most important properties of analytic functions is that, constants aside, they are open as functions (as defined in 14–4.14). This is related to a theorem known as the modulus principle. The modulus (that is, the absolute value) of a nonconstant analytic function with connected domain never achieves its supremum, and achieves its infimum only when it is zero. An easy corollary is the fundamental theorem of algebra.

15–5.1. Definition. Let φ be a function from a topological space S to $\mathbf{R}$. A point $a \in S$ is called a *local weak maximum point* of φ if and only if there is a neighborhood U of a such that, for all $s \in U$,

$$\varphi(a) \geq \varphi(s).$$

It is a *weak maximum point* of φ if the choice $U = S$ will do. A point $b \in S$ is called a *local weak minimum point* if and only if there is a neighborhood V of b such that $\varphi(a) \leq \varphi(s)$ for all $s \in V$. It is a *weak minimum point* of φ if we can take $V = S$.

15–5.2. Theorem. Let f be an analytic function from D to $\mathbf{C}$. Let $a \in D$ and suppose that f is not constant in a neighborhood of a. Then a is not a local weak maximum point of $z \to |f(z)|$. If a is a local weak minimum point of $z \to |f(z)|$, then $f(a) = 0$.

This theorem is known as the *modulus principle*.

Proof. Suppose for a moment that $f(a) = 0$. In this case we must prove only that a is not a local weak maximum point of $z \to |f(z)|$. If it were, we could choose a neighborhood U of a such that $|f(z)| \leq |f(a)| = 0$ for all $z \in U$. Then f would be constant on U, contrary to the hypothesis.

We suppose from now on that $f(a) \neq 0$. Now we must prove that a is neither a local maximum nor a local minimum point for $z \to |f(z)|$. There is no loss of generality and it simplifies the notation somewhat if we assume $f(a) = 1$. (If this is not so, we can instead consider the function $g = f/f(a)$.)

Suppose f is given by

$$f(z) = \sum_{n=0}^{\infty} c_n(z - a)^n \qquad \text{for} \quad |z - a| < \delta. \tag{1}$$

Here $c_0 = f(a) = 1$ by our special assumption. Not all of the coefficients $c_1, c_2, \ldots$ can be zero, for then f would be constant near a. Let k be the least positive integer such that $c_k \neq 0$. Choose a real number θ so that

$$c_k = |c_k| \exp(-i\theta).$$

Since the power series

$$\sum_{n=1}^{\infty} c_{k+n}(z - a)^n$$

has the same radius of convergence as the power series in (1), it defines an analytic function, *a fortiori* a continuous function, which vanishes at a. Hence there is a positive number ϵ such that

$$|z - a| < \epsilon \implies \left| \sum_{n=1}^{\infty} c_{k+n}(z - a)^n \right| < |c_k|.$$

Suppose λ is a positive number such that $\lambda < \delta$, $\lambda < \epsilon$, and $|c_k|\lambda^k < 1$. Put $z_1 = a + \lambda \exp(i\theta/k)$ and $z_2 = a + \lambda \exp(i(\theta + \pi)/k)$. Then

$$|f(z_1)| = \left| 1 + c_k\lambda^k \exp i\theta + \lambda^k \exp i\theta \sum_{n=1}^{\infty} c_{k+n}(z_1 - a)^n \right|$$

$$\geq \left| 1 + |c_k|\lambda^k \right| - \lambda^k \left| \sum_{n=1}^{\infty} c_{k+n}(z_1 - a)^n \right|$$

$$> 1 + |c_k|\lambda^k - \lambda^k|c_k| = 1$$

and

$$|f(z_2)| = \left| 1 + c_k\lambda^k \exp i(\theta + \pi) + \lambda^k \exp i(\theta + \pi) \sum_{n=1}^{\infty} c_{k+n}(z_2 - a)^n \right|$$

$$\leq \left| 1 - |c_k|\lambda^k \right| + \lambda^k \left| \sum_{n=1}^{\infty} c_{k+n}(z_2 - a)^n \right|$$

$$< 1 - |c_k|\lambda^k + \lambda^k|c_k| = 1.$$

Thus, by choosing λ small enough, we can find within any neighborhood of a a point z_1 with $|f(z_1)| > |f(a)|$ and a point z_2 with $|f(z_2)| < |f(a)|$. Hence a is neither a local weak maximum point nor a local weak minimum point for $z \to |f(z)|$.

15-5.3. Corollary. Let f be a nonconstant analytic function with connected domain D. There is no point $a \in D$ for which $|f(a)| = \sup \{|f(z)| \mid z \in D\}$.

Proof. Since f is nonconstant and has connected domain, it is not locally constant at any point. The theorem therefore asserts that it has no local weak maximum points. *A fortiori*, it has no weak maximum point.

15–5.4. Corollary. Let E be a non void compact subset of **C**. Suppose f is a continuous function from E to **C** which is analytic from Int E to **C**. There is a point a of the boundary of E such that

$$|f(a)| = \sup \{|f(z)| \mid z \in E\}.$$

Proof. Since f is continuous and E is compact, there is a point b such that

$$|f(b)| = \sup \{|f(z)| \mid z \in E\}.$$

If b is in the boundary of E, take $a = b$ and we are through. If $b \in$ Int E, let D be the component of Int E containing b. Since **C** is locally connected, D is an open set, and f restricted to D is an analytic function. Since f has a weak local maximum point at b, f must be constant on D by the previous corollary. Now, **C** is connected and $D \neq$ **C**. Therefore, D has a boundary point; let a be one. Now $a \in \bar{D} \subseteq E$, but $a \notin$ Int E because D is closed relative to Int E. Therefore, a is a boundary point of E. Since f is continuous and constant on D, it is also constant on $\bar{D}$. Thus $f(a) = f(b)$ and $|f(a)| = \sup \{|f(z)|\}$ as required.

This form of the maximum modulus principle is often stated: The maximum modulus of an analytic function is attained on the boundary.

15–5.5. Theorem. Let f be an analytic function which is nowhere locally constant. Then f is an open function.

Proof. Let G be an open subset of domain f. We must show that $f(G)$ is an open subset of **C**. Suppose $b \in f(G)$.

Choose $a \in G$ so that $b = f(a)$. Since f is not locally constant, a is an isolated point of $f^{-1}(b)$. Choose a positive number ϵ so small that the closed disk D of radius ϵ about a lies in G and contains no point of $f^{-1}(b)$ other than a itself. Let Γ be the boundary of D, that is, the circle of radius ϵ about a. Since $f(\Gamma)$ is a compact set in **C** which does not contain b,

$$\rho = \inf \{|f(z) - b| \mid z \in \Gamma\} > 0.$$

We shall prove that $f(D)$, and therefore $f(G)$, contains the whole disk

$$E = \{\lambda \mid |\lambda - b| < \tfrac{1}{2}\rho\}.$$

Suppose $\mu \in E$, that is, $|\mu - b| < \rho/2$. Since D is compact, there is a point $z_0 \in D$ such that

$$|f(z_0) - \mu| = \inf \{|f(z) - \mu| \mid z \in D\}.$$

Now, $|f(z_0) - \mu| \leq |f(a) - \mu| = |b - \mu| < \rho/2$. On the other hand, if $z \in \Gamma$, $|f(z) - \mu| \geq |f(z) - b| - |\mu - b| > \rho - \rho/2 = \rho/2$. Comparing, we see that $z_0 \notin \Gamma$. Thus z_0 is an interior point of D. Therefore z_0 is a local weak minimum point for the function $z \to |f(z) - \mu|$. Since $f - \mu$ is analytic and not locally constant at z_0, we conclude that $f(z_0) - \mu = 0$. Hence $\mu \in f(D)$. This proves that $E \subseteq f(D)$, as claimed.

Therefore $f(G)$ is a neighborhood of b. Since b was any point of $f(G)$, $f(G)$ is open.

15–5.6. Corollary. The inverse of an injective analytic function is continuous.

Proof. The inverse of any open, injective function is continuous.

15–5.7. Corollary. An injective analytic function is a homeomorphism from its domain to an open subset of **C**.

15–5.8. The fundamental theorem of algebra. If f is a polynomial function of positive degree (that is, $f(z) = \sum_{n=0}^{k} c_n z^n$ for $z \in \mathbf{C}$, where $c_k \neq 0$ and $k > 0$), there is a complex number z_0 such that $f(z_0) = 0$.

Proof. We shall show that the modulus of f has a minimum point. To simplify the notation we shall assume $c_k = 1$. (If not, consider f/c_k instead.) The theorem is trivial if $k = 1$; so we shall also assume that $k > 1$.

Let $R = 1 + \sum_{n=0}^{k-1} |c_n|$. Since the closed disk of radius R about the origin is compact, there is a point z_0 such that $|z_0| \leq R$ and

$$|f(z_0)| = \inf \{|f(z)| \mid |z| \leq R\}.$$

If $|z| = R$, we have

$$|f(z)| \geq |z^k| - \sum_{n=0}^{k-1} |c_n| \, |z^n| \geq R^k - \sum_{n=0}^{k-1} |c_n| R^{k-1} = R^{k-1}.$$

But $|f(z_0)| \leq |f(0)| = |c_0| < R \leq R^{k-1}$. Therefore, $|z_0| < R$. Hence z_0 is a local weak minimum point for $z \to |f(z)|$. By Theorem 15–5.2, $f(z_0) = 0$.

EXERCISES

1. Let D be a bounded open set in **C**. Suppose that f is a continuous function from $\bar{D}$ to **C** such that f restricted to D is analytic. Given that $|f(z)| = 1$ for all $z \in \bar{D} - D$ and that f does not vanish on D, prove that f is constant on each component of D.

2. Suppose f is an analytic function with bounded domain D. Suppose M is a number with the following property: For every point p of $\bar{D} - D$ and every positive ϵ, there is a neighborhood U of p such that $|f(z)| < M + \epsilon$ for all $z \in U \cap D$. Prove that $|f(z)| \leq M$ for all $z \in D$.

3. Suppose f_1 and f_2 are both analytic functions with the same connected domain. If $z \to |f_1(z)| + |f_2(z)|$ has a weak local maximum point, show that both f_1 and f_2 are constant. Generalize your result.

15–6. THE LOGARITHM

If x is a positive real number, the logarithm of x is usually defined as the unique real number α such that $\exp \alpha = x$. Since $\exp$ is a strictly increasing bijection from **R** to $(0, \infty)$, its inverse, which we shall temporarily denote by L, is a strictly increasing bijection from $(0, \infty)$ to **R**.

If z is a nonzero complex number, we should like to define $\log z$ as the complex number α such that $\exp \alpha = z$. Unfortunately there are many such complex numbers. According to Theorem 15–4.8 all these numbers have the same real part, namely $L(|z|)$, but their imaginary parts differ by multiples of 2π. One way to make a unique choice is to insist that $-\pi < \text{Im} (\log z) \leq \pi$. This makes $\log$ a discontinuous function; for $\log (-1) = \pi i$, but $\text{Im} \left(\log (-1 - \epsilon i)\right) < 0$ for any small positive ϵ. To avoid these discontinuities we shall define $\log$ only on the complement of the negative reals.

Let T be the strip $\{z \mid -\pi < \text{Im } z < \pi\}$. By Theorem 15–4.8, $\exp$ determines a bijection from T to the slit domain $S = \mathbf{C} - \{x \in \mathbf{R} \mid x \leq 0\}$. We shall denote the inverse of this bijection by $\log$. Since $\log$ takes real values on the positive real numbers, $\log$ is an extension of L. By 15–5.6, $\log$ is a continuous function. After a brief digression we shall prove that it is analytic.

For real numbers the exponential addition formula, $\exp a \exp b = \exp (a + b)$, leads immediately to the equation $L(xy) = L(x) + L(y)$. The corresponding formula for $\log$ is not quite so simple. It can happen that $w \in S$ and $z \in S$ but $wz \notin S$, in which case $\log wz$ is not defined. Even if $wz \in S$, $\log wz$ may not be $\log w + \log z$.

Assuming $w, z \in S$, we always have

$$wz = \exp (\log w) \exp (\log z) = \exp (\log w + \log z).$$

We can conclude that

$$\log wz = \log w + \log z \tag{1}$$

provided $\log w + \log z \in T$ or, what is the same thing,

$$-\pi < \text{Im} \log w + \text{Im} \log z < \pi.$$

On the other hand, if $\text{Im} \log w + \text{Im} \log z > \pi$,

$$\log wz = \log w + \log z - 2\pi i,$$

while if $\text{Im} \log w + \text{Im} \log z < -\pi$,

$$\log wz = \log w + \log z + 2\pi i.$$

Often these three equations are summarized by saying that (1) holds "up to multiples of $2\pi i$" or "modulo $2\pi i$."

Now we shall prove that $\log$ is an analytic function. Our proof depends on special facts about $\log$ which enable us to write down its power series explicitly, but the result is much more general. Any continuous function inverse to an analytic function is itself analytic. The easiest proof uses the result of Exercise 1, p. 292, and the fact that differentiability implies analyticity.

15–6.1. Theorem. Suppose f is a continuous function from an open subset G of **C** to **C** such that $\exp \circ f$ is the identity on G. Then f is an analytic function.

Proof. Let a be any point of G and put $b = f(a)$. Let $D = \{z \mid |z - a| < |a|\}$ and define

$$\varphi(z) = b + \sum_{n=1}^{\infty} \frac{(-1)^{n-1}}{na^n} (z - a)^n$$

for $z \in D$. This function φ is analytic. We shall prove that f agrees with φ near a. This will show that f is representable by a power series near a. Since a was chosen arbitrarily in G, this will prove the theorem.

Differentiating the series for φ, we find

$$\varphi'(z) = \sum_{n=1}^{\infty} \frac{(-1)^{n-1}}{a^n} (z - a)^{n-1} = \frac{1}{z},$$

the series being geometric.

Since exp is a continuous function, $\exp^{-1}(D)$ is an open set and every component of $\exp^{-1}(D)$ is open. Let E be the component of $\exp^{-1}(D)$ containing b.

Consider the function ψ defined by

$$\psi(w) = \varphi(\exp w) - w$$

for $w \in \exp^{-1}(D)$. According to the chain rule (15–1.7) ψ is differentiable and its derivative is given by

$$\psi'(w) = \varphi'(\exp w) \exp w - 1 = 0.$$

Therefore ψ is constant on each component of $\exp^{-1}(D)$ by 15–1.12. On E this constant is $\psi(b) = \varphi(\exp b) - b = \varphi(a) - b = 0$. Hence $\varphi(\exp w) = w$ for all $w \in E$. Since f is continuous, there is a neighborhood N of a such that $f(N) \subseteq E$. For $z \in N, f(z) = \varphi(\exp f(z)) = \varphi(z)$ because $\exp \circ f$ is the identity. Thus f and φ agree on a neighborhood of a. As we remarked, this proves the theorem.

15–6.2. Corollary. The function log is analytic.

The logarithm provides us with an interesting example of analytic continuation. Suppose we have found, through the usual arguments with Taylor's series, say, that the real logarithm L can be represented by a power series

$$L(x) = \sum_{n=1}^{\infty} \frac{(-1)^{n-1}}{n} (x - 1)^n \qquad \text{for} \quad 0 < x \le 2.$$

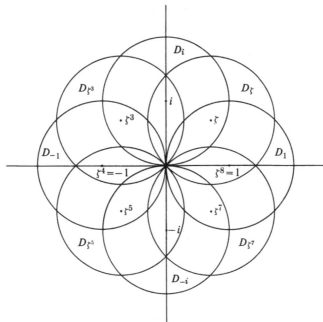

FIGURE 15-4

We want to continue L into the complex plane. Assume we know nothing of the exponential function.

Obviously we begin with the analytic function f_0 defined by

$$f_0(z) = \sum_{n=1}^{\infty} \frac{(-1)^{n-1}}{n} (z-1)^n$$

for $z \in D_1 = \{\lambda \mid |\lambda - 1| < 1\}$. (See Fig. 15-4.)

It is convenient to define D_a, for any complex number $a \neq 0$, to be the open disk with center a and radius $|a|$.

We can expand the function f_0 at any point of D_1. To be definite let us expand it at $\zeta = \frac{1}{2}\sqrt{2} + \frac{1}{2}\sqrt{2}i$. (Then $\zeta^2 = i$, $\zeta^8 = 1$.) The algebra may give us difficulty, but by differentiating with respect to η we find the identity

$$\sum_{n=1}^{\infty} \frac{(-1)^{n-1}}{n} (\xi + \eta)^n = \sum_{n=1}^{\infty} \frac{(-1)^{n-1}}{n} \xi^n + \sum_{n=1}^{\infty} \frac{(-1)^{n-1}}{n} \left(\frac{\eta}{1 + \xi}\right)^n ,$$

provided $|\xi| + |\eta| < 1$. Substituting $\xi = \zeta - 1$ and $\eta = z - \zeta$, we have

$$f_0(z) = f_0(\zeta) + \sum_{n=1}^{\infty} \frac{(-1)^{n-1}}{n\zeta^n} (z - \zeta)^n$$

for $|z - \zeta| < 1 - |\zeta - 1|$. The power series appearing here actually converges for $z \in D_{\zeta}$ to $f_0(z/\zeta)$, as we see directly from the power series for f_0. Hence if we define f_1 on D_{ζ} by

$$f_1(z) = f_0(\zeta) + f_0\left(\frac{z}{\zeta}\right), \qquad (2)$$

then f_0 and f_1 agree on a neighborhood of ζ. Since $D_{\zeta} \cap D_1$ is connected, f_0 and f_1 agree on all of $D_{\zeta} \cap D_1$. This means that

$$f_0(z) = f_0(\zeta) + f_0\left(\frac{z}{\zeta}\right) \qquad (3)$$

for $z \in D_{\zeta} \cap D_1$.

Since $i \in D_{\zeta}$, we can expand f_1 at i. The algebraic problem of finding the new expansion is the same as before thanks to (2), and we can cut it short. Replacing z by z/ζ in (3), we obtain

$$f_0\left(\frac{z}{\zeta}\right) = f_0(\zeta) + f_0\left(\frac{z}{i}\right)$$

for $z/\zeta \in D_{\zeta} \cap D_1$. This condition is equivalent to $z \in D_i \cap D_{\zeta}$. Therefore, adding $f_0(\zeta)$, we get

$$f_1(z) = 2f_0(\zeta) + f_0\left(\frac{z}{i}\right)$$

for $z \in D_i \cap D_{\zeta}$. Now define f_2 on D_i by

$$f_2(z) = 2f_0(\zeta) + f_0\left(\frac{z}{i}\right).$$

Then f_2 is a continuation of f_1. We know that f_0 and f_2 agree on $D_1 \cap D_i$ because both agree with f_1.

We can continue this process by defining

$$f_k(z) = kf_0(\zeta) + f_0\left(\frac{z}{\zeta^k}\right)$$

for $z \in D_{\zeta^k}$. Each of these functions is a continuation of its predecessor.

Because $\zeta^8 = 1$, f_8 and f_0 have the same domain D_1. But $f_8 \neq f_0$. In fact $f_8 = 8f_0(\zeta) + f_0$, and we can easily check from the series for f_0 that $f_0(\zeta) \neq 0$. Thus the continuation has led us back to the starting point, but the continued function is different from the one we started with. This situation is common. What is special about this example is that the continued function f_8 differs from f_0 only by a constant.

Now let us interpret what we have done with the aid of the function log. For each integer n, let $\log_n$ denote the inverse of the function exp restricted to the strip

$$T_n = \{z \mid (n - 1)\pi < \operatorname{Im} z < (n + 1)\pi\}.$$

(The n appearing in $\log_n$ is a subscript only; it has nothing to do with a change of base.) The domain of $\log_n$ is S if n is even and $S^* = \mathbf{C} - \{x \geq 0\}$ if n is odd. Either by Theorem 15–6.1 or from the formula $\log_n z = \log{(-1)^n z} + n\pi i$ we see that $\log_n$ is an analytic function. It follows immediately from the definitions that $\log_{2n}$ and $\log_{2n+1}$ agree on the upper half-plane while $\log_{2n}$ and $\log_{2n-1}$ agree on the lower half-plane.

Because we started with the power series for log at 1, f_0 agrees with $\log = \log_0$ near 1. Since $D_1 \subseteq S$ and D_1 is connected, f_0 agrees with $\log_0$ on all of D_1. Therefore f_1 agrees with $\log_0$ near ζ and, since D_ζ is a connected subset of S, f_1 agrees with $\log_0$ on all of D_ζ. By the same argument, f_2 agrees with $\log_0$ on all of D_i. When we get to f_3, however, the situation is different. Near ζ^3, f_3 agrees with f_2 and therefore with $\log_0$. But $D_{\zeta^3} \cap S$ is not connected; so we can assert only that f_3 agrees with $\log_0$ on that component of $D_{\zeta^3} \cap S$ which contains ζ^3. On the other hand, f_3 agrees with $\log_1$ throughout D_{ζ^3} because $D_{\zeta^3} \subseteq S^* = $ domain $\log_1$. Continuing on around, we find that f_4, f_5, and f_6 represent $\log_1$ on their respective domains, while f_7 represents $\log_1$ only on the component of $D_{\zeta^7} \cap S^*$ which contains ζ^7. But f_6, f_7, f_8, f_9, and f_{10} all represent $\log_2$ on their respective domains.

Since $\log_2 z = 2\pi i + \log z$, we must have $f_8(z) = 2\pi i + f_0(z)$, whence

$$f_0(\zeta) = \tfrac{1}{4}\pi i.$$

The behavior of analytic functions under continuation leads to the consideration of a special kind of relation from $\mathbf{C}$ to $\mathbf{C}$ (i.e., a subset of $\mathbf{C} \times \mathbf{C}$) suggestively called a multiple-valued analytic function. (In spite of the name, a multiple-valued analytic function is not (usually) a function.) A *multiple-valued analytic function* is a relation that can be represented as the union of analytic functions with connected domains each of which can be obtained from any other by successive continuation. If Φ is a multiple-valued analytic function, any analytic function φ with connected domain such that $\varphi \subseteq \Phi$ is known as a *branch* of Φ.

The logarithm is the inverse of the exponential function. Since exp is not injective, the logarithm is not a function. However, the logarithm is a multiple-valued analytic function since it can be represented as the union of the analytic functions, $\{\log_n \mid n \in \mathbf{I}\}$. Each of the functions $\log_n$ is a branch of the logarithm. The one we started with, log, is called the *principal branch* of the logarithm.

Given any analytic function, we can consider all possible ways of continuing it analytically. The union of all the functions arising will be a multiple-valued analytic function (it may turn out to be single-valued) which cannot be further extended; that is, it is not included in any other multiple-valued analytic function. Such a multiple-valued analytic function is called a *complete* analytic function. Every power series with a positive radius of convergence determines a complete analytic function by continuation. The logarithm is easily seen to be a complete analytic function. All entire functions and all rational functions are complete analytic functions which happen to be single valued. So is the function defined by $\sum z^{2^n}$, which cannot be continued at all beyond its original domain. (See Exercise 8. p. 315.)

Temporarily allowing Log to stand for the multiple-valued logarithm, we can put a new interpretation on the addition formula

$$\text{Log } wz = \text{Log } w + \text{Log } z. \tag{4}$$

Here Log w is not determined since Log is not a function. There are infinitely many meanings for Log w. By Log w we mean any one of the numbers α such that $\langle w, \alpha \rangle \in$ Log; that is, any α such that exp $\alpha = w$. The same is true of the other two terms. If we assign meaning arbitrarily to two of the terms, then (4) determines a value for the third term which will be one of its possible meanings. For example, if $w = i$ and $z = -1$, we might take Log w as $\frac{13}{2}\pi i$ and Log z as $-3\pi i$. Then Log $w + $ Log $z = \frac{7}{2}\pi i$, which is one of the possible meanings of Log $wz =$ Log $(-i)$. In this sense (4) is valid for any $w, z \in \mathbf{C} - \{0\}$.

EXERCISES

1. Prove that the logarithm is a complete analytic function.

2. Describe the complete analytic function obtained by continuing the power series

$$\sum_{n=0}^{\infty} \frac{(-1)^n}{n+1}(z-1)^n.$$

3. Suppose Φ is a multiple-valued analytic function such that any two of its branches φ_1 and φ_2 with the same domain differ by some integral multiple of $2\pi i$. Prove that Φ is the logarithm of an analytic function. More precisely, prove that there is an analytic function f with the following property: If φ is a branch of Φ defined at a, there is an integer n such that φ agrees with $\log_n \circ f$ in some neighborhood of a.

15–7. EXPONENTS

We are now in a position to extend the definition of exponents. In Section 10–4 we defined a^x, where $a \in \mathbf{C}$ and $x \in \mathbf{N}^*$, and found the laws of exponents:

$$a^x a^y = a^{x+y}, \tag{1}$$

$$(a^x)^y = a^{xy}, \tag{2}$$

$$(ab)^x = a^x b^x. \tag{3}$$

We then showed that the function could be extended for $a \neq 0$ and $x \in \mathbf{I}$ so that the same laws remain valid. It would be nice to extend the definition further, but unfortunately, without additional restrictions on a, we cannot even extend the domain of x to include $\frac{1}{2}$ and still maintain (1) and (3). For by (1) we must have $(-1)^{1/2}(-1)^{1/2} = (-1)^1 = -1$, and by (3), $(-1)^{1/2}(-1)^{1/2} = 1^{1/2}$; therefore, $1^{1/2}$ must be -1. However, (1) and (3) also imply $1^{1/2} = (1 \cdot 1)^{1/2} = 1^{1/2}1^{1/2} = 1^1 = 1$. We shall restrict a to be real and positive, but allow x to be complex.

Let us sketch briefly the natural method of defining a^x for $a > 0$ and $x \in \mathbf{R}$. For every positive integer n and positive number b there is a unique positive real number y such that $y^n = b$. (See Exercise 2, p. 135.) Denote this number by $\sqrt[n]{b}$.

Because of the uniqueness of roots, we find that $\sqrt[kn]{a^k} = \sqrt[n]{a}$ for any positive integer k. It follows that, for any $m \in I$ and $n \in N, \sqrt[n]{a^m}$ depends only on m/n. Hence we may define a^r for $r \in Q$ by $a^r = \sqrt[n]{a^m}$ for any choice of $m \in I$ and $n \in N$ such that $r = m/n$. We can easily prove that the laws of exponents remain valid for this extension.

Next we prove that the mapping $r \rightarrow a^r$ is strictly increasing from Q to R for each fixed $a > 1$. Moreover, its range is dense in $(0, \infty)$. Then by Theorem 6–5.3 this function can be extended uniquely to be an increasing function from R to $(0, \infty)$. This extended map is $x \rightarrow a^x$. For $a < 1$ put $a^x = 1/(1/a)^x$, and for $a = 1, 1^x = 1$. It is easy to verify that the laws of exponents are still valid.

Superficially it appears that the same construction applies if a is a nonzero complex number, because we can always solve $z^n = b$. But it fails for rational exponents because of the lack of uniqueness.

If we have logarithms at our disposal, the construction of a^x can be shortened to one step. Define

$$a^x = \exp(x \log a)$$

for $a > 0$ and $x \in R$. We can check, first for $x \in N$, then for $x \in Q$, and finally for $x \in R$ that this definition agrees with the natural one.

Now it is obvious, for fixed $a > 0$, how the function $x \rightarrow a^x$ can be extended over C. There can be at most one analytic extension and there is one,

$$z \rightarrow \exp(z \log a).$$

Hereafter, a^z means $\exp(z \log a)$.

The laws of exponents (1) and (3) are valid for this extension, since

$$a^w a^z = \exp(w \log a) \exp(z \log a) = \exp((w + z) \log a) = a^{w+z}$$

and

$$(ab)^z = \exp(z \log ab) = \exp(z(\log a + \log b))$$
$$= \exp(z \log a) \exp(z \log b) = a^z b^z.$$

On the other hand, (2) is not valid. To begin with, $(a^w)^z$ will not even be defined if a^w is not a positive real number. Even if a^w is positive, the formula may fail because $\log a^w$ will not be $w \log a$ unless w is real. However, when w is real

$$(a^w)^z = \exp z(\log a^w) = \exp(zw \log a) = a^{wz}.$$

There is one other case in which $(a^w)^z$ is defined: if $z \in I$. For $n \in N^*, (a^w)^n = a^{wn}$ by induction on n using the first law of exponents. This is true also for negative integers n because $a^{-w} = 1/a^w$.

We could define a^z for any nonzero a as $\exp z\theta$ where θ is some logarithm of a. If a is not negative, the obvious choice would be the principal value $\theta = \log a$. However, one does not use nonintegral exponents with complex numbers without some discussion of precisely what is meant. Even $a^{1/2}$ carries difficulties as we have seen.

It is customary to denote by e the number exp 1, which is easily computed from the original series and found to lie between 2.71828 and 2.71829. From our definitions it is clear that $e^z = \exp z$. The exponential notation is almost always used when the exponent is a simple expression, but the notation exp is preferred in complicated cases like

$$\exp \frac{z^3 + 2z - 1}{z + 1}.$$

EXERCISES

1. Carry out in detail the construction of a^x for positive a and real x by the natural method sketched in the text. Prove also that $\langle a, x \rangle \to a^x$ is a continuous function from $(0, \infty) \times \mathbf{R}$ to $\mathbf{R}$.

2. Let α be any complex number. Show that $(1 + z)^\alpha$, defined as $\exp(\alpha \log(1 + z))$, is analytic and that its expansion at 0 is

$$\sum_{n=0}^{\infty} c_n z^n \quad \text{where} \quad c_0 = 1 \quad \text{and} \quad c_{n+1} = \frac{\alpha - n}{n + 1} c_n.$$

3. Let k be an integer greater than 1. Show that the kth root (the relation inverse to $z \to z^k$ for $z \neq 0$) is a complete analytic function. Let D be an open set in $\mathbf{C} - \{0\}$ which includes the unit circle. Show that there is no branch of the kth root with domain D. Suppose f is a continuous function from a connected open subset G of $\mathbf{C}$ to $\mathbf{C}$ such that $f(z)^k = z$ for all $z \in G$. Prove that f is a branch of the kth root. Show that there are k different branches of the kth root on any nonempty open set which supports a branch of the logarithm.

15-8. GEOMETRIC CONSIDERATIONS

In this section we shall prove some theorems which connect the theory of analytic functions to the geometry of the plane. Theorems of this type might require a considerable excursion into geometry. It depends on how near to the dividing line between analysis and geometry we choose to make the relevant definitions. We shall make the definitions analytical and connect the analysis to the geometry only through pictures and incomplete arguments.

This apparent abandonment of rigor requires some explanation. Geometrical interpretations of complex numbers can give us significant insights into the nature of analytic functions. The fact that the geometrical interpretations are themselves not rigorously derived does not detract in the least from the significance of the insights they may provide. Once insight is acquired it can be carried back over the same unchecked, but apparently reliable, bridges to the firmer ground of pure analysis.

Actually one could set up a rigorous deductive theory of synthetic geometry, make the appropriate definitions, and prove all the geometrical statements we shall need. This, of course, has been done. It is a worthwhile subject to study, but we cannot pursue it here.

When we identify $\mathbf{C}$ with the plane $\mathbf{R} \times \mathbf{R}$ of analytic geometry as usual, the set of complex numbers of absolute value 1 becomes the unit circle. If x is a real number, then exp ix is a point Q on the unit circle (see Fig. 15-5). Taking $P = 1$ and $O = 0$, we would like to prove that the measure of the angle POQ is x.

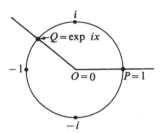

FIGURE 15-5

We begin with the notion of the length of a path in a metric space.

15-8.1. Definition. Let f be a path in a metric space $\langle S, \rho \rangle$ (i.e., a continuous function from $[0, 1]$ to S). For each finite sequence $\{t_k\}$ of real numbers such that $t_0 = 0 < t_1 < t_2 < \cdots < t_n = 1$, compute the sum

$$\sum_{k=1}^{n} \rho(f(t_{k-1}), f(t_k)).$$

If the set of all such sums is bounded, then f is said to be *rectifiable* and the supremum of such sums is called the *length* of the path f. If the set of such sums is unbounded, f is said to be *nonrectifiable*. Sometimes such a path is said to have infinite length.

Note that the length of a path and even its rectifiability depend on the choice of metric for the space S. Thus we are here dealing with a strictly metric property rather than a topological one.

15-8.2. Lemma. If $u, v \in \mathbf{R}$, then

$$|\exp iu - \exp iv| \leq |u - v|.$$

Proof

$$|\exp iu - \exp iv| = \left| \exp i \frac{u+v}{2} \left[\exp i \frac{u-v}{2} - \exp \left(-i \frac{u-v}{2} \right) \right] \right|$$

$$= \left| \exp i \frac{u+v}{2} \right| \left| 2i \sin \frac{u-v}{2} \right|$$

$$= 2 \left| \sin \frac{u-v}{2} \right| \leq |u - v|,$$

the last step by 15-4(19).

15-8.3. Theorem. Let x be a real number. The length of the path $t \to \exp ixt$ is $|x|$.

Proof. Suppose $t_0 = 0 < t_1 < t_2 < \cdots < t_n = 1$. We must consider the sum

$$\sum_{k=1}^{n} \rho(\exp ixt_{k-1}, \exp ixt_k),$$

where ρ is the metric in $\mathbf{C}$, or

$$\sum_{k=1}^{n} |\exp ixt_{k-1} - \exp ixt_k|. \tag{1}$$

The lemma shows that this sum is at most

$$\sum_{k=1}^{n} |xt_{k-1} - xt_k| = |x| \sum_{k=1}^{n} (t_k - t_{k-1}) = |x|.$$

Since this is true for any subdivision, the path is rectifiable and has length at most $|x|$.

Now choose a large integer n and consider the subdivision given by $t_k = k/n$ for $k = 0, 1, \ldots, n$. The sum (1) becomes

$$\sum_{k=1}^{n} \left| \exp ix\frac{k-1}{n} - \exp ix\frac{k}{n} \right| = \sum_{k=1}^{n} 2 \sin \frac{|x|}{2n} = 2n \sin \frac{|x|}{2n}.$$

As $n \to \infty$ the latter approaches $|x|$ (by 15-4(10)). Therefore, the length of the path is at least $|x|$.

To connect this result to the measure of angles, we shall need some discussion of angles. In elementary geometry an angle is usually defined as the union of two rays (half-lines) having a common endpoint but lying in different lines. We want to attach to each angle a positive number, to be called its measure, in such a way that

(i) two angles are congruent if and only if they have the same measure, and

(ii) if a third ray $\overrightarrow{OR}$ (Fig. 15-6) is chosen in the interior of an angle POQ (i.e., $\overrightarrow{OP} \cup \overrightarrow{OQ}$), then

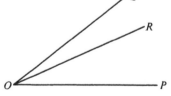

measure POQ = measure POR + measure ROQ.

FIGURE 15-6

It can be shown that such angle-measure functions exist and are unique up to a positive scale factor. Radian measure is an example of such an angle-measure function. It assigns to POQ the ratio, for any circle centered at O, of the length of the arc intercepted by the angle to the radius. We need to know what is meant by the "interior" of an angle and by the "arc intercepted by the angle." The interior of an angle is that one of the two components of its complement which is convex. There are two arcs in the circle having endpoints on $\overrightarrow{OP}$ and $\overrightarrow{OQ}$ and the "arc intercepted" refers to the one meeting the interior of POQ. With this definition the radian measure of an angle is always between 0 and π. By virtue of Theorem 15-8.3 we can say that, if $0 < x < \pi$, the radian measure of POQ is x (where $O = 0$, $P = 1$, and $Q = \exp ix$; see Fig. 15-7). If $0 > x > -\pi$, the radian measure of POQ is $|x|$.

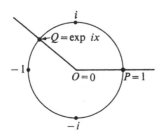

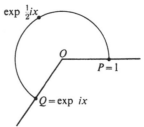

FIGURE 15-7   FIGURE 15-8

If, instead of defining the interior of an angle as the convex complementary component, we allow ourselves to choose either component as the interior, we can have angles exceeding two right angles. Formally, we would define an angle as an ordered pair $\langle \{\overrightarrow{OP}, \overrightarrow{OQ}\}, I \rangle$ consisting of a pair of distinct rays with a common vertex and one of the two components of the complement of $\overrightarrow{OP} \cup \overrightarrow{OQ}$. (Why would it be inappropriate to define an angle as $\langle \overrightarrow{OP} \cup \overrightarrow{OQ}, I \rangle$ or $\overrightarrow{OP} \cup \overrightarrow{OQ} \cup I$?) We can use the same definition of radian measure taking the "arc intercepted" to be the arc meeting I. With this definition the radian measure of an angle is between 0 and 2π. Suppose that $0 < x < 2\pi$, and we name the component containing $\exp \frac{1}{2}ix$ (see Fig. 15-8) as the interior of POQ. Then the radian measure of POQ is again x.

We could now go on to define directed angles and directed angle measure, but even if we did we should be unable to account for angles exceeding four right angles. Directed angles and directed angle measure are intended to describe not a static plane configuration but something more. They describe not only where the sides of an angle are, but also how they got there. We are to conceive of two rays starting in coincidence and one turning with fixed endpoint to a new position. The directed angle measure is to tell us in what direction, and how many times around, the moving ray turned before achieving its final position. This is hard to make precise within the confines of synthetic geometry, so we shall simply jump to a strictly analytic point of view.

A *directed angle* is an ordered pair of rays in $\mathbf{C}$ with a common endpoint. To each directed angle we shall assign a directed angular measure which will be in radians. The direction will be such that $\langle \overrightarrow{01}, \overrightarrow{0i} \rangle$ has measure $+\pi/2$. This direction is usually described as counterclockwise, but this description depends for its validity on the fact that $\mathbf{C}$ is usually represented on paper so that the positive direction appears to be counterclockwise. Our definition of angle is still static.

Although there is now a definite first ray and a definite second ray, there is no description of how the rays got to their relative position. Consequently the measure of an angle will be ambiguous. Every angle will have infinitely many measures differing from one another by multiples of 2π. We begin with the definition of the argument of a nonzero complex number z. This is intended to be the measure of the angle $\langle \overrightarrow{01}, \overrightarrow{0z} \rangle$.

15-8.4. Definition. An *argument* of a nonzero complex number z is the imaginary part of one of the complex numbers α for which $\exp \alpha = z$.

Since any two such complex numbers differ by an integral multiple of $2\pi i$, any two arguments of z differ by an integral multiple of 2π. Although the argument is only a relation, not a function, we shall write, as is traditional, arg z to denote any of the arguments of z. This ambiguous notation has disadvantages; any expression involving arg z must be examined critically to determine exactly what is meant. But there are many situations in which the use of arg simplifies the discussion.

The argument is ambiguous in the same way that the logarithm is ambiguous. In fact, we might write

$$\text{arg } z = \text{Im (Log } z)$$

where Log denotes the multiple-valued logarithm. This equation between ill-defined numbers means that any interpretation of one number is an acceptable interpretation of the other.

We could avoid ambiguity by insisting that arg z should always lie in the interval $(-\pi, \pi]$. Then arg would be a well-defined but discontinuous function. This determination of the argument of z is called the *principal value* of the argument. Except for the negative real numbers this corresponds to the principal branch of the logarithm. Avoiding the discontinuities, we shall write $\text{arg}_0 z = \text{Im (log } z)$. By 15-6.2, arg_0 is a continuous function.

Another way of avoiding ambiguity would be to define arg z as an equivalence class of real numbers. This would give up one of the important properties of the argument, namely, that arg z is a real number which can be combined with other real numbers as usual. When we need an unambiguously defined description of the direction of a complex number z ($\neq 0$), we can use $z/|z|$ and avoid proliferating number systems.

It follows immediately from the definition that

$$z = |z| \exp (i \text{ arg } z) \tag{2}$$

using any determination of arg z. Recalling that the absolute value of a complex number is often called its modulus, we can say that z is determined by its modulus and argument. The reverse is not quite true.

From (2) and a similar equation for w we obtain

$$wz = |w| \, |z| \exp i(\text{arg } w + \text{arg } z),$$

which leads to the equations

$$|wz| = |w|\,|z| \tag{3}$$

and

$$\arg wz = \arg w + \arg z. \tag{4}$$

These equations are often stated: To multiply two complex numbers, multiply their moduli and add their arguments. Formula (3) we have already derived in Chapter 10. Formula (4) connecting indeterminate numbers is like 15–6(4); any determination of two of its terms leads to a correct determination of the third.

Sometimes (4) is written

$$\arg wz = \arg w + \arg z \qquad (\text{modulo } 2\pi),$$

meaning that, whatever the determination of the arguments, this will become an equation if a suitable integral multiple of 2π is added to one side.

If we take polar coordinates in **C**, using the positive real axis as initial ray and measuring angles in the direction of i, the coordinates of z are $|z|$ and $\arg z$. Here the ambiguity of $\arg z$ is the ambiguity of the polar angle with which we must always contend in polar coordinates. When a complex number is written in the form $\rho \exp i\theta$, where $\rho \geq 0$ and θ is real, it is said to be in *polar form*. If $\rho = 0$, the value of θ is irrelevant, of course.

Now let us define the directed angle-measure function μ. Like arg it is really not a function, being determined only modulo 2π. Examination of Fig. 15–9 shows that $\mu(0\alpha, 0\beta)$ ought to be

$$\arg \beta - \arg \alpha = \arg \beta/\alpha.$$

Since angular measure should also be translation invariant, the definition is now determined.

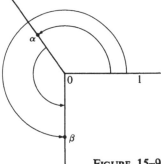

FIGURE 15–9

15-8.5. Definition. The *measure* of the directed angle $\langle \overrightarrow{z_0 z_1}, \overrightarrow{z_0 z_2} \rangle$ is

$$\mu(\overrightarrow{z_0 z_1}, \overrightarrow{z_0 z_2}) = \arg \frac{z_2 - z_0}{z_1 - z_0}.$$

If we think of angles in the most elementary sense as the union of two rays, then the measure of $\overrightarrow{z_0 z_1} \cup \overrightarrow{z_0 z_2}$ is given by the least determination of $|\mu(\overrightarrow{z_0 z_1}, \overrightarrow{z_0 z_1})|$.

Here we should pause to reflect on the famous formula of Euler

$$e^{ix} = \cos x + i \sin x. \tag{5}$$

For us this is almost entirely a matter of definition. For Euler it was not.

To understand the situation we must consider the history of trigonometry. The first trigonometric tables gave the ratio of the chord of a central angle to the radius. The first systematic method of computing tables was given by Ptolemy in the second century. He took the unit angle to be the angle of the equilateral triangle and computed in sexagesimals. Our standard angular unit of 1° is 1/60 of his unit angle, and we still use the smaller sexagesimal subdivisions, minutes (partes minutae) and seconds (partes minutae secundae). His table gives the chord of 1° as $1/60 + 2/60^2 + 50/60^3$, which is correct to three sexagesimal places. The sine (i.e., half the chord of twice the angle) was introduced by Indian mathematicians to simplify the formulas of spherical trigonometry. The sine, like the chord, was a function from the set of geometrical angles (in the sense of our first definition) to the real numbers. The sine is related to sin by the formula

$$\text{sine}\,(\overrightarrow{OP} \cup \overrightarrow{OQ}) = |\sin \mu(\overrightarrow{OP}, \overrightarrow{OQ})|.$$

The tangent first appeared toward the end of the first millenium, A.D., but usable tables were not constructed until the sixteenth century. By the end of the seventeenth century power series expansions had been found for sin, cos, arc sin, and arc tan, but mathematicians continued to think of the arguments of the trigonometric functions as being angles, not numbers. It was Euler who finally codified the theory of the trigonometric functions and showed the advantage of treating them as functions with real arguments, thus setting the stage for many modern developments in analysis.

Originally e^x referred to the natural meaning of the exponent obtained by a limiting process from rational exponents. Newton found the power series expansion for e^x by inverting the series for the logarithm.

Once the series are in hand, the substitution of ix for x leads immediately to (5). At first the formula must have been only a formal curiosity, for the complex numbers were shrouded in mystery and there was no representation of the complex numbers in the geometrical plane. By substituting a complex number where apparently only a real number should go, Euler started the theory of analytic functions of a complex variable.

Thus we can say that the right-hand member of (5) represents the culmination of nearly two thousand years' work in trigonometry, one of the most significant branches of mathematics in the precalculus period because of its applications to astronomy, while the left-hand member marks the beginning of a new and equally significant branch of modern mathematics.

Although the argument function is not well defined, there is a sense in which it is continuous. We shall prove a fact that is intuitively clear. The argument of a complex number can be chosen so that it varies continuously along any path in $\mathbf{C} - \{0\}$.

15–8.6. Theorem. Let f be any path in $\mathbf{C} - \{0\}$. There is a continuous function g from $[0, 1]$ to $\mathbf{R}$ such that, for each t, $g(t)$ is an argument of $f(t)$. Any two such functions differ by a constant integral multiple of 2π. If, for some $t_0 \in [0, 1]$, α is an argument of $f(t_0)$, then there is a unique such function g with $g(t_0) = \alpha$.

Proof. Let $\epsilon = \inf \{|f(t)| \mid 0 \leq t \leq 1\}$. Since f does not take the value 0 and $[0, 1]$ is compact, $\epsilon > 0$. Because f is uniformly continuous, there is a positive integer n such that

$$|t_1 - t_2| < \frac{1}{n} \Rightarrow |f(t_1) - f(t_2)| < \epsilon.$$

The latter inequality implies that $|f(t_1)f(t_2)^{-1} - 1| < 1$, whence $f(t_1)f(t_2)^{-1}$ is not a negative real number and falls in the domain of the continuous function $\arg_0$.

Let β be any argument of $f(0)$. Set

$$h_1(t) = \beta + \arg_0 f(t)f(0)^{-1} \quad \text{for} \quad 0 \leq t \leq \frac{1}{n}$$

and define $h_2, h_3, \ldots, h_n$ recursively by

$$h_{k+1}(t) = h_k\left(\frac{k}{n}\right) + \arg_0 f(t)f\left(\frac{k}{n}\right)^{-1} \quad \text{for} \quad \frac{k}{n} \leq t \leq \frac{k+1}{n}.$$

Each of these functions is continuous. Furthermore, $h_{k+1}(k/n) = h_k(k/n)$ for every k. Hence these functions have a common extension. If g_0 is their least common extension, g_0 is a continuous function from $[0, 1]$ to $\mathbf{R}$. The continuity can easily be established directly, or we can apply the general theorem of Exercise 5, p. 245. Since we can verify by induction on k that $h_k(t)$ is an argument of $f(t)$ for $(k - 1)/n \leq t \leq k/n$, it follows that $g_0(t)$ is an argument of $f(t)$ for $0 \leq t \leq 1$. This proves the existence of one continuous argument function. Having one such function, we can find many others by setting $g_p(t) = 2\pi p + g_0(t)$.

Suppose g and g^* are both continuous functions with the required property. Then for each t, $g(t)$ and $g^*(t)$, being arguments of the same complex number $f(t)$, differ by an integral multiple of 2π. Say $g^*(t) = g(t) + 2\pi q(t)$. Here q is a continuous function from the unit interval to the integers. Since the unit interval is connected, q is constant. Thus g and g^* differ by a constant integral multiple of 2π.

Now suppose α is an argument of $f(t_0)$. Define $g(t) = \alpha - g_0(t_0) + g_0(t)$. Obviously $g(t_0) = \alpha$. Since $\alpha - g_0(t_0)$ is an integral multiple of 2π, $g(t)$ is an argument of $f(t)$ for each t. The uniqueness follows from the result of the last paragraph.

Note that the theorem does not assert that the argument can be chosen continuously on the range of a path. If the path crosses itself (i.e., $f(t_1) = f(t_2)$ for $t_1 \neq t_2$), we may have $g(t_1) \neq g(t_2)$.

Let f be a path in $\mathbf{C} - \{0\}$. Pick any continuous argument function g for f as in the theorem. The number $g(1) - g(0)$ is independent of the choice of g; that is, it depends only on f. It is called the *argument variation* along the path f. Think of the function $t \rightarrow \overrightarrow{Of(t)}$ as a moving ray. It sweeps out an angle as t varies from 0 to 1, sometimes moving forward, sometimes backward. The argument variation along the path f is the measure of this dynamic angle in radians. It is unambiguously determined, and is one of the values of the measure $\mu(\overrightarrow{Of(0)}, \overrightarrow{Of(1)})$ of the corresponding static angle. We leave it to the reader to verify that any other path f^* which determines the same moving ray has the same argument variation as f.

In the special case that f is a closed path $(\text{i.e.}, f(0) = f(1))$, the argument variation along f is an integral multiple of 2π. If it is $2\pi n$, then n is called the *winding number* of the closed path f *about* 0. If f is a path in $\mathbf{C} - \{a\}$, the winding number of f about a is the winding number of $t \rightarrow f(t) - a$ about 0.

We turn now to the local structure of analytic functions. First we show that an analytic function f is a local homeomorphism near any point a of its domain satisfying $f'(a) \neq 0$.

15-8.7. Theorem. Let f be an analytic function defined at a and suppose $f'(a) \neq 0$. There exists a neighborhood U of a on which f is injective.

Proof. To simplify the notation we shall assume $a = 0$. This is no real loss of generality.

Let the power series for f at 0 be $\sum c_n z^n$. Since the power series $\sum_{n=2}^{\infty} n|c_n|z^{n-1}$ has the same radius of convergence as $\sum c_n z^n$, it determines an analytic function which vanishes at 0. Hence there is a positive number ρ such that

$$\sum_{n=2}^{\infty} n|c_n|\rho^{n-1} < \tfrac{1}{2}|f'(0)|.$$

Let U be the disk $\{z \mid |z| < \rho\}$.

Suppose z_1 and z_2 are distinct points of U. For any n,

$$\left|\frac{z_1^n - z_2^n}{z_1 - z_2}\right| = |z_1^{n-1} + z_1^{n-2}z_2 + \cdots + z_2^{n-1}| < n\rho^{n-1},$$

and

$$|f(z_1) - f(z_2) - (z_1 - z_2)f'(0)| = \left|(z_1 - z_2)\sum_{n=2}^{\infty} c_n \frac{z_1^n - z_2^n}{z_1 - z_2}\right|$$

$$\leq |z_1 - z_2|\sum_{n=2}^{\infty} n|c_n|\rho^{n-1} < \tfrac{1}{2}|f'(0)|\,|z_1 - z_2|.$$

Therefore, $|f(z_1) - f(z_2)| > \tfrac{1}{2}|f'(0)|\,|z_1 - z_2|$, whence $f(z_1) \neq f(z_2)$.

15-8.8. Corollary. Suppose f is an analytic function defined at a and $f'(a) \neq 0$. There are open neighborhoods U of a and V of $f(a)$ such that f is a homeomorphism from U to V.

Proof. Take U as in the theorem and let $V = f(U)$. V is open by 15–5.5 and f, restricted to U, is a homeomorphism since it is continuous, open, and injective.

The corollary means that V can be regarded as a distorted copy of U. The topological properties of V and its subsets will be the same as those of U and its subsets, but the geometrical properties may be different. For example, an arc in U will appear as an arc in V, but a straight arc in U will (probably) appear curved in V. We shall investigate the nature of this geometrical distortion.

Consider first the function φ defined by $\varphi(z) = e^{i\theta}z + \beta$, where $\theta \in \mathbf{R}$ and $\beta \in \mathbf{C}$. We shall regard φ as a mapping of the plane into itself. We can represent φ as the composition of two maps $z \to e^{i\theta}z$ and $z \to z + \beta$. The first of these is a rotation of the plane about the origin through an angle θ. (More precisely, through an angle of measure θ. We often ignore this distinction to reduce the complexity of our sentences.) The second is a translation. Both of these transformations preserve distances; hence so does φ. Consequently, φ carries any subset of the plane into a congruent subset.

From elementary geometry we know that such mappings also preserve angles. However, the angles of elementary geometry are not the same as the directed angles we have been considering. In fact, the transformation φ also preserves directed angles. Let $\langle \overrightarrow{z_0z_1}, \overrightarrow{z_0z_2} \rangle$ be a directed angle. Its image is $\langle \overrightarrow{\xi_0\xi_1}, \overrightarrow{\xi_0\xi_2} \rangle$, where

$$\xi_0 = e^{i\theta}z_0 + \beta, \qquad \xi_1 = e^{i\theta}z_1 + \beta, \qquad \text{and} \qquad \xi_2 = e^{i\theta}z_2 + \beta.$$

We have

$$\mu(\overrightarrow{\xi_0\xi_1}, \overrightarrow{\xi_0\xi_2}) = \arg\frac{\xi_2 - \xi_0}{\xi_1 - \xi_0} = \arg\frac{z_2 - z_0}{z_1 - z_0} = \mu(\overrightarrow{z_0z_1}, \overrightarrow{z_0z_2}).$$

Next let ψ be the mapping $z \to \alpha z + \beta$ where $\alpha, \beta \in \mathbf{C}$, $\alpha \neq 0$. We can regard ψ as the composition of the dilation $z \to |\alpha|z$ (a contraction if $|\alpha| < 1$) and the mapping φ (with $\theta = \arg \alpha$). This mapping carries every geometrical figure in the plane into a similar figure and preserves directed angles.

Now suppose f is an analytic function defined at a with $f'(a) \neq 0$. The power series for f begins: $f(a) + f'(a)(z - a) + \cdots$ For values of z near enough to a, the later terms in the power series are negligible; so f is practically indistinguishable from the linear map $z \to f(a) + f'(a)(z - a)$. Hence in a sufficiently small neighborhood of a, f is effectively a similarity transformation. Since an angle is determined in an arbitrarily small neighborhood of its vertex, f must preserve directed angles.

This statement requires a new definition, however, because we have defined angle measure only for angles with straight sides. The idea is to replace a curved line by its tangent, if it has one. We give the definition with somewhat greater generality than usual.

15–8.9. Definition. Let $a \in \mathbf{C}$. A *path away from* a is a continuous function φ from $[0, 1]$ to $\mathbf{C}$ such that $\varphi(0) = a$ but $\varphi(t) \neq a$ for $t > 0$. The path φ away from *a has the direction* ξ if and only if

$$\lim_{t \to 0} \frac{\varphi(t) - a}{|\varphi(t) - a|} = \xi.$$

This is a reasonable definition because we are asking that the ray $\overrightarrow{a\varphi(t)}$ should have a limiting position as $t \to 0$. Although it is not obvious from the definition, the direction of a path φ away from a depends only on range φ, a subset of **C**. Hence it is a geometrical notion. In particular, if the range of φ lies in a ray with a as endpoint, then φ has a direction, namely, the direction of that ray.

15–8.10. Definition. If φ_1 and φ_2 are two paths away from a having directions ξ_1 and ξ_2, respectively, then the directed angular measure from φ_1 to φ_2 is $\mu(\varphi_1, \varphi_2) = \arg \xi_2/\xi_1$.

If range φ_1 and range φ_2 lie on rays $\overrightarrow{az_1}$ and $\overrightarrow{az_2}$, respectively, then $\mu(\varphi_1, \varphi_2) = \mu(\overrightarrow{az_1}, \overrightarrow{az_2})$. Hence this definition is conceptually an extension of the old definition of angle measure. That is why we have chosen to denote it with the same symbol.

15–8.11. Theorem. Let f be an analytic function defined, but not locally constant, at a. Let its power series development at a be

$$f(a) + \sum_{n=k}^{\infty} c_n(z - a)^n,$$

where $c_k \neq 0$. There is a neighborhood U of a with the following property. If φ is a path away from a in U with direction ξ, then $f \circ \varphi$ is a path away from $f(a)$ with direction $c_k \xi^k / |c_k|$.

Proof. We have already shown that a is an isolated point of $f^{-1}(f(a))$ (15–3.8). This means there is a neighborhood U of a such that f does not take the value $f(a)$ in $U - \{a\}$. Now if φ is a path away from a in U, then $f \circ \varphi$ is a path away from $f(a)$.

We must prove that

$$\lim_{t \to 0} \frac{f(\varphi(t)) - f(a)}{|f(\varphi(t)) - f(a)|} = \frac{c_k}{|c_k|} \xi^k,$$

assuming

$$\lim_{t \to 0} \frac{\varphi(t) - a}{|\varphi(t) - a|} = \xi.$$

From the power series we see that

$$f(\varphi(t)) - f(a) = (\varphi(t) - a)^k \sum_{n=k}^{\infty} c_n(\varphi(t) - a)^{n-k}.$$

Since

$$\lim_{t \to 0} \sum_{n=k}^{\infty} c_n (\varphi(t) - a)^{n-k} = c_k \neq 0,$$

the result follows from elementary limit theorems.

15–8.12. Corollary. Under the conditions of the theorem, if φ_1 and φ_2 are paths away from a in U having directions, then

$$\mu(f \circ \varphi_1, f \circ \varphi_2) = k\mu(\varphi_1, \varphi_2).$$

in the following sense: Any interpretation of the right-hand member leads to a correct value for the left-hand member.

Suppose f is an analytic function which is nowhere locally constant on its domain D. Then f' is an analytic function which vanishes at most on a discrete subset E of D. If the point a of the theorem and corollary is in $D - E$, then $k = 1$. This means that the function f preserves angles at such points. A function which preserves angles at a point is said to be *conformal* at that point. Hence we have just proved that a nonconstant analytic function with a connected domain is conformal except on a discrete set. Conversely, a continuous function from an open subset of **C** to **C** which is conformal except on a discrete set is analytic. This is essentially a restatement of the fact that a differentiable function is analytic.

If f' does not vanish anywhere in D, then f is a local homeomorphism at every point. This does not guarantee that f is injective globally, however; hence f need not be a homeomorphism. The function exp is a good example.

This function also provides a simple example of the conformal property. The line Re $z = \lambda$ is carried by exp into the circle of radius exp λ about the origin, while the line Im $z = \mu$ goes into the open-ended ray from the origin having angular position μ. The original lines are perpendicular and, sure enough, the circle and ray cross at right angles. The whole orthogonal network of Cartesian coordinate lines is carried by exp into the orthogonal network of polar coordinate lines.

At points a at which f' vanishes the corollary shows that angles are multiplied by an integer $k > 1$. There isn't room, of course, for angles around $f(a)$ totaling more than 2π; there must be some overlapping. The situation is effectively the same as it would be if we neglected all terms of the power series for f after the first nonconstant term; that is, if we replaced f by

$$z \to f(a) + c_k(z - a)^k.$$

Near a, f is not injective, but k to 1; that is, the inverse image of any point near $f(a)$ contains k points near a. (See Exercise 9.) The interior of each angle exceeding $2\pi/k$ is opened out like a fan to cover a whole neighborhood of $f(a)$. This opening in angle is accompanied by a collapse in radius, since $|f(z) - f(a)| \to 0$ as fast as $|z - a|^k$.

Our results show that any injective analytic function must be conformal at each point of its domain. Such a function is often called a *conformal map*. Conformal maps are useful in the study of Laplace's equation and have received a great deal of attention.

One of the most important theorems in this field is known as Riemann's conformal mapping theorem. If G is a proper open subset of $\mathbf{C}$ and G is homeomorphic to D, the open unit disk, then there is a conformal homeomorphism from G to D. (It is essential that $G \neq \mathbf{C}$.) While we cannot prove this theorem, we can look at a few examples.

Suppose G is the upper half-plane. Then f, defined by

$$f(z) = \frac{z - i}{z + i},$$

maps G conformally onto D.

Suppose H is the first quadrant (i.e., $H = \{z \mid \operatorname{Re} z > 0, \operatorname{Im} z > 0\}$). The mapping $z \to z^2$ carries H conformally onto the upper half-plane. (Note how the "corner" of H is smoothed out.) Since the composition of conformal maps is evidently a conformal map,

$$z \to \frac{z^2 - i}{z^2 + i}$$

carries H conformally onto D.

EXERCISES

1. Define the length of an arc in a metric space. Is it true that the length of a path in a metric space depends only on the range of that path?

2. Suppose that z_1 and z_2 are complex numbers other than 0. Show that

$$\cos \mu(\overrightarrow{0z_1}, \overrightarrow{0z_2}) = \operatorname{Re} \overline{z_1} z_2 / |z_1 z_2|$$

and

$$\sin \mu(\overrightarrow{0z_1}, \overrightarrow{0z_2}) = \operatorname{Im} \overline{z_1} z_2 / |z_1 z_2|.$$

3. Let z_1, z_2, and z_3 be three noncollinear points in $\mathbf{C}$. The directed angles of the triangle $z_1 z_2 z_3$ are $\theta_1 = \mu(\overrightarrow{z_1 z_2}, \overrightarrow{z_1 z_3})$, $\theta_2 = \mu(\overrightarrow{z_2 z_3}, \overrightarrow{z_2 z_1})$, and $\theta_3 = \mu(\overrightarrow{z_3 z_1}, \overrightarrow{z_3 z_2})$, where in each case the determination is made between $-\pi$ and π. Show that $\theta_1 + \theta_2 + \theta_3 = \pm\pi$. Derive the laws of sines and cosines for this triangle.

4. Let f be a path in $\mathbf{C} - \{0\}$. Show that there is a number δ depending only on f such that the argument variation of f is given by $\sum_{k=1}^{n} \arg_0 f(t_k) f(t_{k-1})^{-1}$ for any sequence $t_0 = 0 < t_1 < t_2 < \ldots < t_n = 1$, provided only that $\sup_k (t_k - t_{k-1}) < \delta$.

5. Compute the argument variation along the path f given by

$$f(t) = \cos 2\pi t + it(t - 1)(t - \tfrac{1}{2})^2.$$

6. Give another proof of Corollary 15–8.8 using the result of 14–7.17.

7. Show that every mapping φ of $\mathbf{C}$ into itself which preserves distances has one of the two forms

$$\varphi(z) = \alpha z + \beta \qquad \text{or} \qquad \varphi(z) = \alpha \bar{z} + \beta,$$

where $|\alpha| = 1, \beta \in \mathbf{C}$, but only the first of these types preserves directed angles. (Transformations which reverse directed angles are said to be *anticonformal*.)

8. Show that the direction of a path away from a depends only on its range. That is, prove: If φ_1 and φ_2 are two paths away from a such that range $\varphi_1 = $ range φ_2, and φ_1 has a direction, then φ_2 has the same direction.

9. Suppose that $f(z) = f(a) + \sum_{n=k}^{\infty} c_n(z - a)^n$, where $k > 0$, $c_k \neq 0$. Show that there is neighborhood W of a and an analytic function g with domain W such that $g(a) = 0$, $g'(a) \neq 0$, and $f(z) = f(a) + g(z)^k$ for all $z \in W$.

10. Suppose that g and h are two closed paths in $\mathbf{C} - \{0\}$ such that $g(t)h(t)^{-1}$ is never negative. Prove that g and h have the same winding number about 0.

11. Let φ be a continuous map from $[0, 1] \times [0, 1]$ to $\mathbf{C} - \{0\}$. Show that there is a continuous function ψ from $[0, 1] \times [0, 1]$ to $\mathbf{R}$ such that $\psi(s, t)$ is an argument of $\varphi(s, t)$ for all s, t.

12. A continuous family of paths in a space M is a function $s \to f_s$ from $[0, 1]$ to the set of paths in M such that $\langle s, t \rangle \to f_s(t)$ is continuous. Two closed paths g and h in M are said to be *freely homotopic* in M if and only if there exists a continuous family of closed paths (i.e., f_s is a closed path for each s) such that $f_0 = g$ and $f_1 = h$. Show that "freely homotopic" is an equivalence relation.

Take $M = \mathbf{C} - \{0\}$. If f is a continuous family of paths, show that $s \to$ (argument variation of f_s) is a continuous function. If f is a continuous family of closed paths, show that the winding number of f_s is independent of s. Hence, in order for the closed paths g and h to be freely homotopic, it is necessary that g and h have the same winding number about 0. Show that, conversely, this condition is sufficient.

Suppose F is a polynomial function of degree $n > 0$. Show that for sufficiently large R, the path $g : t \to F(R \exp 2\pi i t)$ has winding number n about 0. Prove the fundamental theorem of algebra by noting that if F did not vanish in $\mathbf{C}$, then g would be freely homotopic in $\mathbf{C} - \{0\}$ to the constant path $h : t \to F(0)$.

ANSWERS AND SOLUTIONS

Page 2

1. The phrases "large integer" and "interesting number" are much too vague. The others are all right.

2. We have no definition of "inhabitant of the United States" complete enough to cover unambiguously all cases which arise in practice. Hence the set of all such inhabitants is not well-defined in a mathematical sense.

Page 5

The sets in question are those sets whose members are drawn from the list $\emptyset$, $\{\emptyset\}$, $\{\{\emptyset\}\}$, $\{\emptyset, \{\emptyset\}\}$. Each of these four sets may or may not be used independently, so there are $2^4 = 16$ possibilities, beginning with $\emptyset$, $\{\emptyset\}$, $\{\{\emptyset\}\}, \ldots$ and ending with $\{\emptyset, \{\emptyset\}, \{\{\emptyset\}\}, \{\emptyset, \{\emptyset\}\}\}$.

Page 14

1. (e), (f), (g), (i), (j), (k), and (l) are true.

2. (b) and (c) are certainly true. None of the others are certainly false.

3. $y \notin B$. If p stands for $x \in A$, q for $y \in B$, and r for $y \in A$, the hypotheses are $(p \text{ and } q) \Rightarrow r$ and $(\text{not } p \text{ or } r) \Rightarrow \text{not } q$. It is probably easiest to examine these statements with truth tables.

	p	q	r	$(p \text{ and } q) \Rightarrow r$	$(\text{not } p \text{ or } r) \Rightarrow \text{not } q$
(1)	T	T	T	T	F
(2)	T	T	F	F	T
(3)	T	F	T	T	T
(4)	T	F	F	T	T
(5)	F	T	T	T	F
(6)	F	T	F	T	F
(7)	F	F	T	T	T
(8)	F	F	F	T	T

Of the eight conceivable cases, (2) is eliminated since we know that $(p \text{ and } q) \Rightarrow r$ is true. Three cases (1), (5), (6) are eliminated since we know that $(\text{not } p \text{ or } r) \Rightarrow \text{not } q$ is true. The remaining four cases are exactly those for which q is false. Hence the conclusion is $y \notin B$.

A formal derivation of this result can be given. The contrapositive form of the second hypothesis is $q \Rightarrow (p \text{ and not } r)$. If we assume q, then p and not-r are consequences. But then the first hypothesis gives r, a contradiction. Therefore q is false. This does not show that $y \notin B$ is the best conclusion we can draw.

For this we note that if we assume only that q is false, then both hypotheses are true. All this means that

$$\text{not } q \Leftrightarrow [((p \text{ and } q) \Rightarrow r) \text{ and } ((\text{not } p \text{ or } r) \Rightarrow \text{not } q)]$$

is a tautology. The truth table argument amounts to a proof of this fact; this method has the tremendous advantage that you need not know in advance what tautology you are trying to prove.

Page 18

1. (a) $(\forall x, y, z) \, (P(x) \text{ and } P(y) \text{ and } P(z)) \Rightarrow (x = y \text{ or } x = z \text{ or } y = z)$
   (b) $(\exists x, y) \, (x \neq y \text{ and } P(x) \text{ and } P(y))$
2. (a) $a \in M$ and $((\exists x) \, \phi(a, x) \text{ and } \phi(x, b))$
   (b) $a \in M$ and $((\exists x) \, \phi(b, x) \text{ and } \phi(x, a))$
   (c) $a \in W$ and $(\exists x, y) \, (\phi(x, b), \phi(y, x), \phi(y, a), \text{ and } x \neq a)$
   (d) $a, b \in W$ and $a \neq b$ and $(\exists x) \, (\phi(x, a) \text{ and } \phi(x, b))$
   (e) $(\exists x, y, z) \, (\phi(x, y), \phi(x, z), y \neq z, \phi(y, a), \text{ and } \phi(z, b))$
   (f) $(\forall y) \, a \neq y \Rightarrow (\text{not-}(\exists x) \, \phi(x, a) \text{ and } \phi(x, y))$
   (g) $a \in M$, $a \neq b$, and $(\exists x \in M)(\exists y \in W) \, (\phi(x, a), \phi(y, a), \phi(x, b), \text{ and } \phi(y, b))$
   (h) $(\forall x)(\exists y \in M) \, \phi(y, x)$
   (i) $(\forall x)(\forall y, z) \, (y \in M, \phi(y, x), z \in M, \text{ and } \phi(z, x)) \Rightarrow y = z$
   (j) $(\forall x)(\exists y, z) \, (y \neq z, \phi(y, x), \phi(z, x), \text{ and } (\forall w) \, \phi(w, x) \Rightarrow (w = y \text{ or } w = z))$
   (k) $(\exists x, a, b) \, (a \neq b, \phi(x, a) \text{ and } \phi(x, b))$
   (l) $(\forall x)(\exists y \in M)(\exists z) \, (\phi(y, z) \text{ and } \phi(z, x))$
   (m) $\text{not-}(\exists x) \, (x \in M \text{ and } (\exists y) \, \phi(x, y) \text{ and } \phi(y, x))$

$\phi(a, b)$ can be expressed as $\psi(a, b)$ and $\text{not-}(\exists x) \, \psi(a, x)$ and $\psi(x, b)$. It is impossible to express $\psi(a, b)$ in terms of $\phi(x, y)$ and quantifiers.

Page 23

The hypothesis is $(\forall x)(\exists y \in M) \, \phi(y, x)$.

(i)	Let a be given.	
(ii)	$(\exists y \in M) \, \phi(y, a)$	by hypothesis;
(iii)	choose b so that $\phi(b, a)$	by (ii);
(iv)	$(\exists y \in M) \, \phi(y, b)$	by hypothesis;
(v)	choose c so that $c \in M$ and $\phi(c, b)$	by (iv);
(vi)	$c \in M$ and $(\exists y) \, \phi(c, y)$ and $\phi(y, a)$	by (iii) and (v);
(vii)	$(\exists z \in M)(\exists y) \, \phi(z, y)$ and $\phi(y, a)$	by (vi);
(viii)	$(\forall x)(\exists z \in M)(\exists y) \, \phi(z, y)$ and $\phi(y, x)$	by (i) through (vii).

Page 25

The relations are all visible in the following diagram:

$$A_3 \subset A_5$$
$$\emptyset = A_8 \subset A_1 \subset A_4 = A_7 \subset A_6 = A_{10} = \mathbf{R}$$
$$A_9 \subset A_2$$

Pages 36–37

1. Use the distributive law twice: $(A \cup B \cup C) \cap (B \cup D) = (A \cap B) \cup (A \cap D) \cup (B \cap B) \cup (B \cap D) \cup (C \cap B) \cup (C \cap D)$. Here all the terms of the form $B \cap X$ are absorbed by $B \cap B = B$, so the result is $(A \cap D) \cup B \cup (C \cap D)$.

2. If we apply the distributive law three times to the left-hand member, we get $(A \cap B \cap C) \cup (A \cap B \cap A) \cup (A \cap C \cap C) \cup (A \cap C \cap A) \cup (B \cap B \cap C) \cup (B \cap B \cap A) \cup (B \cap C \cap C) \cup (B \cap C \cap A)$. Since $A \cap B \cap C$ is absorbed, this reduces to the right-hand member. The dual identity is the same with the members reversed.

3. Let us agree that $A \cup B \cup C$ means $(A \cup B) \cup C$. Then $(A \cup B \cup C)^{\sim} = ((A \cup B) \cup C)^{\sim} = (A \cup B)^{\sim} \cap \tilde{C} = (\tilde{A} \cap \tilde{B}) \cap \tilde{C} = \tilde{A} \cap \tilde{B} \cap \tilde{C}$.

4. $(A - B) \cup C = (A \cap \tilde{B}) \cup C = (A \cup C) \cap (\tilde{B} \cup C) = (A \cup C) \cap (B \cap \tilde{C})^{\sim} = (A \cup C) - (B - C)$.

5. $(A - B) \cup (B - C) \cup (C - A) = (A \cap \tilde{B}) \cup (B \cap \tilde{C}) \cup (C \cap \tilde{A}) = (A \cup B \cup C) \cap (A \cup B \cup \tilde{A}) \cap (A \cup \tilde{C} \cup C) \cap (A \cup \tilde{C} \cup \tilde{A}) \cap (\tilde{B} \cup B \cup C) \cap (\tilde{B} \cup B \cup \tilde{A}) \cap (\tilde{B} \cup \tilde{C} \cup C) \cap (\tilde{B} \cup \tilde{C} \cup \tilde{A})$. Since $A \cup \tilde{A} = B \cup \tilde{B} = C \cup \tilde{C} = U$, the universal set, this is $(A \cup B \cup C) \cap (\tilde{A} \cup \tilde{B} \cup \tilde{C})$. If $A \cap B \cap C = \emptyset$, then $\tilde{A} \cup \tilde{B} \cup \tilde{C} = \tilde{\emptyset} = U$, and the result follows.

6. As shown in the answer to Exercise 5, $(A - B) \cup (B - C) \cup (C - A) = (A \cup B \cup C) \cap (\tilde{A} \cup \tilde{B} \cup \tilde{C})$. By interchanging B and C, we find that this is also $(A - C) \cup (C - B) \cup (B - A)$.

7. $(A \cup (\tilde{B} \cap C))^{\sim} = \tilde{A} \cap (\tilde{B} \cap C)^{\sim} = \tilde{A} \cap (B \cup \tilde{C}) = (\tilde{A} \cap B) \cup (\tilde{A} \cap \tilde{C}) = (B - A) \cup (\tilde{A} - C)$.

8. The first three identities are obvious. $(B \oplus C)^{\sim} = (\tilde{B} \cup C) \cap (B \cup \tilde{C})$. Two applications of the distributive law (8) show that this is $(B \cap C) \cup (\tilde{B} \cap \tilde{C})$. Then $A \oplus (B \oplus C) = (A \cap (B \oplus C)^{\sim}) \cup (\tilde{A} \cap (B \oplus C)) = (A \cap B \cap C) \cup (A \cap \tilde{B} \cap \tilde{C}) \cup (\tilde{A} \cap B \cap \tilde{C}) \cup (\tilde{A} \cap \tilde{B} \cap C)$. Interchanging A and C and using the commutative law, we see that $(A \oplus B) \oplus C = C \oplus (B \oplus A)$ reduces to the same thing. Both sides of the last identity reduce to $(A \cap B \cap \tilde{C}) \cup (A \cap \tilde{B} \cap C)$. The result shows that the subsets of a set form a *ring* with $\oplus$ as the addition and $\cap$ as the multiplication.

Page 37

1. If $X \in \mathfrak{P}(A)$, then $X \subseteq A$. Since $A \subseteq B$, $X \subseteq B$. Thus $X \in \mathfrak{P}(B)$.

2. Use Exercise 1 and induction on n.

3. $2^{2^{2^{\cdot^{\cdot^{\cdot 2}}}}}$, an exponential tower of $(n - 1)$ 2's, for $n > 1$.

4. If there are braces n deep when A is written out in full, then $A \in \mathfrak{P}^{n+1}(\emptyset)$.

Pages 39–40

1. If $\{a, \{a, b\}\} = \{c, \{c, d\}\}$, we could not eliminate the possibility that $a = \{c, d\}$ and $\{a, b\} = c$ on the basis of membership arguments alone. It is true that in this case both $a \in c$ and $c \in a$, and often it is assumed that such circular chains of membership are impossible, but we can prove Theorem 3.3.2 under the other definition without this assumption.

2. If $\langle a, b, c \rangle$ stands for $\{\{a\}, \{a, b\}, \{a, b, c\}\}$, then $\langle x, y, y \rangle = \langle x, x, y \rangle$.

3. If $x \in A \times B$, then $x = \langle a, b \rangle = \{\{a\}, \{a, b\}\}$ for some $a \in A$ and $b \in B$. Obviously, $x \in \mathfrak{P}^2(A \cup B)$. Hence $A \times B \subseteq \mathfrak{P}^2(A \cup B)$.

4. We prove only the third relation. If $\langle x, y \rangle \in (A \times B) \cap (C \times D)$, then $\langle x, y \rangle \in A \times B$, so $x \in A$ and $y \in B$. Similarly, $x \in C$ and $y \in D$. Therefore, $x \in A \cap C$ and $y \in B \cap D$. This shows that $\langle x, y \rangle \in (A \cap C) \times (B \cap D)$. Conversely, if $\langle u, v \rangle \in (A \cap C) \times (B \cap D)$, then $u \in A$, $v \in B$, $u \in C$, and $v \in D$, so $\langle u, v \rangle \in (A \times B) \cap (C \times D)$.

5. If both A and B are not void, choose $a \in A$ and $b \in B$. Then $\langle a, b \rangle \in A \times B$, and the latter is not void. Thus $(A \neq \emptyset \text{ and } B \neq \emptyset) \Rightarrow A \times B \neq \emptyset$. The contrapositive of this is $A \times B = \emptyset \Rightarrow (A = \emptyset \text{ or } B = \emptyset)$. The opposite implication is trivial.

6. This is not valid if $X = \emptyset$, since $\emptyset \times Y = \emptyset = \emptyset \times Z$ for any sets Y and Z. However, if $X \neq \emptyset$ and $X \times Y = X \times Z$, then $Y = Z$. *Proof.* Choose $x \in X$. If $y \in Y$, then $\langle x, y \rangle \in X \times Y$, hence $\langle x, y \rangle \in X \times Z$. Therefore, $y \in Z$. This shows that $Y \subseteq Z$. The opposite inclusion, that $Z \subseteq Y$, is proved in the same way, and therefore $Y = Z$.

Page 49

1. None are injective. f_2 is a restriction of f_4. $f_1 \circ f_2 = \{\langle 1, 1 \rangle, \langle 2, 1 \rangle, \langle 4, 1 \rangle\}$. $f_2 \circ f_3 = \{\langle 0, 1 \rangle, \langle 1, 1 \rangle, \langle 2, 1 \rangle\}$. Both $(f_1 \circ f_2) \circ f_3$ and $f_1 \circ (f_2 \circ f_3)$ are the same as $f_2 \circ f_3$. $\bar{f} = \{\langle \emptyset, \emptyset \rangle, \langle \{0\}, \{2\} \rangle, \langle \{1\}, \{4\} \rangle, \langle \{0, 1\}, \{2, 4\} \rangle\}$.

2. The coordinate projection from $A \times B$ to A is injective if and only if B has at most one element or A is void. It is surjective if and only if B is not void or A is void.

3. It is always injective. It is surjective if and only if A and B are both empty.

4. Suppose $x \in A \times A$. Then $x = \{\{a\}, \{a, b\}\}$ for some $a, b \in A$, and $f^*(x) = \langle f(a), f(b) \rangle = \{\{f(a)\}, \{f(a), f(b)\}\} = \{\bar{f}(\{a\}), \bar{f}(\{a, b\})\} = \bar{\bar{f}}(\{\{a\}, \{a, b\}\}) = \bar{\bar{f}}(x)$. Hence f^* is $\bar{\bar{f}}$ restricted to $A \times A$.

5. We can define a function φ from A to $\mathcal{G}$ by defining, for each $a \in A$, a member $\varphi(a)$ of $\mathcal{G}$. Since a member of $\mathcal{G}$ is itself a function from $\mathcal{F}$ to B, we may define $\varphi(a)$ by giving its values at each point of $\mathcal{F}$. So we put $\varphi(a)(f) = f(a)$ for all $f \in \mathcal{F}$. Now, $f \in \mathcal{F}$ and $a \in A$ imply that $f(a) \in B$, so $\varphi(a)$ is a function from $\mathcal{F}$ to B—that is, a member of $\mathcal{G}$. In symbols, $\varphi = \{\langle a, \{\langle f, f(a) \rangle \mid f \in \mathcal{F}\} \rangle \mid a \in A\}$.

Suppose for a moment that B has more than one element. Then φ is injective but not surjective. If $a_1 \neq a_2$, we can find a function f from A to B such that $f(a_1) \neq f(a_2)$. Then $\varphi(a_1)(f) \neq \varphi(a_2)(f)$, so $\varphi(a_1) \neq \varphi(a_2)$; this proves injectivity. Let b_0 and b_1 be distinct members of B. Let f_0 be the constant function from A to B with value b_0. Define g by $g(f_0) = b_1$, $g(f) = b_0$ for $f \in \mathcal{F} - \{f_0\}$. Then $g \notin$ range φ, because suppose $\varphi(a) = g$. Then $g(f) = \varphi(a)(f) = f(a)$ for all $f \in \mathcal{F}$; in particular, $g(f_0) = f_0(a) = b_0$, a contradiction. (Look carefully at this argument; it is valid even when $A = \emptyset$.)

There remain the possibilities that B is void or has just one element. It will be found that φ is injective when A is void or has just one element (as indeed it must be), but not when A has more than one element. And φ is surjective in all cases except when A is empty and B is not. Checking these facts is really worthwhile.

6. (a) Suppose f is injective. Let $y \in f^{-1}(f(X))$. Then $f(y) \in f(X)$. Choose $x \in X$ so that $f(y) = f(x)$. Since f is injective, $y = x$, whence $y \in X$. Thus $f^{-1}(f(X)) \subseteq X$.

Coupled with (2), this gives $f^{-1}(f(X)) = X$. Conversely, suppose $f^{-1}(f(X)) = X$ for all $X \subseteq A$. Suppose $f(a) = f(b)$. Then $a \in f^{-1}(f(\{b\})) = \{b\}$, whence $a = b$. Thus f is injective.

(b) Suppose f is surjective. Let $y \in Y$. Choose $x \in A$ so that $f(x) = y$. Then $x \in f^{-1}(Y)$ and $y = f(x) \in f(f^{-1}(Y))$. Thus $Y \subseteq f(f^{-1}(Y))$. With (3) this gives $Y = f(f^{-1}(Y))$. Conversely, suppose $f(f^{-1}(Y)) = Y$ for all $Y \subseteq B$. Let $b \in B$ be given. Then $f(f^{-1}(\{b\})) = \{b\}$. Hence $f^{-1}(\{b\})$ is not void. Therefore, $b \in$ range f, and f is surjective.

(c) Suppose f is injective. Let $y \in f(X_1) \cap f(X_2)$. Pick $x_1 \in X_1$ and $x_2 \in X_2$ so that $y = f(x_1)$ and $y = f(x_2)$. Since f is injective, $x_1 = x_2$. Thus $x_1 \in X_1 \cap X_2$, whence $y \in f(X_1 \cap X_2)$. Then (5) implies $f(X_1 \cap X_2) = f(X_1) \cap f(X_2)$. Conversely, suppose $f(X_1 \cap X_2) = f(X_1) \cap f(X_2)$ for all $X_1, X_2 \subseteq A$. Suppose $f(x_1) = f(x_2)$. Then $f(\{x_1\} \cap \{x_2\}) = f(\{x_1\}) \cap f(\{x_2\}) \neq \emptyset$, so $\{x_1\} \cap \{x_2\} \neq \emptyset$. Therefore $x_1 = x_2$. Thus f is injective.

7. By (5) and (3) we have $f(X \cap f^{-1}(Y)) \subseteq f(X) \cap f(f^{-1}(Y)) \subseteq f(X) \cap Y$. Suppose $b \in f(X) \cap Y$. Choose $a \in X$ so that $f(a) = b$. Then $a \in X \cap f^{-1}(Y)$, and therefore $b \in f(X \cap f^{-1}(Y))$. Hence $f(X) \cap Y \subseteq f(X \cap f^{-1}(Y))$. Therefore $f(X \cap f^{-1}(Y)) = f(X) \cap Y$.

Page 50

1. Suppose R and S are functions and let Q be their composition in the sense of 3–4.7. Let $\langle x, y \rangle \in R \circ S$. Then choose z so that $\langle x, z \rangle \in S$ and $\langle z, y \rangle \in R$. Since $\langle x, S(x) \rangle \in S$, the definition of function gives $z = S(x)$, whence $\langle S(x), y \rangle \in R$, and therefore $y = R(S(x)) = Q(x)$. Thus $R \circ S \subseteq Q$. Conversely, if $\langle x, y \rangle \in Q$, then $y = R(S(x))$, so $\langle x, S(x) \rangle \in S$ and $\langle S(x), y \rangle \in R$. Therefore, $\langle x, y \rangle \in R \circ S$. Thus $R \circ S = Q$.

Both $R \circ (S \circ T)$ and $(R \circ S) \circ T$ are

$$\{\langle w, y \rangle \mid (\exists x, z) \langle w, x \rangle \in T, \langle x, z \rangle \in S, \text{ and } \langle z, y \rangle \in R\}.$$

2. Both $\overline{R}(\overline{S}(X))$ and $\overline{(R \circ S)}(X)$ reduce to

$$\{z \mid (\exists x \in X)(\exists y) \langle x, y \rangle \in S \text{ and } \langle y, z \rangle \in R\}.$$

Suppose $\overline{R} = \overline{S}$. We shall prove $R = S$. Let $\langle x, y \rangle \in R$. Then

$$y \in \overline{R}(\{x\}) = \overline{S}(\{x\}),$$

whence $\langle x, y \rangle \in S$. Thus $R \subseteq S$ and the reverse inclusion follows by symmetry.

Page 53

Put $g = \bigcup f_i$. Suppose g is a function. Let i and j be given in I and x in $A_i \cap A_j$. Then $\langle x, f_i(x) \rangle \in g$ and $\langle x, f_j(x) \rangle \in g$. By the definition of function, $f_i(x) = f_j(x)$. Thus (11) holds. Conversely, suppose (11) holds. Obviously, g is a set of ordered pairs. Suppose $\langle x, y \rangle$ and $\langle x, z \rangle$ are in g. Then we can choose i and j in I so that $\langle x, y \rangle \in f_i$ and $\langle x, z \rangle \in f_j$. Then $x \in A_i \cap A_j$, $y = f_i(x)$, and $z = f_j(x)$. By (11), $y = z$, so g is a function. Now, g is an extension of all of the functions f_i. Moreover, if h is any function which is an extension of all the functions f_i, then $g \subseteq h$. Thus g is appropriately called the least common extension of the functions f_i. Clearly, domain $g = \bigcup$ domain f_i and range $g = \bigcup$ range f_i.

Page 54

1. Suppose $f \in \left(\underset{i}{\times} A_i\right) \cap \left(\underset{i}{\times} B_i\right)$. Then f is a function with domain I such that $(\forall i \in I) f(i) \in A_i$ and $(\forall i \in I) f(i) \in B_i$. But this is equivalent to $(\forall i \in I) f(i) \in A_i \cap B_i$. Thus $f \in \underset{i}{\times} (A_i \cap B_i)$. This proves $\left(\underset{i}{\times} A_i\right) \cap \left(\underset{i}{\times} B_i\right) \subseteq \underset{i}{\times} (A_i \cap B_i)$. Since the previous argument is reversible, we obtain also the opposite inclusion.

2. If $f \in \underset{i}{\times} A_i$, then $f(j) \in A_j = B_j \cup C_j$, hence either $f(j) \in B_j$ or $f(j) \in C_j$. Say the former. Now, if $i \in I$ and $i \neq j, f(i) \in A_i = B_i$. Thus $(\forall i) f(i) \in B_i$, that is, $f \in \underset{i}{\times} B_i$. Similarly, if $f(j) \in C_j, f \in \underset{i}{\times} C_i$. Thus $\underset{i}{\times} A_i \subseteq \left(\underset{i}{\times} B_i\right) \cup \left(\underset{i}{\times} C_i\right)$. The opposite inclusion is slightly easier.

3. If $f \in \underset{I \times J}{\times} A_{ij}$, let $\varphi(f)$ be the function $i \to (j \to f(i,j))$. Then $\varphi(f)$ is a function with domain I. Fix an $i \in I$. Since $\varphi(f)(i)$ is a function with domain J and $(\forall j \in J)$ $\varphi(f)(i)(j) = f(i,j) \in A_{ij}$, we have $\varphi(f)(i) \in \underset{j}{\times} A_{ij}$. This shows that

$$\varphi(f) \in \underset{i}{\times} \left(\underset{j}{\times} A_{ij}\right).$$

Thus φ is a function from $\underset{I \times J}{\times} A_{ij}$ to $\underset{i}{\times} \left(\underset{j}{\times} A_{ij}\right)$.

If $\varphi(f_1) = \varphi(f_2)$, then for any i, j, we have $\varphi(f_1)(i) = \varphi(f_2)(i)$ and $\varphi(f_1)(i)(j) = \varphi(f_2)(i)(j)$; the latter is $f_1(i,j) = f_2(i,j)$. Thus $f_1 = f_2$, and φ is injective. On the other hand, if $g \in \underset{i}{\times} \left(\underset{j}{\times} A_{ij}\right)$, then $f(i,j) = g(i)(j)$ defines a function with domain $I \times J$ such that $(\forall i,j) f(i,j) \in A_{ij}$. Then $f \in \underset{I \times J}{\times} A_{ij}$ and $\varphi(f) = g$. Thus φ is surjective.

4. Let φ denote the function $f \to \langle f$ restricted to K, f restricted to $I - K \rangle$. If $f \in \underset{I}{\times} A_i$, then ($f$ restricted to K) $\in \underset{K}{\times} A_i$ and (f restricted to $I - K$) $\in \underset{I-K}{\times} A_i$. Thus φ is a function from $\underset{I}{\times} A_i$ to $\left(\underset{K}{\times} A_i\right) \times \left(\underset{I-K}{\times} A_i\right)$. If $\varphi(f_1) = \varphi(f_2)$, then ($f_1$ restricted to K) = (f_2 restricted to K) and (f_1 restricted to $I - K$) = (f_2 restricted to $I - K$). It follows immediately that $(\forall i) f_1(i) = f_2(i)$; in other words, $f_1 = f_2$. Thus φ is injective. If $\langle g, h \rangle$ is any element of $\left(\underset{K}{\times} A_i\right) \times \left(\underset{I-K}{\times} A_i\right)$, then $f = g \cup h$ is a function in $\underset{I}{\times} A_i$ and $\varphi(f) = \langle g, h \rangle$. Thus φ is surjective.

Pages 58–59

1. Since no singleton appears in the structural set for $\{f, g, h, i, j\}$, it is not isomorphic to either of the others. The bijection $\{\langle a, s \rangle, \langle b, p \rangle, \langle c, t \rangle, \langle d, q \rangle, \langle e, r \rangle\}$ is an isomorphism from the first configuration to the third. There is also a second isomorphism.

2. Since the triplets in the first configuration intersect in a pair while those in the second and third configurations intersect in a singleton, only the second and third configurations can be isomorphic.

3. In the first configuration there is only one member of the structural set containing e. In the other two configurations each member of the basic set appears as an element in at least two members of the structural set.

4. They are isomorphic. There are 64 different isomorphisms. Eight contain the pair $\langle a, s \rangle$, eight contain $\langle a, t \rangle$, etc. One is $\{\langle a, s \rangle, \langle b, t \rangle, \langle c, y \rangle, \langle d, v \rangle, \langle e, u \rangle, \langle f, w \rangle, \langle g, z \rangle, \langle h, x \rangle\}$.

Page 62

1. There are 19.

2. Every Steiner triple system with 7 elements in its basic set is isomorphic to $\langle \{a, b, c, d, e, f, g\}, \{\{a, b, c\}, \{a, d, e\}, \{a, f, g\}, \{b, d, f\}, \{b, e, g\}, \{c, d, g\}, \{c, e, f\}\} \rangle$.

3. Let $\langle A, T \rangle$ be a configuration where $T \in \mathfrak{P}^3(A)$. Think of members of T as *days*, members of a day as *matches*, and members of A as *players*. We write $\exists 1$ for the quantifier *There exists exactly one*, which can be expressed with the usual quantifiers as we noted on page 18. The postulates for a round-robin tournament are as follows.

(i) Every day consists of matches only, and matches are between two players.
$(\forall D \in T)(\forall m \in D)(\exists a, b \in A)\, a \neq b$ and $m = \{a, b\}$.

(ii) Every two distinct players play exactly one match.
$(\forall a, b \in A)\, a \neq b \Rightarrow (\exists 1\, D \in T)\, \{a, b\} \in D$.

(iii) Each player plays exactly one match on each day.
$(\forall a \in A)(\forall D \in T)(\exists 1\, b \in A)\, \{a, b\} \in D$.

4. The obvious first postulate is that every person has exactly two parents, one a man and the other a woman. To rule out the possibility that anyone is his own parent, his own grandparent, his own great-grandparent, etc., we need a much more sophisticated postulate. If we attempt this directly with quantifiers having domain P, we will need infinitely many postulates. However, this difficulty can be avoided by considering that people can be ordered by age. This leads to the following definition.

A genealogy is a configuration $\langle P, \langle M, W, \varphi \rangle \rangle$ where

(i) M and W are complementary subsets of P.

(ii) φ is a subset of $P \times P$. (We shall write $\varphi(p, q)$ instead of $\langle p, q \rangle \in \varphi$.)

(iii) $(\forall p \in P)(\exists m \in M)(\exists w \in W)$

$$\varphi(m, p), \quad \varphi(w, p), \quad \text{and} \quad (\forall q \in P)\ \varphi(q, p) \Rightarrow (q = m \text{ or } q = w).$$

(iv) There exists an ordering of the set P (i.e., a subset B of $P \times P$ such that $\langle P, B \rangle$ is an ordered set in the sense of page 59) such that $(\forall p, q \in P)\, \varphi(p, q) \Rightarrow \langle p, q \rangle \in B$ (more concisely, $\varphi \subseteq B$).

Note that the last postulate involves quantification over $\mathfrak{P}(P \times P)$ instead of merely over P.

Page 64

1. We can letter the members of some triple in the system $\{a, b, c\}$. Choose a fourth element and call it d. The system must contain triples $\{a, d, \ \}$, $\{b, d, \ \}$, and $\{c, d, \ \}$. The blanks must be filled with new names, and we can take them as e, f, and g, respectively. Since there are only seven elements altogether and a must be involved in a triple with f and a triple with g, we must have the triple $\{a, f, g\}$. Similarly, the triples $\{b, e, g\}$ and $\{c, e, f\}$ are forced. Thus we see that every Steiner triple system with seven members is isomorphic to the system $\langle \{a, b, c, d, e, f, g\}, \{\{a, b, c\}, \{a, d, e\}, \{b, d, f\}, \{c, d, g\}, \{a, f, g\}, \{b, e, g\}, \{c, e, f\}\} \rangle$.

2. Choose any two days. The players can be so lettered that these days are $\{\{a, b\}, \{c, d\}, \{e, f\}\}$ and $\{\{a, c\}, \{b, e\}, \{d, f\}\}$. Some third day must contain the match $\{a, d\}$, and on this day e must play against c since he cannot play a or d and he has already played b and f. This day must be $\{\{a, d\}, \{c, e\}, \{b, f\}\}$. Similarly, the days containing the matches $\{a, e\}$ and $\{a, f\}$ are uniquely determined. Thus all five days are determined.

It is important to check that the resulting set of five days is indeed a round-robin tournament. A round-robin tournament can be found for any even number of players.

3. There are 16 such configurations for a fixed basic set; they fall into 10 classes under isomorphism.

4. For a fixed basic set $\{1, 2, 3, 4, 5\}$, there are 120 structures satisfying the given conditions. Of these

20 are isomorphic to $\{\{1, 2, 3\}, \{2, 3, 4\}, \{3, 4, 1\}\}$,
60 are isomorphic to $\{\{1, 2, 3\}, \{2, 3, 4\}, \{3, 4, 5\}\}$,
30 are isomorphic to $\{\{1, 2, 3\}, \{3, 4, 5\}, \{1, 2, 4\}\}$,
10 are isomorphic to $\{\{1, 2, 3\}, \{1, 2, 4\}, \{1, 2, 5\}\}$.

Page 67

1. It is not reflexive, symmetric, or transitive.

2. If A is not void, then $\emptyset$ is a relation in A which is both symmetric and transitive but not reflexive.

3. Suppose $\varphi(a) \cap \varphi(b) \neq \emptyset$. Say $x \in \varphi(a) \cap \varphi(b)$. As shown in 5–1.3, $a\,E\,x$ and $b\,E\,x$. It follows that $a\,E\,b$ and $\varphi(a) = \varphi(b)$. Conversely, $a\,E\,b$ implies $\varphi(a) \cap \varphi(b) = \varphi(a)$, and we know $\varphi(a) \neq \emptyset$ since $a \in \varphi(a)$.

4. There are five partitions of $\{1, 2, 3\}$, shown here grouped in isomorphism classes:

$$\{\{1, 2, 3\}\} \qquad \{\{1, 2\}, \{3\}\}, \{\{2, 3\}, \{1\}\}, \{\{3, 1\}, \{2\}\} \qquad \{\{1\}, \{2\}, \{3\}\}$$

5. To give a formal proof we must translate the hypotheses into formally quantified statements. $\mathcal{P}$ *is a partition of* A becomes

(1) $\mathcal{P} \subseteq \mathfrak{B}(A) - \{\emptyset\}$ and
(2) $(\forall x \in A)(\exists Y \in \mathcal{P})\,(x \in Y$ and $(\forall Z \in \mathcal{P})\,x \in Z \Rightarrow Z = Y)$.

The definition of F is

(3) $F \subseteq A \times A$ and
(4) $(\forall x, y \in A)\,(x\,F\,y \Leftrightarrow (\exists Z \in \mathcal{P})\,x \in Z$ and $y \in Z)$.

The conclusion (a) at the top of page 67 is

(5) $F \subseteq A \times A$,
(6) $(\forall x \in A)\,x\,F\,x$,
(7) $(\forall x, y \in A)\,x\,F\,y \Rightarrow y\,F\,x$, and
(8) $(\forall x, y, z \in A)\,(x\,F\,y$ and $y\,F\,z) \Rightarrow x\,F\,z$.

Here (5) is (3), while (6) and (7) are easy. We give only the proof of (8). It is remarkably long, as is almost any proof carried out in detail.

(9) Let a, b, c be given in A.
(10)   Assume $a\,F\,b$ and $b\,F\,c$.
(11) $a\,F\,b \Leftrightarrow (\exists Z \in \mathcal{P})\,a \in Z$ and $b \in Z$ by (4)
(12) $(\exists Z \in \mathcal{P})\,a \in Z$ and $b \in Z$ by (10) and (11)
(13)   Choose $D \in \mathcal{P}$ so that $a \in D$ and $b \in D$ by (12)
(14)   Choose $E \in \mathcal{P}$ so that $b \in E$ and $c \in E$ Similar to (11), (12), (13)
(15) $(\exists Y \in \mathcal{P})\,(b \in Y$ and $(\forall Z \in \mathcal{P})\,b \in Z \Rightarrow Z = Y)$ by (2)
(16)   Choose $G \in \mathcal{P}$ so that $b \in G$ and
(17) $(\forall Z \in \mathcal{P})\,b \in Z \Rightarrow Z = G$ by (15)
(18) $b \in D \Rightarrow D = G$ by (17)

(19) $D = G$ by (13) and (18)
(20) $E = G$ Similar to the
                                                                    derivation of (19)
(21) $c \in D$ by (14), (20), and (19)
(22) $(\exists Z \in \mathcal{P}) \, a \in Z$ and $c \in Z$ by (13) and (21)
(23) $a \, F \, c$ by (22) and (4)
(24) $(a \, F \, b$ and $b \, F \, c) \Rightarrow a \, F \, c$ by (10) through (23)
(25) $(\forall x, y, z) \, (x \, F \, y$ and $y \, F \, z) \Rightarrow x \, F \, z$ by (9) through (24)

6. There are m equivalence classes.

Page 69

1. We prove only the second statement. Suppose $g(a) = g(b)$. Since the quotient map is onto $\mathcal{Q}$, we can choose x and y so that $a = \varphi(x)$ and $b = \varphi(y)$. Then $f(x) = g(a) = g(b) = f(y)$. By definition x and y are equivalent, so $\varphi(x) = \varphi(y)$, that is, $a = b$. Thus g is injective.

2. (a) Take $F = \{\langle\langle a_1, b_1\rangle, \langle a_2, b_2\rangle\rangle \mid a_1 \, D \, a_2$ and $b_1 \, E \, b_2\}$. It is easy to check that F is an equivalence relation. Denote the quotient map for F by $^-$ written over its argument. We must now check that $\{\langle\overline{\langle a, b\rangle}, \langle\varphi(a), \psi(b)\rangle\rangle \mid a \in A, b \in B\}$ is a bijection from the quotient set of F to $\mathcal{Q} \times \mathcal{R}$.
    (b) Check that $h = \{\langle\langle\varphi(a), \psi(b)\rangle, g(a, b)\rangle \mid a \in A, b \in B\}$ is actually a function.

Page 73

1. The diagonal Δ is obviously both an equivalence relation and a weak order relation. Suppose E is both an equivalence relation and a weak order relation in A. Since E is an equivalence relation, $\Delta \subseteq E$. If $\langle a, b\rangle \in E$, then $\langle b, a\rangle \in E$ since E is symmetric. This is impossible if E is a strong order relation and implies $a = b$ if E is weak; i.e., $\langle a, b\rangle \in \Delta$. Thus $E = \Delta$.

2. $(a \, S \, b$ and $b \, W \, c)$ implies $((a \, S \, b$ and $b \, S \, c)$ or $(a \, S \, b$ and $b = c))$. In either case, $a \, S \, c$.

3. (iii) $\Delta \subseteq W$, (iv) $W \cap W^{-1} = \Delta$, where Δ is the diagonal of $A \times A$.

4. If S and W are corresponding strong and weak order relations such that $S \subseteq R \subseteq W$, then $S = S - \Delta \subseteq R - \Delta \subseteq W - \Delta = S$. Thus $S = R - \Delta$ and $W = S \cup \Delta = R \cup \Delta$. Therefore there is at most one such pair of order relations. But it is easy to check that $R - \Delta$ is a strong order relation. The corresponding weak order relation is $R \cup \Delta$, hence such a pair exists.

Page 75

1. Suppose φ is a weakly order-preserving, injective function from one ordered set $\langle A, S, W\rangle$ to another $\langle B, T, X\rangle$. Suppose $a_1 \, S \, a_2$. Then $a_1 \, W \, a_2$ and $a_1 \neq a_2$. Since φ is weakly order-preserving, $\varphi(a_1) \, X \, \varphi(a_2)$. Since φ is injective, $\varphi(a_1) \neq \varphi(a_2)$. Therefore $\varphi(a_1) \, T \, \varphi(a_2)$. Thus φ is strongly order-preserving.

2. No. Suppose A has more than one member. Then $\emptyset$ is a strong order relation in A, and there are other strong order relations in A. Say S is another, and let W be the corresponding weak order relation. Then the identity function is a strongly order-preserving bijection from $\langle A, \emptyset, \Delta\rangle$ to $\langle A, S, W\rangle$, but it is not an isomorphism.

3. The equivalence relation is $\Delta \cup \{\langle 5, 6\rangle, \langle 6, 5\rangle\}$. The equivalence classes all have just one member except $\{5, 6\}$. The quotient set is $\{\{1\}, \{2\}, \{3\}, \{4\}, \{5, 6\}, \{7\}, \ldots\}$. As an ordered set it is isomorphic to N under the function

$$\{\langle\{1\}, 1\rangle, \langle\{2\}, 2\rangle, \langle\{3\}, 3\rangle, \langle\{4\}, 4\rangle, \langle\{5, 6\}, 5\rangle, \langle\{7\}, 6\rangle, \ldots\}.$$

Page 76

1. No. A could have the trivial order relation as in Exercise 2, page 75.

2. Yes. Suppose φ is a bijective order-preserving function from a linearly ordered set A to an ordered set B. The problem is to prove that φ^{-1} is order-preserving. If $b_1 < b_2$, then since A is linearly ordered we must have either $\varphi^{-1}(b_1) < \varphi^{-1}(b_2)$ or $\varphi^{-1}(b_1) \geq \varphi^{-1}(b_2)$. The latter leads to $b_1 = \varphi(\varphi^{-1}(b_1)) \geq \varphi(\varphi^{-1}(b_2)) = b_2$, which is false. Thus φ^{-1} is order-preserving.

3. Yes. Suppose b_1 and $b_2 \in B$. Choose a_1 and a_2 so that $b_1 = \varphi(a_1)$ and $b_2 = \varphi(a_2)$. Since A is linearly ordered, we must have either $a_1 \leq a_2$ or $a_2 \leq a_1$. Since φ is weakly order-preserving, this gives either $b_1 \leq b_2$ or $b_2 \leq b_1$. Hence B is linearly ordered.

Page 78

1. (3) is false if B is empty and A is an ordered set having a largest element and a smallest element which are different. (3) is true if $B \neq \emptyset$.

2. $\sup B = \bigcup_{Y \in B} Y$, $\inf B = \bigcap_{Y \in B} Y (= X$ if $B = \emptyset)$.

3. $\inf A$ exists if and only if $A \neq \emptyset$. $\sup A$ exists if and only if A is bounded (or finite).

4. (i) If $B \subseteq C$, then any upper bound of C is also an upper bound of B. Hence $\sup C$ is an upper bound of B. Therefore $\sup B \prec \sup C$.

(ii) The hypothesis shows directly that $\sup C$ is an upper bound of B.

5. The commutative laws are completely trivial. By the definition of $\vee$, we have $a \leq a \vee (b \vee c), b \leq b \vee c \leq a \vee (b \vee c),$ and $c \leq a \vee (b \vee c)$. Therefore $a \vee b \leq a \vee (b \vee c)$ and $(a \vee b) \vee c \leq a \vee (b \vee c)$. The opposite inequality follows by similar argument and establishes the associative law for $\vee$. The associative law for $\wedge$ follows in the same way.

Since we have $a \wedge b \leq a \wedge (b \vee c)$ and $a \wedge c \leq a \wedge (b \vee c)$, we have $(a \wedge b) \vee (a \wedge c) \leq a \wedge (b \vee c)$.

It is easy to prove that $(x \vee y) \wedge x = x$ and $x \vee (x \wedge z) = x$. Assume (4) and specialize it with $a = x \vee y, b = x, c = z$. Then

$$(x \vee y) \wedge (x \vee z) = ((x \vee y) \wedge x) \vee ((x \vee y) \wedge z)$$
$$= x \vee ((x \wedge z) \vee (y \wedge z)) = (x \vee (x \wedge z)) \vee (y \wedge z)$$
$$= x \vee (y \wedge z).$$

At the second step we used (4) again together with commutativity. The third step is the associative law. Thus (5) follows from (4). By duality (4) follows from (5).

Pages 81–82

1. $R - \{0\}$ is not complete.

2. Every subset of $\mathfrak{P}(X)$ has a supremum, namely, $\sup B = \bigcup_{Y \in B} Y$.

3. There is a largest equivalence relation in the set A, namely, $A \times A$. The intersection of any family of equivalence relations in A is again an equivalence relation. (If the

family is void, the intersection is taken to be $A \times A$.) Hence every subset of the set of all equivalence relations has an infimum.

4. For each $a \in A$, let $S(a) = \{b \in B \mid b < a\}$. Then, *assuming completeness*, $a = \sup S(a)$; see the proof of 6–5.4. Let a_1 and a_2 be given in A. If $S(a_1) = S(a_2)$, then $a_1 = a_2$. Suppose $S(a_1) \neq S(a_2)$. Say that $S(a_1) - S(a_2)$ is not void. Choose $b_0 \in S(a_1) - S(a_2)$. Let $b \in S(a_2)$. Then $b \geq b_0$ would give $b_0 \in S(a_2)$ from the definition of $S(a_2)$, and this is false; hence $b < b_0$ since B is linearly ordered. Thus, for any $b \in S(a_2)$, $b < b_0 \leq a_1$; that is, a_1 is an upper bound for $S(a_2)$. Therefore $a_1 \geq \sup S(a_2) = a_2$. Similarly, if $S(a_2) - S(a_1) \neq \emptyset$, we can prove $a_2 \geq a_1$. Thus A is linearly ordered.

The counter-example if A is not complete: Let $A = \mathbf{R} \cup \{\theta\}$, where $\theta \notin \mathbf{R}$. Order $\mathbf{R}$ as usual and define $x < \theta \Leftrightarrow x < 0$ and $x > \theta \Leftrightarrow x > 0$ for $x \in \mathbf{R}$. (We have thus put a twin of 0 in $\mathbf{R}$ which is incomparable with 0.) Take $B = \mathbf{R}$. Then B is linearly ordered but A is not, yet B meets every interval in A.

5. (a) Let $\mathfrak{X}$ be a nonvoid subset of $\mathfrak{W}$. Let $V = \bigcap_{X \in \mathfrak{X}} X$. Suppose $a \in A$. Then $(\forall X \in \mathfrak{X}) \langle a, a \rangle \in X$ whence $\langle a, a \rangle \in V$. Suppose $a, b \in A$, $\langle a, b \rangle \in V$, and $\langle b, a \rangle \in V$. Since $\mathfrak{X}$ is not void, we can choose $X \in \mathfrak{X}$. Then $V \subseteq X$, so $\langle a, b \rangle \in X$, $\langle b, a \rangle \in X$. Since X is a weak order relation, $a = b$. Suppose $a, b, c \in A$, $\langle a, b \rangle \in V$, and $\langle b, c \rangle \in V$. For any $X \in \mathfrak{X}$, we have $\langle a, b \rangle \in X$ and $\langle b, c \rangle \in X$; therefore $\langle a, c \rangle \in X$. Thus $(\forall X \in \mathfrak{X})$ $\langle a, c \rangle \in X$; therefore $\langle a, c \rangle \in V$. This proves that $V \in \mathfrak{W}$.

(b) This follows immediately from (i).

(c) Suppose $W \in \mathfrak{W}$ but W is not a linear ordering. Then, say, $u, v \in A$, but $u \neq v$, $\langle u, v \rangle \notin W$, and $\langle v, u \rangle \notin W$. Let $X = W \cup \{\langle x, y \rangle \mid \langle x, u \rangle \in W, \langle v, y \rangle \in W\}$. Then if $a \in A$, $\langle a, a \rangle \in X$. Suppose $a, b \in A$, $\langle a, b \rangle \in X$, and $\langle b, a \rangle \in X$. There are then four cases:

$$\langle a, b \rangle \in W, \langle b, a \rangle \in W;$$
$$\langle a, b \rangle \in W, \langle b, u \rangle \in W, \langle v, a \rangle \in W;$$
$$\langle b, a \rangle \in W, \langle a, u \rangle \in W, \langle v, b \rangle \in W;$$
$$\langle a, u \rangle \in W, \langle v, b \rangle \in W, \langle b, u \rangle \in W, \langle v, a \rangle \in W.$$

In the first case $a = b$; the second, third, and fourth cases lead to the contradiction $\langle v, u \rangle \in W$. Finally, suppose $\langle a, b \rangle \in X$ and $\langle b, c \rangle \in X$. A similar listing of the four cases leads to the conclusion $\langle a, c \rangle \in X$. Thus $X \in \mathfrak{W}$. And $X \supset W$ because $\langle u, v \rangle \in X$ but $\langle u, v \rangle \notin W$. Thus a nonlinear ordering is not maximal. It is easy to prove that a linear ordering is maximal.

(d) Let $Y = \bigcup_{X \in \mathfrak{X}} X$. We verify that Y satisfies the three postulates for a weak order relation. Since $\mathfrak{X}$ is nonvoid, choose $X \in \mathfrak{X}$. Then if $a \in A$, $\langle a, a \rangle \in X \subseteq Y$. Thus Y is reflexive. Suppose $\langle a, b \rangle \in Y$, $\langle b, c \rangle \in Y$. Choose X_1, $X_2 \in \mathfrak{X}$ with $\langle a, b \rangle \in X_1$, $\langle b, c \rangle \in X_2$. Since $\mathfrak{X}$ is linearly ordered by inclusion, either $X_1 \subseteq X_2$ or $X_2 \subseteq X_1$. Hence $X_3 = X_1 \cup X_2 \in \mathfrak{X}$. Since $\langle a, b \rangle \in X_3$ and $\langle b, c \rangle \in X_3$, $\langle a, c \rangle \in X_3 \subseteq Y$. Thus Y is transitive. Finally, suppose $\langle a, b \rangle \in Y$, $\langle b, a \rangle \in Y$. An argument similar to the one just given shows $\langle a, b \rangle \in X_3$, $\langle b, a \rangle \in X_3$, which imply $a = b$; so Y is antisymmetric. This proves $Y \in \mathfrak{W}$.

Page 83

1. We need not assume that A is linearly ordered, because if each two-element set contains its infimum, every two elements are comparable.

2. Yes.

3. Obviously, $<$ is strictly nonreflexive. We must prove that it is transitive and trichotomous. Suppose $g < h$ and $h < k$. Choose x_1 so that

$$g(x_1) < h(x_1) \text{ and } (\forall y \in X) y S x_1 \Rightarrow g(y) = h(y). \tag{1}$$

Choose x_2 so that

$$h(x_2) < k(x_2) \text{ and } (\forall y \in X) y S x_2 \Rightarrow h(y) = k(y). \tag{2}$$

Let x be the smaller of x_1 and x_2 in the ordering of X. If $x = x_1 S x_2$, we have $g(x) = g(x_1) < h(x_1) = k(x_1) = k(x)$; if $x = x_2 S x_1$, then $g(x) = g(x_2) = h(x_2) < k(x_2) = k(x)$; and if $x = x_1 = x_2$, then $g(x) = g(x_1) < h(x_1) = h(x_2) < k(x_2) = k(x)$. In any case, $g(x) < k(x)$. Suppose $y S x$. Then $y S x_1$ and $y S x_2$; hence $g(y) = h(y)$ by (1) and $h(y) = k(y)$ by (2). Therefore $(\forall y \in X) y S x \Rightarrow g(y) = k(y)$. Thus we have proved that $g < k$. Hence $<$ is transitive.

Given g and h in P with $g \neq h$, let $Z = \{x \mid g(x) \neq h(x)\}$. Since Z is not void and X is well-ordered, there is a least element z of Z. It is immediate that $(\forall y \in X) y S z \Rightarrow g(y) = h(y)$. Since $z \in Z$, $g(z) \neq h(z)$. Since $F(z)$ is linearly ordered, either $g(z) < h(z)$ or $h(z) < g(z)$. Correspondingly, either $g < h$ or $h < g$. Thus $<$ is a linear order relation.

Page 85

1. The subsets of S which are the basic sets of subchains are $\{1, 2, 3, 4, 5, 6\}$, $\{2, 3, 4, 5, 6\}$, $\{3, 4, 5, 6\}$, $\{4, 5, 6\}$, and $\emptyset$.

2. T has four subchains, of which $\emptyset$ and $\{1, 3, 4\}$ are isomorphic to $\emptyset$ and $\{4, 5, 6\}$ of S.

3. An isomorphism of S with itself is $\{\langle 1, 4 \rangle, \langle 2, 3 \rangle, \langle 3, 2 \rangle, \langle 4, 1 \rangle\}$.

Page 88

1. The first statement is straightforward. Therefore the union of all subchains of S disjoint from E is the largest subchain disjoint from E.

Suppose T is the largest subchain of S disjoint from the subchain E. Suppose $s \in S - T$ and $f(s) \in T$. If $s \in E$, then $f(s) \in E$ because E is a subchain, contradicting $E \cap T = \emptyset$. Hence $\{s\} \cup T$ is disjoint from E. But $\{s\} \cup T$ is a subchain containing T properly; contradiction. Therefore $s \in S - T \Rightarrow f(s) \in S - T$, and $S - T$ is a subchain.

2. Show that the intersection of any nonvoid family of transitive relations in A is a transitive relation. Since R is included in at least one transitive relation, namely $A \times A$, the intersection of all the transitive relations containing R is the smallest transitive relation containing R.

3. Let T be the reverse of S, i.e., $T = \{\langle a, b \rangle \mid \langle b, a \rangle \in S\}$. Now T is a transitive relation and T contains the reverse of R; i.e., T contains R, since R is symmetric. Therefore $T \supseteq S$ by the definition of S. Taking reverses, we get $S \supseteq T$. Therefore $S = T$, and S is symmetric.

4. The intersection of any nonvoid family of equivalence relations in A is again an equivalence relation. After verifying this, the proof proceeds exactly as the proof of Exercise 2.

5. $f = \{\langle s, t \rangle \mid s < t \text{ and } (\forall u \in S) \text{ not-}(s < u \text{ and } u < t)\}$. Be sure to check that this really defines a function from S to S.

Let T be the subchain generated by E. If $T \neq S$, let s be the least element of $S - T$. If $s \in \text{range } f$, say $s = f(u)$ where $u \in S$, then $u < s$ and so $u \in T$. But T is a subchain, so $s \in T$; contradiction. If $s \notin$ range f, then $s \in E$ and $s \in T$; contradiction. Therefore $T = S$.

Let D be any subset of S which generates S. Since $D \cup$ range f is a subchain of S containing D, we have $S \subseteq D \cup$ range f. Hence $E \subseteq D$. Thus E is the smallest subset of S which generates S.

Page 93

1. Pick any element $a \in S - f(S)$. Let T be the subchain generated by a. Then T is simple.

2. First, let us prove that f is injective. Suppose $s, t \in S$ and $s \neq t$. Since S is linearly ordered, either $s < t$ or $s > t$. By symmetry we may suppose $s < t$. Then $f(s) \leq t$ and $t < f(t)$ by the definition of f; hence $f(s) \neq f(t)$. Thus f is injective.

Let a be the first element of S. Let T be the subchain of S generated by a. As shown in Exercise 1, T is a simple chain. We shall prove $T = S$.

Suppose $s \in S$. If $s = a$, then $s \in T$. Suppose $s \neq a$. Set $U = \{t \in T \mid t < s\}$. Since U is a nonvoid subset of S bounded above by s, it has a largest element u. Then $u < s$. By the definition of f, $u < f(u) \leq s$. Since $u \in T$, $f(u) \in T$; and $u < f(u)$ shows that $f(u) \notin U$. Therefore $f(u) \not< s$. It follows that $f(u) = s$, and hence that $s \in T$. This proves that $S \subseteq T$. But the opposite inclusion is trivial, so $T = S$.

Pages 95–96

1. Suppose $\circ$ is a binary operation in the set A and $*$ is a binary operation in the set B. Then $\langle A, \circ \rangle$ and $\langle B, * \rangle$ are isomorphic if and only if there exists a bijection φ from A to B such that

$$(\forall a_1, a_2 \in A) \quad \varphi(a_1 \circ a_2) = \varphi(a_1) * \varphi(a_2).$$

2. Suppose that b has both a left-identity e and a right-identity f. Then $e = b(e, f) = f$. Thus any left-identity is f and any right-identity is e, so b has just one of each kind and these coincide.

(a)

	a	b	c	d
a	a	b	c	d
b	a	b	c	d
c	c	c	c	c
d	d	d	d	d

Here a and b are left-identities and c and d are left-zeros.

(b)

	a	b
a	a	b
b	a	b

Here a and b are both left-identities and right-zeros.

(c)

	a	b	c
a	a	b	c
b	a	b	c
c	c	c	c

Here a and b are left-identities and c is a zero.

(d)

	a	b	c
a	a	b	c
b	b	b	b
c	c	c	c

Here a is an identity and b and c are left-zeros.

(e) See (b).

3.

	(a)	(b)	(c)	(d)	(e)	(f)	(g)	(h)
associative	yes	no	yes	no	no	yes	yes	yes
commutative	no	yes	yes	no	no	yes	yes	yes
left-identity	no	no	0	2	no	0	no	no
right-identity	all	no	0	no	1	0	no	no
left-zero	all	no	−1	no	−1	no	no	no
right-zero	no	no	−1	2	no	no	no	no

Page 98

1. The rationals and the reals are the only fields.

3. In 8–3.1 (e) it is shown, without using the last part of postulate (iv), that $(\forall x)\, xz = z$. Hence, if $z = u$, we would have for any x, $x = xu = xz = z$, and F would contain only one element. Thus F *has at least two elements* implies $z \neq u$. The opposite implication is trivial.

4. Let $\langle F_1, +, \cdot \rangle$ and $\langle F_2, +, \cdot \rangle$ be fields. They are isomorphic if and only if there exists a bijection φ from F_1 to F_2 such that

$$(\forall a, b \in F_1) \quad \varphi(a + b) = \varphi(a) + \varphi(b) \quad \text{and} \quad \varphi(ab) = \varphi(a)\varphi(b).$$

5. Let G be a subfield of F. Let z and u be the additive and multiplicative identities of F. Let z' and u' be the additive and multiplicative identities of G. Then we have $z' + z' = z' = z + z'$. There is an element $x \in F$ such that $z' + x = z$. Then $z' = z' + z = z' + z' + x = z + z' + x = z + z = z$. Hence the additive identities coincide. Similarly, we have $u'u' = u' = uu'$. Since $u' \neq z' = z$, there is an element $y \in F$ such that $u'y = u$. Then $u' = u'u = u'u'y = uu'y = uu = u$. Thus the multiplicative identities also coincide.

Page 100

2. We omit the verification of the necessity of the given conditions. Assume that they hold. Let a be the restriction of $+$ to $S \times S$ and let m be the restriction of $\cdot$ to $S \times S$. We propose to prove that $\langle S, a, m \rangle$ is a field. If so, it is clearly a subfield of F.

The first problem is to show that a and m are actually binary operations in S; that is, to show that range $a \subseteq S$ and range $m \subseteq S$. These facts are usually expressed by the phrase "S is *closed* with respect to $+$ and $\cdot$." Say $s, t \in S$. Then

$$0 = t - t \in S, \qquad -t = 0 - t \in S, \qquad \text{and} \qquad s + t = s - (-t) \in S$$

by successive applications of (a). Thus range $a \subseteq S$. If $t = 0$, then $st = 0 \in S$. If $t \neq 0$, then $1 = t/t \in S$, $1/t \in S$, and $st = s/(1/t) \in S$ by successive applications of (b). Hence range $m \subseteq S$.

Now we verify the restrictive postulates for $\langle S, a, m \rangle$. The associative, commutative, and distributive laws all transfer automatically to S. Since we have assumed that S contains at least two elements, it contains at least one different from 0. The previous argument shows, therefore, that $0 \in S$ and $1 \in S$. (Be sure to see how the statement $0 \in S$ in the preceding paragraph depends on a temporary assumption.) Hence we may choose $z = 0$ and $u = 1$ in (iv). Then the first and last parts of (iv) follow automatically. The second and third parts are satisfied because we showed that $t \in S \Rightarrow -t \in S$ and $0 \neq t \in S \Rightarrow 1/t \in S$.

The integers are a subset of the field of rational numbers which satisfy (a), (b'), and (c), yet they do not form a subfield.

3. $\phi(0) + \phi(0) = \phi(0 + 0) = \phi(0) = \phi(0) + 0$; hence $\phi(0) = 0$.

$$\phi(a) = \phi((a - b) + b) = \phi(a - b) + \phi(b);$$

hence $\phi(a - b) = \phi(a) - \phi(b)$. If $\phi(1) = 0$, then for any $x \in F$, $\phi(x) = \phi(x \cdot 1) = \phi(x)\phi(1) = \phi(x) \cdot 0 = 0$, that is, range $\phi = \{0\}$. Assume from now on that $\phi(1) \neq 0$. If $x \neq 0$, then $\phi(x)\phi(1/x) = \phi(1) \neq 0$; hence $\phi(x) \neq 0$. If $a \neq b$, then $\phi(a) - \phi(b) = \phi(a - b) \neq 0$, whence $\phi(a) \neq \phi(b)$. Thus ϕ is injective. Also, if $b \neq 0$, $\phi(a) = \phi(b)\phi(a/b)$, whence $\phi(a/b) = \phi(a)/\phi(b)$. In particular, $\phi(1) = \phi(1/1) = \phi(1)/\phi(1) = 1$. Since range ϕ satisfies the criterion of Exercise 2, it is a subfield of G.

Pages 103–104

1. Suppose F is a modular field. Then $0 \in N(F)$. By the definition of $N(F)$, $N(F) - \{0\}$ does not have both properties 8–4.1 (a) and (b). But it does have property (a), so there is an $x \in N(F) - \{0\}$ such that $x + 1 \notin N(F) - \{0\}$. But $x + 1 \in N(F)$ since $N(F)$ has property (b). Hence $x + 1 = 0$ and $-1 \in N(F)$. Consider $A = \{x \mid -x \in N(F)\}$. We have just seen that $1 \in A$. Suppose $a \in A$. Then $-a \in N(F)$, $-1 \in N(F)$, $(-a) + (-1) = -(a + 1) \in N(F)$ by 8–4.3, and $a + 1 \in A$. Thus A has properties (a) and (b). Hence $N(F) \subseteq A$. In other words, $(\forall x) x \in N(F) \Rightarrow -x \in N(F)$. Replacing x by $-x$, we get the opposite implication. Hence $A = N(F)$. Thus $I(F) = N(F) \cup A \cup \{0\} = N(F)$.

2. Suppose $a \in N(F) - \{0\}$. Then $x \to ax$ is a function from $N(F)$ to $N(F)$ by 8–4.3 and it is injective by 8–3.1(d). Since $N(F)$ is finite, this function must also be surjective, so we can choose $b \in N(F)$ so that $ab = 1$. In other words, $1/a \in N(F)$. We have proved $(\forall a \in N(F) - \{0\}) 1/a \in N(F)$. Now $Q(F) = N(F)$ follows from the result of Exercise 1, 8–4.3, and the definition of $Q(F)$.

3. The intersection of any family of subfields of a given field is a subfield. In checking this we find that the verification that the intersection is closed under the field operations is exactly the same as in previous examples of this argument. When we get to postulate (iv) for fields, however, we need the fact that all subfields have the same 0 and 1. If this were not the case, the intersection might be void as far as we can tell.

4. (a) The existence and uniqueness of a function ϕ satisfying (i) and (ii) follow from 8–4.10 and 7–3.8. Then (iii) and (iv) are proved by induction on y.

(b) If $u_1, u_2, v_1, v_2 \in N(F)$ and $u_1 + v_2 = u_2 + v_1$, then $\phi(u_1) - \phi(v_1) = \phi(u_2) - \phi(v_2)$. Hence if $x \in I(F)$ and $x = u - v$ with $u, v \in N(F)$, we can define $\bar{\phi}(x) = \phi(u) - \phi(v)$, this being independent of the choice of u and v. Now $\bar{\phi}$ is an extension of ϕ, and (iii) and (iv) remain valid for $\bar{\phi}$. For clarity we maintain the distinction between ϕ and its extension. It follows from 8–4.6 that domain $\bar{\phi} = I(F)$ and range $\bar{\phi} = I(G)$. Suppose $\bar{\phi}(x) = \bar{\phi}(y)$. Say $x = u_1 - v_1$, $y = u_2 - v_2$, where $u_1, u_2, v_1, v_2 \in N(F)$. Then $\phi(u_1) - \phi(v_1) = \phi(u_2) - \phi(v_2)$, whence $\phi(u_1 + v_2) = \phi(u_2 + v_1)$. Since ϕ is injective, $u_1 + v_2 = u_2 + v_1$, so $x = y$. Thus $\bar{\phi}$ is injective.

(c) If $r_1, r_2, s_1, s_2 \in I(F)$, where $s_1 \neq 0, s_2 \neq 0$, and $r_1 s_2 = r_2 s_1$, then $\bar{\phi}(r_1)/\bar{\phi}(s_1) = \bar{\phi}(r_2)/\bar{\phi}(s_2)$. Therefore if $q \in Q(F)$, say $q = r/s$ with $r, s \in I(F)$, we can define $\psi(q) = \bar{\phi}(r)/\bar{\phi}(s)$ independently of r and s. Now ψ is an extension of ϕ, and (iii) and (iv) remain valid for ψ. It follows from the definition of $Q(F)$ and $Q(G)$ that domain $\psi = Q(F)$ and range $\psi = Q(G)$. Suppose $\psi(q_1) = \psi(q_2)$. Say $q_1 = r_1/s_1$ and $q_2 = r_2/s_2$, where

$r_1, r_2, s_1, s_2 \in I(F)$. Then $\bar{\phi}(r_1)\bar{\phi}(s_2) = \bar{\phi}(r_2)\bar{\phi}(s_1)$ or $\bar{\phi}(r_1s_2) = \bar{\phi}(r_2s_1)$, whence $r_1s_2 = r_2s_1$ and $q_1 = q_2$. Thus ψ is injective. ψ is thus an isomorphism from $Q(F)$ to $Q(G)$ (as fields).

(d) Suppose θ is an isomorphism from $Q(F)$ to $Q(G)$. Using (i) and (ii), we can check first that θ must be an extension of ϕ, then that θ must be an extension of $\bar{\phi}$, and finally that $\theta = \psi$.

Page 107 (top)

Suppose P is the set of all positive elements in some ordering of $Q(F)$. Let $S(F) = \{q \mid (\exists x, y \in N(F)) q = x/y\}$. We shall prove $P = S$, hence that the set of positive elements of $Q(F)$ is uniquely determined, and thus $Q(F)$ can be ordered in only one way. Since $Q(F)$ is a subfield of F, it follows easily from the definition of $N(F)$ that $N(Q(F)) = N(F)$. By 8-5.6 $N(F) = N(Q(F)) \subseteq P$. By 8-5.5(k) and 8-5.1(d), if $x, y \in N(F)$, then $x/y \in P$. Hence $S \subseteq P$. Let $q \in P$. We can write $q = x/y$, where $x, y \in I(F)$. Now $y \neq 0$, so either $y \in N(F)$ or $-y \in N(F)$. Since we have also $q = (-x)/(-y)$ we can arrange that $y \in N(F)$. If $x \notin N(F)$, then $-x \in N(F)$ ($x \neq 0$ because $q \in P$ and therefore $q \neq 0$), so $-q \in S \subseteq P$; but we cannot have both $q \in P$ and $-q \in P$. Therefore $x \in N(F)$ and $q \in S$. This proves $P \subseteq S$.

Now, the isomorphism ϕ from $Q(F)$ to $Q(G)$ was unique, and the construction showed that it takes $N(F)$ onto $N(G)$. Hence it takes $S(F)$ onto $S(G)$. Since $S(F) = P(F) \cap Q(F)$ as shown above and $S(G) = P(G) \cap Q(G)$, ϕ must be order-preserving.

Pages 107–108

1. We shall prove that (c) $\Rightarrow$ (b) $\Rightarrow$ (a) $\Rightarrow$ Archimedean.
Suppose (c). Say $x > 0$. By (c) choose $r \in Q(F)$ so that $0 < r < x$. Put $r = m/n$ where $m, n \in I(F)$. Since $r = (-m)/(-n)$ also, we can assume that $n \in N(F)$. Then $m \leq 0 \Rightarrow m/n \leq 0$, which is a contradiction; so $m > 0$. Therefore $m \in N(F)$ and $1 \leq 1$. Hence $0 < 1/n < m/n < x$. Thus (c) $\Rightarrow$ (b).
Suppose (b). Let $x > 0$ and y be given. If $y \leq 0$, choose $n = 1$; then $nx > y$. If $y > 0$, then $x/y > 0$ and we can choose n so that $1/n < x/y$ by (b), and then $nx > y$. Thus (b) $\Rightarrow$ (a).
Suppose (a). Since $1 > 0$, we have directly $(\forall y)(\exists n \in N(F)) n \cdot 1 > y$, which is the definition of Archimedean. Thus (a) $\Rightarrow$ Archimedean.

2. As we have seen, if $x \in Q(F)$, we can write $x = m/n$ where $m \in I(F)$ and $n \in N(F)$. If $m \leq 0$, then $1 \in N(F)$ and $1 > x$. If $m > 0$, then $m + 1 \in N(F)$ and $m + 1 > x$.

3. Let F be the field of rational functions ordered as at the bottom of page 104. The members of $N(F)$ are the constant functions with positive integral values. Evidently the identity function exceeds all of these, so F is not Archimedean.

Page 116

1. Use induction on r to prove $(\forall r)(\forall p, q) p + r = q + r \Rightarrow p = q$.

2. $p < q \Rightarrow p + r < q + r$ follows immediately from the definition of $<$ together with the associative and commutative laws for $+$. The opposite implication requires the cancellation result of Exercise 1.

3. Suppose $<$ is a strong order relation in S such that $(\forall s) s < f(s)$. Prove by induction on r that $(\forall r)(\forall s) s < s + r$. Hence if $s < u$, choose r so that $u = s + r$; now

it follows that $s < u$. Conversely, suppose $t < u$. Since $<$ is strong, $t \neq u$. Therefore, either $t < u$ or $u < t$. By what we have just proved, the latter implies $u < t$, a contradiction. Thus $t < u$ implies $t < u$. This shows that $<$ and $<$ are the same.

4. A proof can be given similar to that of Theorem 8–5.8. The following proof is quite different.

Let $P(n)$ be the proposition $(\forall T \subseteq S) n \in T \Rightarrow (T$ has a least element$)$. Now $P(e)$ is trivially true. Assume $P(k)$. Let U be any subset of S such that $f(k) \in U$. If $e \in U$, then U has a least element, namely e. Assume from now on that $e \notin U$. Put $T = \{s \mid f(s) \in U\}$. Now $k \in T$, so the inductive assumption implies that there is a least element t of T. Then $f(t)$ is the least element of U. Thus $P(f(k))$ is true. This establishes $(\forall n) P(n)$.

Now if R is a nonvoid subset of S, pick an element $r \in R$. Then $P(r)$ shows that R has a least member. This proves that S is well-ordered.

5. $q \in S_p$ is equivalent to $S_q \subseteq S_p$. Hence $<$ is surely a weak order relation in S. We must prove that $<$ is linear.

We have $S_p \subseteq \{p\} \cup S_{f(p)}$ because $\{p\} \cup S_{f(p)}$ is a chain containing p. Also $S_{f(p)} \subseteq S_p - \{p\}$ because the latter is a chain containing $f(p)$. Hence $S_{f(p)} = S_p - \{p\}$.

Let $P(n)$ be the statement "n is comparable with every member of S." Since $p \in S_e = S$ for all p, e is comparable with every member of S; hence $P(e)$. Assume $P(k)$. Let q be any member of S. If $q \in S_k$ and $q \neq k$, then $q \in S_{f(k)}$. If $q = k$, $f(k) \in S_q$. If $q \notin S_k$, then $k \in S_q$ (by $P(k)$), so $f(k) \in S_q$, since S_q is a subchain. Thus q is comparable with $f(k)$. Since q was arbitrary, $P(f(k))$ holds. This proves $(\forall n) P(n)$ by induction.

That $<$ coincides with $\leq$ follows from Exercise 3, since $(\forall s) s < f(s)$.

6. Use induction to prove successively $(\forall p) p * e = e * p$, $(\forall p, q) f(f(p * q)) = f(p) * q$, and $(\forall p, q) p * q = q * p$.

Page 120

1. Say $\alpha = \overline{\langle p, q \rangle}$. Then $\overline{\langle p, q \rangle} \cdot \overline{\langle p, q \rangle} = \overline{\langle p, q \rangle}$ gives $ppq = qqp$, whence $p = q$ and $\overline{\langle p, q \rangle} = \epsilon$.

2. Let $\Omega(x) = \overline{\langle x, e \rangle}$. There is no difficulty in verifying the required facts. The uniqueness follows from Exercise 1 by induction.

3. There is a bijection φ from S to $N(F)$ such that $\varphi(e) = 1$ and $(\forall s) \varphi(f(s)) = \varphi(s) + 1$. By induction we can prove that $(\forall p, q) \varphi(p + q) = \varphi(p) + \varphi(q)$ and $\varphi(pq) = \varphi(p)\varphi(q)$. If $\langle p, q \rangle \sim \langle r, s \rangle$, then $\varphi(p)/\varphi(q) = \varphi(r)/\varphi(s)$. Hence we can define $\psi(\alpha) = \varphi(p)/\varphi(q)$ where $\langle p, q \rangle \in \alpha$. The formulas $\psi(\alpha \oplus \beta) = \psi(\alpha) + \psi(\beta)$, $\psi(\alpha\beta) = \psi(\alpha)\psi(\beta)$, and $\alpha < \beta \Rightarrow \psi(\alpha) < \psi(\beta)$ can now be verified easily. The last shows that ψ is injective. The range of ψ is $\{x/y \mid x, y \in N(F)\}$, and this was shown in the answer to the exercise at the top of page 107 to be $P(F) \cap Q(F)$.

4. Let $L = \{\langle \alpha, \beta \rangle \mid (\exists w, x, y, z) \langle w, x \rangle \in \alpha, \langle y, z \rangle \in \beta, \text{ and } wz < xy\}$. Suppose $\overline{\langle p, q \rangle} L \overline{\langle r, s \rangle}$. Choose w, x, y, z so that $\langle w, x \rangle \in \overline{\langle p, q \rangle}$, $\langle y, z \rangle \in \overline{\langle r, s \rangle}$, and $wz < xy$. Then $wq = xp$, $ys = zr$, and $xyps = wzqr < xyqr$. Hence $ps < qr$. Hence we have proved $\overline{\langle p, q \rangle} L \overline{\langle r, s \rangle} \Rightarrow ps < qr$. The opposite implication is trivial. That L is a strong order relation now follows easily. It is linear because of the linearity of the order relation in S. Given $\overline{\langle p, q \rangle} L \overline{\langle r, s \rangle}$, choose t so that $ps + t = qr$. Then $\overline{\langle p, q \rangle} + \overline{\langle t, qs \rangle} = \overline{\langle r, s \rangle}$. This shows that $\alpha L \beta \Rightarrow \alpha < \beta$. The reverse implication follows because L is linear.

Page 125

1. Define Ψ by $\Psi(\alpha) = \{\xi \mid \xi < \alpha\}$. Then $\Psi(\alpha) \in \mathcal{P}$. If $\alpha < \beta$, then $\alpha \in \Psi(\beta)$ but $\alpha \notin \Psi(\alpha)$. Since $\mathcal{P}$ is linearly ordered by inclusion, $\Psi(\alpha) \subset \Psi(\beta)$. Thus Ψ is strongly order-preserving. Since $\mathbb{Q}$ is linearly ordered, Ψ is injective.

$\Psi(\alpha) \oplus \Psi(\beta) = \{\xi \oplus \eta \mid \xi < \alpha, \eta < \beta\} \subseteq \{\zeta \mid \zeta < \alpha \oplus \beta\} = \Psi(\alpha \oplus \beta)$. Suppose $\zeta \in \Psi(\alpha \oplus \beta)$, that is, $\zeta < \alpha \oplus \beta$. Then $\alpha\zeta(\alpha \oplus \beta)^* \in \Psi(\alpha)$, $\beta\zeta(\alpha \oplus \beta)^* \in \Psi(\beta)$, and $\zeta = \alpha\zeta(\alpha \oplus \beta)^* \oplus \beta\zeta(\alpha \oplus \beta)^* \in \Psi(\alpha) \oplus \Psi(\beta)$. So $\Psi(\alpha \oplus \beta) \subseteq \Psi(\alpha) \oplus \Psi(\beta)$, and $\Psi(\alpha \oplus \beta) = \Psi(\alpha) \oplus \Psi(\beta)$.

Finally, $\Psi(\alpha) \cdot \Psi(\beta) = \{\xi \cdot \eta \mid \xi < \alpha, \eta < \beta\} \subseteq \{\theta \mid \theta < \alpha\beta\} = \Psi(\alpha\beta)$. Suppose $\theta \in \Psi(\alpha\beta)$. Choose φ so that $\theta < \varphi < \alpha\beta$. Then $\alpha\theta\varphi^* \in \Psi(\alpha)$, $\beta\varphi(\alpha\beta)^* \in \Psi(\beta)$ and $\theta = (\alpha\theta\varphi^*)(\beta\varphi(\alpha\beta)^*) \in \Psi(\alpha) \cdot \Psi(\beta)$. So $\Psi(\alpha\beta) \subseteq \Psi(\alpha) \cdot \Psi(\beta)$, and $\Psi(\alpha\beta) = \Psi(\alpha) \cdot \Psi(\beta)$.

The uniqueness depends on proving that if $A \cdot A = A$, then $A = E$. This can be proved easily from inequalities. It follows that $\Psi(\epsilon) = E$ for any Ψ satisfying $\Psi(\alpha \cdot \beta) = \Psi(\alpha) \cdot \Psi(\beta)$.

2. There exists an isomorphism, say φ, from $\langle \mathbb{Q}, \oplus, \cdot \rangle$ to $\langle P(F) \cap Q(F), +, \cdot \rangle$, and the isomorphism is order-preserving (see Exercise 3, page 120).

If $A \in \mathcal{P}$, then A is nonvoid and bounded in $\mathbb{Q}$; therefore $\varphi(A)$ is a nonvoid, bounded subset of $P(F)$. We can therefore define $\psi(A) = \sup \varphi(A)$. This ψ is an order-preserving bijection from $\mathcal{P}$ to $P(F)$; see Theorem 6–5.3. Note that $\alpha \in A$ implies $\varphi(\alpha) \in \varphi(A)$; therefore $\varphi(\alpha) \leq \psi(A)$.

Suppose $A, B \in \mathcal{P}$. If $\alpha \in A$ and $\beta \in B$, then $\varphi(\alpha \oplus \beta) = \varphi(\alpha) + \varphi(\beta) \leq \psi(A) + \psi(B)$. Since $A \oplus B = \{\alpha \oplus \beta \mid \alpha \in A, \beta \in B\}$, this shows that $\psi(A) + \psi(B)$ is an upper bound for $\varphi(A \oplus B)$. Hence $\psi(A \oplus B) \leq \psi(A) + \psi(B)$.

Suppose $\psi(A \oplus B) < \psi(A) + \psi(B)$. We can choose $c, d \in P(F)$ so that $c < \psi(A)$, $d < \psi(B)$, and $\psi(A \oplus B) < c + d$. By 8–6.2 we can choose $q, r \in Q(F)$ so that $c < q < \psi(A)$ and $d < r < \psi(B)$. Since $q, r \in P(F)$, there exist $\gamma, \delta \in \mathbb{Q}$ with $\varphi(\gamma) = q$, $\varphi(\delta) = r$. Now, $\varphi(\gamma)$ is not an upper bound for $\varphi(A)$, so γ is not an upper bound for A; hence $\gamma \in A$. Similarly, $\delta \in B$. Therefore $\gamma \oplus \delta \in A \oplus B$ and $c + d < q + r = \varphi(\gamma \oplus \delta) \leq \psi(A \oplus B)$; contradiction. Therefore $\psi(A \oplus B) = \psi(A) + \psi(B)$.

The proof that $\psi(A \cdot B) = \psi(A) \cdot \psi(B)$ is very similar.

Page 131

2. We must first show that E is closed with respect to $+$ and $\cdot$. Suppose $e_1, e_2 \in E$. If $d \in D$, then $de_1 = e_1 d$, $de_2 = e_2 d$, $d(e_1 + e_2) = de_1 + de_2 = e_1 d + e_2 d = (e_1 + e_2)d$, $d(e_1 e_2) = (de_1)e_2 = (e_1 d)e_2 = e_1(de_2) = e_1(e_2 d) = (e_1 e_2)d$. Thus $e_1 + e_2 \in E$ and $e_1 e_2 \in E$. The associative and distributive laws and the commutative law for addition transfer from D to E automatically. The commutative law for multiplication in E follows from the definition of E. The multiplicative identity element 1 of D is two-sided, so $1 \in E$. The additive identity element 0 of D turns out to be a two-sided zero element for multiplication, just as in 8–3.1(e). Hence we can choose 0 and 1 as z and u in verifying postulate (iv) for fields. To check the second and third parts, we must verify that if $e \in E$ then the additive inverse of e in D is actually in E and the multiplicative inverse of e in D is actually in E. If $e + x = 0$, and $d \in D$, then $ed + dx = de + dx = d \cdot 0 = 0 \cdot d = ed + xd$, whence $dx = xd$. Thus $x \in E$. If $ey = 1$, then also $ye = 1$ (because, as we remarked, multiplicative inverses are two-sided), hence for any $d \in D$, $yd = ydey = yedy = dy$. Thus $y \in E$.

3. Define addition and multiplication in $\mathbf{R}^4$ as follows:

$$\langle a, b, c, d \rangle + \langle w, x, y, z \rangle = \langle a + w, b + x, c + y, d + z \rangle,$$
$$\langle a, b, c, d \rangle \cdot \langle w, x, y, z \rangle = \langle aw - bx - cy - dz, ax + bw + cz - dy,$$
$$ay - bz + cw + dx, az + by - cx + dw \rangle.$$

Put $R = \{\langle a, 0, 0, 0 \rangle\}$, $i = \langle 0, 1, 0, 0 \rangle$, $j = \langle 0, 0, 1, 0 \rangle$, $k = \langle 0, 0, 0, 1 \rangle$. The verification of the postulates is now routine. The multiplicative inverse of $\langle a, b, c, d \rangle$ is $\langle a/e, -b/e, -c/e, -d/e \rangle$ where $e = a^2 + b^2 + c^2 + d^2$.

Given any division algebra Q satisfying the given conditions, it is readily computed that $\phi(a, b, c, d) = a + bi + cj + dk$ defines a surjection ϕ from $\mathbf{R}^4$ to Q satisfying $\phi(\alpha + \beta) = \phi(\alpha) + \phi(\beta)$ and $\phi(\alpha\beta) = \phi(\alpha)\phi(\beta)$. But just as in the case of fields (Exercise 3, p. 100), any map of division algebras satisfying these conditions is either injective or has range $\{0\}$. Hence ϕ is injective.

Page 134

1. $|w + z|^2 + |w - z|^2 = (w + z)(\bar{w} + \bar{z}) + (w - z)(\bar{w} - \bar{z}) = 2w\bar{w} + 2z\bar{z} = 2|w|^2 + 2|z|^2$.

2. $|x + y + z| \leq |x + y| + |z| \leq |x| + |y| + |z|$. Since we are given that the first and last members are equal, we conclude that equality holds at the second step. Hence there is a nonnegative number r such that $y = rx$ by 10–3.7(h). A similar argument shows that there is a nonnegative number s such that $z = sx$.

Page 135

1. Define $a^1 = a$ and $a^{n+1} = a^n a$. Then 10–4.2(a) and (b) follow by induction on n. For the second part, we must first develop some facts. Suppose $ab = e$. We shall prove that $ba = e$ also. There is an element c such that $bc = e$. Then $ba = bae = babc = bec = bc = e$. If d is any element such that $ad = e$, we have $b = be = bad = ed = d$. Thus there is just one element such that $ab = e$.

Now suppose a^n has been defined for $n \in I$ so that 10–4.2(a) holds and $a^1 = a$. It follows by induction that it must agree with our previous definition for $n > 0$. Since $a^0 a^1 = a^1$, that is, $a^0 a = a$, we must have $a^0 = a^0 e = a^0 ab = ab = e$. Since $a^1 a^{-1} = a^0 = e$, we must have $a^{-1} = b$, the unique inverse of a. It follows by induction that $a^{-n} = b^n$ for $n > 0$. Thus there is at most one way of defining a^n so that 10–4.2(a) holds. To check that this definition actually satisfies 10–4.2(a) requires only the examination of several cases. Once this is established, 10–4.2(b) follows easily, the essential point being that the uniqueness of inverses shows that $(a^n)^{-1} = a^{-n}$.

2. Adapt the proof of Theorem 8–7.4. Take

$$\phi(x) = x + \frac{1}{n(1 + a)^{n-1}} (a - x^n).$$

Page 140

1. Define $\varphi(i) = 2i$ for $i \geq 0$, $\varphi(i) = -2i - 1$ for $i < 0$. Then φ is a bijection from I to N

2. Since $N \times N \sim N$, $(N \times N) \times N \sim N \times N$. Hence $N \times N \times N \sim N$.

3. If we let Z be the smallest fixed point of the function θ, then $h = g^{-1} \circ \psi$.

4. A bijection from the interval $(0, 1)$ to $\mathbf{R}$ is given by $x \to (2x - 1)/x(1 - x)$.

5. Since $\mathbf{R} \supseteq \mathbf{R} - \mathbf{N} \supseteq (0, 1)$, and $\mathbf{R} \sim (0, 1)$ by Exercise 4, $\mathbf{R} \sim \mathbf{R} - \mathbf{N}$ by the Schröder-Bernstein theorem.

6. Let E and O be the sets of even and odd integers respectively. The mapping $\langle A, B \rangle \to A \cup B$ from $\mathfrak{P}(E) \times \mathfrak{P}(O)$ to $\mathfrak{P}(\mathbf{N})$ is a bijection. Therefore $\mathfrak{P}(\mathbf{N}) \sim \mathfrak{P}(E) \times \mathfrak{P}(O)$. The latter is obviously similar to $\mathfrak{P}(\mathbf{N}) \times \mathfrak{P}(\mathbf{N})$.

Page 142

Proof of 11-2.9. Consider the least integer p such that there exists an injection from E to $\mathbf{N}_p$. If φ is such an injection, φ is surjective.

Proof of 11-2.10. This proof is similar to that of 11-2.9.

Proof of 11-2.11 *and* 11-2.12. Suppose B is finite, say card $B = p$, and $A \subseteq B$. Let φ be a bijection from B to $\mathbf{N}_p$. Then φ is an injection from A to $\mathbf{N}_p$; hence card $A \leq p$. If $b \in B - A$, then we can define an injection from A to $\mathbf{N}_{p-1}$ by $\psi(a) = \varphi(a)$ unless $\varphi(a) = p$, and $\psi(a) = \varphi(b)$ if $\varphi(a) = p$; hence card $A < p$. Thus $A \subset B \Rightarrow$ card $A <$ card B. Now 11-2.11 and 11-2.12 follow.

Proof of 11-2.13. Use induction on card B.

Proof of 11-2.14. Use induction on card $\mathfrak{B}$.

Proof of 11-2.15. Use induction on card B.

Proof of 11-2.16. Use induction on the cardinal of the subset.

Pages 142-143

1. Use induction on the cardinal of the ordered set.

2. Assume that a linearly ordered set of cardinal k is isomorphic as an ordered set to $\langle \mathbf{N}_k, <, \leq \rangle$. This is clearly true for $k = 1$. If L is a linearly ordered set of cardinal $k + 1$, let m be its largest element (see 11-2.16). There is an isomorphism φ of $\mathbf{N}_k$ onto $L - \{m\}$. Extend φ by $\varphi(k + 1) = m$.

3. Use induction on the cardinal of A.

4. Let T be the relation defined by the existence of such a function. By induction on k, prove $T \subseteq S$. Then show that T is transitive, hence $S \subseteq T$.

Page 144

1. $\langle m, n \rangle \to m/n$ is a surjection from $\mathbf{N} \times (\mathbf{N} - \{0\})$ to $\mathbf{Q}$.

2. First show that the union of two countable sets is countable. Then use induction on the number of sets in the family.

3. A simple proof using some elementary number theory and some notation we have not yet developed can be given as follows. If A is a finite subset of $\mathbf{N}^*$, put $F(A) = \sum_{a \in A} 2^a$. Then F is a bijection from the set of finite subsets of $\mathbf{N}^*$ to $\mathbf{N}^*$.

To give a formal proof using the theorems and techniques we have already established is quite long. First we define a sequence $\varphi_1, \varphi_2, \varphi_3, \ldots$ of functions inductively. Define φ_1 from $\mathfrak{P}(\mathbf{N}_1)$ to $\mathbf{N}^*$ by $\varphi_1(\emptyset) = 0$, $\varphi_1(\{1\}) = 1$. Define φ_{k+1} from $\mathfrak{P}(\mathbf{N}_{k+1})$ to $\mathbf{N}^*$ by $\varphi_{k+1}(A) = 2^k + \varphi_k(A - \{k + 1\})$ if $k + 1 \in A$ and $\varphi_{k+1}(A) = \varphi_k(A)$ if $k + 1 \notin A$. Now we can prove, by induction on k, that these functions are all injective and that φ_{k+1} is an extension of φ_k. Hence these functions have a least common extension φ which is an injection from $\bigcup_k \mathfrak{P}(\mathbf{N}_k) = \mathfrak{F}$ to $\mathbf{N}^*$.

4. Let φ be a surjection from $\mathbf{N}$ to $\mathbf{Q}$. Define ψ from $\mathbf{R}$ to $\mathfrak{P}(\mathbf{N})$ by $\psi(x) = \{n \mid \varphi(n) < x\}$. Then ψ is an injection from $\mathbf{R}$ to $\mathfrak{P}(\mathbf{N})$.

5. The hypothesis means that every nonvoid subset of S has a greatest member. For each $s \in S$, let $T_s = \{t \mid t \leq s\}$, and put $U = \{s \mid T_s$ is finite$\}$. U is not void, so let u be the greatest member of U. If u is the largest member of S, then $S = T_u$, so S is finite. Suppose u is not the largest member of S. There is a least member v of S greater than u. Then $T_v = T_u \cup \{v\}$, so T_v is finite and $v \in U$. But $v > u$, contrary to the choice of u.

6. For each $x \in L$, let $\psi(x)$ be the least element of $\varphi^{-1}(x)$. Then ψ is an order-preserving injection from L to $\mathbf{N}$. If L is infinite, $\psi(L)$ is an infinite subset of $\mathbf{N}$ and there is an order-preserving bijection f from $\mathbf{N}$ to $\psi(L)$ by 11–3.2. Then $f^{-1} \circ \psi$ is an isomorphism from L to $\mathbf{N}$. If $x, y \in L$ and $x < y$, then $\varphi^{-1}(x)$ is bounded above by any member of $\varphi^{-1}(y)$, so $\varphi^{-1}(x)$ is finite. Thus $\varphi^{-1}(x)$ can be infinite only if x is the largest element of L.

Page 148

$\varphi(2) = 1$. If we assume that $\varphi(k) \geq 1/(k - 1)$, then we find that $\varphi(k + 1) \geq 1/k$, and the construction works out ad infinitum.

Page 155

1. Use the existence of an injection from $\mathbf{N}$ to E. Divide the range of this injection into disjoint infinite sets.

2. Use the existence of an injection from $\mathbf{N} \times \mathbf{N}$ to E. Divide the range of this injection into infinitely many mutually disjoint infinite sets.

3. Let h be a function from $\mathfrak{P}(E)$ to E such that $h(X) \in X$ for every $X \in \mathfrak{P}(E) - \{\emptyset\}$ (this is justified by the axiom of choice). There is a function ψ from $\mathbf{N}$ to E such that

$$\psi(1) = h(E),$$
$$\psi(k + 1) = h(\{e \in E \mid e > \psi(k)\}).$$

Since $\psi(k)$ is not maximal, $\psi(k + 1) > \psi(k)$. It follows immediately that ψ is an order-preserving injection.

4. Suppose S satisfies the ascending-chain condition. If $T \subseteq S$ and T has no maximal elements, then by 11–5.5, there is an order-preserving injection from $\mathbf{N}$ to T. But this contradicts the ascending-chain condition. Conversely, suppose S satisfies the maximum condition. If φ were an order-preserving injection from $\mathbf{N}$ to S, then range φ would contain no maximal element. Note that the axiom of choice does not enter into this half of the argument.

5. If $\{H_i \mid i \in I\}$ is a family of nonvoid sets, choose a well-ordering of $\bigcup_i H_i$, and put $h(i) = $ least element of H_i.

6. Let H_k be the set of all injections from $\mathbf{N}_k$ to E. Since E is infinite, it follows by induction that each $H_k \neq \emptyset$. Let $h \in \bigtimes_k H_k$; then for all k, $h_k \in H_k$. Define φ so that, for all $k \in \mathbf{N}^*$,

$$\varphi(k + 1) = h_{k+1}(i),$$

where i is the least integer not in $h_{k+1}^{-1}(\{\varphi(1), \ldots, \varphi(k)\})$ $(= h_{k+1}^{-1}$ (range φ_k) in the notation of Theorem 11–4.1). Then φ is an injection from $\mathbf{N}$ to E.

Page 159

In what follows P and P' have cardinal $\mathbf{p}$, Q and Q' have cardinal $\mathbf{q}$, etc.

Lemma. If $P \sim P'$, $Q \sim Q'$, $P \cap Q = \emptyset$, and $P' \cap Q' = \emptyset$, then $P \cup Q \sim P' \cup Q'$.

Define $\mathbf{p} + \mathbf{q}$ as card $(P \cup Q)$ where $P \cap Q = \emptyset$. (It is always possible to find sets of given cardinals which are disjoint. For example, $\{1\} \times P$ and $\{2\} \times Q$ are disjoint.) The lemma shows that $\mathbf{p} + \mathbf{q}$ depends only on $\mathbf{p}$ and $\mathbf{q}$ and not on the choice of P and Q. By definition, $\mathbf{q} + \mathbf{p} = \text{card } (Q \cup P) = \mathbf{p} + \mathbf{q}$. If $R \cap (P \cup Q) = \emptyset$, then $(\mathbf{p} + \mathbf{q}) + \mathbf{r} = \text{card } ((P \cup Q) \cup R) = \text{card } (P \cup (Q \cup R)) = \mathbf{p} + (\mathbf{q} + \mathbf{r})$. Thus addition of cardinals is commutative and associative.

Lemma. If $P \sim P'$ and $Q \sim Q'$, then $P \times Q \sim P' \times Q'$.

Define $\mathbf{p} \cdot \mathbf{q}$ as card $(P \times Q)$. Since $P \times Q \sim Q \times P$, $\mathbf{p} \cdot \mathbf{q} = \mathbf{q} \cdot \mathbf{p}$. Since $P \times (Q \times R) \sim (P \times Q) \times R$, $\mathbf{p} \cdot (\mathbf{q} \cdot \mathbf{r}) = (\mathbf{p} \cdot \mathbf{q}) \cdot \mathbf{r}$. Thus multiplication of cardinals is commutative and associative. If $Q \cap R = \emptyset$, then $(P \times Q) \cap (P \times R) = \emptyset$ and $P \times (Q \cup R) = (P \times Q) \cup (P \times R)$. Hence $\mathbf{p} \cdot (\mathbf{q} + \mathbf{r}) = \mathbf{p} \cdot \mathbf{q} + \mathbf{p} \cdot \mathbf{r}$.

If P and Q are sets, let $\mathcal{F}(Q, P)$ denote the set of all functions from Q to P.

Lemma. If $P \sim P'$ and $Q \sim Q'$, then $\mathcal{F}(Q, P) \sim \mathcal{F}(Q', P')$.

Define $\mathbf{p}^{\mathbf{q}}$ as card $\mathcal{F}(Q, P)$. If $Q \cap R = \emptyset$, then $\mathcal{F}(Q \cup R, P) \sim \mathcal{F}(Q, P) \times \mathcal{F}(R, P)$. (The bijection is $f \to \langle f$ restricted to Q, f restricted to $R \rangle$.) Hence $\mathbf{p}^{\mathbf{q}+\mathbf{r}} = \mathbf{p}^{\mathbf{q}} \cdot \mathbf{p}^{\mathbf{r}}$. $\mathcal{F}(R, P) \times \mathcal{F}(R, Q) \sim \mathcal{F}(R, P \times Q)$. (The map is $\langle f, g \rangle \to (r \to \langle f(r), g(r) \rangle)$.) Hence $\mathbf{p}^{\mathbf{r}} \cdot \mathbf{q}^{\mathbf{r}} = (\mathbf{p} \cdot \mathbf{q})^{\mathbf{r}}$. $\mathcal{F}(R, \mathcal{F}(Q, P)) \sim \mathcal{F}(Q \times R, P)$. (The map is $f \to (\langle q, r \rangle \to f(r)(q))$.) Hence $(\mathbf{p}^{\mathbf{q}})^{\mathbf{r}} = \mathbf{p}^{\mathbf{q} \cdot \mathbf{r}}$.

Page 167

3. If $z_n \to \zeta$ under the proposed definition, we could take $\epsilon = 0$ to get $(\exists M \in \mathbf{R})$ $(\forall n > M) |\zeta - z_n| \leq 0$, which is equivalent to $(\exists M \in \mathbf{R})(\forall n > M) z_n = \zeta$; that is, eventually all terms are exactly ζ.

4. Suppose $\{x_n\}$ is a real sequence and ζ is not real. Put $\delta = |\text{Im } \zeta| > 0$. Then $|x_n - \zeta| \geq |\text{Im } (x_n - \zeta)| = |\text{Im } \zeta| = \delta$ for any n. This shows that $x_n \to \zeta$ is false. Without absolute-value signs, the definition of convergence is

$$(\forall \epsilon > 0)(\exists M \in \mathbf{R})(\forall n \geq M) \quad \zeta - \epsilon < x_n < \zeta + \epsilon.$$

5. (a) Bounded and convergent but not monotone. (b) Bounded, but neither convergent nor monotone. (c) Monotone increasing, but unbounded, divergent. (d) Monotone decreasing, bounded, convergent. (e) $x_n = \frac{3}{4}(1 - (-\frac{1}{3})^n)$. Bounded, convergent, but not monotone. In (a), (d), and (e) $M = 1 + 1/\epsilon$ will do, but there are smaller choices in (a) and (e).

6. (a) and (d) converge to $-i$ and 0 respectively.

7. If $|z| \leq 1$, then $|z^n| \leq 1$ for all n, and therefore $\{z^n\}$ is bounded. Suppose $|z| > 1$. Say $|z| = 1 + h$. By induction on n, $|z^n| = |z|^n \geq 1 + nh$; so the sequence $\{z^n\}$ is unbounded, and therefore divergent. If $0 < |z| < 1$, then say $|1/z| = 1 + k$. We have $|z^n| = 1/(|1/z|^n) < 1/(1 + nk)$, and hence $z^n \to 0$. The same result is trivial if $z = 0$.

The sequence $\{z^n\}$ obviously converges if $z = 1$. Conversely, suppose $|z| = 1$ and $z^n \to \zeta$. Let $\epsilon > 0$ be given. Choose k so large that $(\forall n \geq k) |\zeta - z^n| < \epsilon/2$. Then

$|z - 1| = |(z - 1)z^k| = |z^{k+1} - z^k| \leq |z^{k+1} - \zeta| + |\zeta - z^k| < \epsilon.$ Since ϵ was arbitrary, $|z - 1| = 0$, or $z = 1$. Thus we have shown that $\{z^n\}$ is

> bounded and converges to 0 for $|z| < 1$,
> bounded and converges to 1 for $z = 1$,
> bounded and divergent for $|z| = 1$, $z \neq 1$,
> unbounded and divergent for $|z| > 1$.

Page 173

1. Suppose $x_n \to z$. Then $z \geq 0$ by 12–2.6. First, suppose $z \neq 0$. Then $z > 0$. We have $|\sqrt{x_n} - \sqrt{z}| = |x_n - z|/(\sqrt{x_n} + \sqrt{z}) < |x_n - z|/\sqrt{z}$. Given $\epsilon > 0$, choose k so large that $(\forall n > k)\,|x_n - z| < \epsilon\sqrt{z}$. Then $(\forall n > k)\,|\sqrt{x_n} - \sqrt{z}| < \epsilon$. Therefore $\sqrt{x_n} \to \sqrt{z}$.

Now suppose $z = 0$; that is, $x_n \to 0$. Given $\epsilon > 0$, choose k so that

$$(\forall n > k) \quad |x_n| < \epsilon^2.$$

Then $(\forall n > k)\,|\sqrt{x_n}| < \epsilon$. Therefore $\sqrt{x_n} \to 0$.

2. First we prove that $\lim_{n \to \infty} 1/n = 0$. Suppose $\epsilon > 0$ is given. Choose k large enough so that $k > 1/\epsilon$. Then $(\forall n > k)\,|1/n| = 1/n < 1/k < \epsilon$. Hence $\lim_{n \to \infty} 1/n = 0$.

If we write the first expression in the form

$$\frac{1 + 2/n - 1/n^2}{3 - 5/n + 3/n^2},$$

we can conclude from several applications of Theorem 12–2.1 that

$$\lim \frac{n^2 + 2n - 1}{3n^2 - 5n + 2} = \frac{\lim (1 + 2/n - 1/n^2)}{\lim (3 - 5/n + 3/n^2)}$$

$$= \frac{\lim 1 + \lim (2/n) + \lim (-1/n^2)}{\lim 3 + \lim (-5/n) + \lim (3/n^2)} = \frac{1 + 0 + 0}{3 + 0 + 0} = \frac{1}{3}.$$

The second limit is evaluated in a similar manner, this time dividing numerator and denominator by n^3.

In the third limit,

$$\sqrt{n + 1} - \sqrt{n} = \frac{1}{\sqrt{n + 1} + \sqrt{n}} < \frac{1}{2\sqrt{n}}.$$

Applying Exercise 1, we have $\lim (1/\sqrt{n}) = \sqrt{\lim (1/n)} = 0$. Hence $\sqrt{n + 1} - \sqrt{n} \to 0$.

In the last limit, we have

$$n(\sqrt{n^2 + 1} - n) = \frac{n}{\sqrt{n^2 + 1} + n} = \frac{1}{1 + \sqrt{1 + 1/n^2}} \to \frac{1}{2}.$$

3. If $x_n \to z \neq 0$, then

$$\lim (x_n + 4/n - 1/n^2) = z \quad \text{and} \quad \lim (x_n - 5/n + 2/n^2) = z \neq 0.$$

Hence

$$\lim \frac{n^2 x_n + 4n - 1}{n^2 x_n - 5n + 2} = \lim \frac{(x_n + 4/n - 1/n^2)}{(x_n - 5/n + 2/n^2)} = \frac{z}{z} = 1.$$

If $x_n = \alpha/n$, and $\alpha \neq 5$, then

$$\lim_{n \to \infty} \frac{n^2 x_n + 4n - 1}{n^2 x_n - 5n + 2} = \frac{\alpha + 4}{\alpha - 5}.$$

For appropriate choice of α, this expression can take any value except 1. If $x_n = 1/\sqrt{n}$, the limit is 1; and if $x_n = 5/n$, the limit does not exist.

4. Factor the left-hand side of the equation into $x_n(x_n^2 + \frac{1}{2}) = 1/n$. The factor $(x_n^2 + \frac{1}{2})$ is positive; hence x_n is always positive. Also $1/n > x_n/2$. Hence $0 < x_n < 2/n$, and $x_n \to 0$.

Page 175

The function defined by $f(0) = -\infty$, $f(1) = +\infty$, $f(x) = (2x - 1)/x(1 - x)$ for $0 < x < 1$, is an order-isomorphism from $[0, 1]$ to $\mathbf{R}^*$.

Pages 179–180

1. Aside from 5(a), 5(d), and 5(e), which converge, 5(c) converges improperly to $+\infty$, and 5(b) has lim sup $= 1$, lim inf $= -1$. 6(a) and 6(d) converge, and 6(b) converges improperly to ∞.

2. lim sup $1/n = 0$ and $0 \in L$. lim sup $(-1/n) = 0$ and $0 \in U$.

3. (a) Given $\epsilon > 0$, except for a finite set of p, $x_p < \text{lim sup } x_n + \epsilon$ and $y_p < \text{lim sup } y_n + \epsilon$; hence $x_p + y_p < \text{lim sup } x_n + \text{lim sup } y_n + 2\epsilon$. Therefore, lim sup $(x_n + y_n) \leq \text{lim sup } x_n + \text{lim sup } y_n + 2\epsilon$. Since ϵ is arbitrary, the required result follows. If $x_n = (-1)^n$ and $y_n = (-1)^{n-1}$, then we have lim sup $(x_n + y_n) = 0 < \text{lim sup } x_n + \text{lim sup } y_n = 2$.

4. Since lim sup $|a_{n+1}|/|a_n| < 1$, we can choose α so that lim sup $|a_{n+1}|/|a_n| < \alpha < 1$. Then choose k so that $(\forall n \geq k) |a_{n+1}|/|a_n| < \alpha$, or $(\forall n \geq k) |a_{n+1}| < \alpha|a_n|$. By induction on p it follows that $|a_{k+p}| < \alpha^p|a_k|$. But $\alpha^p \to 0$ (Exercise 7, p. 167). Hence $\alpha^p|a_k| \to 0$, and lim $a_n = 0$.

5. Let φ be an order-isomorphism from $\langle \mathbf{R}^*, <, \leq \rangle$ to $\langle S, <, \leq \rangle$. Since suprema and infima are calculated entirely in terms of order, we must have $\varphi(\text{sup } A) = \text{sup } \varphi(A)$ for any set $A \subseteq \mathbf{R}^*$. But sup $\varphi(A)$ is to be calculated in the set S. Although $S \subseteq \mathbf{R}^*$ and the order in S is that order inherited from $\mathbf{R}^*$, it is not necessarily true that suprema calculated in S should agree with those calculated in $\mathbf{R}^*$. (For example, let $S = [0, 1) \cup \{2\}$.) However, when $S = [0, 1]$, it is true. A similar statement holds for infima. Hence

$$\text{lim sup } \varphi(x_n) = \inf_k \text{ sup } \{\varphi(x_n) \mid n \geq k\} = \inf_k \varphi(\text{sup } \{x_n \mid n \geq k\}) = \text{lim sup } x_n.$$

Similarly, lim inf $\varphi(x_n) = \varphi(\text{lim inf } x_n)$. Hence lim x_n exists if and only if lim $\varphi(x_n)$ exists, and if so then $\varphi(\text{lim } x_n) = \text{lim } \varphi(x_n)$.

Note that these considerations have nothing to do with the choice of the isomorphism φ.

Page 182

1. (a), (b), and (d) converge.

2. $\lambda = 0$ and all ρ, or $\lambda \neq 0$ and $|\rho| < 1$.

3. If $\{S_n\}$ denotes the sequence of partial sums of F-$\sum a_n$ and $\{T_n\}$ is the sequence of partial sums of F-$\sum b_n$, then the condition gives $|S_p - S_q| \leq T_p - T_q$ for $p \geq q$. Hence, if $\{T_n\}$ is a Cauchy sequence, so is $\{S_n\}$.

4. If $\limsup |a_{n+1}/a_n| < 1$, then $(\forall n)\, |a_n| < \lambda\rho^n$ for some λ and ρ with $\rho < 1$. (See the answer to Exercise 4, p. 180.) Hence F-$\sum a_n$ converges by the comparison test. On the other hand, if $\liminf |a_{n+1}/a_n| > 1$, then $|a_n| \to +\infty$. In this case it is clear that $\{S_n\}$ does not converge, since $\lim (S_n - S_{n-1})$ does not exist.

5. Using temporarily a familiar notation we have not yet defined, we have for $q \geq p$,

$$\sum_{n=p}^{q} (B_{n+1} - B_n)c_n = B_{q+1}c_q + \sum_{n=p+1}^{q} B_n(c_{n-1} - c_n) - B_p c_p.$$

Let M be an upper bound for $\{|B_n|\}$. Noting that $c_{n-1} - c_n \geq 0$ and $c_n \geq 0$,

$$|S_q - S_{p-1}| \leq M\left(c_q + \sum_{n=p+1}^{q} (c_{n-1} - c_n) + c_p\right) = 2Mc_p.$$

Therefore the sequence $\{S_n\}$ satisfies Cauchy's criterion. We can avoid the use of the indexed sum by proving by induction on q that

$$|S_q - S_{p-1} + B_p c_p - B_{q+1}c_q| \leq M(c_p - c_q)$$

for $q \geq p$, whereupon we obtain the same inequality $|S_q - S_{p-1}| \leq 2Mc_p$.

Page 185

1. Let $x_{4k} = 1$, $x_{4k+1} = 0$, $x_{4k+2} = 2$, $x_{4k+3} = 3$. We have $\liminf x_n = 0$, $\liminf x_{2n} = 1$, $\limsup x_{2n} = 2$, $\limsup x_n = 3$.

2. If the sequence is bounded, then 12–6.6 applies. If not, it contains a subsequence which converges to ∞.

3. If, for some m, the set $A_m = \{n \mid |z_n| < m\}$ is infinite, then $\{z_n\}$ has a bounded subsequence, which in turn has a convergent subsequence. If the set A_m is finite for every m, then $z_n \to \infty$.

4. $\{i, -1, -i, 1\}$, $\{3, i, -1, -2 - i\}$.

5. For every integer k, there are infinitely many rational numbers in the interval $(r, r + 1/k)$. Define f inductively. Let $f(1) = 1$, and let $f(k + 1)$ be the least integer n such that $r < x_n < r + 1/k$ and $n > f(k)$. Then $f(k) \to \infty$ and $x_{f(k)} \to r$.

Page 190

1. $\bigstar(f, A \cup B) = \bigstar(f, A - B) * \bigstar(f, B - A) * \bigstar(f, A \cap B)$,
   $\bigstar(f, A) = \bigstar(f, A - B) * \bigstar(f, A \cap B)$,
   $\bigstar(f, B) = \bigstar(f, A \cap B) * \bigstar(f, B - A)$.

Putting these together, we get the required formula.

3. Define $\prod(f, p, p + 1) = f(p)$, and $\prod(f, p, p + k + 1) = \prod(f, p, p + k)f(p + k)$. Then $\prod(f, p, q)\prod(f, q, q + 1) = \prod(f, p, q + 1)$ and $\prod(f, p, q)\prod(f, q, q + k) = \prod(f, p, q + k)$ by induction on k, using the associative law to prove the inductive step. If S has a two-sided identity element e, define $\prod(f, p, p) = e$. Then (b) holds for $p \leq q \leq r$. Assume further that every element of S has an inverse. This inverse must

be two-sided; see the answer to Exercise 1, p. 135. Define $\prod(f, p, q) = (\prod(f, q, p))^{-1}$ for $p > q$. Formula (b) now holds for all p, q, r. This can be proved directly from the special case $p \leq q \leq r$. For example, if $q \leq p \leq r$, then the special case gives $\prod(f, q, p)\prod(f, p, r) = \prod(f, q, r)$; multiplying by $\prod(f, p, q)$ gives (b) again.

4. Use 12–2.1 and induction on card A.

5. Use the result of Exercise 3(a), p. 179, and induction on card A.

6. Actually we have not defined $\sum_{x \in Y} f(x)$. It means, of course, $\sum_{x \in Y \cap \text{spt} f} f(x)$. If spt f is finite, and $\{Y(i) \mid i \in I\}$ is a family of mutually disjoint sets, then $i \to \sum_{x \in Y(i)} f(x)$ is a function of finite support. Hence $\sum_{i \in I} (\sum_{x \in Y(i)} f(x))$ is effectively finite.

Now suppose that spt f contains just one point x_0. Then $\sum_{i \in I} \sum_{x \in Y(i)} f(x) = \sum_{x \in \bigcup Y(i)} f(x)$, because both sides reduce to $f(x_0)$ or 0, depending on whether or not $x_0 \in \bigcup Y(i)$. Continue by induction on card (spt f) as in the proof of 13–1.4.

Pages 210–211

1. $\sum b_n = \frac{3}{2} \sum a_n$.

2. By induction on k prove that

$$A_k = \sum_{n=1}^{k} \frac{1}{n^2} + \frac{1}{k+1}$$

defines an increasing sequence. Hence $\sum_{n=1}^{\infty} (1/n^2) = \lim_p A_p > A_k$. The other inequality can be proved the same way. Alternatively,

$$\frac{1}{k+1} = \sum_{n=k+1}^{\infty} \left(\frac{1}{n} - \frac{1}{n+1}\right) < \sum_{n=k+1}^{\infty} \frac{1}{n^2} < \sum_{n=k+1}^{\infty} \left(\frac{1}{n-1} - \frac{1}{n}\right) = \frac{1}{k}.$$

3. Prove that the sequence $A_{2k-1} - 1/(4k - 1)$ increases, while $A_{2k} + 1/(4k + 1)$ decreases.

4. We have

$$\frac{2^k}{2^{\alpha(k+1)}} < \sum_{n=2^k+1}^{2^{k+1}} \frac{1}{n^\alpha} < \frac{2^k}{2^{\alpha k}}.$$

Since F-$\sum_k (2^k/2^{\alpha(k+1)})$ diverges for $\alpha \leq 1$, and F-$\sum_k (2^k/2^{\alpha k})$ converges for $\alpha > 1$, the result follows from 13–2.10.

5. Converges for $m \geq 0$. Diverges for $m < 0$. The series clearly converges for $m = 0$. For $m > 0$, we shall prove that the series is dominated by F-$\sum(1/n^{1+m})$, which converges by Exercise 4. The inequality $1 - \alpha h < (1 - h)^\alpha$ for $0 < h < 1$ and $\alpha > 1$ is readily established by differentiation with respect to h. Put $\alpha = m + 1$ and $h = 1/(n + 1)$ to get, for $n > m$,

$$\left|\frac{a_{n+1}}{a_n}\right| = \frac{n - m}{n + 1} < \left(\frac{n}{n + 1}\right)^{m+1}.$$

Then we have $n^{m+1}|a_n|$ is decreasing and hence $a_n < K(n^{-m-1})$.

For $m < 0$, we have $(n + 1)a_{n+1}/na_n = (n - m)/n > 1$. Therefore $n|a_n|$ increases. Since for $n > m$ all the a_n have the same sign, F-$\sum a_n$ converges if and only if F-$\sum |a_n|$ converges. But F-$\sum |a_n|$ dominates $\sum(1/n)$. Hence F-$\sum a_n$ diverges.

6. The series F-$\sum(b_n - a_n)$ has nonnegative terms, hence it either converges or diverges to $+\infty$. Correspondingly, F-$\sum a_n$ either converges or diverges to $-\infty$.

7. Define $b_n = a_n$ if $n \in$ range φ, $b_n = 0$ if $n \notin$ range φ. Then $a_n = b_n + c_n$. Now F-$\sum b_n$ converges to $\sum a_{\varphi(n)}$ by 13–2.8. Hence F-$\sum a_n$ converges if and only if F-$\sum c_n$ converges by 13–2.4. If it converges, $\sum a_n = \sum c_n + \sum b_n = \sum c_n + \sum a_{\varphi(n)}$.

8. This follows immediately from 13–2.10 and the inequalities $|\text{Re } a_n| \leq |a_n|$, $|\text{Im } a_n| \leq |a_n|$, and $|a_n| \leq |\text{Re } a_n| + |\text{Im } a_n|$.

9. This follows from 13–2.23 and the inequalities $|\text{Re } f(x)| \leq |f(x)|, |\text{Im } f(x)| \leq |f(x)|$, and $|f(x)| \leq |\text{Re } f(x)| + |\text{Im } f(x)|$.

10. Let M be a bound for $|g|$ on X. For any finite subset A of X, $\sum_A |f(x)g(x)| \leq M\sum_A |f(x)|$. Hence if f is summable, the sums $\sum_A |fg|$ are bounded, so fg is summable.

11. If $\{f(n) - n\}$ is bounded by k, then

$$\left| \sum_{n=1}^{p} a_n - \sum_{n=1}^{p} a_{f(n)} \right| \leq 2k \max \{|a_j| \mid j \geq p - k\}.$$

If either series converges, the latter expression tends to zero as $k \to \infty$.

12. Let $X = \mathbf{N}$, $f_n(x) = 0$ if $x \neq n$, $f_n(n) = 1$. Then $\lim_n f_n(x) = 0$ for each x, but $\sum_{x \in X} f_n(x) = 1$ for each n.

13. Let $a_{n,n} = 1$, $a_{n-2,n} = -1$ for all n, and $a_{m,n} = 0$ for all other pairs $\langle m, n \rangle$.

14. Let $A_{p,q}$ be defined by (9) and let ζ be the classical double sum. Given $\epsilon > 0$, we can choose M so that $(\forall p, q > M) |\zeta - A_{p,q}| < \epsilon$. Since we assume that the row sums exist, we have for each fixed p,

$$\lim_q A_{p,q} = \sum_{m=0}^{p} \left(\sum_{n=0}^{\infty} a_{m,n} \right).$$

(This is 13–2.4 extended to a finite number of summands.) Hence, for $p > M$,

$$\left| \zeta - \sum_{m=0}^{p} \left(\sum_{n=0}^{\infty} a_{m,n} \right) \right| \leq \epsilon.$$

This shows that $\sum_m \sum_n a_{m,n} = \zeta$.

We can choose the $a_{m,n}$ to obtain any prescribed double sequence $\{A_{p,q}\}$ of rectangular partial sums. If we arrange that $A_{p,q} = (-1)^q/\min(p, q)$, then the classical double sum is 0, but no row sums exist.

15. From the inequality $4ab \leq (a + b)^2$ for any $a, b \in \mathbf{R}$, we obtain

$$\sqrt{(k + 1)(p - k + 1)} \leq \frac{p + 2}{2}.$$

Hence

$$\sum_{k=0}^{p} \frac{1}{\sqrt{k + 1}\sqrt{p - k + 1}} \geq \frac{2(p + 1)}{p + 2}.$$

The terms of the Cauchy product do not approach zero.

16. Consider the double series $\sum_{k=0}^{\infty} \sum_{m=0}^{\infty} (-1)^{mk}(k + 1)z^{(k+1)(m+1)}$. If we sum on m, we get

$$\sum_{k=0}^{\infty} (k + 1)z^{k+1} \sum_{m=0}^{\infty} (-1)^m((-z)^{k+1})^m = \sum_{k=0}^{\infty} \frac{(k + 1)z^{k+1}}{1 + (-z)^{k+1}}.$$

Putting $n = k + 1$, we get the result $\sum_{n=1}^{\infty} (nz^n/(1 + (-z)^n))$.

If we reverse the order of summation, we get

$$\sum_{m=0}^{\infty} z^{m+1} \sum_{k=0}^{\infty} (-1)^k (k+1)((-z)^{(m+1)})^k = \sum_{m=0}^{\infty} \frac{z^{m+1}}{(1+(-z^{m+1}))^2} = \sum_{n=1}^{\infty} \frac{z^n}{(1+(-z)^n)^2}.$$

The reversal is justified provided $\sum_{k=0}^{\infty}\sum_{m=0}^{\infty} (k+1)|z|^{(k+1)(m+1)}$ is convergent. For $|z| < 1$, this sum is

$$\sum_{k=0}^{\infty} (k+1) \frac{|z|^{k+1}}{1-|z|^{k+1}} < \frac{|z|}{1-|z|} \sum_{k=0}^{\infty} (k+1)|z|^k = \frac{|z|}{(1-|z|)^3}.$$

Pages 216–217

2. The odd partial products decrease, the even ones increase.

3. Since F-$\sum_{n=1}^{\infty} (|z|^2/n^2)$ converges, F-$\prod_{n=1}^{\infty} (1 - z^2/n^2)$ converges absolutely for all values of z.

We have

$$\varphi_k(z) = z \prod_{n=1}^{k} \left(1 - \frac{z^2}{n^2}\right) = \frac{(-1)^k z}{(k!)^2} \prod_{n=1}^{k} (z+n)(z-n) = \frac{(-1)^k}{(k!)^2} \prod_{n=-k}^{k} (z-n).$$

Hence

$$\varphi_k(z+1) = \frac{(-1)^k}{(k!)^2} \prod_{n=-k}^{k} (z+1-n) = \frac{(-1)^k}{(k!)^2} \prod_{n=-k-1}^{k-1} (z-n) = \varphi_k(z) \cdot \frac{z+k+1}{z-k}.$$

When $k \to \infty$, $\varphi(z+1) = -\varphi(z)$.

It is easily seen that $\varphi_k(-z) = -\varphi_k(z)$ for all k and z; therefore

$$\varphi(-z) = -\varphi(z).$$

Now we have

$$\varphi(1-z) = -\varphi(z-1) = \varphi((z-1)+1) = \varphi(z).$$

Actually, it can be shown that $\varphi(z) = (1/\pi) \sin \pi z$.

4. $(na_{n+1}/(n-1)a_n) = 1 + (n/(n-1))\epsilon_n$. If F-$\sum|\epsilon_n|$ converges, then

$$\text{F-}\prod_{n=2}^{\infty} \left(1 + \frac{n}{n-1}\epsilon_n\right)$$

is convergent. Hence

$$ma_{m+1} = a_2 \prod_{n=2}^{m} \left(1 + \frac{n}{n-1}\epsilon_n\right)$$

does not tend to zero. Therefore $a_{m+1} \geq \theta/m$, where $\theta > 0$. Thus F-$\sum a_m$ diverges.

If a_{n+1}/a_n is a rational function of n and $n(1 - a_{n+1}/a_n) \to 1$, then $a_{n+1}/a_n = 1 - 1/n + \epsilon_n$, where ϵ_n is a rational functional of n and $n\epsilon_n \to 0$. Then $n^2\epsilon_n$ is bounded, and F-$\sum|\epsilon_n|$ converges.

5. $\prod_{n=1}^{k} (1 + a_n) \geq 1 + \sum_{n=1}^{k} a_n$. Therefore $\lim_{k\to\infty} \prod_{n=1}^{k} (1 + a_n) = \infty$. Evidently F-$\prod(1 - a_n)$ diverges if $a_n \nrightarrow 0$. Hence assume $0 \leq a_n < 1$ for $n \geq p$. Then, since $(\prod_{n=p}^{k} (1 - a_n))^{-1} \geq \prod_{n=p}^{k} (1 + a_n)$, we have $\prod_{n=p}^{k} (1 - a_n) \to 0$ as $k \to \infty$ for all large p.

6. Let $P_k = \prod_{n=0}^{k} (1 + (-1)^n/2n + 1)$. The even partial products decrease to some number α, and the odd partial products increase to some number β. We have

$$\frac{\beta}{\alpha} = \lim \frac{P_{2k+1}}{P_{2k}} = \lim \left(1 - \frac{1}{4k + 3}\right) = 1.$$

7. Let $P_k = \prod_{n=0}^{k} (1 + (-1)^n/\sqrt{n + 1})$. Then

$$\frac{P_{2m-1}}{P_{2m-3}} = \left(1 + \frac{1}{\sqrt{2m - 1}}\right)\left(1 - \frac{1}{\sqrt{2m}}\right)$$

$$= 1 - \frac{1}{\sqrt{2m}\sqrt{2m - 1}}(1 + \sqrt{2m - 1} - \sqrt{2m})$$

$$= 1 - \frac{1}{\sqrt{2m}\sqrt{2m - 1}}\left[1 - \frac{1}{\sqrt{2m} + \sqrt{2m - 1}}\right].$$

Since $1/\sqrt{2m}\sqrt{2m - 1} > 1/2m$ and $(1 - 1/(\sqrt{2m} + \sqrt{2m - 1})) > \frac{1}{2}$, we have

$$P_{2m-1}/P_{2m-3} < 1 - 1/4m.$$

Hence $P_{2m-1} < P_1(1 - \frac{1}{8})(1 - \frac{1}{12}) \cdots (1 - 1/4m)$. Now, F-$\prod(1 - 1/4m)$ diverges to 0; therefore $P_{2m-1} \to 0$. Also, since $P_{2m}/P_{2m-1} \to 1$, $P_{2m} \to 0$. Therefore, because there are no zero factors, F-$\prod(1 + (-1)^n/\sqrt{n + 1})$ diverges to 0.

8. Since F-$\sum a_n$ converges, there is a p such that $|a_n| < \frac{1}{2}$ for $n \geq p$. Then for $m > k \geq p$,

$$\sum_{n=k+1}^{m} a_n - \sum_{n=k+1}^{m} a_n^2 \leq \log P(k, m) \leq \sum_{n=k+1}^{m} a_n - \frac{1}{3} \sum_{n=k+1}^{m} a_n^2.$$

If F-$\sum a_n^2$ diverges, then $\lim_{m\to\infty} \log P(k, m) = -\infty$, and $\lim_{m\to\infty} P(k, m) = 0$.

Suppose F-$\sum a_n^2$ converges. For any $\epsilon > 0$, we can choose $q \geq p$ so that, for $m > k \geq q$,

$$\log (1 - \epsilon) < \sum_{n=k+1}^{m} a_n - \sum_{n=k+1}^{m} a_n^2 \quad \text{and} \quad \sum_{n=k+1}^{m} a_n < \log (1 + \epsilon).$$

Then we shall have $|P(k, m) - 1| < \epsilon$ for $m > k \geq q$, and 13-3.2 applies.

Rearrange the product of Exercise 7. Take the factor $(1 + 1/\sqrt{1})$, then $(1 - 1/\sqrt{2})$, then just enough even factors to bring the partial product above 1, then just enough odd factors to depress the partial product below 1, etc. Now if F-$\prod(1 + b_n)$ is the rearranged product, we shall have $\prod(1 + b_n) = 1$. Since F-$\sum b_n^2$ has all positive terms and is a rearrangement of F-$\sum_{n=0}^{\infty} (1/(n + 1))$, it diverges. Then the result proved in the first part of the exercise shows that F-$\sum b_n$ must diverge; in fact, $\lim_{k\to\infty} \sum_{n=0}^{k} b_n = +\infty$.

Page 222

3. Let $\sum = A + Bi$. Since $|3 + 4i| = 5$, we have

$$\left|\sum_{n=k}^{\infty} \frac{1}{n(3 + 4i)^n}\right| \leq \frac{1}{k} \sum_{n=k}^{\infty} \frac{1}{5^n} = \frac{1}{4k(5^{k-1})}.$$

Therefore

$$\left| \sum_{n=1}^{\infty} \frac{1}{n(3+4i)^n} - \sum_{n=1}^{4} \frac{1}{n(3+4i)^n} \right| < 0.0001.$$

$$\sum_{n=1}^{4} \frac{1}{n(3+4i)^n} = \frac{12-16i}{100} + \frac{1}{2}\frac{-112-384i}{10000} + \frac{1}{3}\frac{-7488-2816i}{1000000}$$

$$+ \frac{1}{4}\frac{-134912+86016i}{100000000}.$$

The real part of this sum is between 0.1115 and 0.1116, and the imaginary part is between -0.1800 and -0.1799. Hence

$$0.1114 < A < 0.1117 \qquad \text{and} \qquad -0.1801 < B < -0.1798.$$

4. $0.6928 < 1 - \frac{1}{2} + \frac{1}{3} - \frac{1}{4} + \frac{1}{5} - \frac{1}{6} + \frac{1}{7} - \frac{1}{15} < \log 2 < 1 - \frac{1}{2} + \frac{1}{3} - \frac{1}{4} + \frac{1}{5} - \frac{1}{6} + \frac{1}{13} < 0.6936.$

Pages 228–229

2. This is trivial if either A or B is empty. So assume $a \in A$, $b \in B$. Then diam $(A \cup B) \le$ diam $A + \rho(a, b) +$ diam B. If $A \cap B \neq \emptyset$, choose $a = b$.

3. Take $\delta = \epsilon - \rho(p, q)$.

4. Define a sequence of spaces recursively by $T_1 = S_1$, $T_{k+1} = T_k \times S_{k+1}$.

5. Given $\epsilon > 0$, choose $a \in A$ so that $\rho(y, a) < \rho(y, A) + \epsilon$. Then $\rho(x, A) \le \rho(x, a) < \rho(x, y) + \rho(y, A) + \epsilon$. Therefore $\rho(x, A) \le \rho(x, y) + \rho(y, A)$. This shows that $\rho(x, A) - \rho(y, A) \le \rho(x, y)$. Interchanging x and y, we get $\rho(y, A) - \rho(x, A) \le \rho(x, y)$. But these two inequalities give $|\rho(x, A) - \rho(y, A)| \le \rho(x, y)$.

6. If $a_1, a_2, \ldots, a_n$ are nonnegative numbers, then

$$\max a_i^2 \le a_1^2 + a_2^2 + \cdots + a_n^2 \le (a_1 + a_2 + \cdots + a_n)^2$$
$$\le n(a_1^2 + a_2^2 + \cdots + a_n^2) \le n^2 \max a_i^2.$$

These are all trivial except the third, which follows from $bc \le \frac{1}{2}(b^2 + c^2)$, since $\sum_{i,j} a_i a_j \le \sum_{i,j} \frac{1}{2}(a_i^2 + a_j^2) = n\sum_i a_i^2$. If $x = \langle x_1, x_2, \ldots, x_n \rangle \in \mathbf{R}^n$ and $y = \langle y_1, y_2, \ldots, y_n \rangle$, the required inequalities follow by putting $a_i = |x_i - y_i|$.

7. If $s \sim t$ means $\rho(s, t) = 0$, then (a), (c), and (d) show that $\sim$ is reflexive, symmetric, and transitive, respectively. If the quotient map is denoted by an overbar, define $\sigma(\bar{x}, \bar{y}) = \rho(x, y)$ and check that σ is well-defined and is a metric for the quotient set. This construction is applied in 14–7.14.

8. The triangle law for τ boils down to two facts. The function $x \to x/(1+x)$ is increasing for $x \ge 0$, and

$$\frac{x+y}{1+x+y} \le \frac{x}{1+x} + \frac{y}{1+y} \quad \text{for} \quad x, y \ge 0.$$

9. Postulates (a), (b), and (c) are trivially satisfied by σ and $\sigma(x, z) = \sup_i \rho_i(x, z) \le \sup_i(\rho_i(x, y) + \rho_i(y, z)) \le \sup_i \rho_i(x, y) + \sup_i \rho_i(y, z) = \sigma(x, y) + \sigma(y, z)$. The function τ is evidently nonnegative and satisfies (a) and (c). To prove the triangle law, let $x, y, z \in S$ be given. Let a positive ϵ be chosen arbitrarily. Choose $i, j \in I$ so that $\rho_i(x, y) < \tau(x, y) + \epsilon$ and $\rho_j(y, z) < \tau(y, z) + \epsilon$. Since I is linearly ordered, we

must have either $i \leq j$ or $j \leq i$. Suppose the former. Then $\tau(x, z) \leq \rho_i(x, z) \leq$ $\rho_i(x, y) + \rho_i(y, z) < \tau(x, y) + \epsilon + \rho_i(y, z) < \tau(x, y) + \tau(y, z) + 2\epsilon$. We obtain the same inequality if $j \leq i$. Since ϵ was arbitrary, we conclude $\tau(x, z) \leq \tau(x, y) + \tau(y, z)$.

10. The four-point metric space $\{a, b, c, d\}$ with metric given by

	a	b	c	d
a	0	α	$\alpha + \beta$	β
b	α	0	β	$\alpha + \beta$
c	$\alpha + \beta$	β	0	α
d	β	$\alpha + \beta$	α	0

where α and β are positive, is not isometric to a subset of $\mathbf{R}$ although each three-point subspace is. A metric space with at least five points with the property that each three-point subset is isometric to a subset of $\mathbf{R}$ is itself isometric to a subset of $\mathbf{R}$.

Pages 231–232

1. Suppose p adheres to E. Say $\{x_n\}$ is a sequence in E with $x_n \to p$. Then $\rho(p, E) \leq$ $\rho(p, x_n)$ for every n. Since $\rho(p, x_n) \to 0$, $\rho(p, E) = 0$. Conversely, if $\rho(p, E) = 0$, we can choose, for each n, $x_n \in E$ so that $\rho(p, x_n) < 1/n$. Then $x_n \to p$, so p adheres to E.

2. Suppose p adheres to the sequence $\{x_n\}$. We can define the function g inductively so that $g(n + 1) > g(n)$ and $\rho(p, x_{g(n)}) < 1/n$. Then $\{x_{g(n)}\}$ is a subsequence converging to p. Conversely, suppose $\{x_{g(n)}\}$ is a subsequence converging to p. Then every ball about p contains infinitely many, in fact all but a finite number, of the terms of $\{x_{g(n)}\}$. Because $g(n) \to \infty$, this implies that every ball about p contains infinitely many of the terms of $\{x_n\}$.

3. Assuming that p is a cluster point of E, we say that $\lim_{s \to p} f(s)$ exists provided there is a $t \in T$ such that

$$(\forall \epsilon > 0)(\exists \delta > 0)(\forall s \in E) \quad 0 < \rho(s, p) < \delta \Rightarrow \rho(f(s), t) < \epsilon. \qquad (*)$$

Note that the value of f at p, if f should happen to be defined at p, has nothing to do with the limit. It is easy to prove that there is at most one t satisfying $(*)$. Hence it is appropriate to denote it by $\lim_{s \to p} f(s)$ when it exists. If p were not a cluster point of E, then every $t \in T$ would satisfy the condition $(*)$.

4. All eight patterns of existence and nonexistence are possible. If the double limit exists, then it is equal to whichever iterated limits may exist. If the double limit does not exist, the two iterated limits may exist and be different.

Page 239

1. If $p, q, r \in S$,

$$\rho(p, q) \leq \rho(p, r) + \rho(q, r) \leq 2 \max \{\rho(p, r), \rho(q, r)\} = 2\theta(\langle p, q \rangle, \langle r, r \rangle).$$

Hence if $p \neq q$, the ball in $S \times S$ of radius $\frac{1}{2}\rho(p, q)$ about $\langle p, q \rangle$ does not meet the diagonal Δ. Hence $\langle p, q \rangle \notin \bar{\Delta}$. Therefore $\bar{\Delta} = \Delta$.

2. Suppose $X \times Y$ is closed in $S \times T$. Let $\{x_n\}$ be a sequence in X converging to p in S. We must prove $p \in X$. Choose $y \in Y$. Then $\{\langle x_n, y \rangle\}$ is a sequence in $X \times Y$

converging in $S \times T$ to $\langle p, y \rangle$. Since $X \times Y$ is closed, $p \in X$. Similarly, Y is closed. The opposite implication follows immediately from 14–2.6.

3. Let $\epsilon > 0$ be given. If $p, q \in \overline{E}$, we can choose $x, y \in E$ so that $\rho(p, x) < \epsilon$ and $\rho(q, y) < \epsilon$. Then $\rho(p, q) \leq \rho(x, y) + 2\epsilon \leq \operatorname{diam} E + 2\epsilon$. This proves $\operatorname{diam} \overline{E} \leq \operatorname{diam} E + 2\epsilon$. Since ϵ is arbitrary, $\operatorname{diam} \overline{E} \leq \operatorname{diam} E$. The opposite inequality is trivial because $E \subseteq \overline{E}$.

4. If $y \notin C(p, \epsilon)$, then $B(y, \delta) \cap C(p, \epsilon) = \emptyset$ where $\delta = \rho(p, y) - \epsilon$. Hence $C(p, \epsilon)$ is closed. If $S = I$ with the usual metric, then $\overline{B(0, 1)} = \overline{\{0\}} = \{0\}$ and $C(0, 1) = \{-1, 0, 1\}$.

5. Let $\langle x, y \rangle \in X \times Y$. Then $B(\langle x, y \rangle, \epsilon) \subseteq X \times Y$ if and only if $B(x, \epsilon) \subseteq X$ and $B(y, \epsilon) \subseteq Y$. If X and Y are both open, an ϵ satisfying the latter condition can always be chosen, so $X \times Y$ is open. Conversely, assume Y is not void and $X \times Y$ is open. For each $x \in X$ we can choose ϵ so that $B(\langle x, y \rangle, \epsilon) \subseteq X \times Y$, whence $B(x, \epsilon) \subseteq X$, and X is open.

6. (a) $p \in E' \Leftrightarrow p \in \overline{E - \{p\}} \Rightarrow p \in \overline{E}$; hence $E' \subseteq \overline{E}$. $p \in \overline{E} - E \Rightarrow E - \{p\} = E$, so $p \in \overline{E - \{p\}} = \overline{E}$; thus $\overline{E} - E \subseteq E'$. Together with $E \subseteq \overline{E}$, these inclusions give $\overline{E} = E \cup E'$. (b) and (c) Suppose every neighborhood of p contains at least two points of E. Then every neighborhood of p contains at least one point of $E - \{p\}$, so $p \in \overline{E - \{p\}}$; i.e., $p \in E'$. Suppose $p \in E'$. Define a sequence by inductive choice. Choose $x_1 \in B(p, 1) \cap (E - \{p\})$. If $x_1, x_2, \ldots, x_{n-1}$ have been chosen, then $B(p, 1/n) - \{x_1, x_2, \ldots, x_{n-1}\}$ is a neighborhood of p, so we can choose x_n in $(B(p, 1/n) - \{x_1, x_2, \ldots, x_{n-1}\}) \cap (E - \{p\})$. Then $\{x_n\}$ is an injective sequence in E, and $x_n \to p$. Now suppose $\{x_n\}$ is an injective sequence in E and $x_n \to p$. Then every neighborhood of p contains infinitely many points in the range of the sequence; a fortiori, it contains at least two points of E. This circular chain of three implications proves (b) and (c). (d) Suppose p adheres to E'. Let $\epsilon > 0$ be given. Choose $q \in B(p, \epsilon) \cap E'$. Every neighborhood of q, in particular $B(p, \epsilon)$, contains at least two points of E and therefore at least one point of $E - \{p\}$. Hence $B(p, \epsilon) \cap (E - \{p\}) \neq \emptyset$. Thus $p \in E'$. This proves that E' is closed.

7. If E is a dense subset of a metric space and $y \in E$, then $E - \{y\}$ is dense unless y is isolated. Hence if there is a minimal dense subset, all of its points are isolated. Conversely, any dense subset of a metric space S must contain all of the isolated points of S. Hence S contains a minimal dense subset if and only if the set of isolated points of S is dense in S, and a minimal dense set is the minimum dense set when such a set exists. This can occur in nondiscrete spaces. For example, let S be the subset of $\mathbf{R}$ consisting of 0 and all points $1/n$ for $n \in \mathbf{N}$. Here 0 is a limit point of isolated points.

Page 245

1. This follows immediately from the inequality of Exercise 5, p. 228.

3. Let π be the projection function from $S \times T$ to S. If G is an open subset of S, then $\pi^{-1}(G) = G \times T$ is open in $S \times T$ by Exercise 5, p. 239 (the cases where T or G is void are trivial). Hence π is continuous. Let H be an open set in $S \times T$ and suppose $x \in \pi(H)$. Choose $y \in T$ so that $\langle x, y \rangle \in H$. There is an $\epsilon > 0$ such that $B(\langle x, y \rangle, \epsilon) = B(x, \epsilon) \times B(y, \epsilon) \subseteq H$. Then $B(x, \epsilon) \subseteq \pi(H)$. This proves that $\pi(H)$ is open. Therefore, π is an open function.

4. The triangle law gives $-\rho(s, p) \le \rho(s, t) - \rho(p, t) \le \rho(s, p)$; hence $|f_s(t)| \le \rho(s, p)$ for all t. A similar calculation shows that

$$|f_r(t) - f_s(t)| = |\rho(r, t) - \rho(s, t)| \le \rho(r, s)$$

for all t, whence $\rho(f_r, f_s) \le \rho(r, s)$. This implies that $s \to f_s$ is continuous. Choosing $t = s$ gives $|f_r(s) - f_s(s)| = \rho(r, s)$, so $\rho(f_r, f_s) = \rho(r, s)$. Thus $s \to f_s$ is an isometry.

5. Let g be the least common extension of the family $\{f_i\}$. If X is any closed subset of T, then $g^{-1}(X) = \bigcup_{i=1}^{n} f_i^{-1}(X)$. Since f_i is continuous, $f_i^{-1}(X)$ is closed relative to E_i; and since E_i is closed, $f_i^{-1}(X)$ is closed in S. Therefore $\bigcup_{i=1}^{n} f_i^{-1}(X)$ is a closed set in S. Hence g is continuous.

Page 249

1. This follows immediately from the inequalities of Exercise 6, p. 228.

2. In the proof of 14–5.3 the choice of δ can now be made independent of q.

3. The family $\{f_i \mid i \in I\}$ of functions from $\langle S, \rho \rangle$ to $\langle T, \sigma \rangle$ is equicontinuous if and only if

$$(\forall p)(\forall \epsilon > 0)(\exists \delta > 0)(\forall i \in I)(\forall q \in S) \quad \rho(p, q) < \delta \Rightarrow \sigma(f_i(p), f_i(q)) < \epsilon.$$

Suppose $\{f_n\}$ is an equicontinuous sequence of functions and $f_n \to g$ pointwise. Let $\epsilon > 0$ and p be given. Choose δ so that

$$(\forall n \in \mathbf{N})(\forall q \in S) \quad \rho(p, q) < \delta \Rightarrow \sigma(f_n(p), f_n(q)) < \epsilon.$$

Then, if $\rho(p, q) < \delta$, we have $\sigma(f_n(p), f_n(q)) < \epsilon$ for all n. Since the metric is continuous, $\sigma(g(p), g(q)) = \lim_n \sigma(f_n(p), f_n(q)) \le \epsilon$. Thus g is continuous.

Suppose $\{f_n\}$ is a sequence of continuous function and that $f_n \to g$ uniformly. Let $p \in S$ and $\epsilon > 0$ be given. By Theorem 14–5.3, g is continuous. Choose δ_0 so that $(\forall q \in S) \rho(p, q) < \delta_0 \Rightarrow \sigma(g(p), g(q)) < \epsilon/3$. Choose k so that

$$(\forall n > k)(\forall s \in S) \quad \sigma(g(s), f_n(s)) < \epsilon/3.$$

Then we have $(\forall n > k)(\forall q \in S) \rho(p, q) < \delta_0 \Rightarrow \sigma(f_n(p), f_n(q)) \le \sigma(f_n(p), g(p)) + \sigma(g(p), g(q)) + \sigma(g(q), f_n(q)) < \epsilon$. For each $n \le k$, we can choose δ_n so that

$$(\forall q \in S) \quad \rho(p, q) < \delta_n \Rightarrow \sigma(f_n(p), f_n(q)) < \epsilon.$$

Hence if $\delta = \min \{\delta_0, \delta_1, \ldots, \delta_k\}$, we have

$$(\forall n \in \mathbf{N})(\forall q \in S) \quad \rho(p, q) < \delta \Rightarrow \sigma(f_n(p), f_n(q)) < \epsilon.$$

This proves that $\{f_n\}$ is equicontinuous.

4. To say that $\{f_n(x)\}$ is a Cauchy sequence for each x means

$$(\forall x)(\forall \epsilon > 0)(\exists k \in \mathbf{N})(\forall n, m > k) \quad |f_n(x) - f_m(x)| < \epsilon.$$

To say that $\{f_n(x)\}$ is a Cauchy sequence uniformly in x means

$$(\forall \epsilon > 0)(\exists k \in \mathbf{N})(\forall n, m > k)(\forall x) \quad |f_n(x) - f_m(x)| < \epsilon.$$

If $\{f_n(x)\}$ is a Cauchy sequence for each x, we can define $g(x) = \lim f_n(x)$. If it is a

Cauchy sequence uniformly in x, then given $\epsilon > 0$, we choose k so that

$$|f_n(x) - f_m(x)| < \epsilon$$

for all m, $n > k$ and all x. Letting $m \to \infty$, we get $|f_n(x) - g(x)| \leq \epsilon$ for all $n > k$ and all x. This shows that $f_n \to g$ uniformly.

Pages 252–253

1. Any two bounded (nonvoid) open intervals in $\mathbf{R}$ or any two semi-infinite open intervals are homeomorphic under a function of the form $x \to ax + b$. The map $x \to (2x - 1)/x(1 - x)$ is a homeomorphism of $(0, 1)$ onto $\mathbf{R}$, and the map $x \to x - 1/x$ is a homeomorphism of $(0, \infty)$ onto $\mathbf{R}$.

2. Any two bounded closed intervals in $\mathbf{R}$ are homeomorphic under a function of the form $x \to ax + b$.

3. Let p be a point in the interior of the triangle. Then radial projection defines a bicontinuous function from the triangle to any circle with center at p.

4. If f is any continuous function on $\mathbf{R}$ which is everywhere positive and periodic with period 2π, then $h: \langle \rho, \theta \rangle \to \langle \rho/f(\theta), \theta \rangle$ (polar coordinates) defines a homeomorphism of the whole plane with itself. If the origin is in the interior of a triangle and $f(\theta)$ is the distance from the origin to the triangle along the θ-ray, then h carries the interior of the triangle onto the interior of the unit disk.

5. The rational numbers have no isolated points; the given set does.

6. If f is a homeomorphism of S_1 onto S_2 and g is a homeomorphism of T_1 onto T_2, then $\langle s, t \rangle \to \langle f(s), g(t) \rangle$ is a homeomorphism of $S_1 \times T_1$ onto $S_2 \times T_2$.

7. Assume that the set of sequences in S which converge in $\langle S, \rho_1 \rangle$ is the same as the set of sequences which converge in $\langle S, \rho_2 \rangle$. If we can prove that a sequence $\{x_n\}$ that converges in both metrics has the same limit in either metric, then ρ_1 and ρ_2 are topologically equivalent by 14–4.3 and 14–6.2. If $x_n \to y$ in $\langle S, \rho_1 \rangle$, then $x_1, y, x_2, y, \ldots$ (How do you write this formally?) is a convergent sequence in $\langle S, \rho_1 \rangle$. Therefore it converges in $\langle S, \rho_2 \rangle$, and hence $x_n \to y$ in $\langle S, \rho_2 \rangle$.

The opposite implication is straightforward.

8. The map $\langle s, f(s) \rangle \to s$, being the restriction of the coordinate projection is continuous in any case. If f is continuous, $s \to \langle s, f(s) \rangle$ is continuous by 14–4.13, and is therefore a homeomorphism. Conversely, if $s \to \langle s, f(s) \rangle$ is a homeomorphism, it is continuous; now f is the composition of $s \to \langle s, f(s) \rangle$ with the second coordinate projection, so it is continuous by 14–4.9.

9. That σ is a metric was shown in Exercise 9, p. 228. Suppose $s_n \to t$ in $\langle S, \rho_1 \rangle$. We are given that then $s_n \to t$ in $\langle S, \rho_2 \rangle$ also. Hence

$$\sigma(s_n, t) = \sup \{\rho_1(s_n, t), \rho_2(s_n, t)\} \to 0;$$

i.e., $s_n \to t$ in $\langle S, \sigma \rangle$. Conversely, since σ dominates ρ, if $\sigma(s_n, t) \to 0$, then $\rho(s_n, t) \to 0$. Thus the identity function from $\langle S, \rho_1 \rangle$ to $\langle S, \sigma \rangle$ is bicontinuous by 14–4.3.

10. Suppose σ is continuous on $S \times S$ in the metric induced by ρ. If $s_n \to t$ in $\langle S, \rho \rangle$, then $\sigma(s_n, t) \to \sigma(t, t) = 0$ by the continuity of σ; that is, $s_n \to t$ in $\langle S, \sigma \rangle$. Since σ dominates ρ, if $\sigma(s_n, t) \to 0$, then $\rho(s_n, t) \to 0$. Thus the identity function from $\langle S, \rho \rangle$ to $\langle S, \sigma \rangle$ is bicontinuous by 14–4.3.

Conversely, suppose σ is topologically equivalent to ρ. Then σ is continuous on $S \times S$ in the metric induced by σ (14–4.21). But $S \times S$ with the metric induced by σ is homeomorphic to $S \times S$ with the metric induced by ρ. Hence σ is continuous in the metric induced by ρ.

It is obvious that the given function σ dominates ρ. To see that it is continuous, note that σ is composed of ρ, f, coordinate projections, absolute value, and arithmetic operations, all of which are continuous.

11. *Hint:* Yes. The proof is on page 394.

Pages 265–266

1. Suppose $x_{g(n)} \rightarrow y$. Let $\epsilon > 0$ be given. Choose k so that

$$(\forall m, n > k) \quad \rho(x_m, x_n) < \epsilon/2.$$

Choose p so that $g(p) > k$ and $\rho(y, x_{g(p)}) < \epsilon/2$. Then $(\forall n > k) \, \rho(y, x_n) \le \rho(y, x_{g(p)}) + \rho(x_{g(p)}, x_n) < \epsilon$. Thus $x_n \rightarrow y$.

2. If $A_k = \sum_{n=1}^{k} \rho(x_n, x_{n+1})$, then

$$\rho(x_n, x_{n+k}) \le A_{n+k-1} - A_{n-1} = |A_{n+k-1} - A_{n-1}|.$$

Since $\{A_n\}$ is a Cauchy sequence in $\mathbf{R}$, $\{x_n\}$ is a Cauchy sequence in S.

3. For each k, let $g(k)$ be chosen so that $(\forall m, n > g(k)) \, \rho(x_m, x_n) < 1/2^k$. Arrange also that g is increasing. Then $\rho(x_{g(n)}, x_{g(n+1)}) < 1/2^n$, so F-$\sum \rho(x_{g(n)}, x_{g(n+1)})$ is convergent by the comparison test.

4. Yes, provided neither is void. A Cauchy sequence $\{s_n\}$ in S lifts to a Cauchy sequence $\{\langle s_n, t \rangle\}$ in $S \times T$.

5. Since the hypothesis is symmetric, we need only prove this one way. Suppose $\langle T, \sigma \rangle$ is complete, f is uniformly continuous from $\langle S, \rho \rangle$ to T, and f^{-1} is continuous from T to S. Let $\{s_n\}$ be a Cauchy sequence in S. Let $\epsilon > 0$ be given. Choose δ so that $(\forall x, y) \, \rho(x, y) < \delta \Rightarrow \sigma(f(x), f(y)) < \epsilon$. Choose k so that $(\forall m, n > k) \, \rho(s_m, s_n) < \delta$. Then $(\forall m, n > k) \, \sigma(f(s_m), f(s_n)) < \epsilon$. Thus $\{f(s_n)\}$ is a Cauchy sequence in T. Since T is complete, say $f(s_n) \rightarrow t$. Then $s_n \rightarrow f^{-1}(t)$ because f^{-1} is continuous. Thus S is complete. Note that we did not use quite all of the hypothesis to prove this half.

6. It was shown in Exercise 4, p. 245, that there is an isometry φ from T to a subset of the set C of all real continuous functions from T to $\mathbf{R}$. Since this space is complete, we can take $S = \overline{\varphi(T)}$ with the metric τ inherited from C.

If we have two such triples $\langle S_1, \tau_1, \varphi_1 \rangle$ and $\langle S_2, \tau_2, \varphi_2 \rangle$. Then $\psi_1 = \varphi_2 \cdot \varphi_1^{-1}$ is an isometry from a dense subset of S_1 to a dense subset of S_2, and $\psi_2 = \varphi_1 \cdot \varphi_2^{-1}$ is an isometry from a dense subset of S_2 to a dense subset of S_1. Since an isometry is clearly uniformly continuous, 14–7.13 applies to show that ψ_1 and ψ_2 can be extended to be continuous functions θ_1 and θ_2 from S_1 to S_2 and from S_2 to S_1, respectively. Now $\theta_1 \circ \theta_2$ extends $\psi_1 \circ \psi_2$, which is the identity on a dense subset of S_2. Hence $\theta_1 \circ \theta_2$ is the identity of S_2. Similarly, $\theta_2 \circ \theta_1$ is the identity of S_1. Thus θ_1 and θ_2 are inverses of each other; since they are easily seen not to increase distances, they are isometries.

7. There is no loss in generality in taking $a = 1$. Define f on the dyadic rationals in $[0, 1]$ inductively. Put $f(0) = p, f(1) = q$. Suppose f has been defined on dyadic rationals

with denominator 2^k so that

$$(\forall m, n) \quad \rho\left(f\left(\frac{m}{2^k}\right), f\left(\frac{n}{2^k}\right)\right) = \frac{|m - n|}{2^k}. \tag{*}$$

Choose $f((2m + 1)/2^{k+1})$ so that

$$\rho\left(f\left(\frac{m}{2^k}\right), f\left(\frac{2m + 1}{2^{k+1}}\right)\right) = \rho\left(f\left(\frac{2m + 1}{2^{k+1}}\right), f\left(\frac{m + 1}{2^k}\right)\right)$$

$$= \frac{1}{2}\rho\left(f\left(\frac{m}{2^k}\right), f\left(\frac{m + 1}{2^k}\right)\right).$$

Condition (*) with k replaced by $k + 1$ is now readily verified. This defines an isometry f from the set of dyadic rationals in $[0, 1]$ to S. Extend f over all of $[0, 1]$ by 14–7.13. It is easy to check that the extension is an isometry.

8. If $x, y \in [\sqrt{a/2}, \infty)$, then

$$\left|\frac{1}{2}\left(x + \frac{a}{x}\right) - \frac{1}{2}\left(y + \frac{a}{y}\right)\right| = \frac{1}{2}\left|1 - \frac{a}{xy}\right| |x - y| \le \tfrac{1}{2}|x - y|.$$

The range of the map is $[\sqrt{a}, \infty) \subseteq S$.

9. Let S be the interval $[x_0 - a, x_0 + a]$ where a is so small that $|f'| \le \lambda < 1$ on S. Then if $x, y \in S$, the mean-value theorem shows that $|f(x) - f(y)| = |f'(\xi)||x - y| \le \lambda|x - y|$ where ξ is between x and y. This inequality, with $y = x_0$, shows that $f(S) \subseteq S$.

Suppose instead $|f'(x_0)| > 1$. Choose b so that $|f'| \ge \mu > 1$ on $J = [x_0 - b, x_0 + b]$. If $y \in J$, the mean-value theorem gives $|f(y) - x_0| \ge \mu|y - x_0|$. Then, if $y, f(y), \ldots,$ $f^{n-1}(y) \in J$, we have $|f^n(y) - x_0| \ge \mu^n |y - x_0|$. Since $\mu^n \to \infty$, it is impossible that $(\forall n)\ f^n(y) \in J$ unless $y = x_0$.

11. It is easy to prove that f has at most one fixed point. Put $x_1 = x, \ldots, x_{n+1} = f(x_n)$. Suppose y adheres to $\{x_n\}$, say $y = \lim_{k \to \infty} x_{g(k)}$. Then $f(y) = \lim_{k \to \infty} x_{g(k)+1}$ and $f(f(y)) = \lim_{k \to \infty} x_{g(k)+2}$, since the condition clearly implies that f is continuous. It also implies that $\{\rho(x_n, x_{n+1})\}$ is a decreasing sequence. Say $\lim_{n \to \infty} \rho(x_n, x_{n+1}) = \theta$. Now

$$\rho(y, f(y)) = \lim_{n \to \infty} \rho(x_{g(n)}, x_{g(n)+1}) = \theta = \lim_{n \to \infty} \rho(x_{g(n)+1}, x_{g(n)+2}) = \rho(f(y), f(f(y))).$$

But $y \ne f(y)$ would imply $\rho(f(y), f(f(y))) < \rho(y, f(y))$. Thus $y = f(y)$, and y is fixed.

Now since $\{\rho(y, x_n)\}$ is a decreasing sequence which has a subsequence decreasing to zero, $\rho(y, x_n) \to 0$. Hence $x_n \to y$.

12. Evidently $\bigcup_n \text{Int } E_n$ is an open subset of S. Suppose it is not dense. Then there is a nonvoid open set G in S with $G \cap \bigcup_n \text{Int } E_n = \emptyset$. Now $G = \bigcup_n (G \cap E_n)$. By the Baire category theorem, not all of the sets $G \cap E_n$ are nowhere dense, i.e.,

$$\text{Int } (G \cap E_n) \ne \emptyset$$

for some n. For this n, we easily find that $G \cap \text{Int } E_n \ne \emptyset$, a contradiction.

Pages 272–273

1. Suppose T is a subspace of a totally bounded space S. Given $\epsilon > 0$, let $\{s_1, s_2, \ldots, s_n\}$ be $\epsilon/2$-dense in S. Choose $t_i \in B(s_i, \epsilon/2) \cap T$ whenever this set is not empty. Then $\{t_i\}$ is ϵ-dense in the subspace T.

2. If X is a finite ϵ-dense set in S and Y is a finite ϵ-dense set in T, then $X \times Y$ is a finite ϵ-dense set in $S \times T$.

3. One implication was established in the proof of 14–8.5, part (B).

Suppose $\langle S, \rho \rangle$ is a totally bounded metric space and let $\{x_n\}$ be a sequence in S. We shall construct a subsequence satisfying Cauchy's criterion.

Let S be divided into a finite number of disjoint sets of diameter less than 1. One of these sets, say S_1, contains infinitely many of the $\{x_n\}$ (i.e., $\{n \mid x_n \in S_1\}$ is infinite). Let S_1 be divided into a finite number of disjoint sets of diameter less than $\frac{1}{2}$. One of them, say S_2, contains infinitely many of the $\{x_n\}$. Continue by induction. Suppose S_k is a set of diameter at most $1/k$ and that $\{n \mid n \in S_k\}$ is infinite. Since S_k is totally bounded, it can be divided into a finite collection of sets $\{T_i \mid 1 \leq i \leq m\}$ each of which has diameter less than $1/(k + 1)$. Then, since $\bigcup_{i=1}^{m} \{n \mid x_n \in T_i\} = \{n \mid x_n \in S_k\}$, at least one of these sets, say $\{n \mid x_n \in T_j\}$, is infinite, and we put $S_{k+1} = T_j$. We thus construct a decreasing sequence of sets $\{S_k\}$ such that, for all k, diam $S_k \leq 1/k$ and $\{n \mid x_n \in S_k\}$ is infinite.

Now let $f(1)$ be the least index such that $x_{f(1)} \in S_1$. Let $f(2)$ be the least index greater than $f(1)$ with $x_{f(2)} \in S_2$. In general, let $f(k)$ be the least index greater than $f(k - 1)$ with $x_{f(k)} \in S_k$.

Now it is clear that $x_{f(k)} \in S_p$ if $k \geq p$. Since diam $S_p < 1/p$, we have $(\forall k, m \geq p)$ $\rho(x_{f(k)}, x_{f(m)}) < 1/p$. Hence $\{x_n\}$ is a Cauchy sequence.

4. To prove this it is sufficient that f be uniformly continuous and surjective. The corresponding result for completeness (Exercise 5, p. 265) is that f must be continuous and f^{-1} uniformly continuous. Neither completeness nor total boundedness is a topological property, but both are uniform properties (that is, if one of two uniformly equivalent spaces is complete or totally bounded, so is the other).

Suppose f is a homeomorphism from S to T. Our results show that if f is uniformly continuous, then we may gain, but cannot lose, total boundedness on passing from S to T. On the other hand, we may lose, but cannot gain, completeness.

5. By passing from a family $\{F_\alpha\}$ of closed sets to the family $\{G_\alpha = S - F_\alpha\}$ of open sets and vice versa and using $\bigcap_{\alpha \in B} F_\alpha = S - \bigcup_{\alpha \in B} G$, we see that

(A) every family of closed sets in S having the finite intersection property has a non-void intersection

is equivalent to

(B) every family of open sets in S containing no finite covering of S is not a covering.

This, in turn, is equivalent to

(C) every open covering of S contains a finite subcovering; i.e., S is compact.

6. If X_β is compact, then $\{X_\alpha \mid X_\alpha \subseteq X_\beta\}$ is a family of closed subsets in the compact space X_β having the finite-intersection property. By Exercise 5, their intersection is not void. Since all the other X_α contain X_β, $\bigcap_{\alpha \in A} X_\alpha \neq \emptyset$.

7. A closed discrete subset of a compact space is itself a discrete compact space and therefore finite. Hence a compact space contains no closed, infinite, discrete subset.

Conversely, suppose S is not compact. Either S is not totally bounded, or S is not complete. If S is not totally bounded, we can find an $\epsilon > 0$ and an infinite sequence $\{x_n\}$ in S with $\rho(x_j, x_k) \geq \epsilon$ for all j, k. (See part (B) of the proof of 14–8.5.) If S is not complete, let $\{x_n\}$ be a Cauchy sequence which does not converge. In either case, the points x_n form a closed, infinite, discrete subset of S.

8. If $\{G_\alpha \mid \alpha \in A\}$ is a covering of $E_1 \cup E_2 \cup \cdots \cup E_k$ by open sets of S, we can find finite subsets B_i of A such that $\{G_\alpha \mid \alpha \in B_i\}$ is a covering of E_i for each i. Then $\{G_\alpha \mid \alpha \in \bigcup_i B_i\}$ is a finite covering of $E_1 \cup E_2 \cup \cdots \cup E_k$. Therefore, the latter is compact by 14–8.6.

9. In view of 14–8.5 we need only show that the completion $\overline{S}$ of S is totally bounded. If X is a finite set $\epsilon/2$-dense in S, then X is ϵ-dense in $\overline{S}$; hence $\overline{S}$ is totally bounded.

10. Suppose f is continuous. We first show that f is a closed subset of $S \times T$, regardless of whether T is compact. Suppose $\langle s, t \rangle$ adheres to f. Then there is a sequence $\{\langle s_n, t_n \rangle\}$ in f such that $s_n \to s$ and $t_n \to t$. But $\langle s_n, t_n \rangle \in f$ means $t_n = f(s_n)$. Then $t = f(s)$ by 14–4.3; so $\langle s, t \rangle \in f$. Thus f is closed.

Now suppose f is a closed subset of $S \times T$ and that T is compact. Suppose $s_n \to s$ in S. We must prove $f(s_n) \to f(s)$. Suppose a subsequence $\{f(s_{\sigma(n)})\}$ converges in T, say to t. Then $\langle s_{\sigma(n)}, f(s_{\sigma(n)}) \rangle \to \langle s, t \rangle$. Since f is closed, $\langle s, t \rangle \in f$. Therefore $f(s) = t$. Thus we have shown that every convergent subsequence of $\{f(s_n)\}$ converges to $f(s)$. Since T is compact, every subsequence of $\{f(s_n)\}$ contains a convergent subsequence. By 14–2.5, $f(s_n) \to f(s)$.

Suppose T is not compact. Then there is a sequence $\{t_n\}$ in T which has no convergent subsequence. Take $S = \{0\} \cup \{1/n \mid n \in \mathbf{N}\}$, and define

$$f = \{\langle 0, t_1 \rangle\} \cup \{\langle 1/n, t_n \rangle \mid n \in \mathbf{N}\}.$$

Then f is a closed subset of $S \times T$, but f is not continuous.

11. Suppose $p \in F \cap G$. We must show that p has a compact neighborhood in $F \cap G$. There is a number $\epsilon > 0$ such that $\overline{B(p, \epsilon)}$ is compact and $B(p, \epsilon) \subseteq G$. Then $\overline{B(p, \tfrac{1}{2}\epsilon)}$ is a compact subset of G and $F \cap \overline{B(p, \tfrac{1}{2}\epsilon)}$, being a closed subset of a compact set, is compact. Now $F \cap \overline{B(p, \tfrac{1}{2}\epsilon)}$ is a compact neighborhood of p in $F \cap G$.

Now let E be a subspace of S which is locally compact in its inherited topology. Let $p \in E$. There is a compact set N which is a neighborhood of p relative to E. There is an open neighborhood H of p in S such that $H \cap E \subseteq N$. Since N is compact, it is closed in S. Therefore $\overline{E} \cap H \subseteq \overline{E \cap H} \subseteq N \subseteq E$ (the first inclusion holds for H open and any E). Thus every point of E has an open neighborhood H in S such that $\overline{E} \cap H \subseteq E$. For each $x \in E$ choose such a neighborhood H_x. Put $G = \bigcup_x H_x$ and $F = \overline{E}$. Then $E = F \cap G$.

12. If U is a compact neighborhood of s in S and V is a compact neighborhood of t in T, then $U \times V$ is a compact neighborhood of $\langle s, t \rangle$ in $S \times T$. It follows immediately that the direct product of two locally compact spaces is locally compact.

13. We shall prove that no function g has a compact neighborhood in S. Construct a sequence $\{f_n\}$ of functions continuous from $[0, 1]$ to $[0, 1]$ as follows: f_1 is zero outside $[\tfrac{1}{3}, 1]$, but $f(\tfrac{1}{2}) = 1$; f_2 is zero outside $[\tfrac{1}{5}, \tfrac{1}{3}]$, but $f_2(\tfrac{1}{4}) = 1$; $\ldots$, f_k is zero outside $[1/(2k + 1), 1/(2k - 1)]$, but $f_k(1/2k) = 1$, etc. Then for $m \neq n$, $\rho(f_m, f_n) = 1$. Therefore this sequence has no convergent subsequence.

If N is any neighborhood of g, then for a suitably small value of α, the sequence $\{g + \alpha f_n\}$ is in N and has no convergent subsequence. Thus N is not compact.

14. Suppose φ is an isometry from a compact metric space S to a proper subset of S. Pick any point x of $S - \varphi(S)$. Since $\varphi(S)$ is compact, it is closed, and $\rho(x, \varphi(S)) = \alpha > 0$.

Now define $x_0 = x$, $x_1 = \varphi(x)$, $x_2 = \varphi(x_1), \ldots, x_{k+1} = \varphi(x_k), \ldots$ Evidently $\rho(x_0, x_n) \geq \alpha$ for all $n > 0$; and since φ is an isometry, $\rho(x_p, x_{n+p}) \geq \alpha$. Thus the sequence $\{x_n\}$ can have no convergent subsequence, contradicting the fact that S is compact.

15. *First proof.* If no such number existed, then for every $n \in \mathbf{N}$, choose x_n so that $B(x_n, 1/n)$ is not contained in any member of the given covering. The sequence $\{x_n\}$ has a convergent subsequence, say $x_{g(k)} \to y$. Choose β so that $y \in G_\beta$, and suppose $B(y, 2\delta) \subseteq G_\beta$. Now k can be so chosen that $\rho(y, x_{g(k)}) < \delta$ and $g(k) > 1/\delta$. This leads to $B(x_{g(k)}, 1/g(k)) \subseteq G_\beta$, contradicting the choice of the x's.

Second proof. For each point $x \in S$, choose $\delta(x) > 0$ so that

$$(\exists \alpha \in A) \quad B(x, 2\delta(x)) \subseteq G_\alpha.$$

Now $\{B(x, \delta(x)) \mid x \in S\}$ is a covering of S and it has a finite subcovering

$$\{B(x, \delta(x)) \mid x \in Y\}.$$

A Lebesgue number is $\epsilon = \min\{\delta(x) \mid x \in Y\}$. If $s \in S$, then for some $x \in Y$, $\rho(x, s) < \delta(x)$; hence $B(s, \epsilon) \subseteq B(x, \delta(x) + \epsilon) \subseteq B(x, 2\delta(x))$. Therefore for some α $B(s, \epsilon) \subseteq G_\alpha$.

16. Suppose $x_n \to y$ in S. Then $\{y\} \cup \{x_n \mid n \in \mathbf{N}\}$ is a compact subset of S. If the restriction of f to every compact subset is continuous, then $f(x_n) \to f(y)$. But this proves that f is continuous.

17. Suppose f is the restriction to E of a continuous function g from S to T. If X is a totally bounded subset of E, then $\bar{X}$ is a compact subset of S (being complete and totally bounded). Therefore g restricted to $\bar{X}$ is uniformly continuous, and f restricted to X, i.e., g restricted to X, is uniformly continuous.

Conversely, suppose that the restriction of f to any totally bounded subset of E is uniformly continuous. By Theorem 14-7.13, f has a unique continuous extension f_X over $\bar{X}$ for each totally bounded subset X of E. If $s \in S$, there is a Cauchy sequence $\{x_n\}$ in E with $x_n \to s$. Regard $\{x_n\}$ as a set X; X is totally bounded. Hence f has a continuous extension f_X over $\bar{X}$. If Y is any other totally bounded set with $s \in \bar{Y}$, then $X \cup Y$ is totally bounded, and $f_Y(s) = f_{X \cup Y}(s) = f_X(s)$. Hence the family $\{f_X\}$ has a common extension over all of S, and this extension is continuous by the result of Exercise 16.

Pages 277–278

1. If $\{G_n\}$ is a countable base for the open sets of S, then $\{T \cap G_n\}$ is a countable base for the open sets of T. See 14-3.13.

2. If X is a countable dense subset of S, then $f(X)$ is a countable subset of T. By 14-4.7, we have $T = f(\bar{X}) \subseteq \overline{f(X)}$, which proves that $f(X)$ is dense.

3. Let U be an open set in $S \times T$ and suppose $\langle s, t \rangle \in U$. There is an $\epsilon > 0$ such that $B(s, \epsilon) \times B(t, \epsilon) = B(\langle s, t \rangle, \epsilon) \subseteq U$. There is an $\alpha \in A$ and a $\beta \in B$ such that $G_\alpha \subseteq B(s, \epsilon)$ and $H_\beta \subseteq B(t, \epsilon)$. Then $\langle s, t \rangle \in G_\alpha \times H_\beta \subseteq U$. This proves that U is a union of sets of the form $G_\alpha \times H_\beta$; hence $\{G_\alpha \times H_\beta\}$ is a base for the open sets of $S \times T$.

4. If $\{G_\alpha \mid \alpha \in A\}$ is a family of mutually disjoint nonvoid open sets in S and X is a countable dense set in S, then for each α we can choose $p(\alpha) \in G_\alpha \cap X$. Then p is an injection of the index set A into X, so A is countable.

5. For any $\epsilon > 0$, $\{B(x, \epsilon) \mid x \in S\}$ is an open covering. Let $\{B(x, \epsilon) \mid x \in Y\}$ be a countable subcovering. Then Y is a countable ϵ-dense set, and the conclusion follows from 14–9.2.

6. A metric space is locally separable if and only if every point has a separable neighborhood (which automatically implies that it has arbitrarily small separable neighborhoods). If D is an uncountable discrete space, $\mathbf{R} \times D$ is a nondiscrete, nonseparable, locally separable space.

$B(x, \varphi(x))$ is the union of a countable sequence $\{B(x, \varphi(x) - 1/n)\}$ of separable subspaces of S; hence it is separable. Since S itself is not separable, this shows incidentally that $\varphi(x) < \infty$.

The function φ is continuous because it satisfies the inequality $|\varphi(x) - \varphi(y)| < \rho(x, y)$. To prove this, note that

$$B(y, \varphi(x) - \rho(x, y)) \subseteq B(x, \varphi(x))$$

and therefore $\varphi(y) \geq \varphi(x) - \rho(x, y)$. By symmetry we also have $\varphi(y) - \varphi(x) \leq \rho(x, y)$.

7. Since each of the functions

$$\frac{|f_n(s) - f_n(t)|}{1 + |f_n(s) - f_n(t)|}$$

is a pseudometric, there is no difficulty in showing that ρ is a pseudometric.

For each n, consider the map $\varphi_n : s \rightarrow \langle f_1(s), \ldots, f_n(s) \rangle$ of S to $\mathbf{R}^n$. Let E_n be a countable set such that $\varphi_n(E_n)$ is dense in range φ_n (since $\mathbf{R}^n$ is separable, such a set exists). The image of $\bigcup_n E_n$ is dense in T. To prove this, it is sufficient to prove

$$(\forall s \in S)(\forall \epsilon > 0) \left(\exists x \in \bigcup_n E_n \right) \rho(x, s) < \epsilon.$$

Let $s \in S$ and $\epsilon > 0$ be given. Choose k so that $1/2^k < \epsilon$. Then choose $x \in E_k$ so that $\sigma(\varphi_k(s), \varphi_k(x)) < \epsilon - 1/2^k$, where σ is the metric of $\mathbf{R}^k$. This means that

$$|f_i(s) - f_i(x)| < \epsilon - 1/2^k \quad \text{for} \quad i = 1, 2, \ldots, k.$$

Then

$$\rho(s, x) < \sum_{n=1}^{k} \frac{1}{2^n} \frac{|f_n(s) - f_n(x)|}{1 + |f_n(s) - f_n(x)|} + \sum_{n=k+1}^{\infty} \frac{1}{2^n}$$

$$< \sum_{n=1}^{k} \frac{1}{2^n} \left(\epsilon - \frac{1}{2^k} \right) + \frac{1}{2^k} < \epsilon.$$

Page 286

1. Under this hypothesis $X \cap G$ is both open and closed in X by 14–3.7 and 14–3.13. Since X is connected, either $X \cap G = X$ or $X \cap G = \emptyset$.

2. The components of a space are always closed (14–10.14). Let E be any one of them and let F be the union of the remaining ones. If there are only finitely many components, F is closed (14–3.2) and E is open.

3. It is easy to check that the set of points at a finite distance from a fixed point is both open and closed. Hence if the metric takes the value ∞, we can construct a nontrivial open and closed subset.

4. Suppose S and T are locally connected metric spaces. Let $\langle s, t \rangle \in S \times T$ and $\epsilon > 0$ be given. There is a connected neighborhood U of s in S with $U \subseteq B(s, \epsilon)$. There

is a connected neighborhood V of t in T with $V \subseteq B(t, \epsilon)$. Then $U \times V$ is a connected neighborhood of $\langle s, t \rangle$ in $S \times T$ with $U \times V \subseteq B(\langle s, t \rangle, \epsilon)$.

Conversely, suppose S and T are not void and that $S \times T$ is locally connected. Let $s \in S$ and $\epsilon > 0$ be given. Choose $t \in T$ and let W be the component of $B(\langle s, t \rangle, \epsilon)$ containing $\langle s, t \rangle$. If π is the coordinate projection of $S \times T$ onto S, then $\pi(W)$ is an open connected neighborhood of s and $W \subseteq B(s, \epsilon)$. This follows from Exercise 3, p. 245, 14–10.19, and 14–10.5. We have proved that S is locally connected.

5. Suppose S and T are pathwise connected metric spaces. Let $\langle s_1, t_1 \rangle$ and $\langle s_2, t_2 \rangle$ be given points of $S \times T$. Let f be a path from s_1 to s_2 in S and let g be a path from t_1 to t_2 in T. Then $x \to \langle f(x), g(x) \rangle$ is a path from $\langle s_1, t_1 \rangle$ to $\langle s_2, t_2 \rangle$ in $S \times T$. Thus $S \times T$ is pathwise connected.

Conversely, suppose S and T are nonvoid and that $S \times T$ is pathwise connected. Given s_1 and $s_2 \in S$, choose $t \in T$ and a path f from $\langle s_1, t \rangle$ to $\langle s_2, t \rangle$ in $S \times T$. Then $\pi \circ f$ is a path from s_1 to s_2 in S, where π is the coordinate projection from $S \times T$ to S.

6. A bounded closed interval is compact, but neither an open interval nor a half-open interval is compact. Therefore a bounded closed interval is not homeomorphic to either an open interval or a half-open interval. To prove that a half-open interval is not homeomorphic to an open interval we must somehow characterize the endpoint in topological terms. If I is an open interval, and p is any point of I, then $I - \{p\}$ is not connected. If H is a half-open interval, and q is its endpoint, then $H - \{q\}$ is connected. It follows that H is not homeomorphic to I.

7. This is trivial for $n = 1$ since S^1 is a circle. For larger n, given $x \ne y$, the intersection of a plane through x and y with S^n is a circle, and they can be connected by a path lying in this circle. To write down an explicit path from x to y in S^n, use vector notation with $\|\langle x_1, x_2, \ldots, x_{n+1} \rangle\| = (x_1^2 + x_2^2 + \cdots + x_{n+1}^2)^{1/2}$. Then

$$t \to \frac{(1 - t)x + ty}{\|(1 - t)x + ty\|}$$

is a path in S^n from x to y unless $x = -y$, in which case the formula fails for $t = \frac{1}{2}$. In this case splice together a path from x to some third vector z and another path from z to y.

8. If f and g are in $B(S)$, then $t \to (1 - t)f + tg$ is a path from f to g, using vector notation (defined in 13–2.16). To check the continuity of this, it is important that $f - g$ be bounded. The existence of this path shows that $B(S)$ is connected. Moreover, if g lies in the ball of radius ϵ about f, then so does the whole path joining f and g. Hence $B(S)$ is locally pathwise connected. Finally, if f and g are in $C(S)$, then the whole path connecting them lies in $C(S)$.

9. **Lemma.** Let G be a connected metric space and let $\{D(\alpha) \mid \alpha \in A\}$ be an open covering of G. For any two points p and q of G, there is a finite sequence $\alpha_1, \alpha_2, \ldots, \alpha_n \in A$ such that $p \in D(\alpha_1), q \in D(\alpha_n)$, and $D(\alpha_i) \cap D(\alpha_{i+1}) \ne \emptyset$ for $1 \le i \le n - 1$.

Proof. The existence of such a chain between p and q is a relation in the set G which is easily seen to be an equivalence relation. (It is the least transitive relation containing the reflexive and symmetric relation $\{\langle r, s \rangle \mid (\exists \alpha \in A) r \in D(\alpha) \text{ and } s \in D(\alpha)\}$. (See Exercise 3, p. 88 and Exercise 4, p. 143.) The equivalence classes of this relation are open. But G has no nontrivial partitions into open sets. Hence there is just one equivalence class. Therefore, any two points can be joined by such a chain.

Returning to the exercise, we take the special case in which G is an open ball in $\mathbf{R}^n$. Let D be an open ball in $\mathbf{R}^n$, and let X be a countable subset of $\mathbf{R}^n$. If p and q are distinct points of $D - X$, we can find uncountably many mutually disjoint segments in D perpendicular to the segment $\overline{pq}$ (here is where $n \geq 2$ comes in). Among these, one, say S, does not meet X. Among the uncountably many arcs $\overline{ps} \cup \overline{sq}$ where $s \in S$, there is one which does not meet X. Hence $D - X$ is pathwise connected.

Now let G be a connected open set in $\mathbf{R}^n$ and let p and q be points of $G - X$. The lemma proves that there is a sequence $p_0 = p, p_1, \ldots, p_n = q$ in $G - X$ and a sequence $D_1, D_2, \ldots, D_n$ of open balls in G such that p_{i-1} and p_i are both in D_i for $1 \leq i \leq n$. (Note that a nonvoid open set $D_i \cap D_{i+1}$ in $\mathbf{R}^n$ must be uncountable; hence $D_i \cap D_{i+1} - X$ is not void.) By the previous argument, p can be connected to q by a path in $G - X$ via $p_1, p_2, \ldots, p_{n-1}$. Thus $G - X$ is pathwise connected. This, of course, implies that it is connected, because each path must lie in a single component of the space.

10. Let $F = \bigcap_A E_\alpha$. F is compact, being a closed subset of a compact set. If F is not connected, then $F = F_1 \cup F_2$, where F_1 and F_2 are disjoint nonvoid sets closed relative to F and therefore closed and compact. Let $G_1 = \{x \mid \rho(x, F_1) < \rho(x, F_2)\}$, $G_2 = \{x \mid \rho(x, F_1) > \rho(x, F_2)\}$ (see 14–1.8 and Exercise 1, p. 245). These are disjoint open sets, and $F_1 \subseteq G_1, F_2 \subseteq G_2$. Since $G_1 \cap E_\alpha$ and $G_2 \cap E_\alpha$ are disjoint nonvoid sets open relative to E_α, while E_α is connected, we conclude that $E_\alpha - (G_1 \cup G_2) = D_\alpha$ is not void. Now $\{D_\alpha\}$ is a family of compact sets linearly ordered by inclusion, so $\bigcap_A D_\alpha \neq \emptyset$ (Exercise 6, p. 272). But $\bigcap_A D_\alpha \subseteq F - (F_1 \cup F_2)$, a contradiction.

11. Let φ be a homeomorphism from $[0, \infty)$ to X. Let $A_n = \varphi([0, n))$ and $B_n = \varphi([n, \infty))$. Since $\varphi([0, n])$ is compact, and therefore closed, $\overline{A}_n = \varphi([0, n])$. Then $S = \overline{X} = \overline{A}_n \cup \overline{B}_n = A_n \cup \overline{B}_n$. But $A_n \cap \overline{B}_n = \emptyset$ since φ is a homeomorphism. Therefore $S - A_n = \overline{B}_n$, and $S - X = S - \bigcup_n A_n = \bigcap_n \overline{B}_n$. Now $\{\overline{B}_n\}$ is a decreasing sequence of compact connected sets; hence $\bigcap_n \overline{B}_n$ is connected (Exercise 10).

12. The set is totally disconnected. Denote it by S. Every line of rational nonzero slope containing one point of $\mathbf{Q} \times \mathbf{Q}$ lies entirely in $\mathbf{R}^2 - S$. If p and q are distinct points of S, then there is such a line L which separates p from q; i.e., $\mathbf{R}^2 - L = G \cup H$, where G and H are open half-planes, $p \in G$, and $q \in H$. Then $S \cap G$ and $S \cap H$ are complementary open subsets of S. Therefore any connected subset of S containing p must be wholly in G (Exercise 1), and so it does not contain q. This shows that no connected subset of S contains more than one point; that is, S is totally disconnected.

If X is the union of all the lines having rational nonzero slope through a point of $\mathbf{Q} \times \mathbf{Q}$, then X is connected. Since $X \subseteq \mathbf{R}^2 - S \subseteq \overline{X} = \mathbf{R}^2$, $\mathbf{R}^2 - S$ is connected by 14–10.4.

Pages 292–293

1. Let $a = f(x)$. Since $f(z) = f(x) + (z - x)\overline{f}(z)$ for all $z \in \text{dom } f$,

$$y = f(g(y)) = a + (g(y) - g(a))\overline{f}(g(y)) \qquad \text{for all } y \in \text{dom } g;$$

hence

$$g(y) = g(a) + \frac{(y - a)}{\overline{f}(g(y))}.$$

Since g is continuous at a and $\overline{f}$ is continuous at $g(a) = x$ with $\overline{f}(x) = f'(x) \neq 0$, this shows that g is differentiable at a and $g'(a) = 1/f'(x)$.

3. (a) Suppose f vanishes at z_0. Let E be a closed disk about z_0 in D. Let

$$M = \sup \{|\varphi(z)| \mid z \in E\}.$$

Let $E_1 \subseteq E$ be a closed disk about z_0 of radius less than $1/2M$. Let z_1 be a point of E_1 at which $|f|$ is largest. Then $|f(z_1)| = |f(z_1) - f(z_0)| \le |z_1 - z_0| |f'(z)| = |z_1 - z_0| |\varphi(z)| |f(z)| \le (1/2M)M|f(z)| = \frac{1}{2}|f(z)|$, where $z \in E_1$. This contradicts the definition of z_1 unless $|f(z_1)| = 0$. Thus f vanishes on a neighborhood of z_0. The set on which f vanishes is therefore an open subset of D. Moreover, because f is continuous, this set is closed as a subset of D. Since D is connected, it is either all of D or empty.

(b) We can choose $z_0 \in D$ (the result is trivial if $D = \emptyset$) and λ so that $g(z_0) = \lambda f(z_0)$. Then $h(z) = g(z) - \lambda f(z)$ defines a solution of the differential equation which vanishes at one point. By part (a), it vanishes everywhere, so $g(z) = \lambda f(z)$ for all z.

Page 298

1. Immediate from 15–2.6.

2. (a) For any z except 0, the test ratio (13–2.15) is $(2n^2 + 1)|z|/(n + 2)$, which tends to ∞ with n. Hence $\rho = 0$. (b) For any z except 0, the test ratio has limit $\frac{1}{2}|z|$, hence $\rho = 2$. (c) $\binom{2n+2}{n+1}/\binom{2n}{n} = (2n + 2)(2n + 1)/(n + 1)^2 \to 4$. Hence $\rho = \frac{1}{4}$. (d) The test ratio method applied directly fails here because, after "closing up" the series to get rid of zero terms, the ratios of consecutive terms are $2|z|$ or $2|z|^2$. This makes it clear that the series diverges for $|z| > \frac{1}{2}\sqrt{2}$ and converges for $|z| < \frac{1}{2}$, but does not determine the actual radius of convergence. However, if we group the terms in threes, the test ratio becomes $4|z|^3$. The grouped series converges therefore if $4|z|^3 < 1$ and diverges if $4|z|^3 > 1$. It is easy to check that the original series converges whenever the grouped series does, and vice versa. Hence $\rho = 2^{-2/3}$. This result can also be obtained from 15–2.6.

3. Let ρ^* be the radius of convergence of $\sum c_n z^n$. If $\sigma > 1/\rho^*$, there is a constant M such that $|c_n| < M\sigma^n$. Then $|a_n| = |c_n - c_{n-1}| \le M(1 + 1/\sigma)\sigma^n$, so $\rho \ge 1/\sigma$. But σ can be chosen arbitrarily close to $1/\rho^*$; hence $\rho \ge \rho^*$.

Inequalities in the other direction depend on whether $\rho > 1$ or $\rho \le 1$. Say $\rho \le 1$. For any $\tau > 1/\rho$, there is a constant K such that $|a_n| \le K\tau^n$. Then

$$|c_n| \le K \sum_{k=0}^{n} \tau^k < K\frac{\tau}{\tau - 1} \cdot \tau^n,$$

so $\rho^* \ge 1/\tau$. Therefore $\rho^* \ge \rho$. This proves that $\rho^* = \rho$ in this case.

Say $\rho > 1$. Then $\sum a_n$ is convergent. Let $b = \sum_{k=0}^{\infty} a_k$. If $b \ne 0$, then $c_n \to b$ and $|c_n|^{-1/n} \to 1$, so $\rho^* = 1$. If $b = 0$, then $c_n = -\sum_{k=n+1}^{\infty} a_k$. For any τ with $1 > \tau > 1/\rho$, choose K so that $|a_k| \le K\tau^k$. Then $|c_n| \le K\tau^{n+1}/(1 - \tau)$, and therefore $\rho^* \ge 1/\tau$. Again $\rho^* = \rho$.

The situation is clarified if we consider the functions f and g defined by the power series

$$f(z) = \sum a_n z^n, \qquad g(z) = \sum c_n z^n.$$

We have $(1 - z)g(z) = f(z)$ on the common part of the domain. If $\rho > 1$ and $f(1) \ne 0$, there is obviously no function g satisfying this relation in any disk of radius larger than 1.

4. Since there are only a finite number of terms different from zero, we may take all sums as being over all of **I**. The result is certainly true for all values of r if $n = 0$. Assume that it is true for $n = p$ and all values of r. Then

$$\sum_k \binom{p+1}{k}\binom{m}{r-k} = \sum_k \binom{p}{k}\binom{m}{r-k} + \sum_k \binom{p}{k-1}\binom{m}{r-k}$$

$$= \binom{p+m}{r} + \binom{p+m}{r-1} = \binom{p+1+m}{r}.$$

This proves the result by induction.

5. Theorem 15–2.10 shows immediately that f' has a derivative at each point of D. Hence, by induction, $f^{(k)}$ is defined on all of D for each k. If

$$f^{(k)}(z) = \sum_{n=0}^{\infty} \frac{(n+k)!}{n!} c_{n+k}(z-a)^n,$$

then

$$f^{(k+1)}(z) = \sum_{n=1}^{\infty} \frac{(n+k)!}{(n-1)!} c_{n+k}(z-a)^{n-1} = \sum_{m=0}^{\infty} \frac{(m+1+k)!}{m!} c_{m+1+k}(z-a)^m.$$

Except for the choice of dummy index, the latter is the required formula for $k + 1$. Since the formula is clearly correct for $k = 0$, it follows for all k by induction.

Pages 307–308

1. Since

$$\frac{1}{z(z-1)(z-2)} = \frac{1}{2z} - \frac{1}{z-1} + \frac{1}{2(z-2)},$$

the power series expansion at a is

$$\frac{1}{2}\sum \frac{(-1)^n}{a^{n+1}}(z-a)^n - \sum \frac{(-1)^n}{(a-1)^{n+1}}(z-a)^n + \frac{1}{2}\sum \frac{(-1)^n}{(a-2)^{n+1}}(z-a)^n$$

$$= \sum \frac{(-1)^n}{2}\left(\frac{1}{a^{n+1}} - \frac{2}{(a-1)^{n+1}} + \frac{1}{(a-2)^{n+1}}\right)(z-a)^n.$$

2. See Section 15–4.

3. Suppose f is analytic at $a = g(b)$ and g is analytic at b. Say $f(z) = \sum_{n=0}^{\infty} c_n(z-a)^n$ for z near a and $g(z) = \sum_{k=0}^{\infty} d_k(z-b)^k$ for z near b. Then

$$f(g(z)) = \sum_{n=0}^{\infty} c_n \left(\sum_{k=1}^{\infty} d_k(z-b)^k\right)^n \tag{1}$$

for z near b. Since the inner power series involves no constant term, there are only a finite number of terms involving $(z-b)^m$ for each m, and these can be collected to give the power series

$$\sum_{m=0}^{\infty} e_m(z-b)^m, \tag{2}$$

where $e_0 = c_0$, $e_1 = c_1 d_1$, $e_2 = c_2 d_1^2 + c_1 d_2$, $e_3 = c_3 d_1^3 + 2c_2 d_1 d_2 + c_1 d_3$, $e_4 = c_4 d_1^4 + 3c_3 d_1^2 d_2 + c_2(2d_1 d_3 + d_2^2) + c_1 d_4$, etc.

To prove that the resulting power series does in fact represent $f \circ g$ near b we must justify the rearrangement of the series (1) to give (2). This can be done in several ways using results of Section 13–2 if we assume that $|z - b|$ is so small that (1) is valid and

$$\sum_{k=1}^{\infty} |d_k| \, |z - b|^k < \rho, \tag{3}$$

where ρ is the radius of convergence of the series for f. Note that, because there is no constant term in the power series in (3), this must be true in some neighborhood of b.

One explicit way is as follows (to simplify notation we assume $b = 0$). By iterating the multiplication of power series we see that for each n there are coefficients $\{e_{n,k}\}$ such that

$$\left(\sum_{k=1}^{\infty} d_k z^k \right)^n = \sum_{k=n}^{\infty} e_{n,k} z^k.$$

Then $f(g(z)) = \sum_{n=0}^{\infty} \sum_{k=n}^{\infty} c_n e_{n,k} z^k$, and this is $\sum_{k=0}^{\infty} (\sum_{n=0}^{k} c_n e_{n,k}) z^k$ provided $\sum_{n=0}^{\infty} \sum_{k=n}^{\infty} |c_n| \, |e_{n,k}| \, |z|^k$ converges (13–2.43). Let

$$\left(\sum_{k=1}^{\infty} |d_k| z^k \right)^n = \sum_{k=n}^{\infty} E_{n,k} z^k.$$

It is easy to prove that $|e_{n,k}| \le E_{n,k}$. Hence the proviso can be replaced by the convergence of $\sum_{n=0}^{\infty} \sum_{k=n}^{\infty} |c_n| E_{n,k} |z|^k$. But this is $\sum_{n=0}^{\infty} |c_n| (\sum_{k=1}^{\infty} |d_k| \, |z|^k)^n$, and we know this converges if (3) is satisfied (with $b = 0$).

4. This is a direct application of Theorem 13–2.31.

5. Let

$$\prod_{n=1}^{k} \left(1 + \frac{z^2}{n^2} \right) = \sum_{m=0}^{\infty} a_{k,m} z^m.$$

(Here $a_{k,m} = 0$ for $m > 2k$.) Evidently, $a_{k,m}$ is nonnegative and increases with k for fixed m. For any positive M and any $p \in N$,

$$\sum_{m=0}^{p} a_{k,m} M^m \le \prod_{n=1}^{k} \left(1 + \frac{M^2}{n^2} \right) \le \prod_{n=1}^{\infty} \left(1 + \frac{M^2}{n^2} \right),$$

the latter product being convergent by 13–3.4. It follows that $a_{k,m} \to b_m$ as $k \to \infty$ and $\sum_{m=0}^{p} b_m M^m \le \prod_{n=1}^{\infty} (1 + M^2/n^2)$. Therefore $\sum_{m=0}^{\infty} b_m M^m$ converges for any M. Hence $\sum_{m=0}^{\infty} b_m z^m$ defines an entire analytic function. Now Exercise 4 gives

$$\prod_{n=1}^{\infty} \left(1 + \frac{z^2}{n^2} \right) = \lim_k \sum_{m=0}^{\infty} a_{k,m} z^m = \sum_{m=0}^{\infty} b_m z^m.$$

6. $f(z) = g(i/z)$, where g is the function defined by the product of Exercise 5.

7. The proof of 15–3.8 is applicable in this case until the connectedness of D is brought in. The set $G = \text{Int} \, f^{-1}(0)$ is an open and closed subset of D. Every component of D is therefore either in G or disjoint from G. (Exercise 1, p. 286.) Hence G is the union of certain components of D. The set $X = f^{-1}(0) - G$ consists of the isolated points of $f^{-1}(0)$. It is therefore a discrete set. Since $f^{-1}(0)$ is closed relative to D and G is open, X is closed relative to D.

8. If such functions exist, we know their power series. Hence they exist if and only if these power series have positive radii of convergence.

(a) no   (b) yes   (c) yes

9. (a) No, because if so, $f(0) = 0$, and $f'(0) = \lim_{n\to\infty} nf'(1/n)$, but $\lim_{n\to\infty} (n(-1)^n/n)$ does not exist.

(b) No, because if so, then $f(0) = 0$, $f'(0) = \lim_{n\to\infty} nf'(1/n) = 0$, and $f''(0) = 2! \lim_{n\to\infty} (n^2 f(1/n))$ (look at the power series). But $\lim_{n\to\infty} n^{1/2}$ does not exist.

10. If $|z| < R$, then

$$f(y) = \sum_{n=0}^{\infty} \frac{1}{n!} f^{(n)}\left(\frac{z}{2}\right)\left(y - \frac{z}{2}\right)^n$$

is valid for $|y - z/2| < R - |z/2|$. But this implies

$$f(z) - f(0) = \sum_{n=0}^{\infty} \frac{1}{n!} f^{(n)}\left(\frac{z}{2}\right)\left(\frac{z}{2}\right)^n - \sum_{n=0}^{\infty} \frac{1}{n!} f^{(n)}\left(\frac{z}{2}\right)\left(\frac{-z}{2}\right)^n,$$

or

$$f(z) = f(0) + 2\sum_{n=0}^{\infty} \frac{1}{(2n+1)!}\left(\frac{z}{2}\right)^{2n+1} f^{(2n+1)}\left(\frac{z}{2}\right).$$

11. We can define g by $g(z) = f(z)/(z-a)$ for $z \neq a$ and $g(a) = f'(a)$. Since $1/(z-a)$ defines an analytic function on $C - \{a\}$, g is evidently analytic on $D - \{a\}$. To see that it is analytic at a, we note that it is represented near a by the power series

$$\sum_{n=0}^{\infty} \frac{f^{(n+1)}(a)}{(n+1)!} (z - a)^n.$$

12. On each component of D, f is either everywhere 0 or vanishes only on a discrete set. Correspondingly, $k_f = \infty$ or $k_f = 0$ except on a discrete set. Thus the set E of points where $k_f \neq 0$ or ∞ meets each component of D in a discrete set. Since D is open, this implies that E itself is discrete.

By looking at the power series at a we see that $f = gh \Rightarrow k_f(a) \geq k_g(a)$ for all $a \in D$. Conversely, assume $k_f \geq k_g$. On each component of D on which g does not vanish everywhere, we can define $h = f/g$ except on the discrete set where g vanishes. If $g(a) = 0$, then

$$f(z) = (z - a)^{k_f(a)} \sum_{n=0}^{\infty} c_n(z - a)^n$$

for z near a, and

$$g(z) = (z - a)^{k_g(a)} \sum_{n=0}^{\infty} d_n(z - a)^n,$$

where $d_0 \neq 0$. Then we can define h near a by

$$h(z) = (z - a)^{k_f(a) - k_g(a)} \left(\frac{\sum_{n=0}^{\infty} c_n(z - a)^n}{\sum_{n=0}^{\infty} d_n(z - a)^n}\right).$$

This expression defines an analytic function since $d_0 \neq 0$.

On components of D on which g does vanish everywhere, take $h(z) = 1$.

386   ANSWERS AND SOLUTIONS

Pages 314–315

1. $\sin \pi/3 = 2 \sin \pi/6 \cos \pi/6 = 2 \sin \pi/6 \sin (\pi/2 - \pi/6)$. Since $\sin \pi/3 \neq 0$, $\sin \pi/6 = 1/2$.

2. Since $\sin (z + \pi/2) = \cos z$, it follows that every period of sin is a period of cos and vice versa. If P is a period of sin and cos, then $\exp iP = \cos P + i \sin P = \cos 0 + i \sin 0 = 1$. Therefore, iP is an integral multiple of $2\pi i$.

3. Suppose $\cos w = \cos z$. Then either $\sin w = \sin z$ or $\sin w = -\sin z$ by (17). Accordingly, either $\exp iw = \exp iz$ or $\exp iw = \exp (-iz)$. Hence either $w - z$ or $w + z$ is an integral multiple of 2π. The converse is clear.

4. Since cos vanishes only for odd multiples of $\pi/2$, the domain of tan is all of $\mathbf{C}$ except the odd multiples of $\pi/2$. Evidently π is a period of tan. Since

$$\sin 2z = \frac{2 \tan z}{1 + \tan^2 z},$$

if P is a period of tan, then $2P$ is a period of sin. Hence all periods of tan are integral multiples of π. Since sin increases strictly and cos decreases on $[0, \pi/2)$ and cos is positive on this interval, tan is strictly increasing on $[0, \pi/2)$. Since $\tan (-z) = -\tan z$, tan also increases strictly on $(-\pi/2, 0]$. Since $\tan \pi/4 = 1$ (from $\sin \pi/4 = \cos (\pi/2 - \pi/4)$), we see that $0 < \tan x/2 < 1$ and $\tan x = 2(\tan x/2)/(1 - \tan^2 x/2) > 2 \tan x/2$ for $0 < x < \pi/2$. This shows that

$$\frac{\tan x}{x} > \frac{2^n}{x} \tan \frac{1}{2^n} x = \frac{\sin x/2^n}{x/2^n} \cdot \frac{1}{\cos x/2^n}.$$

Taking limits, $(\tan x)/x \geq 1$. We obtain strong inequality from

$$\frac{\tan x}{x} > \frac{\tan x/2}{x/2} \geq 1.$$

5. $\cot z = (\cos z)/\sin z$ has domain $\mathbf{C} - \{n\pi\}$ and least period π. $\sec z = 1/\cos z$ has domain $\mathbf{C} - \{(2n + 1)\pi/2\}$ and least period 2π. $\operatorname{cosec} z = 1/\sin z$ has domain $\mathbf{C} - \{n\pi\}$ and least period 2π.

6. Put

$$f_k(z) = \frac{1}{z} + 2z \sum_{n=1}^{k} \frac{1}{z^2 - n^2} = \sum_{n=-k}^{+k} \frac{1}{z - n}.$$

Then

$$f_k(z + 1) = f_k(z) - \frac{1}{z - k} + \frac{1}{z + k + 1}.$$

Since the infinite series converges for each value of $z \in \mathbf{C} - \mathbf{I}$, this shows that $f(z + 1) = f(z)$. Thus f has period 1. To prove that f is analytic, choose a point $a \in \mathbf{C} - \mathbf{I}$. Let k be an integer greater than $|a|$. Then

$$f(z) = f_{k-1}(z) + 2z \sum_{n=k}^{\infty} \frac{1}{z^2 - n^2}.$$

Since f_{k-1} is analytic at a and so is $2z$, it is sufficient to prove that $\sum_{n=k}^{\infty} 1/(z^2 - n^2)$ defines an analytic function on the disk of radius k about the origin. We have

$$\sum_{n=k}^{\infty} \frac{1}{z^2 - n^2} = -\sum_{n=k}^{\infty} \sum_{p=0}^{\infty} \frac{1}{n^{2(p+1)}} z^{2p} = \sum_{p=0}^{\infty} \left(-\sum_{n=k}^{\infty} \frac{1}{n^{2(p+1)}}\right) z^{2p}.$$

The interchange of summations is justified by the fact that

$$\sum_{n=k}^{\infty}\sum_{p=0}^{\infty}\frac{1}{n^{2(p+1)}}|z|^{2p} = \sum_{n=k}^{\infty}\frac{1}{n^2-|z|^2},$$

the latter being convergent for $|z| < k$.

7. $P = \bigcap_{z\in C}\{\lambda\,|\,f(z+\lambda)=f(z)\}$. Thus P is the intersection of closed sets. Hence P is closed. Obviously $0 \in P$. Furthermore, $\lambda, \mu \in P \Rightarrow \lambda - \mu \in P$. Thus P is a closed subgroup of the additive group of C. We shall determine all such subgroups. There are six types as described by (a) through (f).

Suppose first that 0 is an isolated point of P. In this case P is a discrete set: if $\lambda_n \to \mu$ in P, then $\lambda_n - \mu \to 0$ in P; but 0 is isolated, so $\lambda_n - \mu = 0$ for all large n. If 0 is the only point of P, then we have (a). Suppose that $P \neq \{0\}$, say $\lambda_0 \in P$, $\lambda_0 \neq 0$. There are only finitely many points in the discrete compact set $\{\lambda \in P\,|\,|\lambda| \leq |\lambda_0|\}$, so there is an $\alpha \in P$ with $|\alpha| = \inf\{|\lambda|\,|\,\lambda \in P, \lambda \neq 0\}$. Now $n\alpha \in P$ for every $n \in I$. We may have $P = \{n\alpha\}$, that is (b), or there may be other elements in P. In the latter case, there is one, say β, of least absolute value. If β/α were real, we could choose an integer n such that $n < \beta/\alpha < n + 1$. Then $0 < |\beta - n\alpha| < |\alpha|$, contrary to the choice of α. Thus β/α is not real. Evidently $\{m\alpha + n\beta\,|\,m, n \in I\} \subseteq P$. Thus (d) holds unless there is a $\gamma \in P - \{m\alpha + n\beta\}$. We shall prove that this is impossible.

Since β/α is not real, we can write $\gamma = \rho\alpha + \sigma\beta$, where ρ and σ are real. We can choose integers m and n so that $|\rho - m| \leq \frac{1}{2}$ and $|\sigma - n| \leq \frac{1}{2}$. Then $0 \neq \gamma - m\alpha - n\beta \in P$ and $|\gamma - m\alpha - n\beta| = |(\rho - m)\alpha + (\sigma - n)\beta| < \frac{1}{2}|\alpha| + \frac{1}{2}|\beta| \leq |\beta|$. We have a strict inequality in the middle because $(\rho - m)\alpha/(\sigma - n)\beta$ is not real. This contradicts the choice of β.

Now assume that 0 is not isolated in P. Choose a sequence $\{\lambda_n\}$ in $P - \{0\}$ which converges to 0. There are positive integers k_n such that $|k_n\lambda_n| \leq 1$ but $|(k_n + 1)\lambda_n| > 1$. By passing to a subsequence, we can arrange that $k_n\lambda_n \to \alpha$ where evidently $|\alpha| = 1$. Now if $t \in R$, $t\alpha = \lim[tk_n]\lambda_n \in P$, where $[\ ]$ denotes the greatest integer function. Thus $\{t\alpha\,|\,t \in R\} \subseteq P$. Consider now any $\mu \in C$ such that μ/α is not real, and let $P' = P \cap \{t\mu\,|\,t \in R\}$. Arguments similar to those already given (but shorter) show that either $P' = \{0\}$, $P' = \{n\beta\,|\,n \in I\}$, or $P' = \{t\mu\,|\,t \in R\}$, where $\beta \in P' - \{0\}$. These possibilities lead to descriptions (c), (e), or (f), respectively.

Conversely, if P is a closed additive subgroup of C, then the function $z \to \rho(z, P)$ (as in 14–1.8) has P as its set of periods. It follows from 15–3.8 that for a nonconstant analytic function the periods are a discrete set. Hence cases (c) and (e) do not occur for analytic functions. It follows from the modulus principle (Section 15–5) that the case (d) cannot occur for analytic functions. Such a function would have to take its maximum modulus on the compact set $\{\lambda\alpha + \mu\beta\,|\,0 \leq \lambda \leq 3, 0 \leq \mu \leq 3\}$ at an interior point because of the periodicity.

To prove that $\sum(z - m\alpha - n\beta)^{-3}$ defines an analytic function we need the fact (obvious from area considerations) that the number of points of $P = \{m\alpha + n\beta\}$ which satisfy $|m\alpha + n\beta| \leq K$ is bounded by MK^2 for some fixed M and all $K \geq 1$.

Let R be a fixed number > 1. We will prove that $\sum'(z - m\alpha - n\beta)^{-3}$, where $\sum'$ indicates a sum carried out over all $\langle m, n\rangle$ such that $|m\alpha + n\beta| > 2R$, defines an analytic function for $|z| < R$. This is just the problem of justifying the interchange of sums to get

$$\sum{}'\sum_{k=0}^{\infty}\frac{(k+1)(k+2)}{2}\cdot\frac{z^k}{(m+n)^{k+3}} = \sum_{k=0}^{\infty}\left(\frac{(k+1)(k+2)}{2}\sum{}'\frac{1}{(m\alpha+n\beta)^{k+3}}\right)z^k$$

It is legitimate if

$$\sideset{}{'}\sum \sum_{k=0}^{\infty} \frac{(k+1)(k+2)}{2} \cdot \frac{|z|^k}{|m\alpha + n\beta|^{(k+3)}} = \sideset{}{'}\sum (|m\alpha + n\beta| - |z|)^{-3}$$

converges. To estimate the latter sum we divide it into finite blocks $B_1, B_2, \ldots$, where B_j contains all the terms satisfying $2^j R < |m\alpha + n\beta| < 2^{j+1}R$. B_j contains at most $MR^2 4^{j+1}$ terms, all of which are less than $(2^j R - R)^{-3}$. Therefore,

$$\sideset{}{'}\sum (|m\alpha + n\beta| - |z|)^{-3} \le \sum_{j=1}^{\infty} \frac{4M}{R} \cdot \frac{4^j}{(2^j - 1)^3},$$

and the latter converges.

The finite sum involving terms with $|m\alpha + n\beta| \le 2R$ clearly defines an analytic function on all of $\mathbf{C} - P$. Hence $\sum(z - m\alpha - n\beta)^{-3}$ defines an analytic function on all of $\mathbf{C} - P$. Since the series converges as an unordered infinite sum, it is clear that it has the same value as $\sum(z + \alpha - m\alpha - n\beta)^{-3}$ and $\sum(z + \beta - m\alpha - n\beta)^{-3}$. Therefore, $f(z) = f(z + \alpha) = f(z + \beta)$ for all $z \in \mathbf{C} - P$. Hence α and β, and therefore all members of P, are periods of f. It is clear that there are no other periods of f because $|f(z)| \to \infty$ as z approaches any member of P. Moreover, this shows that f is not constant.

8. For $z = \lambda \exp(2\pi k i/2^n)$, we have $f(z) = \sum_{m=0}^{n-1} z^{2^n} + \sum_{m=n}^{\infty} \lambda^{2^n}$. The first of these sums, being finite, determines an entire function, but the second is unbounded as $\lambda \to 1$. Thus f is unbounded on each such ray.

Suppose g is a continuous extension of f with connected open domain E. Let D denote the open unit disk. If D is closed relative to E, then it is both open and closed in E. Since E is connected, $D = E$ and g is not a proper extension of f. If D is not closed relative to E, there is a point ξ on the boundary of D relative to E. Here $|\xi| = 1$ and $\xi \in E$. Since E is open, it contains a point of the form $\exp(2\pi k i/2^n)$. (Such points are dense on the unit circle.) Since g is continuous, it must be bounded near this point, but f is unbounded near this point. This shows that $E \supset D$ is impossible.

Page 318

1. Let E be a component of D. Since E is closed relative to D, $\overline{E} - E \subseteq \overline{D} - D$, and we have $|f(z)| = 1$ for all $z \in \overline{E} - E$. By the maximum modulus principle, $|f(z)| \le 1$ for all $z \in \overline{E}$. Therefore the minimum value of $|f|$ must be taken at a point of E. Since this value is not zero, f is locally constant at this minimum point, by Theorem 15–5.2. Since E is connected, f is constant on E, by 15–3.10.

2. Let $K = \sup \{|f(z)| \mid z \in D\}$. (Conceivably, $K = \infty$, of course.) We must prove $K \le M$. There is a sequence $\{z_n\}$ in D such that $|f(z_n)| \to K$. Since D is bounded, $\overline{D}$ is compact, and we can find a subsequence of $\{z_n\}$ which converges to a point p of $\overline{D}$. By renumbering the subsequence we may as well assume that $z_n \to p$.

Suppose $p \in \overline{D} - D$. For any $\epsilon > 0$, we can find a neighborhood U of p such that $|f(z)| < M + \epsilon$ for $z \in D \cap U$. This gives $|f(z_n)| < M + \epsilon$ for all large enough n, and therefore, $K \le M + \epsilon$. Since ϵ was arbitrary, $K \le M$.

Suppose $p \in D$. Then $|f|$ achieves its maximum at a point of D. By 15–5.2 and 15–3.10, f is constant, say λ, on the component E of D which contains p. Now, E is a bounded, nonvoid open set, and $\mathbf{C}$ is connected. Hence E has a boundary point q relative to $\mathbf{C}$. But E is both open and closed relative to D, so $q \notin D$. Therefore $q \in \overline{D} - D$. For any

$\epsilon > 0$, there is a neighborhood V of q such that $|f(z)| < M + \epsilon$ for $z \in V \cap D$. But $V \cap E$ is not empty, and if $r \in V \cap E$, then $K = |\lambda| = |f(r)| < M + \epsilon$. Again we conclude that $K \leq M$.

3. Let D be the common domain of f_1 and f_2. Suppose $z \to |f_1(z)| + |f_2(z)|$ has a weak local maximum point at z_0. Let λ and μ be complex numbers of absolute value 1 such that $\lambda f_1(z_0) = |f_1(z_0)|$ and $\mu f_2(z_0) = |f_2(z_0)|$ and consider $g = \lambda f_1 + \mu f_2$. We have

$$|g(z)| = |\lambda f_1(z) + \mu f_2(z)| \leq |\lambda| |f_1(z)| + |\mu| |f_2(z)| = |f_1(z)| + |f_2(z)|$$
$$\leq |f_1(z_0)| + |f_2(z_0)| = |g(z_0)| \qquad (*)$$

for all z in some neighborhood U of z_0. Thus $|g|$ has a weak local maximum point at z_0. By 15–5.2 and 15–3.10 we conclude that g is constant on D. This shows that equality holds throughout (*) for every $z \in U$. If g is everywhere 0, then f_1 and f_2 vanish on all of U and therefore on all of D. If g is some other constant ρ, the equality at the second step of (*) implies that $\lambda f_1(z)/\rho$ is real for $z \in U$. Then $\lambda f_1(z)/\rho$ is not open as a function, so it is constant on U by 15–5.5 and therefore it is constant on all of D. This implies that f_1 is constant.

The argument generalizes immediately to $z \to \sum_{k=1}^{n} |f_k(z)|$, and with some additional difficulties to $z \to \sum_{k=1}^{\infty} |f_k(z)|$, provided the latter sum converges for each z in the common domain (from which it can be shown that $g(z) = \sum_{k=1}^{\infty} \lambda_k f_k(z)$ defines an analytic function).

Page 324

1. We must prove that, if f is an analytic function with connected domain D which is a continuation of some branch g of Log, then $f \subseteq$ Log. Say that f agrees with g on the nonvoid open set E. Then $\exp \circ f$ is an analytic function defined on D which is the identity on E. Now 15–3.10 shows that $\exp \circ f$ is the identity on all of D. This shows that $f \subseteq$ Log. The essential point is that $\exp \circ f$ is defined on all of D because $\exp$ is defined on all of C.

2. The given power series represents $(\log z)/(z - 1)$ on its disk of convergence less the point 1. All continuations will represent $f(z)/(z - 1)$ on their domains of definition, where f is some branch of the logarithm, by the principle of permanence of analytic identities. It follows that the complete analytic function is the union of the analytic functions $z \to (\log_n z)/(z - 1)$ (which are defined for $z \in C - [0, \infty)$ if n is odd, for $z \in C - (-\infty, 0] - \{1\}$ if n is even) and the one-element set $\{\langle 1, 1 \rangle\}$.

3. The hypothesis implies that $\{\exp \circ \varphi \mid \varphi \subseteq \Phi\}$ is a family of analytic functions any two of which agree on their common domain. Let f be their least common extension. Then f is analytic by 15–3.5.

Suppose $\varphi \subseteq \Phi$ and $a \in$ domain φ. There is an integer n such that $(n - 1)\pi < \text{Im}\varphi(a) < (n + 1)\pi$, and by continuity we will have also $(n - 1)\pi < \text{Im}\varphi(z) < (n + 1)\pi$ for z in some neighborhood U of a. Together with $\exp(\varphi(z)) = f(z)$, this implies $\varphi(z) = \log_n f(z)$ for $z \in U$.

Page 326

2. The definition shows that $z \to (1 + z)^\alpha$ is an analytic function with domain $C - (-\infty, -1]$. (It is a composition of analytic functions; see Exercise 3, p. 307.) Hence it has a power series development at 0. Since its derivative is $\alpha(1 + z)^{\alpha-1}$, we can calculate the coefficients of this power series directly using 15–3.3.

3. Let Φ be the relation inverse to the kth power function restricted to $\mathbf{C} - \{0\}$. The formula $\varphi_n(z) = \exp\left((1/k)\log_n z\right)$ defines an analytic function φ_n with domain $\mathbf{C} - (-\infty, 0]$ or $\mathbf{C} - [0, \infty)$ according to whether n is even or odd. Since $(\varphi_n(z))^k = z$, $\varphi_n \subseteq \Phi$. It follows from 15–4.10 that $\Phi = \bigcup_{n=1}^{2k} \varphi_n$. Since φ_n and φ_{n+1} agree on an open set, the functions φ_n can all be obtained from one another by continuation. Thus Φ is a multiple-valued analytic function.

Suppose that g is an analytic function with connected domain E which agrees on some open set U with a branch of Φ. Then $g(z)^k = z$ for all $z \in U$. Therefore, $g(z)^k = z$ for all $z \in E$. If $0 \notin E$, it follows immediately that $g \subseteq \Phi$. We must eliminate the possibility that $0 \in E$. Differentiation gives $kg(z)^{k-1}g'(z) = 1$ for $z \in E$. If $0 \in E$, we should have $k \cdot 0 \cdot g'(0) = 1$ (because $k > 1$). This is impossible. This proves that Φ is a complete analytic function.

If we continue a branch of Φ around the unit circle in the positive direction, we come back to a different value of the kth root, the old value multiplied by $\exp(2\pi i/k)$. Hence no branch of the kth root can be defined on a domain including the unit circle. To formalize this argument, let h be a branch of Φ with domain D and define

$$\psi(t) = h(\exp 2\pi it) \exp(-2\pi it/k) \quad \text{for} \quad 0 \leq t \leq 1.$$

This is a continuous function. Since $\psi(t)^k = 1$ for all t, ψ takes values in the discrete set of kth roots of 1. Therefore ψ is constant. But $\psi(0) = h(1)$ and

$$\psi(1) = h(1) \exp(-2\pi i/k) \neq h(1).$$

This argument shows that there cannot even be a continuous function h defined on the unit circle satisfying $h(z)^k = z$. A slight modification shows that the same is true for any other circle surrounding the origin.

Now consider the function f. We have just shown that its domain does not contain any circle surrounding the origin. Since its domain is open, it does not contain 0. Now we prove that f is analytic. For any $a \in G$, we can choose n so that $\varphi_n(a) = f(a)$. On the open set range φ_n the function $z \to z^k$ is injective with inverse φ_n. Since f is continuous, there is a neighborhood V of a in G such that $f(V) \subseteq$ range φ_n. Then for $z \in V$ we have $f(z)^k = z$ and $f(z) \in$ range φ_n, so $f(z) = \varphi_n(z)$. This shows that f agrees with the analytic function φ_n near a. Hence f is analytic.

If ξ is a branch of Log with domain $H \neq \emptyset$, then $f_0(z) = \exp(\xi(z)/k)$ defines a branch of the kth root on H. The functions $f_p = (\exp 2\pi pi/k)f_0$, $p = 1, 2, \ldots, k - 1$, are also branches of the kth root, and they are all different. Suppose θ is any branch of Φ with domain H. Pick $a \in H$. Our earlier considerations show that g agrees with one of the functions f_p near a. Since H is connected (definition of branch), $\theta = f_p$. Thus there are not more than k branches of Φ on H.

Pages 338–339

1. If A is an arc in a metric space, then all paths f which are homeomorphisms from $[0, 1]$ to A have the same length (possibly ∞), and we define this common length to be the length of A. The essential point here is that, if f and g are two homeomorphisms from $[0, 1]$ to A, then $f^{-1} \circ g$ is a homeomorphism from $[0, 1]$ to itself and therefore either an order-preserving or an order-reversing map from $[0, 1]$ to itself.

When we admit paths which are not injective, the length of a path no longer depends only on its range. A noninjective path with range an arc may have length greater than the arc itself.

2. If $\lambda \neq 0$, then cos arg λ and sin arg λ are real, so the equation

$$\lambda = |\lambda| \exp(i \arg \lambda) = |\lambda|(\cos \arg \lambda + i \sin \arg \lambda)$$

shows that cos arg λ = Re $(\lambda/|\lambda|)$ and sin arg λ = Im $(\lambda/|\lambda|)$. Now,

$$\frac{z_2/z_1}{|z_2/z_1|} = \frac{\bar{z}_1 z_2/|z_1|^2}{|z_2/z_1|} = \frac{\bar{z}_1 z_2}{|z_1 z_2|},$$

so the required equations follow directly from the definitions.

3.

$$\theta_1 + \theta_2 + \theta_3 = \arg_0 \frac{z_3 - z_1}{z_2 - z_1} + \arg_0 \frac{z_1 - z_2}{z_3 - z_2} + \arg_0 \frac{z_2 - z_3}{z_1 - z_3}$$

$$= \arg \frac{z_3 - z_1}{z_2 - z_1} \cdot \frac{z_1 - z_2}{z_3 - z_2} \cdot \frac{z_2 - z_3}{z_1 - z_3} = \arg(-1),$$

where $-3\pi < \arg(-1) < +3\pi$. The only values of $\arg(-1)$ in this interval are π and $-\pi$.

By Exercise 2,

$$\frac{\sin \theta_1}{|z_3 - z_2|} = \frac{\text{Im} (\bar{z}_2 - \bar{z}_1)(z_3 - z_1)}{|z_3 - z_2||z_1 - z_3||z_2 - z_1|}.$$

Now Im $(\bar{z}_2 - \bar{z}_1)(z_3 - z_1) = $ Im $(z_3\bar{z}_2 - \bar{z}_1 z_3 - \bar{z}_2 z_1 + |z_1|^2)$. Since $|z_1|^2$ does not contribute to the imaginary part and Im $\bar{w} = -$Im w, we find

$$\frac{\sin \theta_1}{|z_3 - z_2|} = \frac{\text{Im} (z_3\bar{z}_2 + z_1\bar{z}_3 + z_2\bar{z}_1)}{|z_3 - z_2||z_1 - z_3||z_2 - z_1|}.$$

Since the latter is preserved under a cyclic permutation of z_1, z_2, and z_3, we have

$$\frac{\sin \theta_1}{|z_3 - z_2|} = \frac{\sin \theta_2}{|z_1 - z_3|} = \frac{\sin \theta_3}{|z_2 - z_1|}, \tag{1}$$

the Law of Sines. (Note that we have established the Law of Sines for directed angles. To check that this agrees with the ordinary Law of Sines, we must be sure that all angles are positive or all negative. This is clear from (1).)

Set $\alpha = z_3 - z_2$, $\beta = z_1 - z_3$, $\gamma = z_2 - z_1$. Then $|\alpha|^2 = |\beta + \gamma|^2 = |\beta|^2 + |\gamma|^2 + 2\,\text{Re}\,\bar{\gamma}\beta = |\beta|^2 + |\gamma|^2 - 2|\beta||\gamma| \cos \theta_1$. This is the Law of Cosines.

4. The proof of 15–8.6 shows that we need only take δ so that

$$|s - t| < \delta \Rightarrow |f(s) - f(t)| < \inf \{|f(u)| \mid 0 \le u \le 1\}$$

5. Since the path does not cross the positive imaginary axis, the argument variation is zero.

6. We may assume $a = 0$ and $f'(0) = 1$, because we can replace the more general function g by f where $f(z) = g(z + a)/g'(a)$. Say $f(0) = b$. We will construct neighborhoods U of 0 and V of b such that f is a homeomorphism of U onto V.

We have $f(z) = z + \psi(z)$ where $\psi(z) = \sum_{n=2}^{\infty} c_n z^n$. Using an estimate essentially the same as that in 15–8.7, we can find a positive number ρ such that

$$|\psi(z_1) - \psi(z_2)| < \tfrac{1}{2}|z_1 - z_2| \tag{1}$$

for z_1, $z_2 \in D$, where $D = \{z \mid |z| < \rho\}$. This implies that f is an injection on D, since $f(z_1) = f(z_2)$ leads to $\psi(z_1) - \psi(z_2) = z_2 - z_1$, which contradicts (1) if $z_1 \neq z_2$.

Now we proceed as in 14–7.17. Because we have restricted the domain of the inequality (1), we must show that, for y near b, the function $z \to y - \psi(z)$ maps some closed (and therefore complete) neighborhood W of 0 into itself. Take $V = \{y \mid |y - b| < \frac{1}{3}\rho\}$ and $W = \{z \mid |z| \le \frac{2}{3}\rho\}$. For $y \in V$ and $z \in W$, $|y - \psi(z)| \le |y| + |\psi(z) - \psi(0)| < \frac{1}{3}\rho + \frac{1}{2}|z - 0| \le \frac{2}{3}\rho$, so $y - \psi(z) \in W$. By 14–7.16, there is a continuous map φ of V into W such that $\varphi(y)$ is the fixed point of the contraction $z \to y - \psi(z)$. This means that $\varphi(y) + \psi(\varphi(y)) = y$, that is $f(\varphi(y)) = y$, for $y \in V$.

Since φ is continuous and f is injective on $\varphi(V)$, f is a homeomorphism from $\varphi(V)$ to V. All that remains therefore is to show that $\varphi(V)$ is a neighborhood of 0. Suppose it is not. Choose a sequence $\{z_n\}$ in $D - \varphi(V)$ such that $z_n \to 0$. Then $f(z_n) \to f(0) = b$, and eventually we have $f(z_n) \in V$. Since f is injective on D, $z_n \in \varphi(V)$; contradiction.

It should be remarked that once we find that f is injective on D, it follows from 14–8.15 that f is a homeomorphism when restricted to any compact subset of D. From this it follows that f is a homeomorphism from G to $f(G)$ for any open set G with $\overline{G} \subseteq D$. It does not follow easily that $f(G)$ is an open set, however. We have proved this in two different ways using analytical facts about f, but it is actually true for any continuous and injective f. The general theorem is: If G is an open subset of $\mathbf{R}^n$ and f is a continuous injection from G to $\mathbf{R}^n$, then f is a homeomorphism from G to $f(G)$ and $f(G)$ is an open set in $\mathbf{R}^n$.

7. Suppose ψ is a mapping of D into itself which preserves distances, and $\psi(0) = 0$, and $\psi(1) = 1$. For each $\lambda \in \mathbf{C}$ there are at most two points z with $|z| = |\lambda|$ and $|z - 1| = |\lambda - 1|$, namely, λ and $\bar{\lambda}$ (there is only one point if $\lambda \in \mathbf{R}$). Since ψ is distance-preserving, $\psi(\lambda)$ must be either λ or $\bar{\lambda}$. By considering also $|\psi(\lambda) - \psi(i)| = |\lambda - i|$, we see that, if $\psi(i) = i$, then $\psi(\lambda) = \lambda$ for all λ, and if $\psi(i) = -i$, then $\psi(\lambda) = \bar{\lambda}$ for all λ.

If φ is a distance-preserving map of $\mathbf{C}$ into itself, then

$$\psi(z) = \frac{\varphi(z) - \varphi(0)}{\varphi(1) - \varphi(0)}$$

is another distance-preserving map with $\psi(0) = 0$ and $\psi(1) = 1$. Accordingly, $\varphi(z) = \alpha z + \beta$ or $\varphi(z) = \alpha \bar{z} + \beta$, where $\beta = \varphi(0)$ and $\alpha = \varphi(1) - \varphi(0)$.

In the second case the image of the angle from $\overrightarrow{z_0 z_1}$ to $\overrightarrow{z_0 z_2}$ has measure

$$\arg \frac{\bar{z}_2 - \bar{z}_0}{\bar{z}_1 - \bar{z}_0} = -\arg \frac{z_2 - z_0}{z_1 - z_0}.$$

8. We shall take $a = 0$ to simplify the notation. Suppose φ_1 has direction ξ. Let $\epsilon > 0$ be given and choose $\delta > 0$ so that

$$t < \delta \Rightarrow \left| \frac{\varphi_1(t)}{|\varphi_1(t)|} - \xi \right| < \epsilon.$$

Now $\varphi_1([\delta, 1])$ is a compact, therefore closed, set which does not contain 0. Hence there is an $\eta > 0$ such that $t < \eta \Rightarrow \varphi_2(t) \notin \varphi_1([\delta, 1])$. Since range $\varphi_2 \subseteq$ range φ_1, this shows that $t < \eta \Rightarrow \varphi_2(t) \in \varphi_1([0, \delta))$. Thus

$$t < \eta \Rightarrow \left| \frac{\varphi_2(t)}{|\varphi_2(t)|} - \xi \right| < \epsilon.$$

Hence φ_2 has the direction ξ also.

9. Let $h(z) = \sum_{n=0}^{\infty} c_{n+k}(z-a)^n$. Then $f(z) = f(a) + (z-a)^k h(z)$. On a sufficiently small neighborhood U of c_k there is a branch φ of the kth-root function. Put $g(z) = (z-a)\varphi(h(z))$ for $z \in W = h^{-1}(U)$.

10. Suppose h^* is a continuous function from $[0, 1]$ to $\mathbf{R}$ such that $h^*(t)$ is an argument of $h(t)$ for all t. Then $g^*(t) = h^*(t) + \arg_0 g(t)h(t)^{-1}$ defines a continuous function g^* from $[0, 1]$ to $\mathbf{R}$ such that $g^*(t)$ is an argument of $g(t)$ for all t. Hence the winding number of g is $(g^*(1) - g^*(0))/2\pi = (h^*(1) - h^*(0))/2\pi$, the winding number of h.

11. Let ϵ be a lower bound for $|\varphi|$. Since φ is uniformly continuous, we can choose n so that $|\varphi(s_1, t_1) - \varphi(s_2, t_2)| < \epsilon$ if $|s_1 - s_2| < 1/n$ and $|t_1 - t_2| < 1/n$. This inequality implies that $\varphi(s_1, t_1)\varphi(s_2, t_2)^{-1} \in$ domain $\arg_0$. Let β be any argument of $\varphi(0, 0)$ and define the functions $h_{j,k}$ by recursion as follows:

$$h_{1,1}(s, t) = \beta + \arg_0 \varphi(s, t)\varphi(0, 0)^{-1}$$
$$\text{for } 0 \le s \le 1/n, 0 \le t \le 1/n,$$
$$h_{j+1,1}(s, t) = h_{j,1}(j/n, 0) + \arg_0 \varphi(s, t)\varphi(j/n, 0)^{-1}$$
$$\text{for } j/n \le s \le (j+1)/n, 0 \le t \le 1/n,$$

and

$$h_{j,k+1}(s, t) = h_{j,k}(j/n, k/n) + \arg_0 \varphi(s, t)\varphi(j/n, k/n)^{-1}$$
$$\text{for } j/n \le s \le (j+1)/n, k/n \le t \le (k+1)/n.$$

The functions $\{h_{j,k} \,|\, 1 \le j \le n, 1 \le k \le n\}$ have a common extension ψ which is the function required.

12. Put $I = [0, 1]$. We shall abbreviate the function $t \to \varphi(s, t)$ by $\varphi(s, \cdot)$ and the function $s \to \varphi(s, t)$ by $\varphi(\cdot, t)$. In these terms "g is freely homotopic to h" means "there exists a function φ from $I \times I$ to M such that $\varphi(0, \cdot) = g$, $\varphi(1, \cdot) = h$, and $\varphi(\cdot, 0) = \varphi(\cdot, 1)$.

Let g, h, and k be closed paths in M. First, g is freely homotopic to g because we can define $\varphi(s, t) = g(t)$ for all s, t. Thus "freely homotopic" is reflexive. Suppose g is freely homotopic to h. Choose φ as above and define $\psi(s, t) = \varphi(1 - s, t)$. Then ψ is continuous, $\psi(0, \cdot) = h$, $\psi(1, \cdot) = g$, and $\psi(\cdot, 0) = \psi(\cdot, 1)$. Thus "freely homotopic" is symmetric. Suppose also that h is freely homotopic to k. Let θ be a continuous function from $I \times I$ to M such that $\theta(0, \cdot) = h$, $\theta(I, \cdot) = k$, and $\theta(\cdot, 0) = \theta(\cdot, 1)$. Then $\eta_1(s, t) = \varphi(2s, t)$ for $0 \le s \le \frac{1}{2}$, $t \in I$, and $\eta_2(s, t) = \theta(2s - 1, t)$ for $\frac{1}{2} \le s \le 1$, $t \in I$, define two continuous functions η_1 and η_2 which have a common extension η to $I \times I$ because $\eta_1(\frac{1}{2}, \cdot) = h = \eta_2(\frac{1}{2}, \cdot)$. This common extension is continuous by Exercise 5, p. 245. Since $\eta(0, \cdot) = g$, $\eta(1, \cdot) = k$, and $\eta(\cdot, 0) = \eta(\cdot, 1)$, g is freely homotopic to k. Thus "freely homotopic" is transitive.

Now let f be a continuous family of closed paths in $\mathbf{C} - \{0\}$. By Exercise 11 there is a continuous function ψ from $I \times I$ to $\mathbf{R}$ such that $\psi(s, t)$ is an argument of $f_s(t)$ for all s, t. Since $(s \to (\text{argument variation of } f_s)) = \psi(\cdot, 1) - \psi(\cdot, 0)$, it is clearly continuous. If the f_s are all closed, the argument variation of f_s is always 2π times an integer. Since $[0, 1]$ is connected, $s \to (\text{argument variation of } f_s)$ is constant. Hence if g and h are freely homotopic in $\mathbf{C} - \{0\}$ they have the same winding number about zero.

Suppose g and h are two closed paths in $\mathbf{C} - \{0\}$ with the same winding number. Choose continuous functions g^* and h^* from I to $\mathbf{R}$ such that $g^*(t)$ is an argument of $g(t)$ and $h^*(t)$ is an argument of $h(t)$ for all t. Put

$$\varphi(s, t) = |g(t)|^{1-s}|h(t)|^s \exp i((1 - s)g^*(t) + sh^*(t)).$$

This defines a continuous function φ from $I \times I$ to $\mathbf{C} - \{0\}$. We find that $\varphi(0, \cdot) = g$ and $\varphi(1, \cdot) = h$ by direct calculation. Furthermore,

$$\varphi(s, 1)\varphi(s, 0)^{-1} = \exp i((1 - s)(g^*(1) - g^*(0)) + s(h^*(1) - h^*(0)))$$
$$= \exp i(g^*(1) - g^*(0)) = 1$$

for all s, the second step because g and h have the same winding number, that is $g^*(1) - g^*(0) = h^*(1) - h^*(0)$, and the third because g is closed so $g^*(1)$ and $g^*(0)$ are arguments of the same number. This proves that g and h are freely homotopic.

Taking R as in 15–5.8 and using the same estimates, we see that

$$\frac{F(R \exp 2\pi it)}{(R \exp 2\pi it)^n}$$

is never negative for $t \in I$. Hence by Exercise 10, if

$$g(t) = F(R \exp 2\pi it),$$

then g has the same winding number as $t \to R^n \exp 2\pi int$, namely, n. Now let $\varphi(s, t) = F(sR \exp 2\pi it)$. If range $\varphi \subseteq \mathbf{C} - \{0\}$, then $g = \varphi(1, \cdot)$ would be freely homotopic to the constant map $\varphi(0, \cdot)$ and therefore have the same winding number. But obviously a constant map has winding number 0. Therefore, $0 \in$ range φ and F has a root.

Page 253

11. Let S be the set of nonnegative rational numbers. Let α be a positive irrational number. If x is a positive rational number, then there is a unique $n \in I$ such that $2^n\alpha < x < 2^{n+1}\alpha$. Define a function φ from $\mathbf{Q}$ to S by

$$\varphi(0) = 0,$$
$$\varphi(x) = 2^n x \qquad \text{if} \quad 2^n\alpha < x < 2^{n+1}\alpha,$$
$$\varphi(x) = -2^{n+1}x \qquad \text{if} \quad -2^{n+1}\alpha < x < -2^n\alpha.$$

This is readily seen to be a homeomorphism; the only point at which there is any question is 0. Actually, any countable nonvoid metric space without isolated points is homeomorphic to $\mathbf{Q}$.

INDEX OF SYMBOLS
AND SPECIAL NOTATIONS

$\Rightarrow$, $\Leftarrow$, $\Leftrightarrow$, implication, 14
$\forall$, universal quantifier, 16
$\exists$, existential quantifier, 16
$=$, equals, 2
( ), parentheses, 3
   functional values, 41
   in exponent, derivative, 291
(,), open interval, 279
[ ], brackets, 3
[,] closed interval, 279
(,], [,), half-open intervals, 279
{ }, set formation, 24
   notation for a sequence, 160
$\in$, membership, 3
$\notin$, nonmembership, 3
$\subseteq$, $\supseteq$, set inclusion, 5
$\subset$, $\supset$, proper set inclusion, 5–6
$\emptyset$, null set, 3–4
$\bigcup$, union of a family of sets, 51
$\cup$, union of sets, 29
$\bigcap$, intersection of a family of sets, 51
$\cap$, intersection of sets, 29
$-$, set difference, 34
   subtraction, 99
$\sim$, set complement, 34
$\langle\ ,\ \rangle$, $\langle\ ,\ ,\ \rangle$, ordered pair, ordered
   triple, etc., 38
$\circ$, composition of functions, 44
   composition of relations, 50 Ex. 1
$\rightarrow$, convergence, 161, 229
   defining a function, 43
$\times$, direct product, 53
$\times$, direct product, 39
$'$ (prime), derivative, 238
$\overline{}$ (overbar), conjugate, 132
   closure, 233

$<$, $>$, $\leq$, $\geq$, order in **R**, 132
$|\ |$, absolute value, 133
$\sqrt{}$, square root, 109
$+$, addition, 97
$\cdot$, multiplication, 97
0, zero, 97
1, multiplicative unit, 97
2, 3, 4, 5, 6, 7, 8, 9, digits, 220
∞, $+\infty$, $-\infty$, infinity, 173–175
$\aleph_0$, $\aleph_1$, . . . , cardinal numbers, 156
arg, $\arg_0$, argument, 330
c, cardinal of the continuum, 157
C, set of complex numbers, 132
card, cardinal of a set, 142
cos, cosine function, 310
diam, diameter of a set, 228
e, a number about 2.718, 326
exp, exponential function, 308
F-$\prod$, formal infinite product, 212
F-$\sum$, formal infinite sum, infinite
   series, 191
glb, greatest lower bound, 78
I, set of integers, 132
$I(\ )$, subset of a field, 101
Im, imaginary part of a complex
   number, 132
inf, infimum, 78
Int, interior of a set, 237
lim, limit in metric spaces, 230
   limit in **R** or **C**, 166
$\lim_{x \to p}$, limit, 287
$\overline{\lim}$, $\underline{\lim}$, superior and inferior limits, 175
lim inf, inferior limit, 175
lim sup, superior limit, 175
log, logarithm, 319
lub, least upper bound, 78
max, maximum, 78

min, minimum, 78
N, set of positive integers, 132
N*, set of nonnegative integers, 132
N_k, segment of the integers, 141
$N(\ )$, subset of a field, 101
$P(\ )$, set of positive elements in an
   ordered field, 105
$\mathfrak{P}(\ )$, power set, 37
$\prod$, extension of multiplication, 189,
   190 Ex. 3
  infinite product, 212
π, a number about 3.1416, 311

Q, the set of rational numbers, 132
$Q(\ )$, prime subfield of a field, 102
R, the set of real numbers, 132
R*, the set of extended real numbers, 174
Re, real part of a complex number, 132
$\sum$, extension of addition, 188
  effectively finite sum, 190
  infinite sum, 191
sin, sine function, 310
spt, support of a function, 189
sup, supremum, 78

INDEX

Absolute convergence, of infinite
    products, 215
  of infinite series, 195
Absolute value of a number, 133
Addition, of cardinals, 156–157, 159 Ex.
  of natural numbers, 113
  of positive real numbers, 123
  of rational numbers, 117
  of real numbers, 126
Adherent point, 231
  to a sequence, 231 Ex. 2
a-function, 152
Alternating series, 182 Ex. 5
Analytic continuation, 304 ff, 320 ff
Analytic functions, 287 ff
  arithmetic combinations, 300
  defined by power series, 301
  definition, 298
  local structure, 334 ff
And, 9–10
Angles, 328 ff
Anticonformal, 339 Ex. 7
Antisymmetric relation, 70
Arc, 284
Archimedean ordered field, 107
Arcwise connected space, 284
a-relation, 152
Argument, of a complex number, 330
  of a function, 41
  variation along a path, 333
Arithmetic combinations, of analytic
    functions, 300
  of continuous functions, 243–244
Ascending chain condition, 155 Ex. 4
a-set, 152
Associative-commutative law, 186–188,
    204–205

Associative law, for cardinals, 156
  general, 186–188
  for sets, 30, 52
Associative operation, 94
Associativity, of direct products, 39, 48
  of functions, 44
  of relations, 50 Ex. 1
Assumptions, 20, 23
a-subset, 158
Attractive fixed point, 262
Axiom, 59, 153
  of choice, 148 ff, 154
Axiomatic set theory, 151 ff

Baire category theorem, 263
Ball in metric space, 226
Base, for open sets, 274
  for open sets in a direct product, 277
    Ex. 3
Beman, W. W., 84, 128
Biconditional, 14
Bicontinuous function, 250
Bijective function, 45
  continuous function on a compact
    space is a homeomorphism, 271
Binary combinations of sets, 29
Binary operations, 94
Binary relation, 50
Binomial coefficients, 295
Binomial theorem, 296
Bolzano-Weierstrass property, 269
Bolzano-Weierstrass theorem, 184
Bound, greatest lower, 77
  least upper, 77
  lower, 77
  upper, 77
Bound variable, 27

Boundary of a set, 238
Bounded, 77
  above, 77
  below, 77
  complex-valued function, 134
  function, 77, 228
  sequence, 161
  set in metric space, 228
  subset of **C**, 134
  totally, 267
Boundedness of a continuous function on
    a compact space, 271
Brackets, 3
Branch of a multiple-valued function, 323

Canonical bijection of direct product, 48
Canonical models, 63
Cantor's paradox, 154
Cardinal (numbers), 155 ff
  of the continuum, 157
  of **R**, 218
  of a separable metric space, 275
  of a set, 142
Cardinal exponents, 157
Cartesian product of two sets, 39
Categorical postulate system, 63
Category of a set, 264
Cauchy criterion for convergence, of
    infinite products, 213
  in metric space, 253
  in **R** or **C**, 180–181
Cauchy product of two series, 209
Cauchy sequence, 181, 253
Center, of a division algebra, 131 Ex. 2
  of a power series, 293
Chains, 84 ff
Characteristic of a field, 102–103
Characteristic function of a set, 157, 200
Circle of convergence of a power series,
    294
Circular functions, 310 ff
Class, 2
Classical double sum, 208
Classification, 63
Closed ball, 239 Ex. 4
Closed function, 243
Closed interval, 279–280
Closed set, 232
Closed subset of a subspace, 234

Closure, 233
Cluster point, 231
Codomain of a function, 45
Coefficients of a power series, 293
Cohen, 154, 158
Collection, 2
Commutative laws, for cardinal
    arithmetic, 156
  general, 186–188
  for sets, 30, 52
Commutative operation, 94
Compact spaces, 266 ff
Comparable elements in an ordered set,
    75
Comparison test for convergence of
    infinite series, 196, 182 Ex. 3
Compatible, 104
Complement, 34
Complete analytic function, 323
Complete metric space, 253 ff
Complete ordered field, 108 ff
Complete ordered set, 79 ff
Completeness of space of continuous
    functions, 255
Completion of a metric space, 259
Complex exponents, 325
Complex numbers, 129 ff
Components, of open sets in a locally
    connected space, 283
  of a space, 282
Composition, of functions, 44
  of relations, 50 Ex. 1
Compound symbols, 3
Compute to k decimals, 221
Conceptually equivalent postulate systems,
    61, 66, 71
Conditional connective, 13–14
Conditionally convergent series, 195
  rearrangement of, 198
Configuration, 55
  notation, 72
Conformal, 337–338
  mapping theorem, 339
Conjugate complex numbers, 132
Connected, locally, 282
  spaces, 278 ff
  subsets of **R**, 280
Connectives, logical, 8 ff
Consistency of postulates, 26 (footnote), 62

Constant function, 42
Continuous function, 239 ff
  on a connected space, 279
  definition, 240
Contraction, 262
  fixed-point theorem, 261
Contrapositive, 19
Convergence, of infinite products, 212
  of infinite series, 191
  in metric spaces, 229 ff
  in **C**, 161
Convex metric space, 265 Ex. 7
Coordinate projections, 47, 54
Coordinates in a direct product, 39, 54
Corollary, 60
Cosine function, 310 ff
Cosines, law of, 338 Ex. 3
Countable axiom of choice, 154 Ex. 7
Countable base for the open sets, 274
Countable set, 143
Counterimage of a set, 47
Covering, 266
Criteria for the existence of limits in
    R and **C**, 180 ff

Decimal representation of numbers, 220
Decreasing sequence, 161
Dedekind, 84, 121, 128
DeMoivre, 129
DeMorgan's laws, 35, 52
Dense, ε-dense, 267
  set, 239
Denumerable set, 143
Derivative, 288
Derived set, 239 Ex. 6
Diagonal of a direct product, 39
Diameter of a set in a metric space, 228
Difference of two sets, 34
  symmetric, 37 Ex. 8
Differentiability, of analytic functions,
    299
  of compositions, 290
  of power series, 297
  of rational functions, 290
Differentiation, 287 ff
Digital representatives, 220
Direct product, of a family of sets, 53
  of two metric spaces, 225
  of two sets, 39

Directed angle, 329
  between two paths away from a point,
    336
Direction of a path away from a point,
    335
Disconnected, 278
Discrete space, 238
Disjoint sets, 34
Disk of convergence of a power series, 294
Distance function, 223
Distributive lattice, 78 Ex. 5
Distributive law, for cardinal arithmetic,
    156
  for direct products, 30 Ex. 4
  for sets, 30, 52
Divergence, of infinite products, 212
  in metric spaces, 229
  in **R** or **C**, 161
  to zero, 212
Division algebra, 131 Ex. 2
Domain, of a function, 41
  of a logical variable, 15
  of a relation, 50
Dominated convergence theorem, 206
Dominates (sets), 136
  (series), 196
Double limit, 208
Double sequence, 207, 231 Ex. 4
Double series, 207 ff
Double sum, 208
Doubly periodic, 314 Ex. 7
Duality, of interior and closure, 237
  of intersection and union, 35–36, 52
  of open and closed sets, 236
Dummy index, 166
Dummy variables, 26 ff
  conventions concerning, 162–163
Dyadic, 220

Effectively finite, 190
Empty set, 3
Entire functions, 302
Enumerable set, 143
ε-dense, 267
Equality, 2
Equicontinuity, 249 Ex. 3
Equivalence classes, 66
Equivalence relations, 65 ff
Equivalent postulate systems, 61

Eudoxus of Cnidos, 121
Euler, 129, 331
Exclusive or, 10
Existential quantifier, 16
Exponential form of a number, 220
Exponential function, 308 ff
Exponents, in cardinal arithmetic, 157
   integral, 134–135
   nonintegral, 324 ff
Extended binary combinations of sets, 30
Extended complex plane, 178
Extended direct products, 39
Extended real number system, 174
Extension, of a binary operation to finite sets, 186
   of a continuous function, 273 Ex. 17
   of a function, 43
   least common, 53 Ex.
      continuity of, 245 Ex. 5
   of a uniformly continuous function, 258

Factorial function, 295
Factoring functions, 67 ff
False propositions, 9
Family, 48
Field, 94 ff
   complete ordered, 108 ff
   definition, 96
   ordered, 104
   of rational functions, 104
Finite intersection property, 272 Ex. 5
Finite member of **R***, 173
Finite set, 141
First element, of an ordered pair, 37
   of a simple chain, 89
First-order predicate calculus, 8
First principle of induction, 87
Fixed point of a function, 81
Formal fractions, 116 ff
Formal infinite product, 212
Formal infinite sum, 191 ff
Fractions, 116 ff
Freely homotopic paths, 339 Ex. 12
Function, 40 ff
   defined by formula, 42
   definition, 41
   spaces, 226

Fundamental theorem of algebra, 318, 339 Ex. 12

Generating subset of a chain, 86
Gödel, 26, 62, 153, 154, 158
Graph, of a function, 40
   of a relation, 49
Greatest lower bound, 77
Group, 135 Ex. 1

Half-open intervals in **R**, 280
Harmonic series, 196
Heine-Borel property, 267, 269
Heine-Borel theorem, 270
Holomorphic, 298
Homeomorphism, 249 ff
   of a compact space, 271
Homotopic paths, 339 Ex. 12
Hypothesis of the continuum, 158

Idempotent laws for sets, 30
Identity, element, 95
   function, 41
   map, 46
If . . . , then . . . , 13
Image of a set, 46
Imaginary part of a complex number, 132
Implies, 14
Improper convergence, 178
Inclusion, 5
   map, 46
Inclusive or, 10
Incomparable elements in an ordered set, 76
Inconsistent postulate systems, 62
Increasing sequence, 161
Index set, 48
Indexed direct product, 53
Indexed intersection, 51
Indexed union, 51
Induced map, of a direct product, 47
   of the power set, 46
Induction, first principle, 87
   second principle, 83, 87
   transfinite, 83
Inductive definition of functions, 89 ff, 144 ff
Inductive proof, 83, 87
Inferior limits, 175

Infimum, 77
  in **R**, 174
Infinite metric, 224
Infinite products, 212
Infinite series, 191 ff, 182 Ex.
Infinite set, 141
Infinity, 173–175
Injective function, 44
Integers, 132
Interior, point, 236
  of a set, 237
Intermediate-value theorem, 281
Intersection, of a family of sets, 51
  of two sets, 29
Intervals, 279
  connectedness, 280
Into, 42
Intrinsic property of a configuration, 57
Intuitionists, 8
Inverse image, of a function, 45
  of a set, 47
Isolated point, 238
Isometric, 224
Isomorphic, 55
Isomorphism, of chains, 85
  of configurations, 55
  of ordered fields, 105
  of ordered sets, 73
  of simple chains, 91
Isotone, 73

Join of sets, 30
Juxtaposition notation, 30

Knaster fixed-point theorem, 81

Lattice, 78 Ex. 5
Laws of sines and cosines, 338 Ex. 3
Least common extension of a family of
    functions, 53 Ex.
Least subsystems, 85–86
Least upper bound, 77
Lebesgue convergence theorem, 206
Lebesgue number of a covering, 273
    Ex. 15
Left-identity element, 95
Left-zero element, 95
Lemma, 60

Length of a path, 327
Lexicographic ordering, 83 Ex. 3
Limit, arithmetic combinations, 168
  definition, 166, 230
  in metric spaces, 229 ff
  in **R**, 160 ff
Lindelöf covering theorem, 275
Linear order relation, 75
List notation for sets, 4, 25
Local structure of analytic functions,
    334 ff
Locally, 272
  compact, 272 ff
  connected, 282 ff
  pathwise connected, 284
  separable, 277 Ex. 6
Logarithm, 319 ff, 215
  principal branch, 323
Logic, 7 ff
Logical connectives, 8 ff
Lower bound, 77

m-adic rational, 219
m-adic sequence, 218
map, 41, 45
  of ordered sets, 73 ff
Mapping, 41
Mathematical induction, 83, 87
Mathematical logic, 7
Maximal, 76
Maximum, 76
  condition, 83
  point, 315
Measure of a directed angle, 331
Meet of sets, 30
Membership, 3
Metric, 223
  continuity of, 244
  infinite-valued, 224
  in R^n or C^n, 224–226
  space, 223 ff
Metrizable space, 251
Minimal, 76
Minimum, 76
  condition, 83
  point, 315
Möbius function, 196
Model for a postulate system, 62

Modular field, 102
Modulus principle, 315
Modus ponens, 19
Monomial function, 244
Monotone, 73
  sequence, 161
Moore-Smith theorem, 256
Multiple-valued analytic function, 323
Multiplication, of cardinals, 156, 157, 159 Ex.
  of natural numbers, 114
  of positive real numbers, 124
  of rational numbers, 119
  of real numbers, 126

Natural boundary of a function, 315 Ex. 8
Natural map associated with an equivalence relation, 66
Natural numbers, 88–89, 151, 153
Natural projection in direct products, 54
Necessary condition, 14
Negative, 104
Neighborhood of a point, 238
Neutral element, 95
Newton, 332
Nondense set, 239
Nonnegative, 104
Nonpositive, 104
Nonrectifiable path, 327
Not, 9
Nowhere dense set, 239
n-sphere, 226
Null set, 3

One-element set, 4
One-to-one, 45
Only if, 14
Onto, 42
Open covering, 266
Open function, 243
Open intervals, 279
Open sets, 235
Open subsets of **R**, 283
Operator, 41
Or, 9–10
Order, of listing the members of a set, 4
  in the natural numbers, 114
  in the positive real numbers, 122

in the rational numbers, 119–120
in the real numbers, 127
relations, 70 ff
in subsets, 72
Ordered field, 104 ff
  complete, 108 ff
Ordered pair, 38
Ordered set, 72
Ordered triples, etc., 38
Ordering a set, 72
Order-preserving map, 73
Order-reversing map, 73
Ordinary induction, 87

Parentheses, 3
Partial order, 75
Partial products, 212
Partial sums, 191
Partition of a set, 66
Path, 284
  away from a point, 335
  components of a space, 284
  length of, 327
Pathwise connected space, 284
Peano, 62
Peano's axioms for the natural numbers, 89
Period, 312, 314 Ex. 7
Periodic function, 312
  doubly, 314 Ex. 7
Permanence of analytic identities, 311
Placeholder, 15, 42
Point, 223
Pointwise, 199
  convergence, 247–248
Polar form of a complex number, 331
Polynomial function, 244
Positive, 104
  integers, 132
  real numbers, 121 ff
Postulates, 59, 153
Power series, 239 ff, 210
Power set, 37
Precompact space, 272 Ex. 9
Prime subfield of a field, 102
Principal branch of the logarithm, 323
Principal value of the argument, 330
Pringsheim's theorem, 211 Ex. 14

Product, direct, of two sets, 39
  of sets (old term), 29
Proof, 19 ff
Proper set inclusion, 5–6
Proper subset, 6
Proposition, 61
  in logic, 8
Propositional function, 15
Propositional scheme, 15
Pseudometric, 259, 228 Ex. 7
Ptolemy, 332
Purely imaginary, 132
Pythagoreans, 121

Quantification, 16
Quantified proposition with null domain, 17
Quantifiers, 16
Quaternary relation, 50
Quaternion algebra, 131 Ex. 3
Quotient map of an equivalence relation, 66
Quotient set, 66

Raabe's test, 197, 216 Ex. 4
Radius of convergence of a power
  series, 294
  formula for, 295
Range, of a function, 41
  of a relation, 50
Ratio, 121
  test, 196, 182 Ex. 4
Rational function, 244
Rational numbers, 116 ff, 132
Real numbers, 126 ff
Real part of a complex number, 132
Rectifiable path, 327
Redundant postulates, 61
Reflexive relation, 65
Region of convergence of a power
  series, 294
Regular function, 298
Relation, 49 ff
Relative consistency, 63
Remainder of a series after k terms, 193
Representative of an equivalence class, 66
Represents, power series represents a
  function, 298
Repulsive fixed point, 262

Restriction of a function, 43–44
Riemann, 196, 338
Riemann hypothesis, 196
Riemann sphere, 178
Right-identity element, 95
Right-zero element, 95
Round-off, 221
Rule of trichotomy, 75
Russell's paradox, 25, 140, 153, 156

Schröder-Bernstein theorem, 138
Second element of an ordered pair, 37
Segments of the integers, 141
Self-contradictory postulate system, 62
Semigroup, 188
Separability of the space of continuous
  functions, 276
Separable spaces, 273 ff
Sequence, 160
Set, 1 ff
  complements, 34
  empty, 3
  formation, 24
  inclusion, 5
  intersection of, 29, 51
  null, 3
  union of, 29, 51
  universal, 34
Set-theoretic paradoxes, 25
Similarity of sets, 136
Simple chain, 88
Simple ordering, 75
Sine, curve, 285–286
  function, 310 ff
  law of sines, 338 Ex. 3
Space, of continuous functions, 226,
  248, 276
  completeness, 255
  separability, 276
  metric, 223 ff
  topological, 250
Sphere, 226
Square roots in complete ordered
  fields, 108–109
Standard English numerals, 220
Standard m-adic representation of a
  number, 219
Standard m-adic sequence, 218
Steiner triple system, 62 Ex. 2

Strictly decreasing, 161
Strictly increasing, 161
Strictly monotone, 161
Strong order relation, 70
Strongly order-preserving, 73
Structure, 55
Subchain, 84
Subcovering, 266
Subfield, 98
Subsequence, 182
Subset, 5
Subspace of a metric space, 224
Successor function, 84
Sufficient condition, 14
Sum, by columns of a double series, 208
   by rows of a double series, 208
   of f over X, 201
   of an infinite series, 191
   of sets (old term), 29
Summable function, 201
Superior limit, 175
Superlative order words, 76
Superset, 5
Support, 189
Supremum, 77
   in $\mathbf{R}$, 174
Surjective, 42
Symmetric difference of two sets, 37
   Ex. 8
Symmetric relation, 65

Tautologically equivalent formulas and
      propositions, 12
Tautologies, 10 ff
   and set identities, 31 ff
   computation of, 11
Taylor's series, 300
Terms, of an infinite series, 191
   of a sequence, 160
Ternary operation, 94
Ternary relation, 50
Theorem, 60
Topological properties, 224, 251
Topological space, 250
Topologically equivalent metrics, 250
Topology, 250, 252
Total ordering, 75

Totally bounded, 267
Totally disconnected set, 282
Transfinite induction, 83
Transformation, 41
Transitive relation, 65
Triangle law, 223
Trichotomy, 75
Truth table, 11
Two-valued logic, 8

Unary operation, 84
Uncountable set, 143
Uniform, 245 ff
   continuity, 246
      of a continuous function on a
         compact space, 272
   convergence, 247–248
   equivalence of metric spaces, 265
      Ex. 5
Uniqueness, of limits, 229
   of power series, 299
Unit element, 95
Universal quantifier, 16
Universal set, 34
Unordered finite sums, 186 ff
Unordered infinite sums, 199 ff
Upper bound, 77

Value of a function, 41
Variable, dummy, 26 ff
   logical, 15
Vector space, 199
Venn diagrams, 32 ff

Wallis' product, 216 Ex. 2
Weak maximum or minimum point of a
      function, 315
Weak order relation, 70
Weakly monotone, 161
Weakly order-preserving, 73
Weierstrass, 302
Well-ordered set, 82
Well-ordering theorem, 154, 155 Ex. 6
Whitney, 29
Winding number of a closed path, 334

Zero element, 95